Modern Fluoropolymers

Wiley Series in Polymer Science

Series Editor:

Dr John Scheirs
ExcelPlas Australia
P. O. Box 102
Moorabbin 3189
Victoria
AUSTRALIA

Modern Fluoropolymers
High Performance Polymers for Diverse Applications

Forthcoming titles:

Polymer Recycling
Science, Technology and Applications

**Preparation, Properties and Technology of
Metallocene-based Polyolefins**

COMMERCIAL FLUOROPOLYMERS AND MANUFACTURERS

AF	Teflon AF	DuPont
	Cytop	Asahi Glass
ECTFE	Halar	Ausimont
ETFE	Tefzel	DuPont
	Hostaflon ET	Hoechst
	Aflon COP	Asahi Glass
	Neoflon	Daikin
FEP	Teflon FEP	DuPont
	Hostaflon TFA	Hoechst
	Algoflon	Ausimont
	Neoflon AP	Daikin
FKM	see VDF-HFP	
MFA	Hostaflon TFA	Hoechst
	Algoflon	Ausimont
PCTFE	Kel-F	Dyneon
	Aclon	Allied Signal
	Voltaflef	Elf Atochem
	Daiflon	Daikin
PFA	Teflon PFA	DuPont
	Hostaflon PFA	Hocchst
	Neoflon	Daikin
PFPE	Fomblin	Ausimont
	Krytox	DuPont
	Denum	Daikin
PTFE	Teflon PTFE	DuPont
	Fluon	ICI
	Hostaflon PTFE	Hoechst
	Algoflon	Ausimont
	Fluon	Asahi-ICI
	Polyflon	Daikin
PVDF	Kynar	Atochem USA
	Hylar	Ausimont
	Foraflon (now Kynar)	Elf Atochem
	Solef	Solvay
	Neoflon	Daikin
	KF	Kureha Chem
PVF	Tedlar	DuPont
VDF-HFP	Viton A	DuPont Dow Elastomers
	Fluorel	Dyneon
	Tecnoflon	Ausimont
	Dai-El	Daikin
VDF-HFP-TFE	THV	Dyneon
	Fluorobase T	Ausimont
	Viton B	DuPont Dow Elastomers
	Dai-El	Daikin
TFE-PMVE-CSM	Kalrez	DuPont Dow Elastomers
TFE-PP	Aflas	Dyneon
VDF-CTFE	Kel-F 3700	Dyneon
VDF-HPFP	Tecnoflon SL	Ausimont
VDF-HPFP-TFE	Tecnoflon T	Ausimont
VDF-HFP-TFE-CSM	Viton G	DuPont Dow Elastomers
	Fluorel	Dyneon
VDF-TFE-CSM	Viton GLT	DuPont Dow Elastomers

JOHN NEWTON

Modern Fluoropolymers

High Performance Polymers for Diverse Applications

Edited by

John Scheirs
ExcelPlas Australia, Mooraboin, Victoria, Australia

WILEY SERIES IN POLYMER SCIENCE

JOHN WILEY & SONS
Chichester • New York • Weinheim • Brisbane • Singapore • Toronto

Reprinted March 1998

Other Wiley Editorial Offices

John Wiley & Sons, Inc., 605 Third Avenue,
New York, NY 10158-0012, USA

VCH Verlagsgesellschaft mbh, Pappelallee 3,
D-69469 Weinheim, Germany

Jacaranda Wiley Ltd, 33 Park Road, Milton,
Queensland 4064, Australia

John Wiley & Sons (Asia) Pte Ltd, 2 Clementi Loop #02-01,
Jin Xing Distripark, Singapore 129809

John Wiley & Sons (Canada) Ltd, 22 Worcester Road,
Rexdale, Ontario M9W 1L1, Canada

Library of Congress Cataloging-in-Publication Data

Modern fluoropolymers : high performance polymers for diverse
applications / edited by John Scheirs.
       p.   cm.
    Includes bibliographical references and indexes.
    ISBN 0-471-97055-7 (alk. paper)
    1. Polymers.   2. Organofluorine compounds.   I. Scheirs, John.
TA455.P58M54   1997
620.1´92 — dc21   96-49974
                                                              CIP

British Library Cataloguing in Publication Data

A catalogue record for this book is available from the British Library

ISBN 0-471-97055 7

Typeset in 10/12pt Times by Laser Words, Madras, India
Printed and bound in Great Britain by Biddles Ltd, Guildford, Surrey
This book is printed on acid-free paper responsibly manufactured from sustainable forestation,
for which as least two trees are planted for each one used for paper production.

Contents

List of Contributors

DR V. ARCELLA — *Fluoropolymers R&D, Ausimont SpA, via Lombardia 20, 20021 Bollate (MI) Italy*

DR P. AVAKIAN — *DuPont Central R & D Experimental Station, PO Box 80356, Wilmington, DE 19880-0356 USA*

DR B. BANKS — *NASA Lewis Research Center, 21000 Brookpark Road, MS-302-1, Cleveland, OH 44135, USA*

DR S. BOWERS — *DuPont Dow Elastomers SA, Geneva, Switzerland*

DR R. F. BRADY, JR. — *Naval Research Laboratory, Chemistry Division, Coatings Section, Washington DC, 20375-5342, USA*

DR D. M. BREWIS — *Institute of Surface Science and Technology, Department of Physics, Loughborough University, Loughborough LE11 3TU, Leicestershire, UK*

DR W. H. BUCK — *DuPont Fluoroproducts, Wilmington, DE 19880-0323, USA*

PROFESSOR P. E. CASSIDY — *Department of Chemistry, Southwest Texas State University, 601 University Drive, San Marcos, TX 78666-4616 USA*

DR N. N. CHUVATKIN — *ORGSTEKLO Company, 606006 Dzerzhinsk, Russia*

MR P. CUMMINGS — *Precision Polymer Engineering, Clarendon Road, Blackburn, BB1 9SS, UK*

DR R. FERRO — *Ausimont SpA, Tecnoflon, Technical Service and Development, via Lombardia 20, 20021 Bollate (MI), Italy*

DR J. W. FITCH III   *Department of Chemistry, Southwest Texas State University, 601 University Drive, San Marcos TX 78666-4616, USA*

DR K. HINTZER   *Dyneon GmbH, Werk Gendorf, D-84504 Burgkirchen, Germany*

DR G. S. HOOVER   *Pelmor Laboratories, Inc., 401 Lafayette Street, Newtown, PA 18940, USA*

DR D. E. HULL   *Fluorothermoplastic Business Unit, Dyneon, Building 220-10E-10, St Paul, MN 55144-1000, USA*

DR R. A. IEZZI   *Elf Atochem North America Inc., Technical Polymers R & D, 900 First Avenue, PO Box 61536, King of Prussia, PA 19406-0936, USA*

DR B. V. JOHNSON   *Fluorothermoplastic Business Unit, Dyneon, Building 220-10E-10, St Paul MN 55144-1000, USA*

DR D. L. KERBOW   *DuPont, Washington Works, PO Box 1217, Parkersburg, WV 26102-1217, USA*

DR P. R. KHALADKAR   *DuPont de Nemours & Co, Engineering — L1315, PO Box 80840, Wilmington, DE 19880-0840, USA*

DR E. KHOSRAVI   *IRC in Polymer Science & Tech, University of Durham, Durham, DH1 3LE, UK*

DR G. LÖHR   *Dyneon GmbH, Werk Gendorf, D-84504 Burgkirchen, Germany*

DR B. J. LYONS   *22 Hallmark Circle, CA 94025-6683, USA*

DR J. B. MARSHALL   *DuPont Dow Elastomers LLC, Wilmington, DE 19809, USA*

DR I. MATHIESON   *Institute of Surface Science and Technology, Department of Physics, Loughborough University, Loughborough LE11 3TU, Leicestershire, UK*

DR M. T. MAXSON   *Dow Corning Corporation, Mail No. CO41D1, Midland, Michigan 48686-0994, USA*

DR A. W. NORRIS   *Dow Corning Corporation, Mail No. CO41D1, Midland, Michigan 48686-0994, USA*

DR M. J. OWEN   *Dow Corning Corporation, Mail No. CO41D1, Midland, Michigan 48686-0994, USA*

DR I. Yu. PANTELEEVA — *ORGSTEKLO Company, 606006 Dzerzhinsk, Russia*

DR I. M. POZZOLI — *Ausimont, Melt Processable Fluoropolymers, Viale Lombardia 20, 20021 Bollate (MI), Italy*

DR P. R. RESNICK — *DuPont Fluoroproducts, Fayetteville, NC 28302, USA*

DR I. P. RODRICKS — *Fluorothermoplastic Business Unit, Dyneon, Building 220-10E-10, St Paul, MN 55144-1000, USA*

MR E. W. ROSS — *Pelmor Laboratories, Inc., 401 Lafayette Street, Newtown, PA 18940, USA*

DR J. SCHEIRS — *ExcelPlas Australia, PO Box 102, Moorabbin, 3189 VIC. Australia*

MR D. SEILER — *Marketing Manager Fluoropolymers, Elf Atochem North America, Inc., 2000 Market Street — 21st Floor, Philadelphia, PA 19103-3222, USA*

DR T. SHIMIZU — *DAIKIN Industries Ltd, R & D Department, 1-1 Nishi-Hitosuya, Settsu, Osaka 566, Japan*

DR J. B. STALEY — *Fluorothermoplastic Business Unit, Dyneon, 220-10E-10, St Paul, MN 55144-1000, USA*

DR G. STANITIS — *Ausimont USA, Inc., Crown Point & Leonards Lane, PO Box 26, Thorofare, NJ 08086, USA*

DR H. W. STARKWEATHER, JR. — *DuPont Central Research and Development, Experimental Station, Wilmington, DE 19880-0356 USA*

DR N. SUGIYAMA — *Research Centre, Asahi Glass Company, Ltd, Hazawa-cho, Kanagawa-ku, Yokohama, 221, Japan*

DR T. TAKAKURA — *Research Center, Asahi Glass Co., Ltd, 1150 Hazawa-cho, Kanagawa-ku, Yokohama 221, Japan*

DR M. TATEMOTO — *Daikin Industries Ltd, Settsu, Osaka 566, Japan*

DR C. TOURNUT — *Elf Atochem, Centre de Recherche Rhône-Alpes, PO Box 63, 69493 Pierre Benite Cedex, France*

DR A. VAN CLEEFF — *DuPont Dow Elastomers LLC Wilmington, DE, 19809, USA*

DR G. VITA — *Ausimont, Melt Processable Fluoropolymers, Viale Lombardia 20, 20021 Bollate (MI), Italy*

Series Preface

The Wiley Series in Polymer Science aims to cover topics in polymer science where significant advances have been made over the past decade. Key features of the series will be developing areas and new frontiers in polymer science and technology. Emerging fields with strong growth potential for the twenty-first century such as nanotechnology, photopolymers, electro-optic polymers etc. will be covered. Additionally, those polymer classes in which important new members have appeared in recent years will be revisited to provide a comprehensive update.

Written by foremost experts in the field from industry and academia, these books place particular emphasis on structure-property relationships of polymers and manufacturing technologies as well as their practical and novel applications. The aim of each book in the series is to provide readers with an in-depth treatment of the state-of-the art in that field of polymer technology. Collectively, the series will provide a definitive library of the latest advances in the major polymer families as well as significant new fields of development in polymer science.

This approach will lead to a better understanding and improve the cross fertilisation of ideas between scientists and engineers of many disciplines. The series will be of interest to all polymer scientists and engineers, providing excellent up-to-date coverage of diverse topics in polymer science, and thus will serve as an invaluable ongoing reference collection for any technical library.

Dr John Scheirs
Series Editor
June 1997

Preface

The last book on fluoropolymers was published in 1972 (edited by Dr. Leo Wall) and while it was a comprehensive overview of fluoropolymer development at that time, many of the polymers described therein have failed to reach commercial maturity. Moreover, the last 25 years have seen the introduction of numerous new fluoropolymers and fluoroelastomers and these developments have widened considerably the scope and applications of fluorine-containing polymers.

This book presents an overview of modern fluoropolymers with an emphasis on structure/property behaviour and their diverse fields of application. The incorporation of fluorine into both organic and inorganic polymers confers hydrophobicity, enhanced thermal, chemical and oxidative stability, reduced adhesion, increased solubility, improved biocompatibility and increased gas permeability. The introduction of fluorine in the form of hexafluoroisopropylidene groups can also impart solubility and processability to otherwise largely intractable polymers such as polyimides and polybenzimidazoles (see Chapter 8).

From a historical perspective, fluoropolymer developments have always been closely linked to military and strategic applications. For example, PTFE from the Du Pont pilot plant which began in 1943 was utilized almost immediately at Oak Ridge, Tennessee, in equipment for separating the isotopes of uranium for the first atomic bomb. PTFE also found military use as a nose cone cover for artillery shells in World War II. Even before this, extensive work was done on a partially fluorinated polymer, PCTFE, in order for it to be used in the Manhattan atomic bomb project. In a series of parallel developments, PCTFE was independently developed in Germany by the chemical giant IG Farben. Then in the early 1940s, PVF came into commercial production in the US while pilot-plant production of PTFE was carried out in the UK at ICI in the period 1944–47.

In 1948, another polymer which began with a promising future was polyfluoroprene (the polymer of 2-fluoro-1,3-butadiene). It was officially the first commercial fluoroelastomer though it was surpassed by the development of Kel-F (VDF–CTFE) in 1954 by the US Army and Viton VDF–HFP-type fluoroelastomers in 1958 since these showed vastly superior properties. The emerging aerospace industry in the early 1950s spurred the development of these

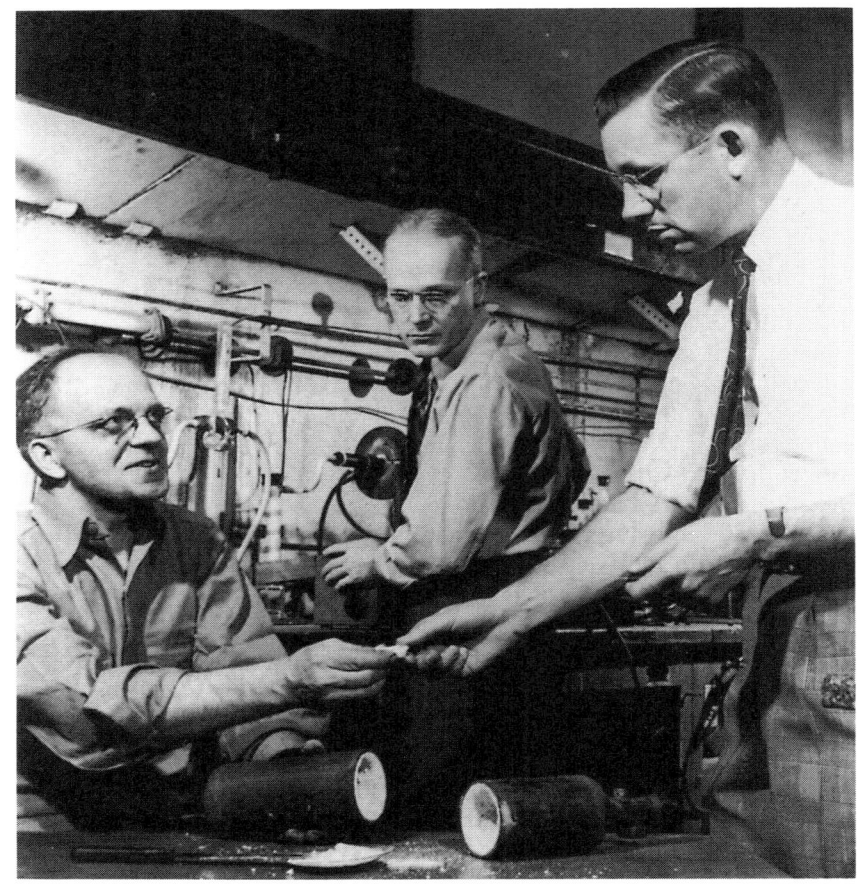

Discovery of PTFE (Teflon®) inside a cylinder of tetrafluoroethylene (TFE) gas by Dr Roy Plunkett (right) in the DuPont Laboratories in 1938. (Courtesy of DuPont)

fluoroelastomers in response to the critical need for rubbers with better heat and fuel resistance to be used in seals and hoses for military jet engines.

Later other fluoroelastomers were developed by the US Air Force for seals on their experimental supersonic aircraft. While the use of fluoropolymers in military and strategic applications still continues to drive fluoropolymer research, see for example submarine radome coatings (Chapter 6), F-111 fuel tank seals (Chapter 20), lubrication seals for B-series strategic bombers (Chapter 19), perfluoropolyether liquids as lubricants for bearings in UF_6 enrichment plants as well as for satellites (Chapter 24), canopies for supersonic fighter aircraft (Chapter 9) and heat-resistant shields for satellites and spacecraft coatings that are resistant to atomic oxygen (Chapter 4), there are also a myriad of terrestrial applications.

Apart from high-tech military applications, fluoropolymers touch our daily lives in a number of diverse ways, from the breathable Gortex® material in sporting and outdoor wear based on fibrillar PTFE to Scotchgard® coatings (fluorinated acrylic esters polymers) which impart water and oil repellency. In addition, fluoroelastomer tubes and seals are now extensively used in the fuel systems of cars due to their excellent resistance to methanol and low permeation (see Chapter 19). Another diverse application area of fluoropolymers is as lubricants (liquid perfluoropolyethers) in computer hard-drives to prevent wear between the head and the spinning hard disk. Fluorinated polymer lubricants are even used to lubricate the delicate mechanisms of Rolex® watches. Furthermore, on prestige buildings and skyscrapers, fluoropolymer coatings provide excellent long-term durability since they show no sign of fading or cracking even after 25 years of continuous outdoor exposure (see Chapter 14 and Chapter 29). Moreover, their low surface energy means that streaks of dirt and adherent dust are simply washed off by rain.

Cable insulation is a rapidly expanding market area for fluoropolymers, given the increasing demand for high-speed data transmission cables for Internet and optical communications. Fluoropolymers are specified here because of their high temperature and fire resistance. The so-called 'plenum cables' that run in air-conditioning ducts and in the open spaces between ceilings and floors of offices are an application that can only be met by fluoropolymers by virtue of their low flame-spreading and low smoke-generating behaviour in a fire situation. Hook-up wires for trains and planes also represent another large application area for crosslinked fluoropolymers (see Chapter 18).

Amorphous perfluoropolymers are presently making major inroads as plastic optical fibres because of their high transparency in the near infra-red. In contrast, the presence of vibrational C–H overtone absorptions in the near infra-red spectrum of hydrocarbon polymers limits their use as optical fibres due to high signal losses by attenuation. Their low refractive index also makes fluoropolymers useful materials for fibre optic cladding applications. In addition, the hydrophobicity of fluoropolymers is advantageous for their use as optical fibres, since water

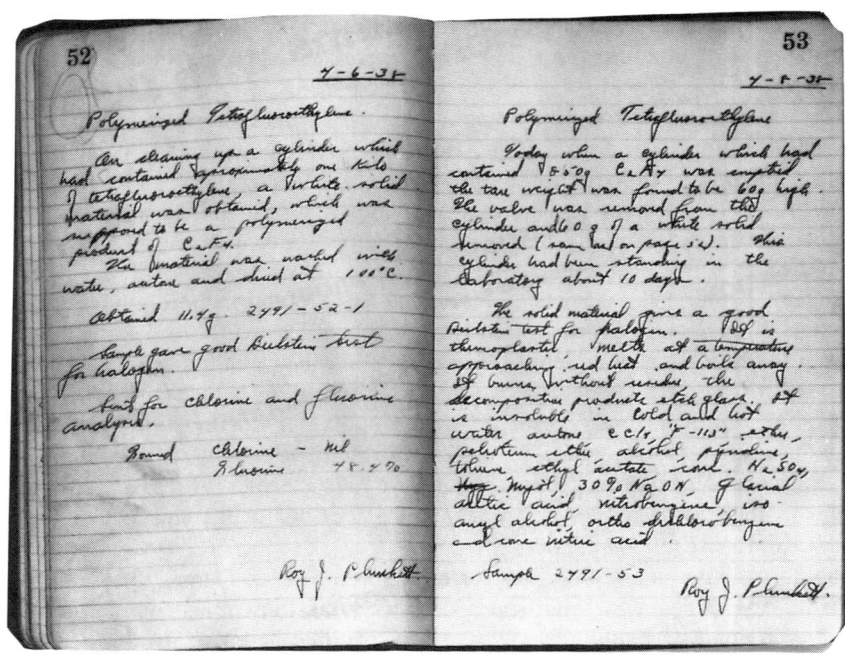

Laboratory notebook of Dr Roy Plunkett in which the discovery of Teflon® was first recorded on April 6, 1938. (Courtesy of DuPont)

absorption significantly increases losses in the near infra-red region used by the transmitting lasers.

Fluoropolymers are the only materials capable of withstanding the combination of high temperatures and high acidity encountered in flue gas desulphurizers in coal-burning power plants. Thus, in many cases, fluoropolymers perform in applications where no other materials can. In fact, lesser performing materials could prove hazardous or at the very least lead to increased maintenance and downtime if they were used in place of fluoropolymers (e.g. in the case of seals used in deep oil exploration).

Components such as O-rings, gaskets, seals, etc. usually represent the weak points in an engineering application and it is in these critical areas that fluoropolymers are frequently specified. Though fluoropolymers are used in very small volumes relative to commodity polymers their function is unique and instrumental to the overall performance of the application in which they are used. As a result, the chemical processing industry is heavily reliant on fluoropolymers and owes much of its development to their availability (see Chapters 16 and 25).

Fluoropolymer coatings, by virtue of their excellent UV resistance, have great potential to upgrade the service life of cheaper engineering plastics such as ABS (which on its own has very little resistance to weathering). This demonstrates the concept where the high-performance FP is used as a protective veneer on a cheaper substrate that provides mechanical integrity. This idea is also behind the wide use of fluoropolymer liners in steel pipes in chemical processing applications.

The excellent stability of fluoropolymers in the UV is becomingly increasingly important now that further miniaturization of electronic integrated circuitry is creating the need to reduce the wavelength of the light source used in resist manufacture from near-and mid-UV (300–450 nm) to far-UV (<250 nm). In particular, KrF excimer lasers (248 nm) are used currently for RAM fabrication and it is at such low wavelengths that most other polymers degrade.

Fluoropolymers are also characterized by their outstanding purity, and from this property numerous applications are derived. Their inertness combined with that fact they contain no additives, makes fluoropolymers the logical choice for high-purity pipes and fittings for the semiconductor and pharmaceutical industries.

Thus, from humble beginnings as a waxy deposit in a cylinder of refrigerant gas, certain grades of fluoropolymers now have the distinction of being the most expensive polymers commercially available (e.g. Teflon® AF) and combine the chemical inertness of PTFE with optical transparency and processability. The high cost of fluoropolymers is the result of complex manufacturing operations with multistep and exacting syntheses. For example, chapters on PFA (Chapter 11), Teflon® AF (Chapter 22) and Cytop® (Chapter 28) give an appreciation on the sequences of reactions involved. Even though the fluoropolymer product slate is now replete with a diverse range of commercially available fluoropolymers, new ones continue to emerge. Examples of these include, MFA and THV, both of which were commercialized only in the past few years (see Chapters 21 and 13 respectively).

As with any family of polymers, fluoropolymers by and large share common strengths but also common deficiencies. However, within this large group of polymers, significant variability does exist in structure and properties. The challenge is no longer simply specifying a 'fluoropolymer for the job' over a non-fluorinated polymer, but rather to judiciously select the most suitable fluoropolymer grade. With this in mind the opening chapter is designed to convey an appreciation of the elements involved in selecting fluoropolymers and fluoroelastomers to avoid in-service failure and to enable more informed choices in material specification.

John Scheirs

About the Editor

John Scheirs obtained a Ph. D. in Applied Science from the University of Melbourne, Australia. Subsequently, he worked as a development chemist in a development/technical service laboratory for the Exxon-Mobil polymer joint venture in Melbourne. In 1990, he spent one year as a guest researcher at Sussex University in the UK developing new, sensitive techniques for studying polymer degradation. From 1994 to 1996 he has been engaged in research in laboratories in Italy, Belgium and France. Dr Scheirs has authored over 50 scientific papers, including eight encyclopedia chapters and has given presentations at ACS, IUPAC and ANTEC symposia. He is on the editorial board of *Polymer Degradation and Stability*. Dr Scheirs is a member of the Society of Plastic Engineers (SPE), American Chemical Society (ACS), Institute of Materials (formally the PRI) and the Royal Australian Chemical Society (RACI). Current affiliation is with ExcelPlas Australia, a polymer consulting company. In the past two years he has worked on projects concerning the durability of fluoropolymers, particularly polyvinylidene fluoride and perfluoropolyethers.

1

Structure/Property Considerations for Fluoropolymers and Fluoroelastomers to Avoid In-service Failure

J. SCHEIRS
ExcelPlas Australia

1 INTRODUCTION

While fluoropolymers exhibit a range of outstanding properties such as chemical resistance and high-temperature stability, they do not always offer a panacea. Their expansion into increasingly aggressive environments such as those encountered in chemical processing, oil wells, motor vehicle engines, nuclear reactors and space applications are presenting a challenge to conventional fluoropolymers.

The reputation of fluoropolymers for being of high purity, resistant to all chemicals and having stabilities well beyond temperatures at which their hydrogenated counterparts degrade, has led product designers to specify fluoropolymers for some of the most extreme applications. For example, the established performance of PTFE has led people to believe that PTFE offers a universal solution to new chemical resistance problems. Pushing their performance properties to the limit without sufficient in-service reliability or testing has led to inevitable failures, some of which have become well documented.

Fluoropolymers, though endowed with many unique properties, possess some deficiencies relative to many engineering plastics; such as, poorer mechanical properties, higher permeabilities and high cost. Specific shortcomings of some commercial fluoropolymers are shown in Table 1.1. Judicious material selection

Modern Fluoropolymers. Edited by John Scheirs
© 1997 John Wiley & Sons Ltd

Table 1.1. List of Achilles' heels of common fluoropolymers

Fluoropolymer	Achilles' heel
PTFE	Ionizing radiation, creep
PVDF	Ketones, strongly alkaline solutions
PCTFE	High processing temperatures
FEP	Fatigue, poor high temp. properties
ETFE	Elevated temperatures + oxygen
VDF–HFP	Amines, cryogenic temperatures
PFA	Low heat deflection temp.
PVF	Aluminium + sunlight

and in-service testing are necessary precursors to avoid catastrophic and costly failures.

Thus, it is important to have an understanding of the mechanisms involved in fluoropolymer and fluoroelastomer deterioration as well as an appreciation of the structure/properties limitations within this class of polymers. This chapter seeks to address those situations in which failure of fluoropolymer components may occur, and to highlight some of the important considerations when selecting fluoropolymers for diverse applications.

2 STRUCTURE/PROPERTY RELATIONSHIPS

2.1 FLUOROPLASTICS

An understanding of how basic polymer structure dictates performance properties is fundamental to the proper selection of a fluoropolymer for a particular application.

The PTFE chain adopts a slowly twisting helix that comprises 13 CF_2 groups every 180° turn such that every main chain bond is rotated 20° from the

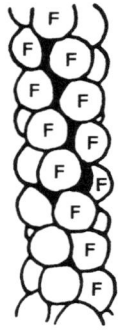

Figure 1.1. Schematic of the PTFE molecular helix. The molecule can be viewed as a nearly perfect cylinder with an outer sheath of fluorine atoms over a central core of carbon atoms (adapted from Koo [**116**], 1972)

next. This is a thermodynamically favoured configuration, rather than a planar zigzag (as with polyethylene), because of the mutual repulsion of the adjacent fluorine atoms. The helical twisting conformation of PTFE molecules results in an almost perfect cylinder comprised of an outer sheath of fluorine atoms enveloping a carbon-based core (Figure 1.1). This morphology is conducive for PTFE molecules to pack like parallel rods. Individual cylinders can, however, slip past one another and this accounts for the tendency of PTFE to cold flow. Figure 1.2 shows atomic force micrographs of PTFE molecules which clearly demonstrates this packing behaviour. In contrast to PTFE, PVF is comprised of alternating hydrogen atoms which being smaller than fluorine atoms permit molecular chains to pack in a planar zigzag fashion.

The PTFE chain is incredibly stiff due to the mutual repulsion of fluorine atoms which tends to inhibit the bending of the chain backbone. The extremely high molecular weight of PTFE results in a melt viscosity about a million times higher than that of most other polymers ($10^{10}-10^{12}$ Pa . s). The very high melt viscosity of PTFE even suppresses normal crystal growth. For instance, the crystallinity of PTFE when polymerized is ~90 per cent and the melting point is around 340 °C; however, after fabrication, even with slow cooling from the 'melt', the crystallinity rarely reaches 70 per cent and the melting point reduces to approximately 328 °C. This is consistent with the observation that an increase in the molecular weight (MW) of PTFE, lowers the heat of crystallization and hence the degree of crystallinity considerably [1].

FEP is a copolymer of TFE and HFP so it can be viewed as PTFE with an occasional side group of CF_3 attached. The CF_3 groups may be regarded as defects which are incorporated into the crystallites and thereby lead to a decrease in melting point. The side groups also act as impediments to the polymer chains slipping past one another, thus reducing cold flow due to creep relaxation.

PFA is a copolymer of TFE and PMVE (perfluoro(methyl vinyl ether)) ($CF_2=CFOCF_3$) in an approximately 100 : 1 mole ratio. This small amount of comonomer is sufficient to reduce crystallinity of PTFE and confer improved toughness. Also, the creep limitations of PTFE are suppressed in PFA due to increased side chain entanglement. The longer perfluoropropoxy side group in PFA has a greater effect on reducing crystallinity than the perfluoromethyl side groups in FEP. This is the reason why only such a small amount of PMVE is copolymerized with TFE to produce PFA.

ETFE polymers are stiffer and have greater tensile strength than PTFE, FEP and PFA. This is because ETFE molecules adopt a planar zigzag configuration with there being strong intermolecular attractive forces between the CF_2 groups on one chain and the CH_2 groups on an adjacent chain. ETFE also exhibits exceptionally low creep as a consequence of the electronic interchain interaction that occurs between the bulky $-CF_2-$ groups of one chain and the smaller $-CH_2-$ groups of an adjacent chain [2]. Both PVDF and ETFE have chains of alternating hydrogenated and perfluorinated units that can interpenetrate during

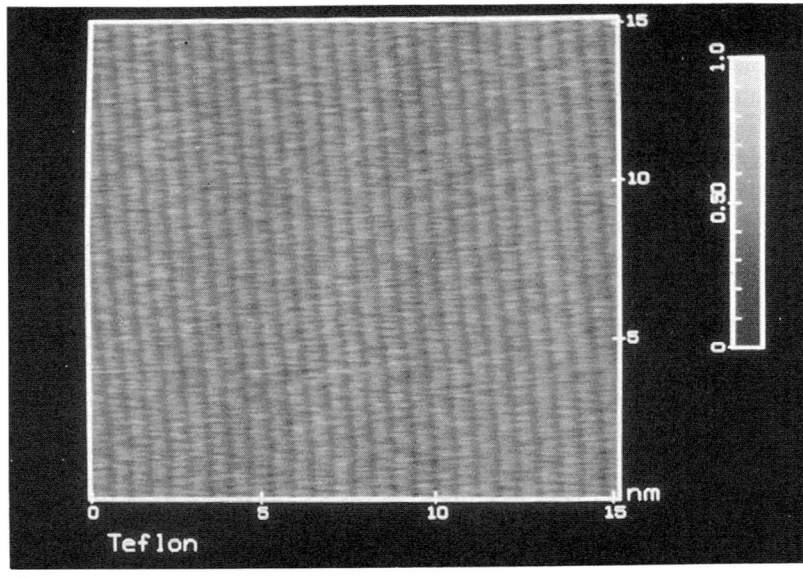

(a)

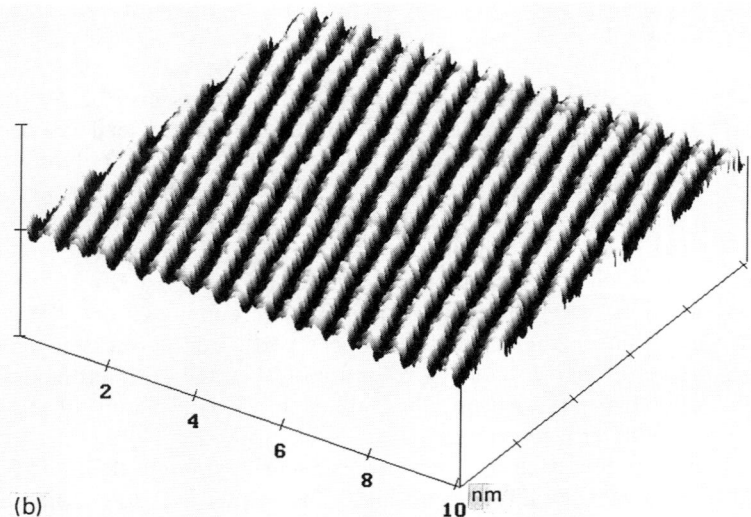

(b)

Figure 1.2. Atomic force photomicrographs of the surface of Teflon: (a) plan view (courtesy of Professor Paul Smith Materials, Department, Institute of Polymers, Swiss Federal Institute of Technology, Zurich); (b) oblique view (courtesy of Julius Vancso, Faculty of Chemical Technology, University of Twente, Netherlands)

crystallization with the larger CF_2 groups on one chain 'interlocking' with the smaller CH_2 groups on another chain.

Fully fluorinated polymers such as PTFE or FEP, on the other hand, exhibit no interpenetration, so the chains are able to slip past each other [2]. ETFE also has exceptional abrasion resistance due to the high interchain attractive forces. Because these attractive forces diminish with increasing temperatures, ETFE has good moulding characteristics such as its low melt flow which allows filling of thin sections and gives cycle times comparable to other engineering polymers. Similarly, extrusion and moulding performance of ECTFE is comparable to that of HDPE.

PCTFE has a low melting point but is more difficult to extrude than FEP. The presence of the chlorine atom promotes interchain attractive forces thus improving the mechanical properties of PCTFE relative to PTFE. PCTFE has superior creep resistance, together with greater hardness and tensile strength compared to PTFE. Also, with PCTFE, the presence of the chlorine atom, which has a greater atomic radius than fluorine, hinders the close packing that is possible in PTFE. This results in a lower melting point and a reduced propensity for polymer crystallization [3]. Similarly, ECTFE has good stiffness due to the interchain attraction supplied by the incorporation of the chlorine substituent.

2.1.1 Mechanical Properties

Figures 1.3 and 1.4 show that mechanical performance properties of fluoropolymers can be ranked into two categories based on whether the polymers are fully fluorinated or contain hydrogen atoms in their structures. Generally, the fluoropolymers with hydrogen in their structure have about 1.5 times the strength of fully fluorinated polymers and are about twice as stiff. Fully fluorinated polymers, on the other hand, exhibit higher maximum service temperatures and greater elongation.

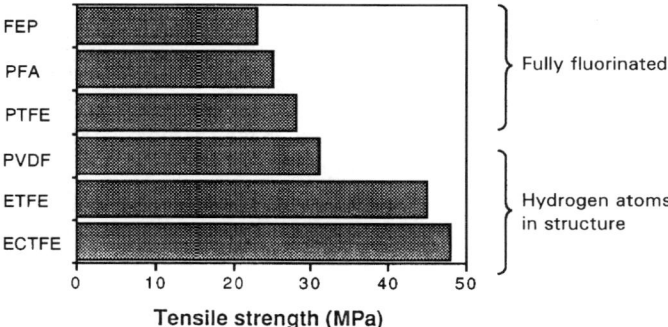

Figure 1.3. Tensile strength values for commercial fluoropolymers (according to ASTM D638; data from Imbalzano [11], 1991)

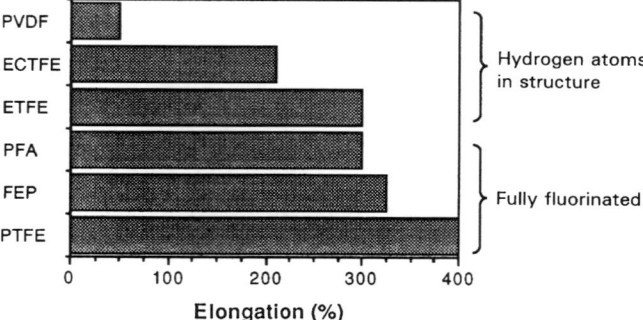

Figure 1.4. Elongation values for commercial fluoropolymers (according to ASTM D638; data from Imbalzano [11], 1991). Note: values are nominal and can vary by 20 per cent

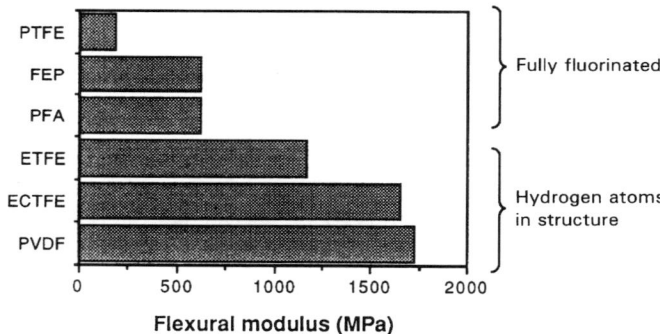

Figure 1.5. Flexural modulus values for commercial fluoropolymers (according to ASTM D790; data from Imbalzano [11], 1991)

PVDF has one of the highest flexural moduli amongst the fluoropolymers (Figure 1.5). Its high modulus, however, can be intentionally decreased (i.e. flexibility increased) by copolymerizing with HFP (hexafluoropropylene) (<15 per cent). Two such grades are Solef® 11010 and Kynar® Flex 2800 both of which have increased impact strength and elongation but decreased modulus. ECTFE and ETFE also possess relatively high modulus values due to interchain attractive forces. PTFE, FEP and PFA display low stiffness (despite the rigidity of their molecular chains) because of their very low intermolecular attractive forces.

2.1.2 Optical Properties

FEP and PFA represent the second-generation fluoropolymers which despite being melt processable are, however, crystalline (between 50 and 70%). This leads to poor optical properties (low clarity) and an inability to be dissolved thereby making the preparation of optically thin coatings difficult. Amorphous fluoropolymers (such as Teflon® AF) represent the third generation of fluoropolymers. They incorporate a dioxole ring whose bulky cyclic structure hinders crystallization. As a result, the polymer has exceptionally high clarity and excellent optical properties. In fact, Teflon® AF has one of the lowest refractive indices of any plastic [4]. In addition, the polymer has high transparency across the entire spectrum from the UV to the near infra-red making it an excellent candidate material for fibre optics and optical devices.

2.2 FLUOROELASTOMERS

Fluoroelastomers are predominantly based on VDF together with other structures which disrupt the crystallinity which homopolymer PVDF would otherwise achieve. These elastomers thus exploit the favourable properties conferred by short sequences of VDF, with the added benefit of low or negligible crystallinity.

In contrast to the rod-like microstructure of PTFE in which the elementary fibrils are approximately 6 nm wide and the molecular chains are all extended, elastomers based on VDF–HFP and TFE–VDF–HFP possess fine particles 16–30 nm in diameter [5]. Figure 1.6 is a transmission electron micrograph showing the emulsion particles of a VDF–HFP elastomer. From this micrograph a fine textured structure can just be discerned on the surface of the elastomer particles. Figure 1.7 is a higher magnification image of the same elastomer. Here the fine structure in the form of spherical particles, is clearly visible. The properties of VDF–HFP elastomers such as their resilience and flexibility can be related to these spherical domains being interconnected. The diameter of these fine particles has been found to be proportional to the MW of the elastomer [5].

2.2.1 Compression Set

Fluoroelastomers find primary application in seals where their role is to prevent leakage after prolonged compression. Fluoroelastomers are rarely used in applications where they are in tension, more often than not they are in compression. Compression-set resistance is a fundamental property of elastomers for O-ring fabrication since a proper seal can only be maintained when the O-ring attempts to revert to its original uncompressed dimensions. Compression-set resistance can be optimized by selecting fluoroelastomers with high MW and/or those crosslinked with a bisphenol-cure system. Another cure system which gives excellent high-temperature compression-set resistance is the peroxide cure system in which free radicals attack a reactive bromine atom attached to the chain.

Compression-set resistance is also strongly dependent on the type of fillers used. The best general-purpose filler for fluoroelastomers is MT Black since it provides an excellent balance of compression set and heat resistance. However, if superior high-temperature compression-set resistance is required then coal fines

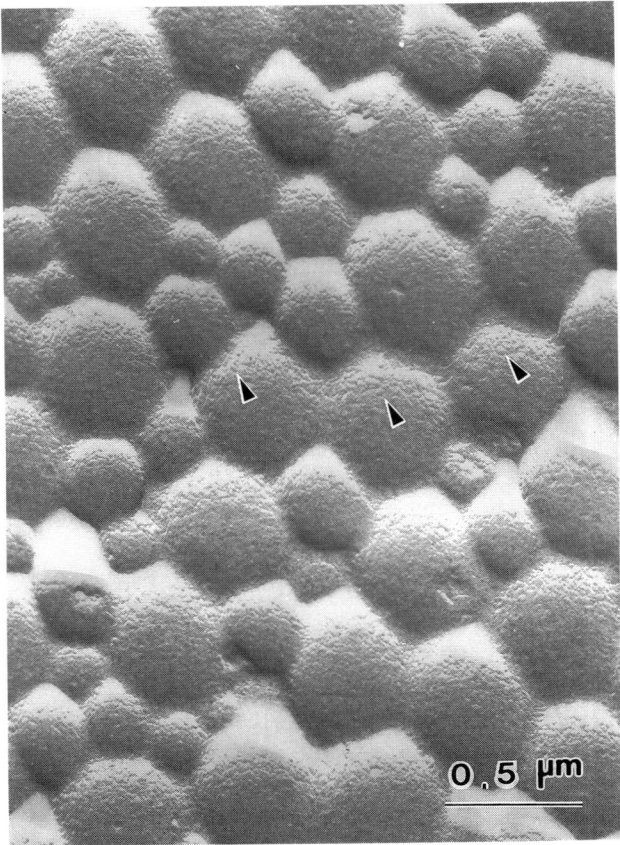

Figure 1.6. Transmission electron micrograph of emulsion particles of VDF–HFP using the surface replication and Pt–Pd shadowing technique. Note the textured microstructure as indicated by the arrows (courtesy of S. Yamaguchi [5], Daikin Industries, Japan)

can be used, but there will be a sacrifice in the processing and tensile/elongation properties since coal dust is less reinforcing then MT Black. Another carbon filler known as SRF Black can be used to give high strength and high modulus compounds, although in the case of peroxide-cured fluoroelastomers mould sticking may occur. In the case of non-black fillers, barium sulphate gives the best compression set; however, it has poorer tensile properties than MT Black. For improved tensile properties, diatomaceous silica should be used [6].

In addition to compression-set resistance, elongation is a another important property to be considered for fluoroelastomeric seals and O-rings since some elongation may occur during installation of the O-ring.

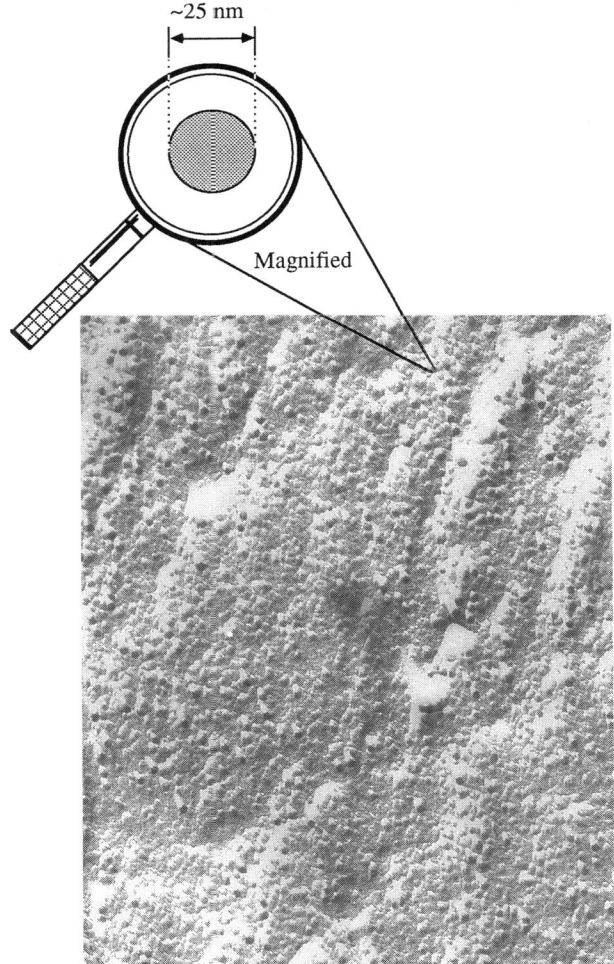

Figure 1.7. Higher magnification micrograph of the VDF–HFP elastomer shown in Figure 1.6. The spherical sub-particles which are visible represent the molecular microstructure of the fluoroelastomers. The elastomeric properties of fluoroelastomers can be related to this fine structure (courtesy of S. Yamaguchi [5], Daikin Industries, Japan)

3 MECHANICAL FAILURE MODES

While the mechanical properties of fluoropolymers are often a secondary consideration in their selection criteria, their tensile, flexural and creep properties are nevertheless of fundamental importance in any engineering application.

3.1 TENSILE FAILURE

Stress–strain data on fluoropolymers as a class indicate that these materials perform much better in compression than in tension. For example, depending on its crystallinity, PTFE has a tensile yield strength of about 12.4 MPa (30 per cent strain) at 23 °C; in compression, however, this value can be as high as 25.5 MPa (25 per cent strain).

The mechanical properties of fluoropolymers are only modest when compared to engineering or even commodity plastics. Recently the failure of a PVF component in a rocket engine has been reported to be responsible for the malfunctioning of an Atlas rocket [7]. This was attributed to the low tensile strength of PVF which is about half that of PET.

3.2 CREEP

Certain fluoropolymers suffer from deformation under load which is termed creep or 'cold flow'. PTFE and FEP are the fluoropolymers which are most susceptible to creep. Careful design consideration must be given to articles fabricated from PTFE and FEP, particularly products intended for service under conditions of continuous stresses. PFA exhibits lower creep than both PTFE and FEP due to increased side chain entanglement.

The creep limitations of PTFE can be overcome, in part, through the incorporation of specific fillers such as carbon or bronze powder. While these dramatically improve the deformation under load characteristics of PTFE, the electrical insulating properties suffer. Also, bronze powder and other fillers based on transition

Table 1.2. Izod impact strengths for filled and unfilled fluoropolymers (data from Waterman [50], 1994)

Polymer	Impact strength (J/cm notch)
ECTFE	No break
ETFE	No break
PFA	No break
FEP.	No break
PFA + 20% glass fibre	7.0
ETFE + 30% glass fibre	4.0
ECTFE + glass fibre	3.7
ETFE + 30% carbon fibre	2.4
PTFE/PPS alloy	2.21
PTFE + 30% glass fibre	2.2
PVDF	2.1
FEP + 20% glass fibre	2.0
PVF	1.8
PCTFE	1.4
PTFE + 15% graphite	1.4
PTFE	1.3–2.1
PVDF + 20% carbon fibre	1.2

metals can catalyse the thermal degradation of PTFE [8]. The creep resistance of filled PTFE can be improved further by inert gas sintering which leads to a decrease in the degree of porosity.

The addition of fillers to improve creep resistance can lead to a reduction in elongation. In the case of ETFE, the addition of 25 per cent glass fibre doubled its tensile strength and creep resistance, but caused a 25 times reduction in elongation. In general, the addition of fillers leads to a decrease in the impact strength of the fluoropolymer (Table 1.2). Some glass-filled PTFEs are also prone to discoloration. Figure 1.8 shows the fracture surface of glass-filled PTFE.

PTFE is widely used for gasket manufacture because of its wide-ranging chemical resistance. However, PTFE is subject to cold flow, even when glass

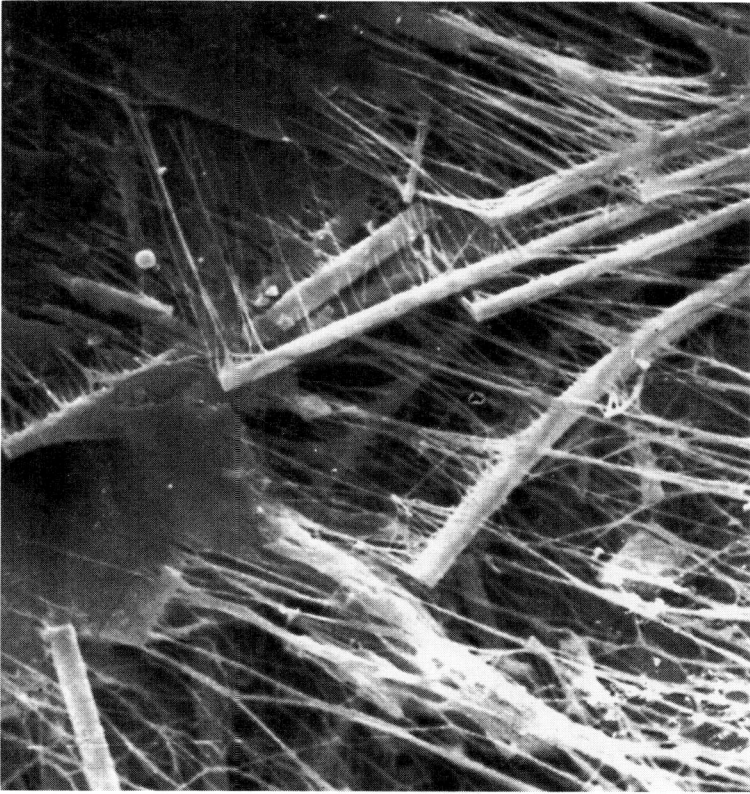

Figure 1.8. Ductile fracture in a glass-fibre reinforced PTFE component. Note: the fibrillation of PTFE demonstrates its excellent adhesion to the glass fibres (from Engel [118], *Atlas of Polymer Damage* reproduced by permission of Carl Hanser Verlag)

reinforced. In order to maintain a reliable seal with PTFE gaskets it is necessary to totally enclose them (Figure 1.9) or to use a metal containment ring (see Figure 1.10).

Alternatively, an alloy of PTFE and polyphenylene sulphide is available which is less susceptible than PTFE to creep and cold flow. These PPS–PTFE alloys find application in packings, bushings and seals, although the material is in limited availability outside the USA. Alloys comprising 65 wt% of PTFE, 20 wt% of polyether sulphone and the balance graphite are also available and these possess

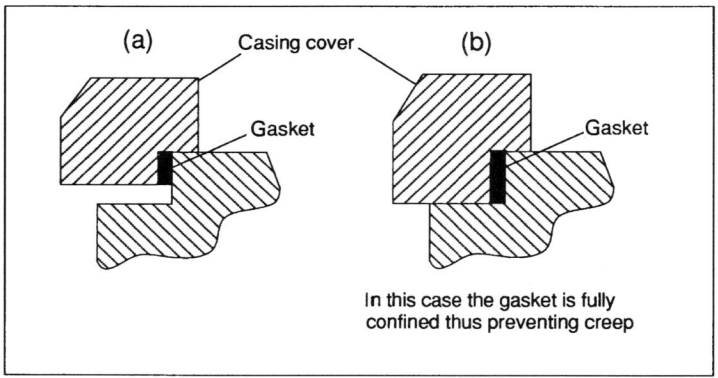

Figure 1.9. (a) PTFE gasket can undergo 'cold-flow' leading to premature seal failure; (b) improved design prevents creep of gasket

Figure 1.10. Examples of steel back-up rings around PTFE gaskets to limit the cold-flow of the polymer (courtesy of Dow Plastic-Lined Piping Products [31])

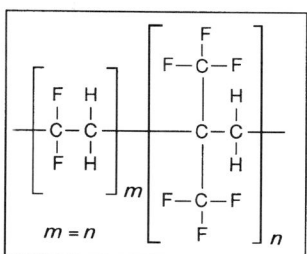

Figure 1.11. An alternating copolymer of VDF and HFIB (hexafluoroisobutylene). Note that the bulky trifluoromethyl groups shield the protons along the backbone against chemical or oxidative attack. The trifluoromethyl groups also impart outstanding creep resistance to this polymer

good creep resistance as well as abrasion resistance and can be used for pistons and bearings [9].

PTFE having a highly fibrillated, expanded structure confers an improved resistance to creep and is thus a suitable material for leak-resistant gaskets. This expanded PTFE, commercially available as Gore-Tex® GR, is produced by exploiting the unique fibrillation capability of dispersion-polymerized PTFE and is manufactured by W. L. Gore and Associates Inc.

An alternating copolymer of HFIB (hexafluoroisobutylene) and VDF has been prepared which has outstanding creep resistance together with excellent mechanical and chemical properties. This polymer has been patented by Allied Chemicals [10] under the designation CM-1. From Figure 1.11 it can be seen that the two trifluoromethyl groups of the HFIB unit shield the protons effectively from thermal and oxidative attack and confer the creep resistance, while the methylene groups in the main chain lead to strong interchain forces with CF_2 and CF_3 groups on adjacent chains.

3.3   FATIGUE RESISTANCE

The flex life of the polymer is a critical property in applications such as bellows and flexible hoses. ETFE has outstanding resistance to fatigue failure. It can withstand between 6 and 12 million folding cycles before failure (according to ASTM D2176). In contrast, FEP can only withstand up to 80 000 cycles before failure. PFA is marginally better than FEP in terms of fatigue resistance, being able to withstand up to half a million cycles. PTFE can withstand more than 1 million cycles of folding [11]. The flex life of the polymer is strongly dependent on crystallinity in the case of PTFE. However, in the case of PFA, increased crystallinity seems to have little effect on the flex life [12]. PVDF possesses relatively high crystallinity (68%) and exhibits exceptional ability to withstand repeated flexure and fatigue.

3.4   WEAR

Despite its very low coefficient of friction, unfilled PTFE wears rapidly and is unsatisfactory for use in high-speed bearings. Wear performance of PTFE can be improved through the use of special fillers; however, certain fillers such as molybdenum disulphide can sensitize the polymer towards thermal degradation at temperatures approaching its melting point. The wear resistance of PTFE can be improved 250 times by the addition of bronze powder.

3.5   CRACKING

The presence of voids in PTFE can also affect its fracture toughness and mechanical properties since a crack front will preferentially propagate through a system of voids. The specific gravity of PTFE for optimum mechanical properties is 2.15–2.23 g/cm^3.

Microcracks can also be introduced in PTFE during its fabrication. Heating PTFE to temperatures above 327 °C causes it to turn to a translucent gel state. PTFE can be fabricated when in this gel state to produce flared pipes and flange faces. However, during this operation PTFE is very prone to cracking due to its sensitivity to mechanical shock. Care must be taken during PTFE forming operations to prevent unnecessary jarring or impact. Cracks initiated during forming of PTFE pipe liners can cause premature pipe failure if the defect is undetected during installation.

3.6   STRESS CRACKING

Certain combinations of chemicals and mechanical load can induce stress cracking in fluoropolymers, especially PVDF. In fabrication by injection moulding, slow cooling optimizes the crystallization process and relaxes stresses but leads to uneconomical cycle times. In general, the broader the molecular weight distribution (MWD) of the PVDF, the greater their resistance to stress cracking [13]. Annealing cycles will also provide increased stress crack resistance by relieving residual stresses. Ideally, FEP components also should be annealed to prevent stress cracking in service. Heating FEP at 260 °C and allowing it to cool slowly results in stress relief.

The stress crack resistance of ETFE varies inversely with its processability. There are several grades of ETFE available and those with improved processability can be plagued by lower stress crack resistance.

Rotomoulding is a popular fabrication method for fluoropolymers. Articles made from fluoropolymers by rotomoulding must be cooled in a uniform manner to avoid the creation of residual stresses which are detrimental to the performance of the part. The mouldings should first be cooled in ambient air, then by a water spray and finally with a stream of air. In an attempt to produce more economical cycle times, some manufacturers may employ a more severe cooling regime and thus introduce built-in stresses that can accelerate in-service part failure [12].

4 CHEMICAL RESISTANCE

4.1 FLUOROPLASTICS

4.1.1 PTFE

Despite all the drawbacks associated with PTFE such as poor processability, potential voiding, creep relaxation and permeability it is still the polymer of choice when it comes to handling aggressive acids (e.g. HF, H_2SO_4) at high temperatures (e.g. 280–450 °C). No other polymer can match its resistance under this combination of conditions. The excellent chemical and thermal stability of PTFE can be accounted for by its C−F bond energy being one of the highest known (481 kJ/mol), whereas its interchain attractive forces ranks among one of the lowest (~3 kJ/mol). Although PTFE exhibits outstanding chemical resistance, it can absorb limited amounts of compounds which contain little or no hydrogen [14].

4.1.2 FEP

The structure of FEP resembles that of PTFE, except that a perfluoromethyl group replaces a fluorine atom on each repeat unit. The presence of this bulky side group tends to distort the highly crystalline structure of the PTFE chain and results in a higher amorphous fraction. Since organic solvents permeate through the amorphous phase (the crystallite regions not being permeable), it would be expected that FEP is more susceptible to permeation than PTFE. However, since FEP resins are melt processable they are void free and permeation can only proceed via molecular diffusion, unlike in PTFE where both molecular diffusion, and porous transport are active. Overall, its permeation characteristics are slightly superior to those of PTFE. Table 1.3 shows that the permeability of various

Table 1.3. Permeability of FEP to chemicals at 23 °C. Note: FEP being essentially non-polar is not permeated significantly by ketones (in contrast to PVDF); however, it is quite permeable to aliphatic hydrocarbon solvents (data according to ASTM E96-35T using 24 μm film thickness)

Permeant	Permeability constant[a]
Ethyl acetate	0.27
Acetone	0.37
Benzene	0.75
Ethanol	1.61
Toluene	5.38
Water	8.14
Dipentene	23.50
Decane	112.18

[a] Units, $mol/(m.s.Pa) \times 10^{15}$.

solvents through FEP is a function of their polarity, with ketones showing a low tendency to permeate and aliphatic hydrocarbons such as decane having high permeation constants. It should be noted, however, that FEP is somewhat susceptible to permeation by H_2SO_4 (98%) despite the acid being quite polar [15].

4.1.3 PVDF

At first glance the properties of ETFE and PVDF would seem to be similar; however, the alternating units of CH_2 and CF_2 in PVDF induce a dipole that makes the polymer susceptible to attack by highly polar solvents such as ketones and amides. Some highly polar solvents such as dimethylacetamide

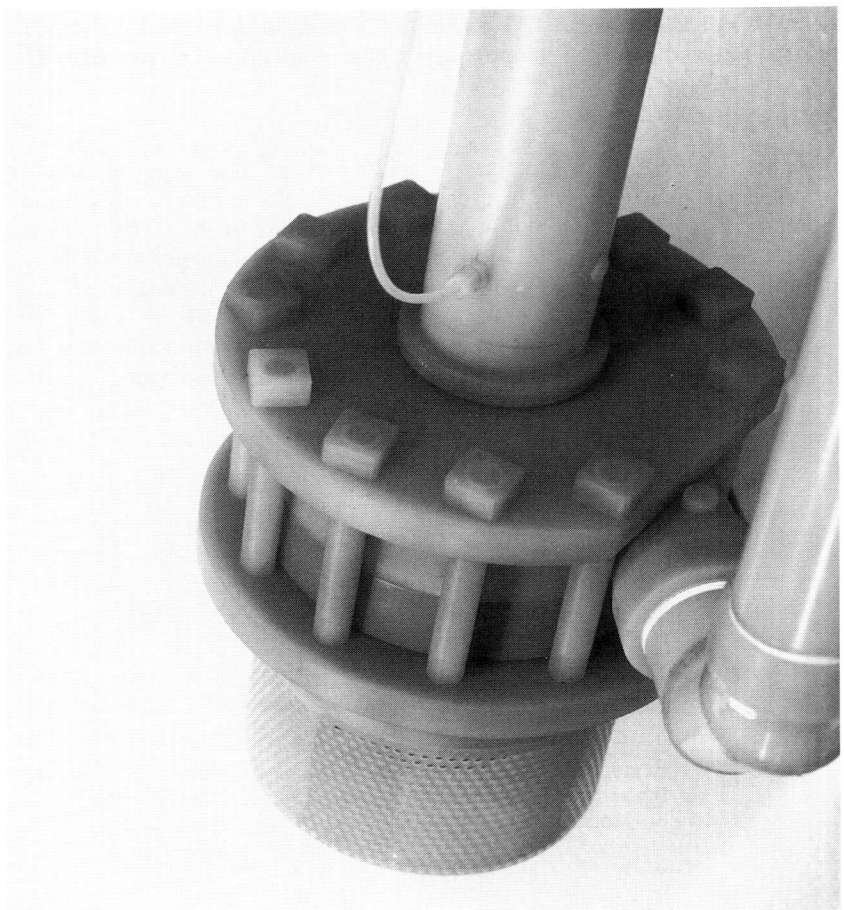

Figure 1.12. Example of an all-fluoropolymer construction used in the chemical process industry (courtesy Vanton Pump & Equipment Corp.)

and dimethylformamide dissolve PVDF at elevated temperatures while organic amines and other strong nucleophiles cause discoloration and embrittlement. In contrast, ETFE has no known solvents at room temperature. PVDF also has poor tolerance to methylene chloride and pyridine while fuming sulphuric acid leads to sulphonation [3]. PVDF is not recommended for flue gas desulphurizers due to condensation of $H_2 SO_4$ which occurs within the desulphurizer; instead, PTFE is the preferred choice for such applications.

PVDF components should thus not be used in contact with ketones, strong bases, amines and esters. These can cause swelling, softening and even dissolution depending on the conditions. In fact, certain ketones and esters (e.g. isophorone and dimethyl phthalate) act as latent solvents for PVDF dispersions and coatings. Generally in the case of ketones, the longer the chain and the greater the degree of branching, the lower the solvating power of the ketone for PVDF [13]. PVDF does, however, have excellent resistance to bromine and is the material of choice in that industry (Figures 1.12 and 1.13).

4.1.3.1 Alkaline Degradation

PVDF can degrade severely in the presence of strongly alkaline liquids. This is one of the main failings of PVDF. It can break down extensively on exposure to NH_4OH after a relatively short period of time (e.g. within two months). The degradation of PVDF by NH_4OH occurs through a dehydrofluorination mechanism and leads to carbon–carbon double bonds. The α crystalline form of PVDF is more susceptible to dehydrofluorination because of its less sterically hindered conformation. PVDF from different manufacturers can vary in their relative proportions of α, β and γ crystalline forms. Thus, some grades containing a higher α crystalline content are inherently more susceptible to thermal degradation and discoloration [16]. Aqueous LiOH has also been found to lead to dehydrofluorination and oxidation of PVDF [17]. In fact, a mixture of potassium hydroxide (20%) solution with ethanol and acetone (1 : 9) is an effective dehydrofluorination system for PVDF and is used in the preparation of carbynes [18].

It has been found that PVDF forms a brittle layer when exposed to a 50% NaOH solution. This brittle layer subsequently cracks under an applied tensile stress (Figure 1.14). This cracking is not considered to be environmental stress cracking as this requires the simultaneous action of tensile stress and a cracking medium [19].

There have been reported failures (by cracking) of PVDF pipes and tank liners in 10% sodium hydroxide in the temperature range 52–80 °C [20]. Van Tilberg (1993) has suggested that PFA would offer superior performance over PVDF for liners used in chemical road tankers [21]. The enhanced chemical resistance of PFA relative to PVDF was demonstrated by exposing PFA to 70% nitric acid, 98% sulphuric acid and 50% sodium hydroxide for up to three months. After this time no signs of chemical attack or crazing were discernible [21]. The only significant effect on the properties of PFA was a marked increase in elongation

Figure 1.13. Pump for handling liquid bromine constructed almost entirely from PVDF (courtesy Vanton Pump & Equipment Corp.)

after 28 d exposure at 80 °C in all three solutions, although no detectable changes occurred in tensile strength or ductility. The increase in ductility with the absence of loss of strength suggests some degree of plasticization of the PFA results [21].

It has been shown that the weld region of welded PVDF sheets is attacked by a 50% NaOH solution in preference to the unwelded sections. The weld regions

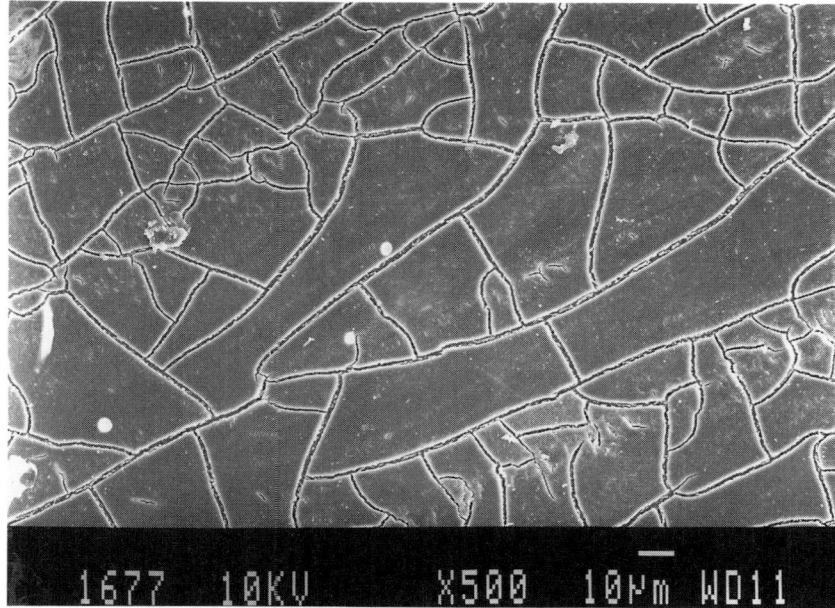

Figure 1.14.   Cracking of PVDF sheet after 28 days' exposure to 50% sodium hydroxide at 80°C (courtesy of Industrial Research Ltd, Auckland, New Zealand and reproduced by permission of Rapra Technology Ltd, Shropshire, UK.)

thus show greater deterioration (cracking and discoloration) than the surrounding material (Figure 1.15). Also, cracks form perpendicular to the weld line. Weight measurements have shown a weight gain of 0.12% in the unwelded panel and 0.37% in the welded panel [22].

4.1.3.2 Discoloration

PVDF despite its outstanding thermal and photooxidative stability is nevertheless often prone to discoloration problems. PVDF is widely used in ultrapure water applications and in the food processing industry where its discoloration is perceived as a contamination problem. PVDF discolours due to a dehydrofluorination reaction leading to conjugation in the backbone akin to that which occurs in PVC and EVA. The problem has been observed with PVDF pipes carrying hot ultrapure water [23], PVDF pipes and fittings exposed to UV sterilization lamps [24] and PVDF pipes in prolonged contact with aqueous ammonium hydroxide [16] or sodium hydroxide [25].

While the discoloration in water pipes can appear quite severe, it has been shown that only 0.1% of the PVDF is affected even when it appears totally black [26]. As a result the mechanical properties of the PVDF discoloured by hot

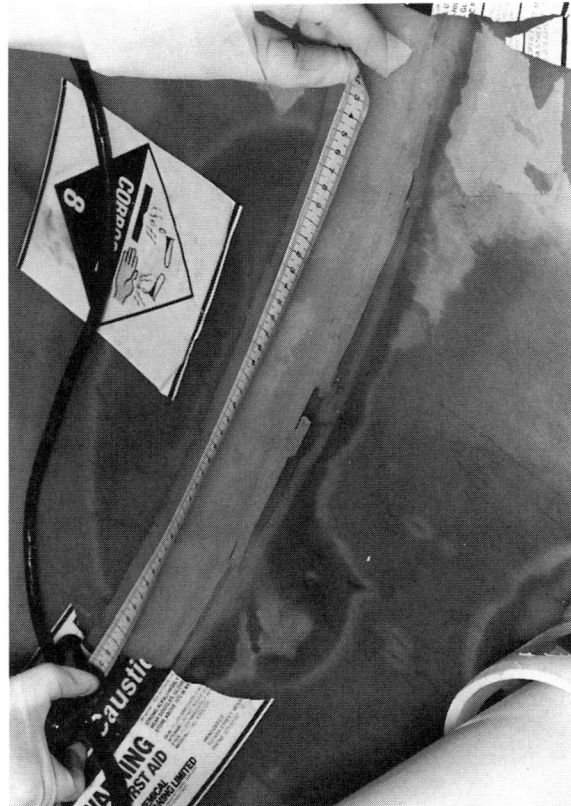

Figure 1.15. Cracking and discoloration of a PVDF liner at the weld region (courtesy of Industrial Research Ltd, Auckland, New Zealand, and reproduced by permission of Rapra Technology Ltd, Shropshire, UK).

water remain essentially unaffected. However, this is not the case where PVDF is discoloured by alkaline solutions. It has been shown that PVDF discolours from a natural, semitransparent state to a dark chocolate brown colour following exposure to ammonium hydroxide for two months [16] and that a drastic decrease in toughness accompanies this colour change.

4.1.3.3 Effect of Temperature

When determining the resistance of fluoropolymers to aggressive chemicals, the effects of temperature need to be considered. For instance, PVDF exposed to concentrated sulphuric acid at 40 °C does not cause any noticeable deterioration, however, raising the temperature to 100 °C for a limited period of time causes severe discoloration [22]. Empirical chemical resistance data for PVDF obtained from pump manufacturers shows that PVDF is attacked by acetone at room temperature. At 66 °C, glacial acetic acid, aniline, butylamine, benzaldehyde, ethyl acetate and ethyl ether may affect PVDF, while at 100 °C, nitric acid, sulphuric acid (conc.), cyclohexanol and phenol can also affect PVDF.

4.1.3.4 Testing

A method for determining the chemical resistance of PVDF has been published by Fisher [27] which involves sequential immersion of the test specimen in the selected chemical solution after which the sample is washed, dried and exposed again. This method is seen to better simulate actual tanker lining conditions. Also, it should be recognized that for some chemicals, the vapour phase exerts a greater attack on the fluoropolymer than the liquid phase. To this end ASTM C868 has been developed for cell testing of fluoropolymer linings which give chemical resistance data for the three-phases; liquid, interface and vapour as occurs in actual usage.

4.1.3.5 Antagonism Between Fluoroelastomers and Glass

It has been found that glass-lined vessels are susceptible to environmental degradation when exposed to hydrobromic acid in the presence of fluoropolymers [28]. For example, PVDF parts, fittings and joints are not recommended for use in the presence of HBr in contact with glass equipment since hydrogen fluoride which is evolved as a result of the decomposition of the fluoropolymer attacks glass liners and vessels.

4.1.4 PCTFE

In contrast to the detrimental effects it has on PVDF, sodium hydroxide (50%) exerts negligible effects on PCTFE. PCTFE is also very resistant to concentrated nitric acid, hydrochloric acid and sulphuric acid. However, ketones (e.g. acetone and MEK) cause significant swelling of the PCTFE which plasticizes the polymer and leads to a ~5% weight increase. Toluene, carbon tetrachloride and ethylene oxide also plasticize the polymer and lead to an increase in flexibility. Certain chemicals such as gasoline can cause an amber discoloration of PCTFE. Silicones tend to induce stress cracking. PCTFE is also attacked by nitrogen tetroxide, highly chlorinated–fluorinated solvents and chlorine gas [29].

4.1.5 Solvent Permeation

4.1.5.1 Liner Collapse

Fluoropolymers are widely used as liners for metal pipes in the chemical process industry for conveying corrosive chemicals. Fluoroplastic-lined pipe combines the mechanical strength and rigidity of a metallic pipe with the corrosion resistance and durability of a plastic liner. However, under certain conditions, failure of these linings can occur. Failure is due to the fact that some fluoropolymers are susceptible to permeation by certain industrial chemicals [30]. Any vapours that permeate between the liner and pipe become trapped, and the build-up of pressure together with the formation of corrosion products can induce the liner to collapse. Interlayer pressure build-up usually manifests itself as a single-lobe collapse (Figure 1.16). A three-lobe collapse as in Figure 1.17 is characteristic of a vacuum failure which is the result of the combined effect of negative pressure exerted inside the lined pipe and a softened liner.

Pipes for PTFE liners usually have spiral diamond grooves to allow permeation vapours to vent at the flange. Often fluoropolymer-lined pipe is insulated

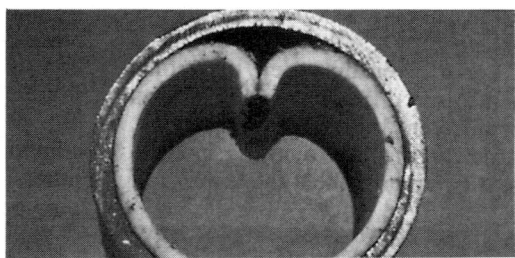

Figure 1.16.   Single-lobe collapse of PTFE liner due to chemical permeation through the liner (courtesy of Dow Plastic-Lined Piping Products [31])

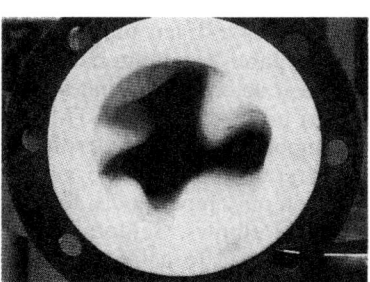
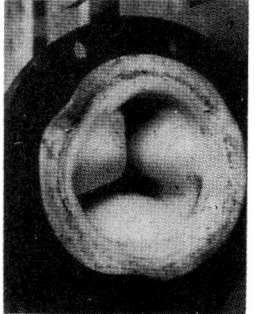

Figure 1.17.   Three-lobe collapse of PTFE liner due to a combination of chemical permeation and internal vacuum conditions (courtesy of Dow Plastic-Lined Piping Products [31])

to prevent freezing, conserve energy or to maintain a particular process temperature. If the insulation is installed improperly, it may block pathways for venting permeants. Under conditions which favour high permeation rates, blockage of the vent path may lead to collapse of the fluoropolymer liners and premature failure of the lined component [31].

4.1.5.2 Blisters

Blistering of the plastic liner can be another precursor to permeation failure. This happens when the permeant vapour partially diffuses through the wall of the plastic liner and then condenses due to falling temperatures. Once the plastic-lined pipe is reheated, the trapped liquid vaporizes creating a blister on the surface of the liner [30]. This process can also cause microvoids in the plastic liner which can increase the likelihood of crack initiation.

The occurrence of such failures can be minimized by properly considering permeation factors, careful fluoropolymer selection and the provision of adequate venting. Pipe manufacturers must incorporate venting systems in pipes and fittings to allow permeating chemical vapours to escape. For example, PTFE, PFA and FEP lined systems require that all pipes and fittings have vent holes or vent grooves; PVDF lined systems require venting only below a minimum liner thickness and ECTFE and ETFE require no venting according to ASTM F 423-91. While vent grooves and weep holes solve the problem of venting the liquid permeants, it makes the problem worse for gaseous permeants since it establishes a partial differential pressure across the liner, potentially increasing the permeation rate [32].

4.1.5.3 Temperature and Polarity Effects

Fluoropolymer liners are most often used for transporting aggressive chemicals, up to a ceiling temperature of 240°C. Above this temperature, chemical degradation of the liner and chemical permeation become significant factors. While PVDF is not recommended for contact with NaOH concentrations greater than 5%, ECTFE does handle caustic solutions (both dilute and concentrated) very well up to temperatures of 115°C [33]. One of the main uses of ECTFE is to produce rotomoulded tanks for handling and storage of acids such as nitric and hydrochloric acid. ECTFE is often used as fluoropolymer linings in tankers because of its chemical inertness combined with its low vapour permeability. Conversely, PVDF handles 96% sulphuric acid and chlorinated benzenes much better than ECTFE.

The life expectancy of FP liners can be shortened by extreme temperature, pressure, thermal cycling and specific chemicals. Higher temperatures increase the overall permeation by increasing the free volume of the polymer thus promoting diffusion and enhancing chemical solubility. Furthermore, permeation rates are strongly dependent on temperature, often increasing at an exponential

Table 1.4. Polarities of some common fluoropolymers and chemicals. Note: for maximum chemical resistance the quantitative difference between the polarity of the fluoropolymer and that of the chemical should be as high as possible (adapted from Buxton [30], 1994)

Polymer	Polarity	Chemical	Polarity
PTFE	0	Benzene	0
PFA	~0.1	Hydrochloric acid	1.1
ETFE	~0.5	Sulphuric acid	~2.0
ECTFE	~0.9	Nitric acid	2.2
PVDF	~1.4	MEK	~2.8

rate with increasing temperature. The similarity between the polarities for the fluoropolymer and the permeant chemical is the critical factor influencing whether significant permeation will occur. Higher permeation rates are shown for polymer–permeant pairs having the same polarity. For example, methyl ethyl ketone (MEK) permeates faster through ETFE or ECTFE (both of which are polar) than through PTFE (which is non-polar). The same holds for non-polar pairs also, for example, benzene permeates through PTFE roughly 10 times faster than water, even though benzene is a much larger molecule [30]. Table 1.4 shows the polarities of some common fluoropolymers and chemicals.

Permeation rates are not only affected by the type of fluoropolymer but also by the quality of the processed polymer. For example, improper sintering of PTFE will lead to poorly coalesced particles and microvoids thus increasing the rate of permeation. In addition, incorrect particle size and PTFE crystallinity will also affect the permeation rate.

4.1.6 Gas Permeability

Fluoropolymer tubing such as PFA is being used in the semiconductor industry for supplying high-purity gases. Recent work has shown that PFA is inferior to PCTFE (Kel-F) as a gas barrier [34]. In fact, the rate of permeability of oxygen is 70 times greater in PFA than for PCTFE, while that of water is 10 times greater for PFA than for PCTFE.

4.2 FLUOROELASTOMERS

The chemical stability of a fluoroelastomer is a function of its fluorine content. For example, as the fluorine content of a fluoroelastomer is increased from 66 to 70% the volume swell in certain media drops from 96 to only 10%. However, as the fluorine content of the polymer is increased so too is cost, due to the preparation and refinement of the expensive perfluorinated monomers. Table 1.5 shows the effect of increasing the fluorine content of the fluoroelastomer on its degree of swell in both benzene and hydraulic fluid.

Table 1.5.   Effect of fluorine content on solvent swell of fluoroelastomers (data from Schroeder [6], 1987)

Fluoroelastomer	% Fluorine	Volume swell (%)	
		Benzene (21 °C)	Aircraft hydraulic fluid (121 °C)
VDF–HFP	65	20	171
VDF–HFP–TFE	67	15	127
VDF–HFP–TFE–CSM	69	7	45
TFE–PMVE–CSM	71	3	10

4.2.1 VDF–HFP Elastomers

Copolymers of VDF and HFP have excellent resistance to oils, fuels and lubricants as well as to aliphatic and aromatic hydrocarbons. However, they are attacked by low MW esters, ethers, ketones and amines which is a legacy of the VDF in their structure.

4.2.1.1 Acid Resistance

VDF-based fluoroelastomers generally have good resistance to strong acids. For example, Viton®-type fluoroelastomers remain tough and elastic even after prolonged exposure at 150 °C in anhydrous hydrofluoric acid or chlorosulfonic acids. However, problems can occur when the elastomers are formulated with acid acceptors based on oxides or hydroxides of alkali metals (e.g. magnesium oxide or calcium hydroxide). In this case, acids can react directly with these additives and cause surface cracks and failure. Special grades of VDF-based fluoroelastomers are available which are formulated with lead-based acid acceptors and they offer high acid resistance. For example, VDF–HFP–TFE elastomer compounded with MgO swells by 61% in fuming nitric acid compared to a swell of only 24% when the fluoroelastomer is formulated with PbO.

One of the largest current applications for fluoroelastomers is in expansion joints of flue ducts of desulfurizers used in coal-fired plants. In this application the elastomer needs to resist both high temperatures and highly acidic flue gas steams. The most effective fluoroelastomer in this application is VDF–HFP–TFE with a peroxide cure package.

4.2.1.2 Steam Resistance

A major application area for Viton®-type fluoroelastomers is extruded profiles, hoses and seals where resistance to hot water and steam is required. Peroxide-cured fluoroelastomers offer higher resistance to hot water and steam than their counterparts cured by diamines or bisphenol compounds. Swelling of peroxide-cured Viton® GF is very low in steam and hot aqueous media; e.g. swelling by only ~3% after three weeks at 162 °C. For peroxide curing, special cure-site monomers containing a labile halogen, such as bromide or iodide must be incorporated into the polymer. Also, together with the peroxide, activators such as triallyl isocyanurate are added to enhance peroxide curability. The peroxide used for both Viton GF and Viton GLT is 2,5-dimethyl- 2,5-bis(*tert*-butyl peroxy)hexane (Luperco® 101-XL) while 2,5- dimethyl- 2,5-bis(*tert*-butylperoxyhexyne) (Luperco® XL) is used in the case of Viton GH. While all Viton G-type elastomers have excellent resistance to degradation in steam they can fail prematurely if the steam is contaminated with oil or corrosion inhibitors [35].

Since no water is generated during peroxide curing, the elastomers can be cured under mild pressure (or even atmospheric pressure) without the problem of air bubbles and voids that afflicts diamines or bisphenol-cured vulcanates. In contrast, unpressurized processing of diamines or bisphenol-cured fluoroelastomers can result in mouldings with an undesirable spongy structure. The water resistance of fluoroelastomers can be increased further by using lead oxide acid scavengers instead of calcium hydroxide.

4.2.1.3 *Amines and Other Strong Nucleophiles*

VDF–HFP fluoroelastomers can be embrittled when used in certain hostile environments containing a strong nucleophilic compound. The following applications in particular may cause VDF–HFP elastomers to fail.

4.2.1.3.1 Corrosion inhibitors Fluoroelastomers designated as FKM such as Viton® A are widely used for seals. However, it has been shown that oil additives can reduce the life of these sealing compounds. One study showed that aromatic amine additives containing the imide groups (−NH−) reduce the elongation of FKM rubber by up to 75% (Figure 1.18) [36]. One of the most serious problems with fluoroelastomer seals is their failure in oilfield applications where downtime can result in significant lost revenue. Seals frequently fail due to loss of elasticity as a result of an interaction with amine additives present in the lubrication oils.

Amine additives are added to oil as oxidation inhibitors and corrosion inhibitors. The interaction of amines with fluoroelastomers proceeds by the same mechanism which is used to cure (vulcanize) fluoroelastomers. In the first stage, dehydrofluorination produces double bonds in the polymer backbone. Because of their electron-deficient character these double bonds are susceptible to nucleophilic substitution by amines. The amine promotes crosslinking between these unsaturated sites which increases the crosslink density of the fluoroelastomer, reduces rubber elasticity and increases hardness. As a result the seals fail to perform adequately. It has been determined that disecondary butyl *p*-phenylenediamine (DBP) and ethoxy dihydrotrimethyl quinoline (EDTQ) (both oxidation inhibitors for oil) are the amine additives that exert the most powerful crosslinking effect on fluoroelastomers [37].

It has been demonstrated that aged oil was less aggressive to fluoroelastomer seals than fresh oil [38]. Recently it has been shown that water formed during the oxidation of oil reduces the detrimental effect of amines on the fluoroelastomers

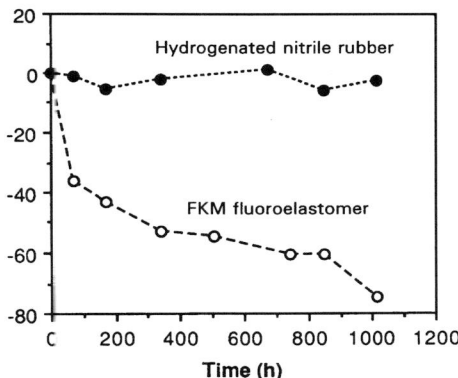

Figure 1.18. Effect of immersion of fluoroelastomer FKM (68% fluorine) and hydrogenated nitrile rubber in motor oil containing amine additives on the elongation properties of the rubbers (adapted from Dinzburg [37], 1995)

[37]. The mitigating effect of water is due to hydrolysis of the $-CF=NH-$ group formed by interaction of the amine with the fluoroelastomer.

4.2.1.3.2 Oil stabilizers Fluoroelastomer engine components (based on VDF–HFP) such as valve-stem seals and oil seal packing rings have been found to crack when exposed to engine oil at high temperatures. These failures have been attributed to extensive crosslinking and embrittlement by metal dithiocarbamate additives present in the engine oil [38]. These dithiocarbamate additives, which serve as antioxidants in the motor oil, also act as accelerators in the vulcanization of elastomers. While not actually increasing the rate of dehydrofluorination of elastomers, these additives promote crosslinking which leads to increased hardening and cracking of the rubber components. For example, fluoroelastomers aged in oil containing zinc diamyldithiocarbamate were found to lose 75% of their elongation after 28 d at 163 °C.

4.2.1.3.3 Drilling fluids Fluoroelastomers used in oil drilling are being exposed to both high temperatures, high pressures as well as a complex cocktail of gases, drilling fluids and amine corrosion inhibitors that are used. Oil wells can also present a very acidic environment to fluoropolymers since a mixture of hydrochloric acid and hydrofluoric acid is pumped into the well to dissolve the mud layers in the geological stratum. One of the main failure modes of fluoroelastomeric components is swell, since a relatively small degree of swelling can lead to a dramatic reduction in physical properties [39].

Methanol is often injected into a well during oil drilling to suppress the formation of hydrates which arise due to the presence of water in the well.

Furthermore, the combination of water, carbon dioxide and hydrogen sulphide creates a highly acidic environment which promotes corrosion, therefore organic bases (amines) are added to neutralize the system. Both methanol and carbon dioxide can swell fluoroelastomers while amines initially harden the rubber by excessive crosslinking and ultimately can dissolve the fluoroelastomer [39].

4.2.1.3.4 Chloramine Chloramine is used by some water distribution companies to chlorinate drinking water. It has been found that a changeover from free chlorine sterilization to chloramine caused the failure of a number of elastomeric O-rings and washers in the water distribution network. It was determined that chloramine solutions are particularly aggressive towards all types of elastomers. Research work has shown that Viton® A exhibits swells of 9.2 and 18% due to water absorption when exposed to aqueous chloramine solutions with pH values of 5.5 and 8 respectively [40].

4.2.1.4 Fuels

Fluoroelastomers are finding increased use for under-the-bonnet automotive applications where both high temperatures and highly aromatic unleaded petrol present a challenging environment to other elastomers (such as nitrile rubber or epichlorohydrin rubber). In particular, fluoroelastomers show excellent resistance to 'sour' petrol (i.e. fuel containing peroxides) and they have low permeation values thus minimizing loss of fuel to the atmosphere.

Care must also be exercised when selecting fluoroelastomers for applications involving contact with methanol or methanol containing fuels. Methanol, because of its high polarity can swell some VDF-based fluoroelastomers. It is necessary to select a VDF-based polymer with a fluorine content exceeding 69.5% if swelling problems are to be avoided. Such elastomers include VDF–HFP–TFE and VDF–PMVE–TFE. In particular, Kalrez®-type polymers which typically have fluorine contents around 73% exhibit outstanding resistance to methanol and methanol–fuel blends.

The increased use of cleaner burning fuels such as methanol–fuel blends and gasoline containing tertiary butyl ether has meant that material selection for fuel lines and seals has to be re-evaluated. Fluorosilicones are also good candidates for applications involving immersion in methanol–gasoline blends. The pendant trifluoropropyl group imparts outstanding solvent resistance making fluorosilicones the elastomers of choice for aviation fuels such as methanol–gasoline blends [41]. However, extensive work has shown that significant differences in fuel resistance exist between different fluorosilicones [41].

4.2.1.5 Organic Solvents

Since Viton® is used for protecitve gloves and clothing, the permeation of organic solvents through this polymer has received close attention [42]. The resistance of

Viton® fluoroelastomers to permeation by aliphatic ketones and esters is not so good and this has been related back to their fluorine content [43]. For example, it was found that the extent of swelling (i.e. absorption) of Viton® fluoroelastomers shows a dependence on the percentage of fluorine content according to the sequence Viton® 1430 > Viton® 1141 > Viton® 1144 > Viton® 1433 where Viton® 1433 (having a fluorine content of 70%) shows the lowest absorption while Viton® 1430 with 66% fluorine exhibits the highest absorption for both esters and ketones.

4.2.1.6 Chlorofluorocarbons

The failure of fluoroelastomeric seals in fire suppressant storage systems has prompted a detailed investigation [44] into the compatibility of fluoroelastomers with fire extinguishing fluids such as HFC-125, HFC-227 and FC-218. It was found that fluoroelastomer seals are severely swelled by perfluorocarbons and chlorofluorocarbon liquids. Even TFE–PMVE–CSM (i.e. Kalrez®) which has renowned chemical resistance undergoes a 40% weight gain when exposed to Freon 113 for 10 d at 25 °C. Thus, care must be exercised when specifying or using perfluoroelastomers in sealing applications with Freon®-type liquids or gases. Seals capable of withstanding liquid or vapour HCFC s have recently been reported [45] based on blends of PFA and CTFE–VDF (e.g. Dei-el Perfluoro GA-55 and Kel-F 3700).

4.2.1.7 Ozone and Chlorine

Exposure of Viton® gaskets to ozone leads to a 20% reduction in the elongation, whereas ozone exposure affects PTFE gaskets in a beneficial manner by increasing both the tensile strength and its elongation [46]. Chlorine, on the other hand, causes a 6 and 11% reduction in the tensile strength of both Viton® and PTFE gaskets respectively.

4.2.2 Perfluoroelastomers

Copolymers of perfluoromethyl vinyl ether and TFE offer virtually unmatched resistance to all classes of chemicals except fluorinated solvents. While these perfluorinated elastomers have excellent chemical resistance to a broad spectrum of aggressive chemicals they are adversely affected by hydraulic fluid (especially those containing phosphoric esters), diethylamine and fuming nitric acid resulting in swelling of the elastomer by 41, 61 and 90% respectively.

4.2.3 TFE–PP

Fluoroelastomers based on TFE–PP (Aflas®) have a relatively low fluorine content (54%) and thus show limited chemical resistance to aromatic hydrocarbons. However, the absence of VDF in the structure of this elastomer does make it resistant to highly polar solvents such as ketones which are the Achilles' heel of fluoroelastomers containing VDF. Recent work has shown that VDF–HFP elastomers uptake four times the amount of liquid acetone (105%) of TFE–PP elastomers [47]. In addition, TFE–PP fluoroelastomers display exceptional resistance to dehydrofluorination and embrittlement by organic bases such as amines. The higher resistance of TFE–PP fluoroelastomers to amine-induced degradation compared with VDF-based fluoroelastomers is because the activated hydrogen atoms of the VDF unit are not present in TFE–PP. Given this, they are usually crosslinked with a peroxide curing system unless, of course, a cure site monomer (e.g. glycidyl ether) is incorporated into the structure.

While TFE–PP elastomers also have good resistance to steam and hot acids they show extensive swelling in chlorinated solvents such as carbon tetrachloride, trichloroethylene and chloroform with swelling of 86, 95 and 112 per cent respectively after 7 days at 25 °C. Curiously, a relatively high degree of volume swelling (71%) is also seen in acetic acid [48].

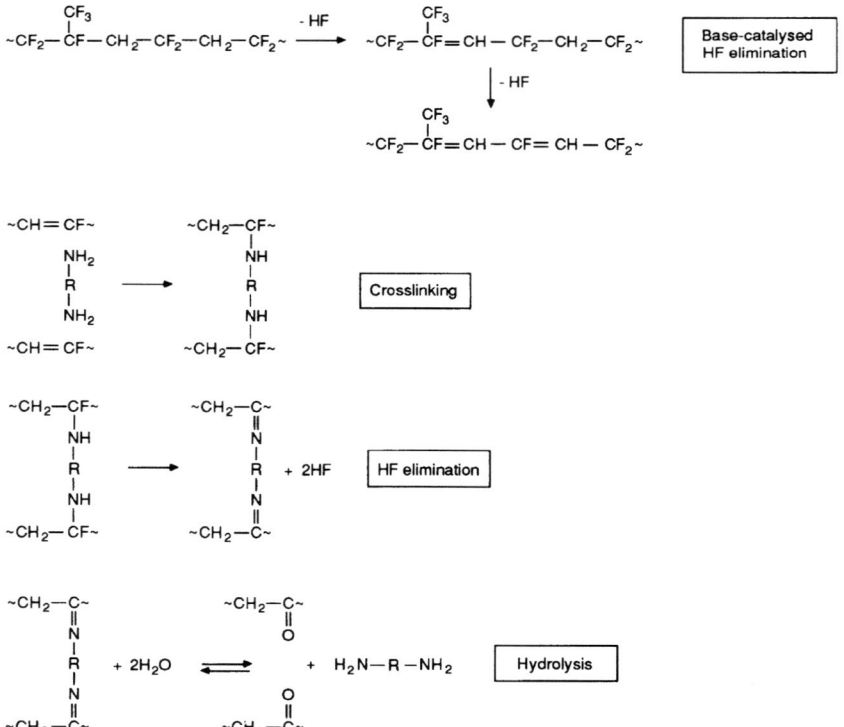

Figure 1.19. Mechanism of diamine cure of VDF–HFP elastomers. Also shown is the mechanism of hydrolytic breakdown of the crosslinks

4.2.4 Chemical Resistance of Crosslinks

There are three main systems employed for crosslinking of perfluoroelastomers: namely, peroxide cure, diamine cure and bisphenol cure. The type of cure system influences the chemical resistance of the fluoroelastomer due to the different chemical reactivity of the crosslinks. For example, the bisphenol-cured perfluoroelastomers show poor compatibility with glacial acetic acid, and volume swells of up to 32% have been reported after just 70 h exposure [49]. Bisphenol crosslinks are also easily hydrolysed. Peroxide-cured fluoroelastomers exhibit superior steam and acid resistance as compared with bisphenol cured systems. It should also be noted that bisphenol cured systems are extremely sensitive to contaminants during compounding since as little as 0.1 pph of sulphur can completely inhibit a bisphenol cure system.

Fluoroelastomers based on VDF–HFP can be crosslinked with a diamine ($H_2N-R-NH_2$) to give diimine crosslinks as shown in Figure 1.19. However, elastomers crosslinked in this way are susceptible to hydrolysis which regenerates the original diamine, leads to the formation of ketone groups in the elastomer and results in loss of desirable properties. Hydrolysis of VDF–HFP vulcanates has been confirmed by the observation that much of the diamine present in the polymer can be extracted by long-term immersion of the sample in water [3].

5 TEMPERATURE PERFORMANCE

5.1 MAXIMUM SERVICE TEMPERATURE

A mistake that is often made with PTFE is that since it has such excellent stability at high temperatures, it is automatically assumed that its mechanical properties are also maintained at elevated temperatures. This is not the case and in fact the opposite is true. The mechanical properties of PTFE undergo a dramatic decline as the temperature is raised. Both the tensile strength and elastic modulus of PTFE exhibit a severe temperature dependency (Figures 1.20 and 1.21 respectively).

Exacerbating this situation, PTFE undergoes an increase in crystallinity when held at temperatures in the crystallization zone (307–327 °C) which leads to a further decrease in tensile strength, impact strength and resistance to cyclic fatigue. Expanded PTFE (Gore-Tex® GR), on the other hand, because of its highly fibrillated structure maintains useful properties at temperatures as high as 315 °C for limited periods.

Like PTFE, FEP also lacks mechanical strength at elevated temperatures and is therefore best suited for ambient temperature applications. At room temperature, the mechanical properties of PFA and FEP are comparable; however, differences between PFA and FEP become significant as the temperature is increased, with PFA being superior. PFA can be used up to 260 °C while FEP should strictly be restricted to temperatures below 200 °C (Figure 1.22) [12].

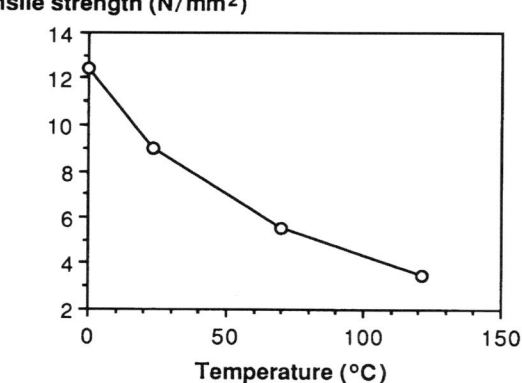

Figure 1.20. Tensile strength of PTFE as a function of temperature (data from DuPont Teflon brochure [8])

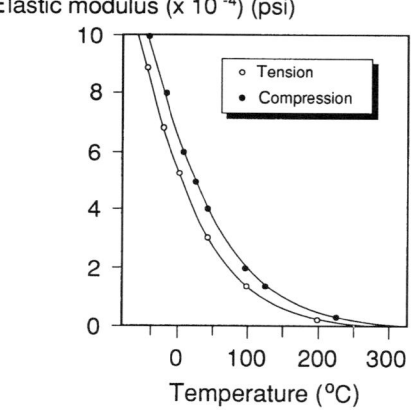

Figure 1.21. Dependency of the elastic modulus of PTFE on temperature

When considering the maximum service temperature of PVF and PVDF one must also consider the effect that structural defects play on the melting point and heat deflection temperatures. For instance, PVF and PVDF contain varying amounts of head-to-head irregularities (up to 12% for PVF) as well as chain branching (determined by polymerization temperature) both of which hold an inverse relationship with melting temperature and heat deflection temperature.

As in the case of PTFE, PCTFE shows a decline in its mechanical properties at temperatures higher than ambient (Table 1.6). In particular, the tensile strength of PCTFE diminishes by approximately 93% as the temperature is increased from room temperature to 200 °C [50].

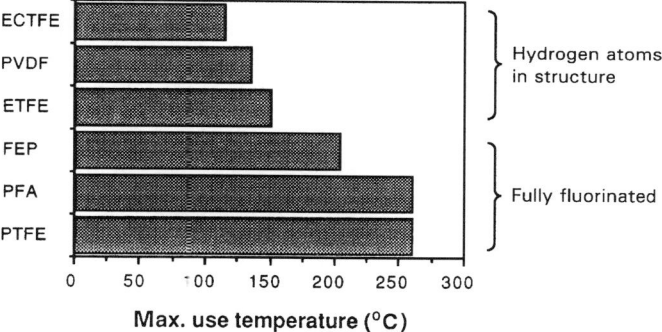

Figure 1.22. Maximum service temperatures for commercial fluoropolymers (according to UL-746B; data from Imbalzano [11], 1991)

Table 1.6. Effect of temperature on the tensile strength of PCTFE

Temperature (°C)	Tensile strength (MN/m²)
23	32
100	11
200	2

5.1.1 Heat Distortion Temperature

Table 1.7 shows the heat distortion temperature (HDT) values of various fluoropolymers. The HDT of ETFE is very low at just 70 °C; however, the addition of glass or carbon fibre fillers give remarkable improvements in this property, increasing the HDT by 240%. It is interesting to note that in the case of PFA even with 20% glass fibre reinforcement it has a HDT of only 82 °C. The alloy of PTFE/PPS has the highest HDT of any fluoropolymer-based material.

5.2 MINIMUM SERVICE TEMPERATURE

PTFE has excellent low-temperature properties and can be used down to −200 °C, in contrast to this PCTFE should only be used down to about −40 °C. Similarly, PVDF has a glass transition temperature (T_g) of about −40 °C, thus this is also the lowest service temperature for this polymer. Although PVDF can also exhibit some ductility limitations at very low temperatures, recent work has overcome this by blending in minor amounts of fluorosilicone elastomer, thereby improving the low temperature impact properties of PVDF without compromising the melting point and crystallinity of the polymer [51].

The low-temperature flexibility of fluoroelastomers is dictated mainly by their glass transition temperature which in turn depends on the freedom of motion of

Table 1.7. Heat distortion temperatures for various fluoropolymers (data from Waterman [50], 1994)

Polymer	Heat Distortion Temperature ($^{\circ}$C)
PTFE/PPS	>260
ETFE + 30% carbon	241
ETFE + 30% glass	238
ECTFE + glass fibre	200
FEP + 20% glass fibre	158
PVDF	85–148
PVDF + 20% carbon	140
PTFE	120
PTFE + 15% graphite	95
PVF	82
PFA + 20% glass fibre	82
ECTFE	75
ETFE	70

Note: HDT measurements obtained at 1.8 MN/m^2

segments of the polymer chain. If there is good flexibility and rotation of chain segments, then the fluoroelastomers will possess a correspondingly low T_g and will exhibit good low-temperature properties. The brittle point of a fluoroelastomer can be determined using ASTM D-2137.

The importance of low-temperature performance of elastomeric seals is underscored by the case history of the Challenger disaster. Shuttle engineers from Morton Thiokol (a NASA contractor) had in fact expressed concern prior to launch about the O-rings on the shuttle's solid rocket boosters stiffening in the cold and thereby losing their ability to effect an integral seal.

Figure 1.23 shows the useful temperature range for commercial fluoroelastomers. VDF–HFP fluoroelastomers represent the cornerstone of the fluoroelastomer industry, but unfortunately with a T_g of only −20 °C they exhibit poor low-temperature properties. New fluoroelastomers with improved low-temperature properties are being developed to remedy the deficiencies of current VDF–HFP-type fluoroelastomers. Fluoroelastomers based on VDF copolymerized with perfluoroalkyl vinyl ethers (e.g. PMVE) have superior low-temperature properties; however, these elastomers are also much more expensive than their VDF–HFP counterparts. Figure 1.24 shows the structure of Aftex® which is a copolymer of ethylene and a complex perfluorovinyl ether.

Plasticizers have been investigated as a means of improving the low-temperature flexibility of fluoroelastomers, but since they are generally less stable than the fluoroelastomer itself, they tend to decrease the overall thermal stability of the elastomer and accelerate deterioration. An exception is Technoflon® FOR which is purported to contain a stable fluorinated amide plasticizer which improves low-temperature flexibility and suppresses the brittleness point without sacrificing other physical properties. Peroxide-cured fluorocarbon elastomers can

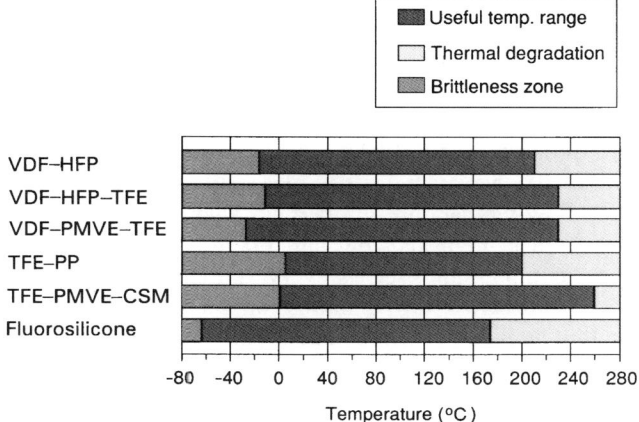

Figure 1.23. Useful service temperature ranges for commercial fluoroelastomers

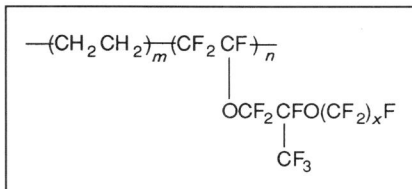

Figure 1.24. Structure of Aftex® which is a copolymer of ethylene and a complex perfluorovinyl ether

also be blended with fluorosilicones to improve low-temperature properties, but at the expense of high-temperature stability and solvent resistance.

The US Air Force employs a stringent testing criteria to specify fluoroelastomers for hydraulic seals on their aircraft. It is designed to simulate service conditions experienced in aircraft hydraulic systems and involves cycling the temperature from 163 to −54 °C while hydraulic fluid pressure on one side of the seal is cycled from 50 to 1500 psi. Test results on VDF–HFP elastomers show that those crosslinked with a dihydroxy aromatic system (denoted as E-60C) seal at −29 °C but leaked at −34 °C, while the low compression set Viton GLT materials sealed at −45 °C but leaked at −54 °C [35].

In TFE–PP copolymers, the TFE : PP molar ratio is about 55 : 45 with ≈70% of TFE units alternating with the PP units forming tetrads. The polymer is almost amorphous. While TFE imparts improved thermal stability to a fluoroelastomer it has a negative effect on the low-temperature flexibility. For example, TFE–PP elastomers exhibit good high-temperature properties but poor low-temperature performance since their T_g ($-2\,°C$) is quite high for an elastomer. Hence, these materials should not be specified in sealing applications at subzero temperatures.

Figure 1.25 shows the brittleness temperatures of various fluoroelastomers compared to nitrile rubber. Nitrile rubber, despite having very good resistance to oils and alcohols, cannot compete with fluoroelastomers in terms of low-temperature properties. While Viton® GF has a brittleness temperature of $-30\,°C$, fluorosilicones remain flexible until $\approx -60\,°C$. Together with their outstanding low-temperature properties, fluorosilicones also show low compression set, both at high and low temperatures, as shown in Figure 1.26. Low-temperature compression set is very important for low-temperature sealability [52]. Obviously, a low compression set is favourable for adequate resilience in sealing applications. In contrast, both nitrile rubber and Viton® GF have low-temperature compression sets approaching 100%, reflecting their low resilience at $-30\,°C$.

Fluorosilicones have outstanding low-temperature properties because the oxygen atoms in the backbone bestow a high degree of chain flexibility and permit a high level of mobility of chain segments. However, since they contain less fluorine than fluorocarbon elastomers they possess inferior high-temperature stability and solvent resistance. The aircraft and aerospace industries employ more fluorosilicone than fluorocarbon elastomers due to the extreme low-temperature demands (down to $-65\,°C$) that high-altitude flight places upon seals and O-rings.

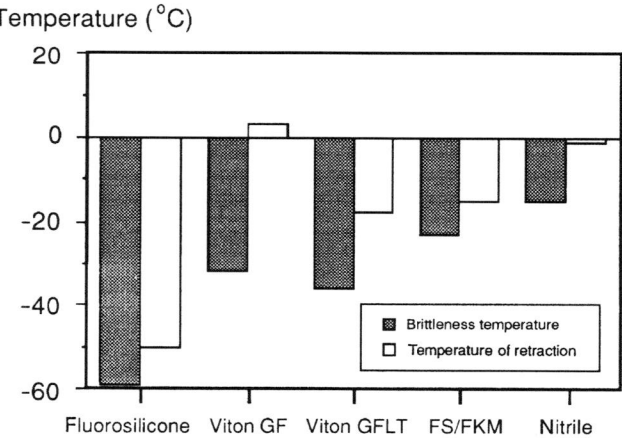

Figure 1.25. Brittleness temperatures and temperatures of retraction of fluoroelastomers with nitrile rubber being shown for comparison. FS/FKM is a fluorosilicone/fluorocarbon blend. Note that only the fluorosilicone rubber passes the $-40\,°C$ temperature requirement established by automotive design engineers (data from Fiedler *et al.* [52], 1990)

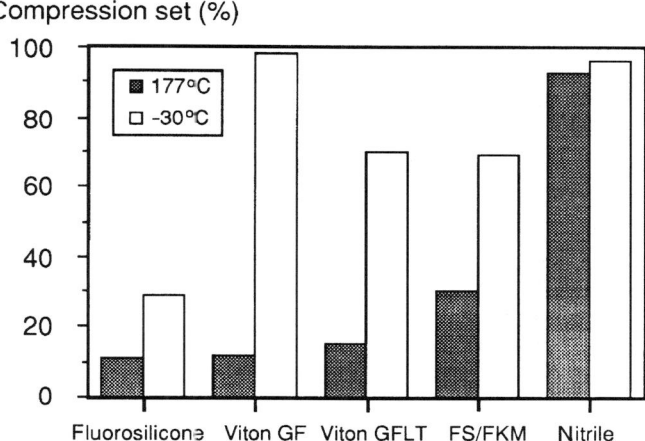

Figure 1.26. Compression set of fluoroelastomers after 22 h at −30 °C and 177 °C. Nitrile rubber is shown for comparison. Note that the fluorosilicon shows the best low-temperature compression set resistance as indicated by a compression set of only 28% (data from Fiedler *et al.* [52], 1990)

The Wright Patterson Air Base laboratories of the US Air Force have developed a thermally stable fluorosilicone based on poly(fluoroalkylarylenesiloxanylene) known as FASIL. Seals for supersonic aircraft which fly at velocities of Mach 2–3 must be able to tolerate temperatures up to 315 °C in contact with fuel. Fluorosilicones with the chemical structure shown in Figure 1.27 are used by the US Air Force for this application. In contrast, seals produced from polysulphide rubbers can only withstand continuous service temperatures of 120 °C [48].

The F-111 fighter bombers use fluorosilicones as fuel tank sealants which exploit their low-temperature flexibility and resistance to jet fuels [53]. In

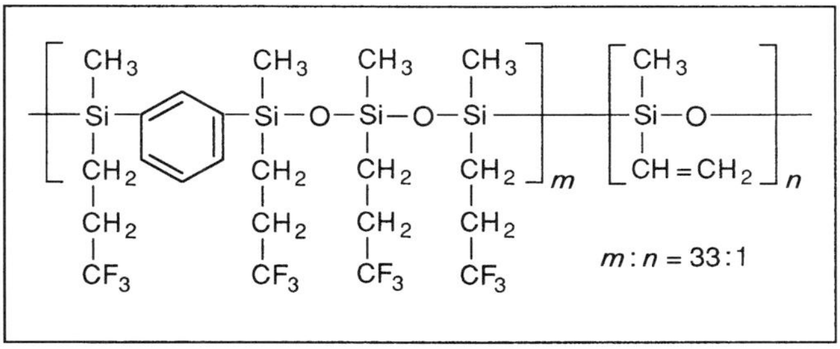

Figure 1.27. Poly(fluoroalkylarylenesiloxanylene) (FASIL) developed by the US Air Force for seals for supersonic aircraft where it can tolerate temperatures up to 315 °C and exposure to jet fuel

addition, many commercial airliners use fluorosilicone putty to seal their wing structures, since all the fuel is carried in the wing. It is important to note, however, that most fluorosilicones that are cured at room temperature employ an acetoxy cure system that evolves acetic acid on curing. The acetic acid can accelerate corrosion of steel structures and wiring installations.

As fluorosilicones cannot evolve substantial quantities of HF they cannot cause stress cracking of titanium, and furthermore they demonstrate good adhesion to titanium and aluminium. Given this combination of factors (together with their resistance to temperature extremes and JP-4 jet fuel), they are the elastomer of choice for military aircraft which employ titanium and aluminium superstructures. The future usage of these fluorosilicone elastomers is set to increase as larger supersonic passenger aircraft become a reality. There is still some development to go, however, since the material for this application must be able to withstand 60 000 hours (6.7 years) at 177 °C. Moreover, Mach 2.7 planes can develop a skin temperature of about 232 °C. The fact is that the speed of these high-speed civil transports may well be dictated by the temperature capabilities of the fuel tank sealants [54].

A group of fluorine-containing elastomers, which are little known at present, but nevertheless can be obtained commercially, are the fluoroalkoxyphosphazenes. They share many similarities with fluorosilicones such as good low-temperature resistance but also experience their limited thermal stability. Fluorophosphazenes have a backbone comprised solely of alternating phosphorus and nitrogen atoms (Figure 1.28) hence these elastomers are more

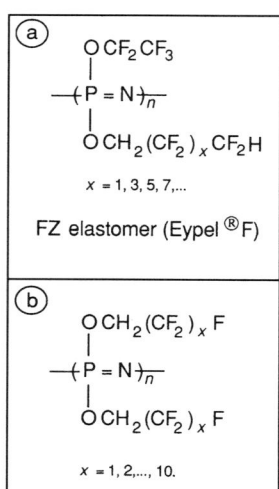

Figure 1.28. Fluoroalkoxyphosphazene elastomers have backbones comprised solely of alternating phosphorus and nitrogen atoms which give them excellent low-temperature properties

strictly referred to as phosphonitrile fluoroelastomers. This structure confers better low-temperature properties than fluorocarbon elastomers (down to below −65 °C) but inferior high-temperature performance (i.e. they can only be used up to 175 °C). They are also similar to fluorosilicones in their fluid resistance but are mechanically tougher, making them more suitable for dynamic applications. In addition, they have excellent vibration damping characteristics and are liquid oxygen compatible. Overall the preferred application of fluorophosphazenes stems from their very wide operating temperature range and their good damping properties and flexural fatigue resistance over this temperature range. They have been marketed by Ethyl Corporation since 1983 under the Eypel™F trade name.

5.3 THERMAL EXPANSION

A factor that is often overlooked when selecting polymers for specific applications is their dimensional stability in response to temperature variation. In the case of PTFE, a marked change in specific volume occurs around 19 °C with the volume increasing up to 1.8% around this temperature region (Figure 1.29). It is important that any PTFE components intended for use above room temperature that have been fabricated below this temperature, or vice versa, are carefully checked for tolerance compliance. Furthermore, the final service temperature of the article must be accurately determined [8]. FEP resins, unlike PTFE, do not exhibit a

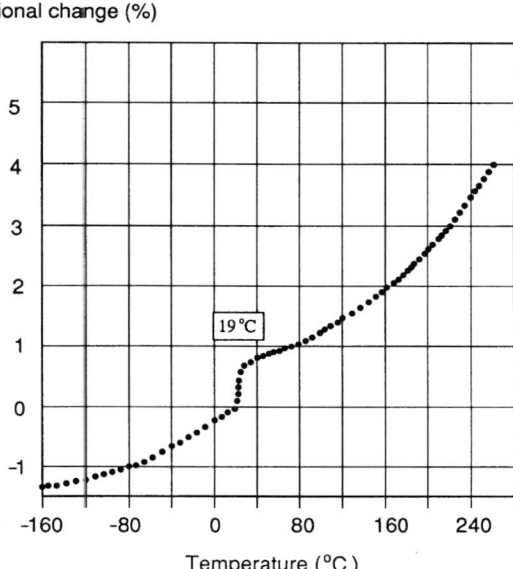

Figure 1.29. Uniaxial thermal expansion of PTFE as a function of temperature. Note: transition zone at 19 °C leads to a large step change in volume

marked change in volume around room temperature, since they lack a first-order transition in this temperature range.

The high degree of thermal expansion of the fully fluorinated thermoplastics makes it difficult to machine components to close tolerances. PTFE exhibits the highest coefficient of thermal expansion of fluoropolymers (100×10^{-6} K^{-1}), while FEP is marginally lower (80×10^{-6} K^{-1}). These values are high when compared to engineering plastics (which have values in the range 50–60×10^{-6} K^{-1}). On the other hand, other fluoropolymers such as PCTFE and ETFE have values of 60×10^{-6} and 40×10^{-6} K^{-1} respectively.

Perfluoroelastomers, due to their fully fluorinated nature, share high a degree of thermal expansion. The linear coefficient of thermal expansion of perfluoroelastomers is of the order of 50% higher than for fluoroelastomers. For example, as the temperature is raised from ambient to 316 °C, Kalrez® perfluoroelastomers can undergo a 21% increase in volume. In applications such as O-rings in a confined situation, extrusion of the O-ring may occur [55].

6 ELECTRICAL PROPERTIES

The electrical properties of PTFE are outstanding and are even superior to those of polyethylene. The very low dielectric strength of PTFE is the result of its highly symmetrical chain structure in which all electrical dipoles (i.e. C—F bonds) are equally balanced. The excellent electrical properties of PTFE can be impaired, however, by the presence of voids and microcracks. The pressure applied during fabrication is crucial to ensuring optimum electrical properties. Too low a pressure gives voids, while too high a pressure causes microcracks [8]. The electrical properties of PTFE are also adversely affected by ionizing radiation which causes an increase in both the dielectric constant and the dissipation factor.

The excellent dielectric properties of PTFE present a limitation in applications where PTFE is used as a tube or as a lining in steel pipe for the transport of corrosive and inflammable liquids/gases. In such applications it does suffer from a tendency to accumulate and generate static electrical charges. This is a safety concern with PTFE since a spark caused by static discharge could initiate an explosion or fire [50].

As would be expected from their similar structures, PTFE and FEP have almost identical electrical insulating properties. Figure 1.30 shows the dependence of dielectric loss (ε'') on temperature for PTFE and FEP (16.8 wt% HFP) [56]. The excellent dielectric properties of PTFE are demonstrated by its low and almost invariant dielectric constant. The magnitude of the dissipation factor is greater for FEP than for PTFE because the molecular structure of FEP is asymmetrical. It is also evident that the electrical properties of FEP change with temperature, and this dependency should be considered when designing insulators for use at temperatures other than room temperature.

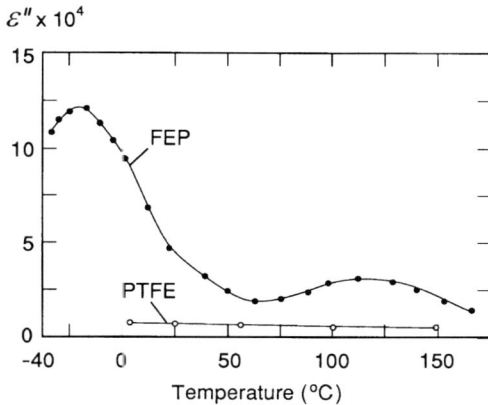

Figure 1.30.   Dependence of dielectric loss (ε'') on temperature for PTFE and FEP (16.8 wt% HFP) (data from Bur [56], 1972)

The new amorphous fluoropolymers (such as Teflon AF) have even better electrical properties than PTFE and have the lowest dielectric constant of any known organic polymer. But it should be noted, however, that they also rank as the most costly commercial polymers.

PVDF is generally recognized not for its excellent electrical properties but rather for its unusual electrical properties. PVDF can be crystallized in a form where the majority of hydrogen atoms reside on one side of the main chain and the fluorines on the other, which induces a polarity that manifests itself by piezoelectricity (i.e. it changes its polarization in response to mechanical stress) [2]. Since normal melt processing of PVDF will produce a small fraction of this polar form, the electrical properties of PVDF are inferior to those of other fluoropolymers. The d.c. and a.c. dielectric breakdown strength of PVDF are 770 MV/m and 233 MV/m respectively. The d.c. strength is comparable to capacitor grade PP; however, the a.c./d.c. ratio is significantly lower than that of other polymers [57].

Since ETFE cannot form the polar conformation that PVDF adopts, it has superior electrical properties to PVDF. The dielectric constant of PVDF ranges from 8.4 to 6.4 as the field frequency ranges from 60 to 1 million Hz. In contrast, the dielectric constant of ETFE is 2.6 and remains approximately constant over the frequency range [58]. Although the electrical properties of ETFE are superior to PVDF its dielectric constant is still high when compared to PTFE (2.6 versus 2.1 at 10^6 Hz). Despite this it is, nevertheless, extensively used for insulating electrical wiring because of its favourable mechanical properties especially its resistance to 'cut-through' and creep [59]. PFA, on the other hand, displays excellent electrical properties, but it loses its dielectric strength in the presence of a corona discharge since polar oxygenated species form on its surface.

PCTFE has poor electrical insulating properties due to its unsymmetrical structure which results from replacement of a fluorine atom with a chlorine atom.

6.1 TRACKING RESISTANCE

Tracking is the result of the high-voltage leakage currents arcing on the surface of a polymeric insulator. The sparks generate temperatures of 450–600 °C on the polymer surface and develop carbonaceous tracks which extend in a dendritic fashion. The tendency of a polymer to undergo tracking is related to its char formation during thermal degradation and can be assessed by thermogravimetry (TG).

Many fluoropolymers (especially those which are only partially fluorinated, e.g. PVDF and ECTFE) possess poor tracking resistance, while PTFE is tracking resistant. This resistance to tracking is by virtue of the non-charring nature of PTFE which in turn is due to its tendency to depolymerize at elevated temperatures. The tracking resistance of PTFE (and related perfluorinated polymers, PFA, MFA and FEP) is thus outstanding because it leaves no carbonized path. PVDF, on the other hand, readily forms char at temperatures above 500 °C as reflected by its plateau in a TG experiment (Figure 1.31). Thermal decomposition of PVDF proceeds via a chain stripping reaction which yields ≈40% of a polyacetylenic char (Figure 1.32). It is this polyenic char that is deposited as a result of tracking.

The resistance of a polymer to tracking can be quantified by measuring the threshold voltage to induce tracking. The initial tracking voltages for PVDF,

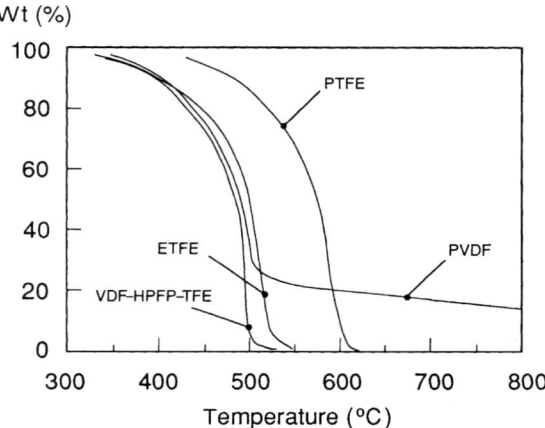

Figure 1.31. Thermogravimetric profiles of selected fluoropolymers in air under a heating rate of 40 °C/min. Note: the formation of char as indicated by the formation of a plateau in the weight loss curve of PVDF reflects its poor resistance to electrical tracking

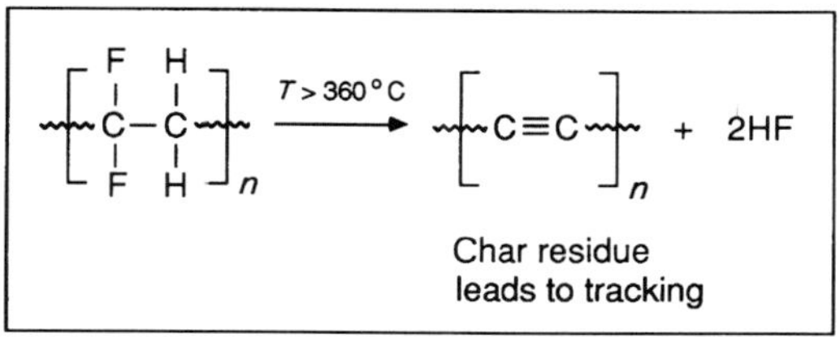

Figure 1.32. Mechanism of char formation which occurs during electrical tracking of PVDF

VDF–HPFP–TFE, VDF–HFP and ETFE are 1.5, 1.5, 2.25 and 3.0 kV respectively. PTFE does not have a tracking onset voltage, though it does suffer from surface erosion.

6.2 MICROELECTRONIC APPLICATIONS

Fluoropolymers find application in electronic devices such as capacitors and relays. Newer developments involve the use of fluoropolymers in fast microelectronic devices. These devices consist of multilayer sandwiches of low capacitance dielectrics (e.g. fluoropolymers) and low resistivity metals (e.g. copper) [60]. A problem however, lies in the lack of metal adhesion to fluoropolymers. In fact, fluoropolymers are notoriously difficult and problematic to metallize. This is related both to their low surface energy and the presence of outer surface contamination. While surface oxidation can aid the adhesion process, the type and distribution of surface groups is not clear. Using sophisticated surface techniques, it has been shown that the extrusion process is responsible for surface oxidation of PFA due to the reaction of the hot extrudate with the atmosphere immediately after exiting the film-forming die [61].

Various strategies have been devised to increase the surface energy of fluoropolymers in order to promote adhesion. Among these, wet-chemical modification of the polymer with a reducing agent has been widely investigated. Recently the effect of a mild reducing agent (benzoin dianion in DMSO) on PTFE, TFE, FEP, Teflon AF and PCTFE was studied by Hung and Burch [62]. They found the order of adhesive strength was: PCTFE > PFA > FEP > PTFE. Teflon AF was found to be totally inert towards this wet chemical treatment. Other methods for improving the metal–polymer adhesion of fluoropolymers involves low-pressure plasma treatment [63]. For further reading on surface treatments to promote adhesion and bondability of fluoropolymers see Chapter 7 by Brewis and Mathieson [64].

Good adhesion of fluoroelastomers to metals is also important for many applications. Generally diamine-cured fluoroelastomers provide better adhesion to metal than bisphenol and peroxide-cured systems. Calcium oxide acid acceptors can assist in providing good metal adhesion.

7 PURITY

7.1 ULTRA-PURE WATER

Fluoropolymers are the materials of choice for ultra-pure water distribution in the semiconductor industry since they contain no additives (such as process aids or heat stabilizers) and are not susceptible to hydrolysis (Figure 1.33).

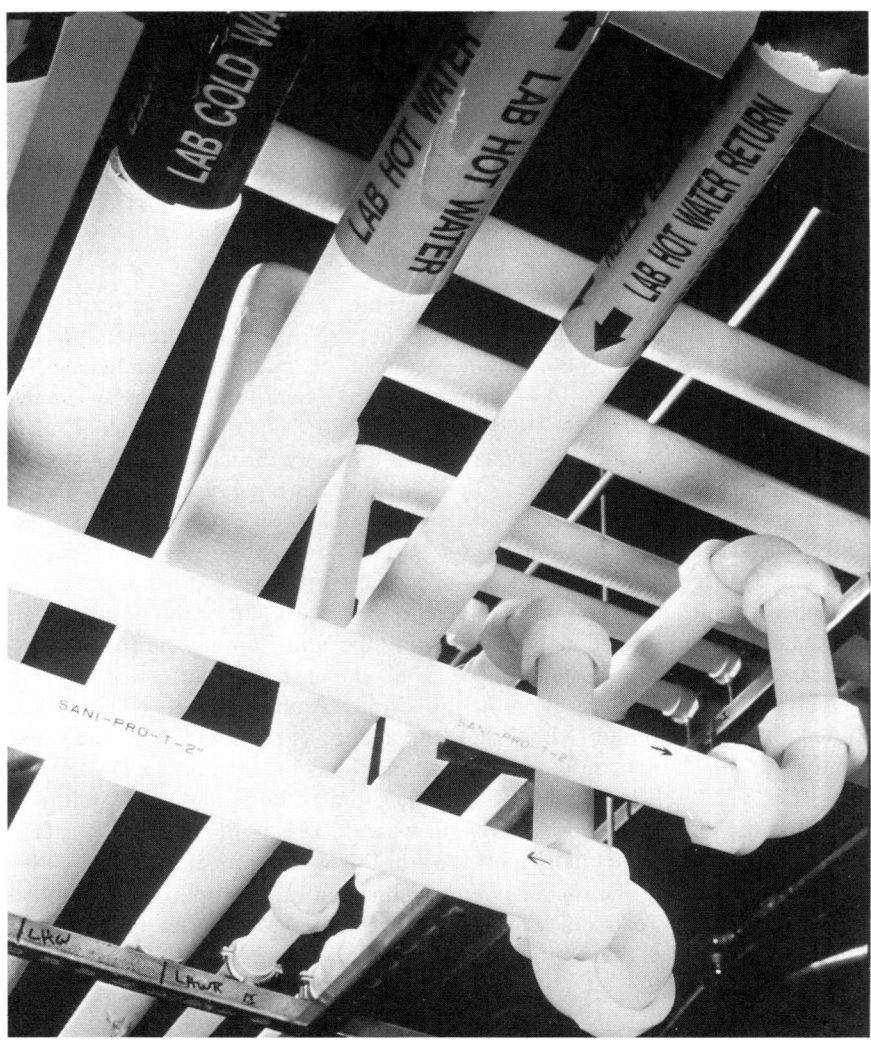

Figure 1.33. Ultra-pure water distribution is the largest growth industry of PVDF in the 1990s (courtesy of D. Seiler, Atochem North America)

However, polymer end groups generated during polymer initiation, chain transfer and polymer moulding steps can contribute impurities to high-purity water. The presence of organic species in ultra-pure water is especially undesirable for silicon wafer processing.

For example, the acid fluoride groups ($-COF$) of some fluoropolymers can be hydrolysed to produce HF which can corrode processing equipment and damage sensitive components such as silicon wafers stored in contact with PFA. Unwanted reactive end groups can be removed by exposure of the polymer to elemental fluorine to convert the end groups to stable CF_3 groups. Teflon PFA 440 is an example of such a stabilized polymer. Such stabilized PFA polymers do, however, exhibit lower tensile properties [65].

Following testing of the entire fluoropolymer family, it has been found that PFA and FEP polymers release negligible amounts of total organic hydrocarbons (Figure 1.34). As described, PFA is treated with elemental fluorine to convert reactive end groups to CF_3 and thus it imparts a very low degree of contamination [66]. ECTFE was found to release marginally more organic impurities. PVDF contributes a much higher level of total organic carbon (TOC) than the above polymers, due to its higher polarity and hence wettability [67]. Furthermore, the end groups of PVDF are predominantly *tert*-butyl or methyl groups and are thus reactive. Also, sometimes there are residues of chemicals used during polymerization such as the water-soluble initiators (e.g. *tert*-butyl peroxide), surfactants (e.g. potassium perfluorocarboxylic acid) and 'seeding chemicals'. For instance, methyl cellulose is used as a 'seed' in the PVDF polymerization process [24] and a residue of this can act as a nutrient source for bacteria. In addition to direct contamination of ultra-pure water by PVDF, the dehydrofluorination of PVDF can cause fluoride ions to contaminate ultra-pure water and this can cause

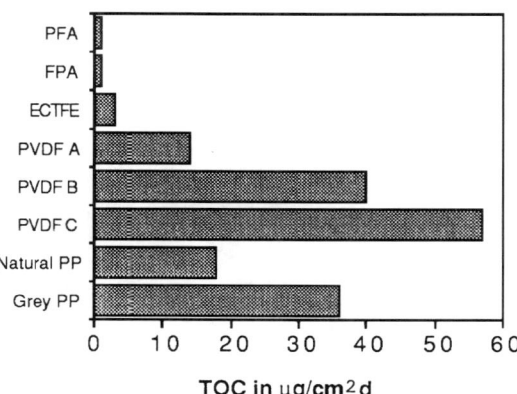

Figure 1.34. Amount of total organic hydrocarbons released from various plastic piping materials in contact with high-purity water at 23 °C (adapted from Sinha [67], 1991)

damage to silicon wafers in direct contact with it even at fluoride levels as low as 1 ppm [68].

The purity of fluoropolymer pipes is also influenced by the size of the internal weld bead since it can cause turbulent flow leading to 'hang-ups'. Recently, a system has been devised to minimize the internal weld bead by inserting silicone rubber bellows inside the pipe during welding [69]. Other methods aim at avoiding internal weld bead formation by the use of inner pipe adaptors placed in the ends of the pipes to be bonded. The inner pipe adaptors meet with a reverse taper near the line to be bonded [70].

7.2 OUTGASSING

Outgassing of volatile compounds is another mode by which fluoropolymers can contaminate ultra-pure water (Figure 1.35). The extent of outgassing from a range of commercial fluoropolymers has been quantified by headspace analysis, using gas chromatography coupled with mass spectrometry after heating the samples (2.5 g) for 2 h at 70 °C [71]. It was found that PFA, PTFE and PCTFE showed negligible outgassing while ECTFE, ETFE and THV all outgassed appreciably. In the case of ECTFE, relatively high levels of 2-ethyl hexanoic acid were measured along with lower levels of a phthalate ester. ETFE outgassed 1,1,2 trichlorotrifluoroethane (CFC-113) and diethylphthalate. The level of outgassing for ECTFE and ETFE was found to be higher for pellets than finished products, indicating that

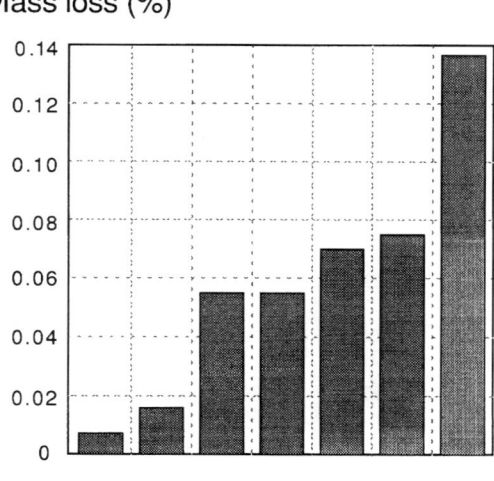

Mass loss (%)

Figure 1.35. Mass loss from various fluoropolymers pellets at 125 °C and 10^{-3} Torr (according to ASTM F1227; adapted from Mikkelsen, Alberg and Prestidge [71], 1995)

these volatile compounds evaporate during the moulding process. Analysis of the THV polymer before and after melt processing showed considerable outgassing from the fabricated part of tetrafluorohexane, presumably a thermal degradation product. PCTFE was found to outgas to a very small extent with the principal volatile being chlorotrifluoroethylene (CTFE), the monomer.

8 SURFACE PROPERTIES

8.1 COEFFICIENT OF FRICTION

Fluoropolymers are characterized by their low frictional behaviour, a property which is essential in their application as bearings. The coefficient of friction (COF) is related to the polymer's molecular cohesion and the hardness, and is independent of contact area and the sliding velocity. The exceptionally low COF of PTFE is a function of its low surface energy and low molecular cohesion; these in turn are due to the smooth molecular profile of the PTFE chain. The COF of a fluoropolymer increases as the fluorine atom is replaced by chlorine or hydrogen atoms (that is, as one goes from PTFE to PCTFE to PVDF and to PVF) [72]. Partially fluorinated polymers thus have a higher coefficient of friction than both FEP and PTFE.

8.2 WEAR

The wear rate of PTFE is strongly dependent on temperature. It is the increased strength of PTFE at low temperatures that decreases its wear. It should be noted that sliding speed also has a pronounced effect on the rate of wear of unfilled PTFE. Table 1.8 shows the wear factors for PTFE filled with various fillers. It is evident that fillers dramatically improve the wear resistance of PTFE. While fillers do impart outstanding wear resistance to PTFE they should be specified with caution since they can adversely affect chemical resistance and impact strength. Table 1.9 shows some common fillers for PTFE together with their advantages and limitations. The limitations of filled PTFE materials stem mainly from the lower chemical resistance of the filler. For example, glass-filled PTFE is not compatible with HF, while the bronze-filled PTFE grades should not be used in

Table 1.8. Wear factors for various filled PTFE compounds (data from Waterman [50], 1994)

Filler	% By volume	Wear factor
PTFE	—	7000–1400
PTFE + glass fibre	22	5.5
PTFE + glass fibre/MoS$_2$	12.2/2.3	5
PTFE + carbon/graphite	26.3	4
PTFE + bronze/MoS$_2$	23.8/2.3	4

Note: wear factor units are: 10^{-8} cm^3 s/cm kg h.

Table 1.9. Common fillers for PTFE and their advantages and limitations (data from Waterman [50], 1994)

Filler	Loading (vol.%)	Advantages	Limitations
Glass	22	A loading of about 22% (by volume) of glass fibre is optimum for PTFE, although loadings up to 36% can be incorporated. Increasing glass content improves creep resistance and chemical resistance (particularly to mineral acids)	Glass-filled PTFE is suitable as a bearing material for low pressure-valves. However, at high loads and high speeds the rate of wear increases and the risk of scoring the shaft is much increased. Glass-filled PTFE should not be used in contact with HF
Bronze	25–30	PTFE/bronze materials make excellent bearings where they are superior to PTFE/glass at high speeds due to its greater thermal conductivity. They also perform well against softer mating materials where PTFE/glass would cause high wear	Bronze-filled PTFE is not as thermally stable as unfilled PTFE due to the destabilizing effect exerted by copper in the bronze. Higher wear rate than PTFE–glass especially in water. Furthermore, bronze-filled composites are not suitable for applications involving contact with foodstuffs
Bronze/ graphite	40	PTFE–bronze and graphite filler composites have very good wear resistance both under wet and dry conditions. They have a lower rate of wear than bronze-filled PTFE, especially in aqueous environments	Bronze-filled PTFE is not as thermally stable as unfilled PTFE due to the destabilizing effect exerted by copper in the bronze
Powdered coke	28	PTFE–powder coke exhibits very good chemical resistance with reasonable wear and creep resistance. It is recommended for applications involving HF where glass-filled PTFE would be attacked. They are also the material of choice for piston rings in oil-free compressors	PTFE filled with powdered coke undergoes higher wear than glass-filled PTFE, but is less likely to score shafts
Graphite	15	PTFE–graphite composites possess the lowest coefficient of friction of all filled PTFE grades. They are recommended where reasonably good deformation resistance and/or wear resistance are required together with a low coefficient of friction. They also find application in pipe for handling flammable/corrosive liquids/gas since the graphite filler dissipates static electricity	Higher wear rates than glass-filled PTFE, but does not cause scoring of shafts

the presence of strong acids such as 50% H_2SO_4, conc. HCl, and conc. HNO_3. Also, bronze-filled PTFE should not be used in contact with mercury since an amalgam will form which may cause property deterioration.

9 THERMAL DEGRADATION

9.1 FLUOROPLASTICS

Figure 1.36 shows the relative thermal stability of selected fluoropolymers as determined by thermogravimetry. ETFE exhibits the lowest thermal stability due to the ethylene linkages in its structure. FEP, despite being fully fluorinated, shows some thermal instability due to its high degree of $-CF_3$ branching. Next, in stability is PFA which possesses a stable, highly shielded ether group in its structure. PTFE exhibits the highest stability, as one would expect given its regular $-CF_2-$ structure.

9.1.1 PTFE

PTFE gives a monomer yield of 95% compared with just 26% for PCTFE [73]. Depropagation (unzipping) to form monomer competes with chain radical transfer reactions, and the dominant pathway is dependent on the structure of the fluoropolymer. The high bond strength of the C−F bonds makes depolymerization the dominant mechanism. In partially fluorinated fluoropolymers, on the other hand, the lower bond energies of C−H and C−Cl bonds increase the likelihood of chain transfer reactions.

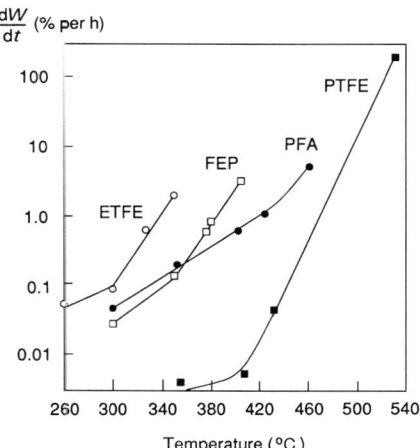

Figure 1.36. Thermal stability of various fluoropolymers by measuring rate of weight loss as a function of temperature (adapted from Baker and Kasprzak [86], 1993)

9.1.2 FEP

The thermal stability of FEP is considerably lower than that of PTFE. At temperatures above 200 °C the properties of FEP rapidly decline as a result of degradation reactions. FEP undergoes a bimodal degradation process; the first step involves preferential elimination of HFP from the backbone at a rate four times faster than the depolymerization of PTFE. Then in a distinct secondary process, the remaining backbone undergoes decomposition at the same rate as that of PTFE decomposition [74].

9.1.3 PFA

The improved thermal stability of PFA compared with FEP can be attributed to the ether group in the side chain which serves as a spacer and eliminates steric strain at the branching point. However, PFA (e.g. Teflon PFA 340) has been found to undergo degradation during processing or use at high temperatures as a result of the presence of reactive end-groups (e.g. $-COF$ and $-CH_2OH$). This degradation leads to crosslinking reactions and an increase in the molecular weight disribution (MWD) when the unstable end-groups decompose to form radicals which then undergo radical recombination reactions [65].

9.1.4 PVDF

The thermal stability of PVDF is inferior to PTFE but much superior to PVF and PCTFE. PVDF has been found to undergo sensitized thermal decomposition by certain additives such as silica, titanium dioxide and antimony oxide. These inorganic compounds catalyse thermal decomposition of the polymer at temperatures above 375 °C. It has also been reported that PVDF polymerized using $K_2S_2O_8$ as the initiator is less stable than PVDF produced using a peroxide initiator [75].

9.1.5 PVF

PVF films discolour at high temperatures, but retain considerable strength after heat ageing at 217 °C [75]. Unlike, PTFE, the fluorine-deficient PVF does not yield appreciable amounts of monomer during pyrolysis. Instead, HF is the major product of PVF thermal degradation. PVF heated in air has lower stability and gives a higher yield of HF than in nitrogen. In air, HF loss occurs at about 350 °C, followed by backbone cleavage around 450 °C. Benzene is also a major degradation product of PVF and is formed by chain scission and subsequent cyclization [75]. Although PVF is much less thermally stable than PVDF, it is ranked ahead of PVDF as a fire safety material [76].

9.1.6 ETFE

HF evolution accompanies ETFE degradation which is autocatalytic as in the case of PVDF. Iron and transition metal salts can accelerate the degradation of

ETFE via dehydrofluorination and oligomer formation. Surprisingly, copper salts act as thermal stabilizers [2].

Crosslinked ETFE is widely used for high stability electrical insulation for trains and aeroplanes. However, recent work by McDonnell Douglas Corp. has shown that it does not satisfy all the requirements expected for insulation, such as property retention after prolonged exposure at elevated temperatures [77]. Workers have found that crosslinked ETFE insulation turns yellow after just a few days at 220 °C, and after two months' ageing the insulation had turned brown. It was concluded that irradiation of ETFE improves the mechanical properties of ETFE, but at the expense of thermal stability.

The oxidative stability of ETFE has been related to the oxidative degradation of tandem ethylene linkages. For example $-CF_2-CH_2-CH_2-CH_2-CH_2-CF_2-$ is less oxidative stable than $-CF_2-CF_2-CH_2-CH_2-CF_2-CF_2-$ because the shielding effect provided by the fluorine atoms does not extend over more than one C—C bond length so that the methylene groups near the centre of the tetramethylene sequence have almost the same susceptibility to oxidative attack as those in polyethylene.

9.1.7 ECTFE

ECTFE has comparable thermal stability to ETFE. ECTFE can be thermally stabilized by the addition of an ionomer which considerably reduces dehydrofluorination and dehydrochlorination reactions, and suppresses polymer discoloration [78].

9.1.8 PCTFE

PCTFE is susceptible to degradation at temperatures as low as 250 °C. In fact, there is only a narrow margin between its degradation temperature and its processing temperature (see Figure 1.37). Careful temperature control is thus necessary during the melt processing of PCTFE since the processing temperatures (230–290 °C) approach the decomposition temperature of the polymer. The degradation of PCTFE proceeds by chain scission and leads to the formation of terminal unsaturation as shown in equation (1) [79].

$$\sim CF_2-CClF-CF_2-CClF-CF_2-CClF\sim \longrightarrow \sim CF_2-CF=CF_2 + CCl_2F-CF_2-CClF\sim \tag{1}$$

PCTFE shows a marked tendency towards becoming brittle as a result of thermal degradation which is a consequence of it increasing in crystallinity. The propensity for PCTFE to embrittle with high-temperature use can be offset by polymerizing the CTFE with low levels of VDF (typically less than 5%).

9.2 THERMAL STABILITY OF FLUOROELASTOMERS

Fluoroelastomers represent one of the most thermally stable elastomer systems currently available. They have widespread use in such demanding applications as

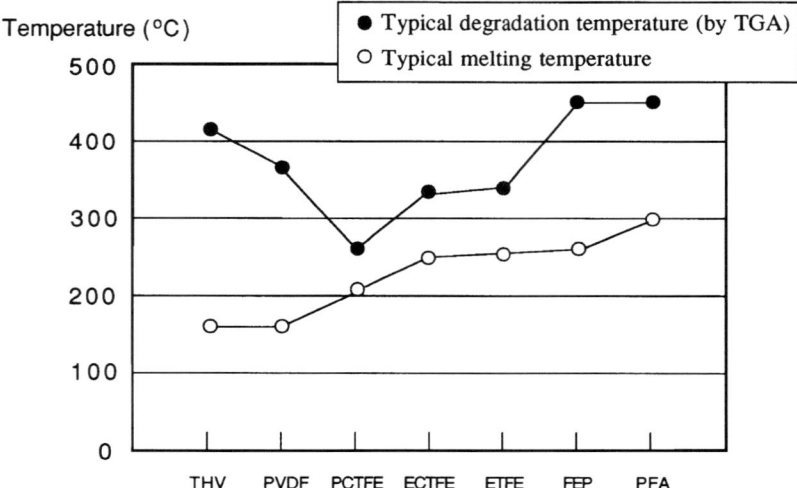

Figure 1.37. Plot of typical melting temperatures and typical degradation temperatures of commercial fluoropolymers. Note the sensitivity of PCTFE to thermal degradation during processing (data courtesy of 3M)

fuel tank sealants where the in-service temperature range can vary from −54 to 177 °C. In simplistic terms, the thermal stability of a fluoroelastomer is a function of its fluorine content. For instance, Kalrez® (a copolymer of TFE and PMVE (perfluoromethyl vinyl ether)) is the only commercially known fully - fluorinated fluoroelastomer and as a result it maintains thermal stability to beyond 300 °C. Moreover, this perfluoroelastomer becomes more elastic rather than embrittled with heat ageing. Viton® and Fluorel® both contain hydrogen atoms in their structure and as a result their thermal stability is lower than Kalrez®. The long-term maximum service temperature for Viton® is 215 °C as compared with 315 °C for Kalrez®. In addition, it has been shown that heating Viton® A at 150 °C results in the formation of unsaturation and that metal oxides promote this dehydrofluorination reaction at even lower temperatures [80]. Copolymers of VDF and CTFE well known under the Kel-F® trade name, have inferior heat stability when compared with VDF–HFP elastomers and have upper long-term use temperatures of only 200 °C.

9.2.1 HPFP Fluoroelastomers

Fluoroelastomers containing hydropentafluoropropylene (HPFP) were pioneered by Montedison in Italy (since HFP-based elastomers were under patent protection), but it has now been realized that elastomers such as VDF–HPFP and VDF–HPFP–TFE (known as Tecnoflon® SL and Tecnoflon® T respectively) have limited thermal stability when compared to their analogues containing HFP. This is because of their lower fluorine content.

9.2.2 Fluorosilicones

Fluorosilicones are in an interesting thermodynamic state where they are actually in equilibrium with their cyclic trimer and tetramer (Figure 1.38). However, it is the oligomers, not the polymer, which are thermodynamically favoured. This is also the reason why during manufacture the polymerization is stopped before equilibrium is reached. Thermal degradation of fluorosilicones can occur by a reversion mechanism where heat shifts the equilibrium towards the tetramer. Thus, the polymer breaks down to form the cyclic tetramer (a thermodynamically stable compound). This reaction is accelerated by basic compounds such as the polymerization catalyst (KOH). This degradation pathway is more dominant in the absence of oxygen since oxidative crosslinking becomes competitive in oxygen-containing atmospheres [53].

It can be seen from the structure of the most common fluorosilicone PMTFPS (poly[methyl (3,3,3-trifluoropropyl) siloxane]) (Figure 1.38) that the fluorine is only incorporated at the terminus of the pendant group. This is because of the marked tendency of α or β fluoroalkyl substituents to undergo thermal cleavage. The hydrocarbon spacer provided by two methylene groups gives optimum thermal stability and the γ fluorosubstituent is provided by copolymerizing with 3,3,3-trifluoropropene.

9.2.3 Thermal Instability of Cure Sites

Ideally for maximum thermal stability, crosslinking of fluoroelastomers should occur by controlled chain extension without the introduction of weak links or potential sites for degradation. Unfortunately in practice this is not feasible and crosslinking of fluoroelastomers is associated with decreased stability compared

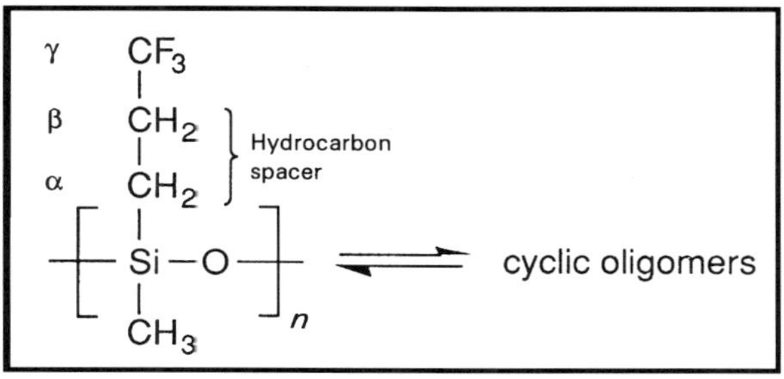

Figure 1.38. Structure of poly[methyl (3,3,3-trifluoropropyl)siloxane] (PMTFPS). Fluorosilicones are in equilibrium with their cyclic trimer and tetramer and this γ-substituted fluoroalkyl gives optimum thermal stability

with the original fluoropolymer. In fluoropolymers these sites for crosslinking have to be created either by incorporation of comonomers containing active groups (known as cure site monomers) or by dehydrofluorination to introduce unsaturation into the backbone [81].

The incorporation of cure sites into a TFE–PMVE introduces thermal instability as indicated by heat ageing experiments. TFE–PMVE containing a perfluorophenoxy group as a cure site monomer has been shown to have half the stability of a TFE–PMVE copolymer without the cure site. However, as stated above, fully fluorinated elastomers such as TFE–PMVE–CSM still have far better heat stability than hydrofluorinated elastomers such as VDF–HFP. For instance, TFE–PMVE–CSM can be used continuously at 316 °C with its elongation properties effectively remaining constant. Curatives based on the triazine structure have been shown to provide the most thermally stable crosslinks for Kalrez® perfluoroelastomers.

9.2.4   HF Evolution

Because of their chemical structure there is the possibility of elimination of hydrogen fluoride (HF) from the fluoroelastomer main chain at elevated temperatures. This is analogous to the splitting out of hydrogen chloride from PVC. If the fluoroelastomers are used as seals or sealants in contact with metals such as titanium then the evolution of HF may cause stress corrosion cracking [82]. It is also known that the degradation of fluoropolymers in air yields a higher concentration of hydrolysable fluorine compounds than in nitrogen (or inert atmosphere).

Fluoroelastomers are often formulated with acid acceptors such as the following: CaO, Mg(OH)$_2$, Ca(OH)$_2$, CaCO$_3$, MgCO$_3$ and LiCO$_3$. The use of these inorganic acid scavengers to reduce HF elimination must be balanced against their deleterious effects on the polymer's physical properties.

The thermal stability of fluoroelastomers has been studied in detail by Cox and Wright [83]. It was found that VDF–HFP was the most thermally stable elastomer system of those studied, while VDF–CTFE elastomers showed considerably lower thermal stability due to the presence of CTFE.

9.2.5   Limiting Oxygen Index

Many applications employ fluoropolymers as seals, bellows and the like in oxygen-enriched atmospheres (i.e. O$_2$ concentration >21%) such as breathing apparatus, gas distillation units and rocket fuel systems. In such applications it is important to select fluoropolymers that possess high oxygen indices to avoid situations of enhanced combustibility of the elastomer. The limiting oxygen index (LOI) is a measure of the combustibility of the polymer by measuring the lowest concentration of oxygen which will support combustion of the polymer and is measured in accordance to ASTM D2863. The fully fluorinated polymers such as PTFE, FEP and PFA have LOI values all exceeding 95%. On the other hand, partially fluorinated polymers possess LOI that are considerably lower. PVF has the lowest LOI of any fluoropolymer with a value of 22.6%. ETFE has a LOI of 30% which is quite low, while that of PVDF is marginally higher with a LOI of 44%. ECTFE has a relatively high LOI of 64% which is characteristic of chlorinated materials. ECTFE chars while ETFE drips under fire conditions. PVF chars rather than burns despite its low LOI and thus finds extensive application in aircraft interiors.

Because of its combination of good properties at cryogenic temperatures and high oxygen index (i.e. 100), PCTFE meets the NASA liquid oxygen impact compatibility requirements under MSFC106A, and the flammability requirements in accordance with NHB 8060.1C. It thus finds widespread use in the shuttle, orbiter and the space station (in construction) for oxygen rich environments because of these factors and its very low degree of outgassing.

Fluoroelastomers based on VDF and CTFE (such as Kel-F) cured with hexamethylenediamine carbamate or bisphenol A have oxygen indices in the range 48–60 and are rated as V-0 by Underwriters Laboratory. Fluorophosphazene elastomers exhibit very good LOI values since they char on burning rather than dripping or flowing, and thus they have potential in wire and cable insulation and for flame-resistant closed-cell insulating foams.

The oxygen index values of fluoroelastomers are extremely sensitive to compounding variables as well as formulation [84]. For example, lead oxide-cured HFP–VDF fluoroelastomers (e.g. Viton B) have oxygen indices of 100%. Curing with MgO lowers the LOI to 55, while calcium carbonate-filled HFP–VDF elastomers may have a LOI of only 42. The 3M grade denoted as Fluorel L-3206-6 are extensively used by NASA and has been shown to have excellent resistance to ignition under the most severe conditions encountered in space exploration [48].

PVDF has long been used as a valve seating material for oxygen cylinders. However, occasional failures resulting in ignition have been reported for PVDF valve seats [85]. It was found using TG that the onset of thermal decomposition is much lower in pure oxygen than in air. In fact, the onset of thermal degradation is reduced to 315 °C in pure oxygen.

9.2.6 Toxicity

Fluoropolymers by virtue of their inert nature are among the safest synthetic polymers known. However, during high-temperature use and reprocessing of fluoropolymers the possibility exists for the formation of perfluoroisobutene (PFIB) which because of its extremely high toxicity is a cause for concern even in very small concentrations. The purported 2 h lethal dose of PFIB for rats is just 1 ppm. It has been shown by workers at DuPont that FEP (Teflon FEP) can form PFIB at temperatures as low as 380 °C [86]. The branched structure of FEP makes it more susceptible to forming PFIB compared with linear fluoropolymers. In the case of PTFE, higher temperatures are required (e.g. 525 °C).

The use of PTFE powder as a flame-retardant additive in engineering plastics to impart non-dripping characteristics during combustion has been questioned due to the possible risk of PFIB generation [87]. However, it is now known that pyrolysis of perfluoropolymers alone produces PFIB and that co-pyrolysis with any hydrogenated material produces little, or usually no PFIB. It is believed that this is due to the rapid addition of hydrogen radicals in the smoke to the fluoro-carbon radical intermediates formed by fluorocarbon degradation, thus preventing recombination of fluorocarbon radicals to yield PFIB (private correspondence: D. Kerbow, Du Pont).

Another major pyrolysis product of fluoropolymers is carbonyl difluoride (COF_2) which is also highly toxic hence its other name 'fluophosgene'. Exposure to carbonyl difluoride can develop into a condition termed 'polymer fume fever' which has its onset some hours after exposure and presents influenza-like symptoms. When ETFE, FEP or PFA are heated to 400 °C in air, carbonyl difluoride is the principal gas evolved [86]. This product is quickly hydrolyzed to hydrogen fluoride according to equation (2). HF besides its toxicity can also attack metal and glass surfaces.

$$COF_2 + H_2O \longrightarrow CO_2 + 2HF \qquad (2)$$

The toxicity of smoke from full-scale fires involving perfluoropolymers has been studied by Clarke et al. [88].

When heated above 150 °C, fluorosilicones can evolve a toxic fluorinated alde-hyde (3,3,3-trifluoropropionaldehyde) (TFPA). The evolution of TFPA is usually very low and obviously its yields are dependent on temperature, sample size and oxygen concentration; however, like PFIB, its high toxicity necessitates the need to minimize human exposure particularly during curing of the elastomers where temperature >150 °C may be employed [53].

Interestingly, the toxicity of the combustion products of fluorophosphazene elastomers is much lower than those of fluoroelastomers and fluorosilicones, and even considerably less than for many other commonly used fire-resistant elastomers [89].

10 RADIATION DEGRADATION

10.1 GAMMA IRRADIATION

Fluoropolymers, while collectively being quite susceptible to radiation-induced degradation, respond differently with regard to whether chain scission or crosslinking occurs. On exposure to ionizing radiation, polymers that are completely halocarbon such as PTFE or PCTFE suffer rapid loss of MW, while fluoropolymers that contain hydrogen (e.g. PVDF) rapidly crosslink into a network (Figure 1.39). Copolymers of TFE–HFP represent an intermediate group where crosslinking and chain scission both occur, although crosslinking

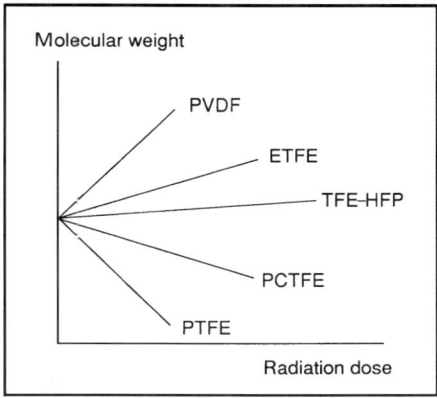

Figure 1.39. Schematic showing the effect of radiation on the MW of various fluoropolymers

predominates to a limited extent. TFE copolymers with PFVE undergo chain scission accompanied by a reduction in their mechanical properties [90].

10.1.1 PTFE

Since the introduction of nuclear power plants it has been known that PTFE is very susceptible to radiation-induced degradation. In 1954 it was shown that exposure of PTFE to a large radiation dose in an atomic reactor caused it to be converted from a solid polymer with a melting temperature of 327 °C to a wax melting at 100°C [91]. Interestingly, this shortcoming is now being exploited by converting PTFE scrap into lubricants. In practical terms, PTFE loses 50% of its original tensile strength after exposure to as little as 1 Mrad (10^4 Gy) of radiation in air.

It has been demonstrated that carbon-filled PTFE is far more resistant to ionizing radiation than unfilled PTFE [92]. This has been attributed to the lower rate of oxygen diffusion in the filled system. Electron spin resonance spectroscopy has identified that peroxy radicals are involved in the radiation-induced chain scission processes. This is consistent with the observation that the radiation resistance of PTFE is reduced by a factor of 40 times in the presence of oxygen [93].

Although oxygen is effectively excluded in thick sections of PTFE due to diffusion limitations, the effect of ionizing radiation is nevertheless profound due to the buildup of enormous pressure by entrapped gaseous radiolysis products such as TFE which are unable to diffuse out rapidly enough [94].

The poor stability of PTFE towards radiation has actually been utilized to create novel fasteners for medical devices that degrade during gamma sterilization so that the apparatus is easily disassembled and discarded [95].

10.1.2 FEP

The radiation tolerance of FEP is up to 10 times greater than that of PTFE. The primary effect of radiation on FEP is chain scission with the rate of chain scission being much higher in oxygen than in nitrogen.

10.1.3 PFA

The radiation resistance of PFA is superior to that of PTFE; however, since it is fully fluorinated, chain-scission is the dominant pathway and a decline in its physical properties is inevitable. Figure 1.40 shows that the tensile strength of PFA decreases steadily with increasing exposure to ionizing radiation.

10.1.4 ETFE

ETFE resins show good retention of tensile properties when exposed to low-level ionizing radiation. This is due to the fact that competing degradation mechanisms, namely chain scission and crosslinking, are occurring at an approximately equal rate so the net MW change is small. Figure 1.41 shows that the radiation resistance of ETFE is superior to both PTFE and FEP. For this reason, ETFE is used as cable insulation in nuclear energy installations [2]. However, at higher radiation doses, the relative elongation of ETFE is severely affected (Figure 1.42) as indicated by its tensile elongation dropping sharply when irradiated. From Figure 1.42 it is clear that the change in mechanical properties is much more pronounced in the presence of air [96].

The crosslinking of ETFE which occurs when it is exposed to low doses of ionizing radiation (either gamma or electron beam) can be utilized to advantage

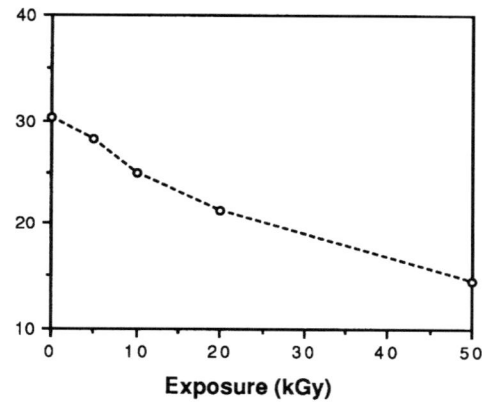

Figure 1.40. Effect of radiation on the tensile strength of PFA (data from Gangal [12], 1980)

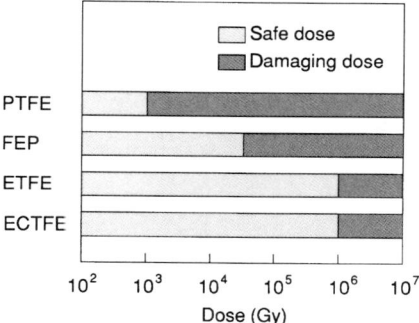

Figure 1.41. Radiation resistance of common fluoropolymers used as cable insulating materials (data from Schönbacher and Stolarz-Izycka [117], 1979). Note: the replacement of fluorine with hydrogen as in the case of ETFE and ECTFE leads to improved radiation resistance. 10^4 Gy = 1 Mrad

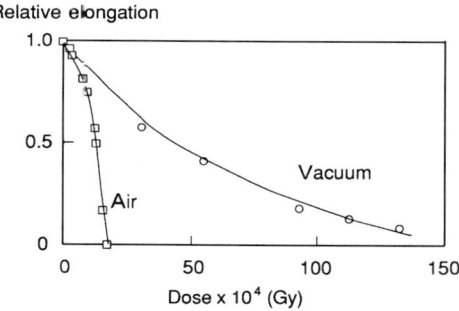

Figure 1.42. The effect of γ-irradiation on the relative elongation of ETFE at 70 °C. Note: samples irradiation in air and vacuum with doses of 5.8×10^3 Gy/h and 7.7×10^3 Gy/h respectively (adapted from Clough, Gillen and Dole [96], 1991)

since it imparts improved high-temperature properties such as resistance to cut-through by a hot soldering iron. While regular ETFE is only rated for continuous use at 150 °C, radiation-crosslinked ETFE wire insulation is recommended for continuous service at 200 °C.

10.1.5 PVDF

PVDF and HFP–VDF predominantly undergo crosslinking when exposed to low-level gamma radiation (up to 20 Mrad) as indicated by the increased gel fraction with increasing dose [90,97]. It is important to note the distinction here between PVDF and PTFE. PVDF undergoes a property improvement due to crosslinking (Figure 1.43) while in PTFE crosslinking does not occur. Crosslinking in PVDF

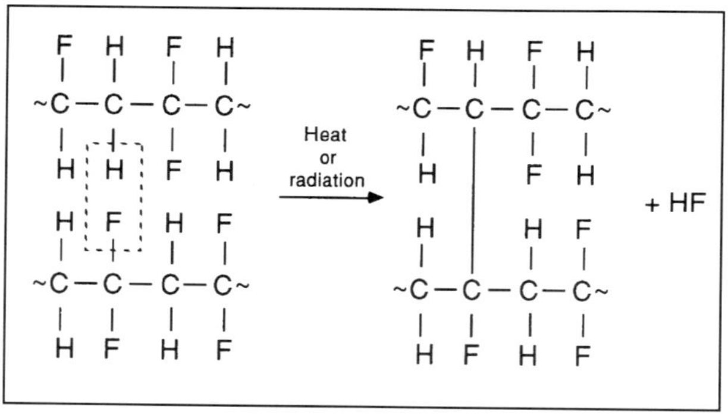

Figure 1.43. Crosslinking mechanism of PVDF which can be activated either by heat or ionizing radiation. Note: this type of crosslinking is not possible in PTFE, therefore chain scission is the preferred route of degradation

is usually accompanied by an increase in tensile strength of the polymer and a decrease in both the degree of crystallinity and melting point. The overall resistance of PVDF to nuclear radiation is very good. The tensile strength of PVDF is virtually unaffected after 1000 Megarads of gamma radiation in vacuum. The impact strength and elongation are slightly reduced due to crosslinking effects. PVDF components are used in plutonium reclamation plants [13]. In fact, PVDF sheets filled with gadolinium oxide have radiation shielding ability and can be used for lining containers of nuclear waste [98]. However, it has been shown that chemical changes and discoloration do occur at relatively low doses [99]. Degradation of PVDF by gamma rays also leads to the formation of an absorption at 1714 cm^{-1} which is attributed to the formation of double bonds due to dehydrofluorination; however, mechanical properties are largely retained.

10.1.6 VDF–HFP

The radiation-induced degradation of a VDF–HFP copolymer was studied by Zhong and Sun [100]. They found that exposure to radiation resulted in loss of weight and that a linear relationship exists between the dose and the loss of weight. A weight loss of 0.4% corresponds to a dose of about 650 kGy at 150 °C. A linear relationship has also been observed for PTFE where the weight loss is directly proportional to the square of the irradiation dose [101]. TFE–PP based elastomers show better resistance to gamma irradiation than Viton® elastomers [102].

10.1.7 PVF

Like VDF-based polymers, PVF undergoes crosslinking and forms a gel fraction when exposed to ionizing radiation. The tensile strength and resistance to etchants of PVF increases with radiation dose while the melting point and crystallinity decrease [90].

10.1.8 PCTFE

Little work has been done on the effect of radiation on PCTFE, but it is known that the radiation resistance of PCTFE is superior to that of other fluoropolymers.

10.2 ELECTRON BEAM

Electron beam irradiation of solid PTFE induces extensive chain scission and leads to a drastic MW reduction (by up to six orders of magnitude) to give micropowders [103]. A small business has grown around this process to produce micronized PTFE which finds use as an additive in inks, coatings and in thermoplastics to improve rub resistance and lubricity [104]. In contrast, PVDF undergoes crosslinking as a result of electron beam irradiation and the tensile and elongation properties remain essentially unchanged even up to 20 Mrad [13].

Electron beam radiation causes most fluoropolymers to crosslink due to free radical reactions. Copolymers of VDF–HFP (e.g. Viton® A) exhibit a greater tendency to crosslink than PVDF [97]. Certain N-containing chemicals act as sensitizers and increase the degree of crosslinking for a given radiation dose. For example, *m*-phenylenemaleimide and triallyl cyanurate have been found to act as sensitizers with the former exerting the greatest influence. The electron beam-induced crosslinking of VDF–HFP and TFE–VDF–HFP has been used to advantage to produce heat shrinkage tubes.

Work on the electron-beam irradiation of PVDF films has found that low doses cause no change in physical properties of the polymer while high doses (e.g. 10^4 Gy) cause extensive chain scission, 82% increase in gel fraction and a 75% decrease in tensile strength [105].

In the case of PVF, it shows good resistance to low doses of electron beam radiation retaining its strength at doses of 3×10^5 Gy and only starts to show embrittlement at doses of 10^7 Gy. For comparative purposes, PTFE is degraded by doses more than 10 times lower.

11 UV STABILITY

As a group, fluoropolymers generally show exceptional intrinsic UV stability. This is due to the fact that the strength of the C–F bond makes them able to resist pure photolysis and because fluoropolymers generally contain no light-absorbing chromophores either in their structure or as impurities. PTFE, for example, is completely stable to outdoor weathering and no chemical or physical changes have been observed after 30 years of continuous exposure in Florida.

The outdoor durability of a fluorinated coating is directly related to its fluorine content [106]. Figure 1.44 demonstrates the effect of fluorine content of a coating on the outdoor durability as assessed by gloss retention [106].

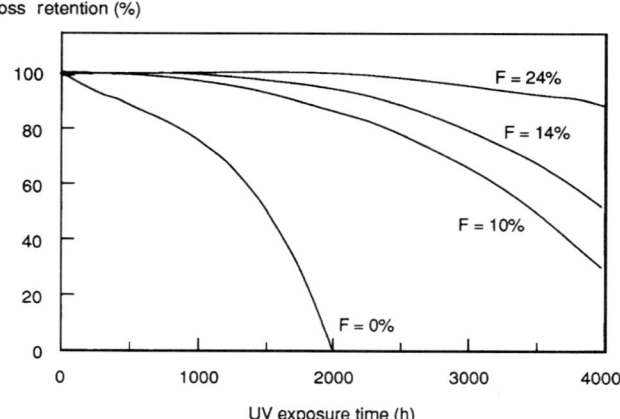

Figure 1.44. The dependence of the UV resistance of a fluoropolymer coating (as indicated by gloss retention) on its fluorine content

While UV degradation of some fluoropolymers does take place outdoors it occurs over very long time scales (e.g. 5–10 years) and only very sensitive analytical techniques can detect this low level of degradation in shorter time periods [107]. For instance, X-ray photoelectron spectroscopy (XPS) can be used to detect the increase in the oxygen concentration of the surface of the fluoropolymer due to photooxidation processes. The high sensitivity of XPS allows chemical changes to be detected in PVDF after four years of outdoor exposure [108]; however, no significant changes in gloss or colour can be measured. Similarly, electron spin resonance spectroscopy (ESR) has been used to detect radicals in weathered PVDF [109].

Despite their excellent resistance to UV degradation, there are nevertheless reports in the literature regarding the UV instability of certain fluoropolymers. However, attention must be paid to the wavelength of UV light employed (since $\lambda < 300$ nm are most damaging) in order to make accurate determinations regarding their sensitivity to UV radiation.

FEP polymers absorb UV radiation only at wavelengths below 180 nm which makes them susceptible to UV degradation only by far UV in space. As it happens, FEP is used as a thermal control blanket for space vehicles. These blankets comprises a 25 μm film of FEP which is metallized on one side. Adams and Garton [110] found that after short-term exposure ($\sim$7 h) to a 150 nm UV source, FEP film showed changes in the IR spectrum indicating the formation of vinylic unsaturation (due to HF elimination), ketonic carbonyls and acid fluoride groups. The defluorination of FEP when exposed to far-UV radiation is consistent with the work of George et al. [111] who showed by ESR that FEP forms a range of fluorinated radicals when exposed to far-UV radiation.

Unlike completely fluorinated polymers (e.g. FEP, PFA), the degradation of PCTFE is greatly accelerated by UV light [112]. Copolymers of TFE and HFP have been reported to undergo scission and crosslinking during UV irradiation [113]. Also, copolymers of TFE and trifluoronitrosomethane are readily degraded under UV irradiation [114] with the evolution of COF_2, $CF_3N=CF_2$ and $(CF_3)_2NH$.

12 CASE STUDIES OF FLUOROPOLYMER FAILURE INVESTIGATIONS

Case Study No. 1

Problem: PVDF tubing used in the food process industry showed signs of cracking.

Analysis: Chemical analysis revealed that the tube was stress cracking due to the action of sodium hydroxide. Improper annealing during manufacture also probably contributed to failure.

Solution: ECTFE was used to replace PVDF since ECTFE has superior resistance to caustic solutions (e.g. NaOH) especially at elevated temperatures.

Case Study No. 2

Problem: Non-stick PTFE panels applied to a ship's hull as a non-fouling surface became encrusted with barnacles.

Analysis: Microscopic examination of the panels showed a high degree of porosity in which the marine microorganisms were able to secrete their adhesive-like compounds and thus anchor themselves. The porosity was a consequence of the panels being fabricated by a sintering process, since PTFE is not melt processable or able to be solvated.

Solution: A PVDF dispersion coating was applied which is virtually void free.

Case Study No. 3 (case history from Fluoroware, Minnesota, USA)

Problem: PVDF pipes in a vitamin E manufacturing plant carrying hydrochloric acid (the catalyst) were cracking and leaking acid.

Analysis: PVDF was being chemically attacked by HCl.

Solution: PFA was used to replace PVDF since PFA is not affected by hydrochloric acid.

Case Study No. 4

Problem: PTFE sheet used in the mining industry was showing grooves and surface defects.

Analysis: Microscopic investigation showed microfibrils of PTFE, indicating wear of the polymer.

Solution: ETFE was used to replace PTFE since ETFE has outstanding abrasion resistance.

Case Study No. 5 (adapted from ref. 115)

Problem: Metallized PVF films used in solar collection systems became dull and embrittled after long-term solar exposure.

Analysis: Infra-red analysis of the film showed the formation of absorption bands at 1340 cm^{-1} and 1740–1790 cm^{-1} indicating the formation of $-CH=CHF$ groups as a result of scission and disproportionation reactions. The enhanced photodegradation of the Al-backed PVF is attributable to the reaction being driven towards the favourable formation of very stable aluminium fluoride complexes.

Solution: ECTFE was used to replace PVF since ECTFE has superior resistance to metal-catalysed photodegradation.

Case Study No. 6

Problem: Silicon wafers for semiconductor manufacture became cloudy after storage in a PFA wafer holder.

Analysis: Surface analysis of the wafers indicated that the acid fluoride end-groups present in the polymer caused micro-etching of the wafers.

Solution: The PFA was replaced with a chemically modified PFA which contains no reactive end groups (such as Teflon PFA 440).

ACKNOWLEDGEMENTS

The author is indebted to Dr Dewey Kerbow (DuPont, Experimental Station, Wilmington) and Kenneth Heffner (ElectroChem Engineering & Mfg. Co., Emmaus, Pennsylvania) for their helpful comments on the manuscript. The author would like to extend special thanks to G. Julius Vancso, Faculty of Chemical Technology, University of Twente, The Netherlands for the atomic force micrograph depicting the molecular chains of Teflon and Sadaatsu Yamaguchi, Masayoshi Tatamoto and Tetsuo Shimizu of Daikin Industries Ltd, Research and Development, Japan, for contributing electron micrographs of fluoroelastomers.

The author would also like to gratefully acknowledge the following persons for contributing information for this chapter: L. William Buxton (Du Pont, Wilmington, Delaware, USA), Dale K. Heffner (Electro Chemical Engr. & Mfr., Emmaus, Pennsylvania, USA), Roger L. Clough (Sandia National Laboratories, Albuquerque, New Mexico, USA), Marc Tavlet (European Organization for Nuclear Research, Geneva, Switzerland), George Black (Vanton Pump and Equipment Corp., New Jersey, USA), John Kalnins (Dow Chemical Company, Bay City, Michigan, USA), Marty Burkhart (George Fischer Piping

Systems, Texas, USA) and Bruce Smart (DuPont, Experimental Station, Wilmington, USA).

REFERENCES

1. Suwa, T., Takehisa, M. and Machi, S., *J. Appl. Polym. Sci.*, **17**, 3253 (1973).
2. Kerbow, D. L., 'Ethylene-tetrafluoroethylene copolymer resins' in *Modern Fluoropolymers: High Performance Polymers for Diverse Applications*, (ed. J. Scheirs), Wiley, London (1997).
3. Saunders, K. J., *Organic Polymer Chemistry*, Chapman & Hall, London, 2nd edn, Chapter 7, p. 149 (1983).
4. Resnick, P. R. and Buck, W. H. 'Teflon® AF amorphous fluoropolymers' in *Modern Fluoropolymers: High Performance Polymers for Diverse Applications*, (ed. J. Scheirs), Wiley, London (1997).
5. Yamaguchi, S. and Tatemoto, M. 'Fine structure of fluorine elastomer emulsions', *Sen'i Gakkaishi*, **50**, 414 (1994).
6. Schroeder, H., 'Fluorocarbon elastomers' in *Rubber Technology*, (ed. M. Morton), Van Nostrand Reinhold, Melbourne, Australia, p. 410 (1987).
7. Anon., *Materials World*, February, p. 61 (1996).
8. Teflon PTFE brochure, DuPont, (1990).
9. Tanaka, N., Japanese Patent, JP 07 331 011 (1995), *Chem. Abs.*, **124**, 204 560w.
10. Minhas, P. S. and Petrucelli, F., *Plast. Eng.*, **33**, 60 (1977).
11. Imbalzano, J. F., *Chemical Engineering Progress*, April, p. 69 (1991).
12. Gangal, S. V., 'Tetrafluoroethylene-perfluorovinyl ether copolymers', *Kirk-Othmer: Encyclopedia of Chemical Technology*, Wiley Vol. 11, p. 671 (1980).
13. *Kynar Technical Brochure*, Atochem North America, 1990 (available from Atochem, Plastics Department, Three Parkway, Philadelphia, PA 19102, USA).
14. Starkweather, H. W., *Macromolecules*, **10**, 1161 (1977).
15. Gangal, S. V., 'Fluorinated ethylene-propylene polymers', *Kirk-Othmer: Encyclopedia of Chemical Technology*, John Wiley & Sons, Vol. 11, p. 657 (1980).
16. Burkhart, M., Story, B., Wallace, R. M. and Patrick, B., *Microcontamination*, **10**, 27 (1992).
17. Crowe, R. and Badyal, J. P. S., *J. Chem. Soc., Chem. Commun.*, **14**, 958 (1991).
18. Kudryavtsev, Yu. P., Evsyukov, S. E. and Babaev, V. G., *Izv. Akad. Nauk, Ser. Khim.*, **5**, 1223 (1992), *Chem. Abs.*, **118**, 7486b.
19. Stirling, C. D., Van Tilberg, V. S. M. and Miller, N. A., *Polymers & Polymer Composites*, **1**, 167 (1993).
20. Hoa, S. V. and Ouellete, P., *Polym. Eng. Sci.*, **23**, 202, 1983.
21. Van Tilburg, V. S. M., Stirling, C. D. and Miller, N. A., *Polymers & Polymer Composites*, **1**, 121 (1993).
22. Lepoutre, P. A., Stirling, C. D. and Van Tilberg, V. S. M., *Corros. Australasia*, **15**, 9 (1991).
23. Burkhart, M., Wermelinger, J. and Klaiber, F., *Microcontamination*, **13**, 27 (1995).
24. Pate, K. T., McIntosh, R., Hanselka, R., *Ultrapure Water*, **7**, 49 (1990).
25. Kise, H. and Ogata, H., *J. Polym. Sci.*, **21**, 3443 (1983).
26. Vanderveken, Y., *Solef PVDF Coloration*, Solvay Technical Publication, Brussels, Belgium, p. 3 (1993).
27. Fisher, A. O., *Managing Corrosion with Plastics*, Vol. VII, pp. 223–236 (1986).
28. Itzhak, D. and Peled, P., *Materials Letters*, **17**, 312 (1993).

29. Aclar® PCTFE brochure (available from Allied Signal Inc., PO Box 2332, Morristown, NJ 07962-2332, USA).
30. Buxton, L. W. and Henthorn, G. V., *Chem. Eng.*, **101**, 133 (1994).
31. *Dow Plastic-lined Piping Products Engineering Manual*, 1994 (available from Dow Chemical Co., Bay City, Michigan, 48706, USA).
32. Hall, N. L., 'Using fluoropolymers to resist permeation of corrosives', *Chem. Processing*, May (1994)
33. Kane, R. D., *Chemical Engineering Progress*, August, (1992), p. 76.
34. Ma, C., Haider, A. M. and Shadman, F., *IEEE Trans. Semicond. Manuf.*, **6**, 361 (1993).
35. MacLachlan, J. D., *Polym. Plast. Technol. Eng.*, **11**, 41 (1978).
36. Nersasian, A., 'The effect of lubricating oil additives on the properties of fluorohydrocarbon elastomers', STLE 1979 Annual Meeting Preprint No. 79-AM-3C-3 (1979).
37. Dinzburg, B. N., *Lubrication Engineering*, October, p. 796 (1995).
38. Abu-Isa, I. A. and Trexler, H. F., *Rubber Chem. and Tech.*, **58**, 326 (1985).
39. De Costobadie, J. and Page, N. M., *Proc. Int. Conf. Offshore Mech. Arct. Eng.*, 12th, 891 (1993) *Chem. Abs.*, **122**, 217 907y.
40. Reiber, S., *Rubber World*, **210**, 38 (1994).
41. Virant, M. S., Fiedler, L. D., Knapp, T. L. and Norris, A. W., 'The effect of alternative fuels on fluorosilicone elastomers', SAE Technical Paper 910102, presented at International Congress and Exposition, Detroit, Michigan, February, (1991).
42. Zellers, E. T. and Guo-Zheng, Z., *J. Appl. Polym. Sci.*, **50**, 531 (1993).
43. Aminabhavi, T. M. and Munnolli, R. S., *Canad. J. Chem. Engin.*, **72**, 1047 (1994).
44. McKenna, G. B., Horkay, F., Verdier, P. H. and Waldron, W. K., *Nat. Instit. Standards Tech. Spec. Publ.*, **890**, 201 (1995) *Chem. Abs.* **124**, 347 561h.
45. Martinez, F. V. and Boufakhreddine, N. F., US Patent 5 459 202, 1995 (to Du Pont).
46. Greene, A. K., Vergano, P. J., Few, B. and Serafini, J. C., *J. Food Engineering*, **21**, 439 (1994).
47. Wang, P. and Sung, N-.H., *Polym. Mater. Sci. Eng.*, **69**, 372 (1993).
48. Elias, H-, G. and Vohwinkel, F., *New Commercial Polymers — 2*, Gordon and Breach, New York (1986).
49. Logothesis, A. L., 'Chemistry of fluorocarbon elastomers', *Progress in Polym. Sci.*, **14**, 251 (1989).
50. Waterman, N. A. and Ashby, M. F. (ed.) 'Fluoroplastics' in *Elsevier Materials Selector*, Vol. 3, Elsevier Applied Science, London, p. 1703 (1994).
51. Mascia, L., Pak, S. H. and Caporiccio, G., *Polym. International*, **35**, 75 (1994).
52. Fiedler, L. D., Knapp, T. L., Norris, A. W. and Virant, M. S., 'Effect of methanol/gasoline blends at elevated temperature on fluorosilicone elastomers', presented at the Society of Automotive Engineers International Congress and Exposition, Detroit, Michigan, USA, February (1990).
53. Maxson, M. T., Norris, A. W. and Owen, M. J., 'Fluorosilicones' in *Modern Fluoropolymers: High Performance Polymers for Diverse Applications* (ed. J. Scheirs), Wiley, London (1997).
54. Hergenrother, P. M., *Trends of Polym. Sci.*, **4**, 104 (1996).
55. Marshall, J. B., 'Kalrez®-type perfluoroelastomers — synthesis, properties and applications' in *Modern Fluoropolymers: High Performance Polymers for Diverse Applications*, (ed. J. Scheirs) Wiley, London (1997).
56. Bur, A. J. 'Dielectric properties of fluorine-containing polymers' in *Fluoropolymers*, (ed. L. Wall), Wiley-Interscience, Chapter 15, p. 475 (1972).

57. Jow, T. R., and Cygan, P. J., *Conf. Rec. IEEE Int. Symp. Electr. Insul.*, p. 181 (1992).
58. Anderson, B. C. and Uschold, R. E., 'Fluoropolymers' in *Encyclopedia of Material Science and Engineering*, Vol. 3 (ed. M. B. Bever), p. 1809 (1986).
59. Avakian, P. and Starkweather, H. W. Jr, 'Dielectric properties of fluoropolymers' in *Modern Fluoropolymers: High Performance Polymers for Diverse Application*, (ed. J. Scheirs), Wiley, London (1997).
60. Sacher, E., *Progress in Surf. Sci.*, **47**, 273 (1994).
61. Piyakis, K., Sacher, E., Domingue, A., Pireaux, J. J., Leclerc, G., Bertrand, P. and Lhost, J. B., *Applied Surface Science*, **84**, 227 (1995).
62. Hung, M. H. and Burch, R. R., *J. Appl. Polym. Sci.*, **55**, 549 (1995).
63. Neagu, E. and Neagu, R., *Applied Surface Science*, **64**, 231 (1993).
64. Brewis, D. M. and Mathieson, I., 'Dielectric properties of fluoropolymers' in *Modern Fluoropolymers: High Performance Polymers for Diverse Applications*, (ed. J. Scheirs), Wiley, London, 1997.
65. Feiring, A. E., Imbalzano, J. F. and Kerbow, D. L., *Trends in Polym. Sci.*, **2**, 26 (1994).
66. Imbalzano, J. and Kerbow, D., US Patent 4 943 658 (1988) (to DuPont).
67. Sinha, D., *Chemical Engineering Progress*, October, p. 84 (1991).
68. Goodman, J. and Andrews, S., *Solid State Technology*, **33**, 65 (1990).
69. Lueghamer, A., German Patent DE 4 422 372 (1996). *Chem. Abs.*, **124**, 148 139b.
70. Fujiwara, K. and Uratani, M., Japanese Patent JP 08 11 217 (1996), *Chem. Abs.*, **124**, 262 801m.
71. Mikkelsen, K. J., Alberg, M. J. and Prestidge, J. K., *Microcontamination*, June (1995).
72. Seymour, R. B., *Engineering Polymer Sourcebook*, McGraw-Hill, New York, p. 216.
73. Madorsky, S. L., *Thermal Degradation of Organic Polymers*, Interscience, New York (1964).
74. Carlson, D. P. and Schmiegel, W., 'Fluoropolymers, Organic', *Ullmann's Encyclopedia of Industrial Chemistry*, Vol. A11, Verlag Chem., Weinheim, p. 39 (1988).
75. Nguyen, T., *Rev. Macromol. Chem. Phys.*, **C25**, 227 (1985).
76. Kourtides, D. A. and Parker, J. A., *Soc. Plast. Eng., Tech. Pap., ANTEC'77*, **23**, 322 (1977).
77. Morelli, J. J., Fry, C. G., Grayson, M. A., Lind, A. C. and Wolf, C. J., *J. Appl. Polym. Sci.*, **43**, 601 (1991).
78. Chen, C-. S. and Chapoy, L. L., Eur. Patent Appl. 683 204 (1995) *Chem. Abs.*, **124**, 119 077z.
79. Iwasaki, M. *et al. J. Polym. Sci.*, **25**, 377 (1957).
80. Novikov, A. S., Galil, F. A., Slovokhotova, N. A. and Dyumaeva, T. N., *Vysokomol. Soedin.*, **4**, 423 (1962).
81. Paciorek, K. J., 'Chemical crosslinking of fluoroelastomers' in *Fluoropolymers*, (ed. L. Wall), Wiley-Interscience, Chapter 10, (1972).
82. Wright, W. W., 'The evolution of hydrogen fluoride from hydrofluoro-elastomers' in *Developments in Polymer Degradation — 6*, (ed. N. Grassie), Elsevier Applied Science, London, Chapter 1, p. 1 (1985).
83. Cox, J. M. and Wright, B. A. and Wright, W. W., *J. Appl. Polym. Sci.*, **8**, 2935, 2951 (1964).
84. Johnson, P. R., Pariser, R. and McEvoy, J. J., *Rubber Age*, **107**, 29 (1975).
85. Hirschler, M. M., *Eur. Polym. J.*, **18**, 463 (1982).
86. Baker, B. B. and Kasprzak, D. J., *Polym. Degrad. Stab.*, **42**, 181 (1993).

87. Scheirs, J. and Camino, G. in *Recycling of PVC and Mixed Plastic Waste*, (ed. F. P. La Mantia), ChemTec Publishing, Ontario, p. 167 (1996).
88. Clarke, F. B., Van Kuijk, H., Valentine, R., Makovec, G. T., Seidel, W. C., Baker, B. B., Kasprzak, D. J., Bonesteel, J. K., Janssens, M. and Herpol, C., *J. Fire Sciences*, **10**, 488 (1992).
89. Alarie, Y. C., Lieu, P. J. and Magill, J. H., *J. Combustion Toxicology*, **8**, 242 (1981).
90. Rosenberg, Y., Siegmann, A., Narkis, M. and Shkolnik, S., *J. Appl. Polym. Sci.*, **45**, 783 (1992).
91. Charlesby, S., *Nucleonics*, **12**, 18 (1954).
92. Chipara, M. D. and Chipara, M. I., *Polym. Degrad. Stab.*, **37**, 67 (1992).
93. Clough, R., 'Radiation-resistant polymers' in *Encyclopedia of Polymer Science and Engineering*, Vol. 13, Wiley, New York, p. 667 (1988).
94. Florin, R. E., 'Radiation chemistry of fluorocarbon polymers' in *Fluoropolymers*, (ed. L. Wall), Wiley-Interscience, Chapter 11 (1972).
95. Franetzki, M., German Patent, DE 4 440 597 (1996) *Chem. Abs.*, **125**, 18129h.
96. Clough, R. L., Gillen, K. T. and Dole, M., 'Radiation resistance of polymers and composites' in *Irradiation Effects on Polymers*, (ed. D. W. Clegg and A. Collyer), Elsevier Applied Science, London, Chapter 3, p. 95 (1991).
97. Klier, I., Strachota, S. and Vokal, A., *Plasty Kauc.*, **28**, 172 (1991); *Chem. Abs.*, **117**, 252 422e; Klier, I. and Vokal, A., *Radiat. Phys. Chem.*, **38**, 457 (1991).
98. Hutton, L. T. and Lavanga, D. J., Eur. Patent EP 722 173 (1996) *Chem. Abs.*, **125**, 125 932m.
99. Kawano, Y. and Soares, S., *Polym. Degrad. Stab.*, **35**, 99 (1992).
100. Zhong, X. and Sun, J., *Polym. Degrad. Stab.*, **36**, 239 (1992).
101. Charlesby, A. and Davison, W. H. Y., *Chem. Ind.*, 232 (1957).
102. Ito, M., *Radiat. Phys. Chem.*, **47**, 607 (1996).
103. Lunkwitz, K., Brink, H. J., Handte, D. and Ferse, A., *Radiat. Phys. Chem.*, **33**, 523 (1989).
104. E-beam Services Inc. 146 Du Pont Street, Plainview, NY 11803, USA.
105. Suther, J. L. and Laghari, J. R., *J. Mater. Sci. Lett.*, **10**, 786 (1991).
106. Scheirs, J., 'Fluoropolymer coatings; new developments' in *Polymeric Materials Encyclopedia*, (ed. J. Salamone), CRC Press, Florida, USA, p. 2498–2507 (1996).
107. Scheirs, J., Burks, S. and Locaspi, A., *Trends in Polym. Sci.*, **3**, 74 (1995).
108. Sjostrom, C., Jernberg, P. and Lala, D., *Mater. Struct.*, **24**, 3 (1991).
109. Okamoto, S. and Ohya-Nishiguchi, H., *J. Jpn. Soc. Colour Mater.*, 63, 392 (1990); *Chem. Abs.*, **114**, 25 835m.
110. Adams, M. R. and Garton, A., 'Far-ultraviolet degradation of selected polymers' in *Polymer Durability; Degradation, Stabilization and Lifetime Protection*, (ed. R. L. Clough, N. C. Billingham and K. T. Gillen) Advances in Chemistry Series 249, American Chemical Society, Chapter 10, p. 139 (1996).
111. George, G. A., Hill, D. J. T., O'Donnell, J. H., Pomery, P. J. and Rasoul, F. A., in *LDEF — 69 Months in Space*, (ed. A. S. Levine), NASA Conference Publication 3194, National Aeronautics and Space Administration, Washington, DC, p. 867 (1993).
112. Wall, L. A. and Straus, S., *J. Research (National Bureau of Standards)*, **65**, 227 (1961).
113. Bowers, G. H. and Loveloy, E. R., *Ind. Eng. Chem. Prod. Res. Dev.*, **1**, 89 (1962).
114. Shultz, A. R., Knoll, N. and Morneau, G. A., *J. Polym. Sci.*, 62, 211 (1962).
115. Smith, D. M., Welch, W. F., Graham, S. M., Chughtai, A. R. and Schissel, P., *Solar Energy Mater.*, **19**, 111 (1989).

116. Koo, G. P., 'Structure and mechanical properties of fluoropolymers', in *Fluoropolymers*, (ed. L. Wall), Wiley-Interscience, New York, Chapter 16, p. 507 (1972).
117. Schöbacher, H. and Stolarz-Izycka, A., *CERN Report 79-04*, (1979) (available from CERN, TIS Division, CH-1211 Geneva, Switzerland).
118. Engel, L., Klingele, H., Ehrenstein, G. W. and Schaper, H., *An Atlas of Polymer Damage*, Carl Hanser Verlag, Munich (1981), p. 168.

2

Fluorocarbon Elastomers

V. ARCELLA
Fluoropolymers R & D – Ausimont, Italy

R. FERRO
Fluoroelastomers Technical Service & Market Development – Ausimont, Italy

1 INTRODUCTION

Fluoroelastomers are synthetic copolymers designed for demanding service applications in hostile environments, characterized by broad temperature range and contact with aggressive chemicals. Their most important properties have been related to the structure of the backbone leading to the following general classification: fluoroinorganic elastomers; fluorocarbon elastomers.

In fluoroinorganic elastomers, such as fluorosilicone [1] and fluorophosphazene [2] elastomers, polymeric chains are comprised of inorganic repeating units having fluorinated organic pendant groups. The most important properties of this class of products are the retention of tensile properties at high temperature and the exceptional low temperature flexibility [3]. The introduction of fluorine improves the thermal stability and solvent resistance. Fluorocarbon elastomers are the most common fluoroelastomers, characterized by carbon–carbon linkages in the polymer backbone. Fluorinated monomers (the starting materials to prepare these polymers) can be divided in two important different classes: vinylidene fluoride (VF_2)-based fluorocarbon elastomers and tetrafluoroethylene (TFE)-based fluorocarbon elastomers (perfluoroelastomers). The development of VF2-based fluorocarbon elastomers, described in this chapter, were initiated for military interest. In the early 1950s the M. W. Kellog Company in a joint project with the US Army developed the first commercial fluorocarbon elastomer: the copolymer of VF2 and chlorotrifluoroethylene (CTFE), available in 1957 under the trade name of Kel-F [4]. The most common fluorocarbon elastomer, the copolymer of VF2 and hexafluoropropylene (HFP), was developed on a commercial scale by Du Pont, and became available in 1955 under the trade name of VITON A [5]. Since that time other important VF2-based fluorocarbon elastomers have

Modern Fluoropolymers. Edited by John Scheirs
© 1997 John Wiley & Sons Ltd

Table 2.1. Historical development of fluorocarbon elastomers

Structure	Year	Company
$(CH_2-CF_2)(CF_2-CF)$ Cl	1955	3M
$(CH_2-CF_2)(CF_2-CF)$ CF_3	1957	DuPont
$(CH_2-CF_2)(CF_2-CF)(CF_2-CF_2)$ CF_3	1960	DuPont
$(CH_2-CF_2)(CHF-CF)$ CF_3	1960	Montecatini Edison
$(CH_2-CF_2)(CHF-CF)(CF_2-CF_2)$ CF_3	1960	Montecatini Edison
$(CH_2-CF_2)(CF_2-CF)(CF_2-CF_2)$ OCF_3	1975	DuPont
$(CF_2-CF_2)(CF_2-CF)$ OCF_3	1975	DuPont
$(CF_2-CF_2)(CH_2-CH)$ CH_3	1977	Asahi Glass
$(CF_2-CF_2)(CH_2-CH)(CH_2-CF_2)$ CH_3	1990	Asahi Glass 3M
$(CH_2-CH_2)(CF_2-CF)(CH_2-CF_2)$ CF_3	1992	Ausimont

been developed with the aim of improving their application performance, in particular their chemical resistance against polar solvents and strong bases, and their low-temperature flexibility. Table 2.1 gives an illustration of the historical development.

2 CHEMICAL STRUCTURE AND RELATED PROPERTIES

Homopolymers of VF2, TFE and hydrogenated ethylene (E) respectively are highly crystalline polymers. The copolymerization of these monomers with suitable comonomers, such as CTFE, HFP, methylvinylether (MVE) and propylene (PP) gives amorphous materials with good elastomeric properties.

All these suitable monomers have the similarity of a pendant group attached to the vinyl functionality which 'disturbs' and prevents the crystallization of the macromolecular chain. For an amorphous structure they have to be copolymerized in appropriate amounts depending on the comonomers used, i.e. their ability to modify the mechanism of crystalline domain formation. Monomer combinations of important commercially available fluorocarbon elastomers have been already reported [6]. New monomer combinations constitute today's commercial fluoroelastomers, as reported in Table 2.2. In the case of VF2/HFP copolymers, completely amorphous polymers are obtained when the amount of HFP is higher than 19–20% on a molar base [7]. The triangular diagram of Figure 2.1 shows roughly the elastomeric region for VF2/HFP/TFE terpolymer systems. Commercially, VF2-based fluorocarbon elastomers have been, and still are, the most successful among fluoroelastomers. Despite their higher cost compared to other speciality rubbers, they are often preferred for many applications thanks to their exceptional thermal stability and resistance to oxidation and chemical degradation, although under some conditions they can be inferior to other materials. The higher thermal and chemical stability of fluoroelastomers can be mostly related to the high bond energy of the $C-F$ bond, and, due to the presence of fluorine, also to the higher bond energy of the $C-C$ and $C-H$ links [8]. They show outstanding resistance to aromatic and aliphatic hydrocarbons, such as fuels and oils. So they are not replaceable in many high-temperature automotive and aerospace applications, when in contact with such fluids. However, due to the polar character of the main VF2 monomer, they are highly swelled when in contact with polar solvents, such as low molecular weight

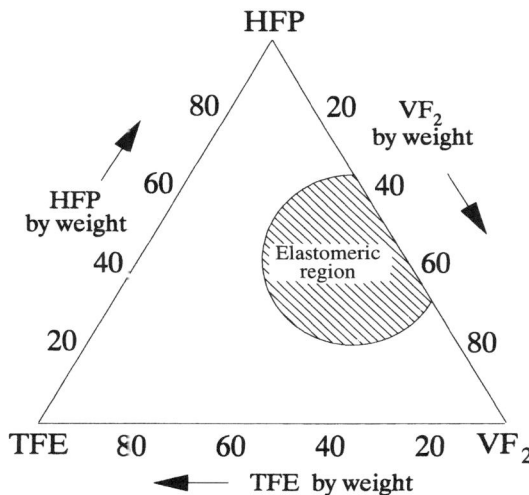

Figure 2.1. VF$_2$/HFP/TFE compositions

Table 2.2. Elastomeric combinations of different monomers

'Basic' monomer \ 'Elastomeric' monomer	$CF_2=CF-CF_3$	$CHF=CF-CF_3$	$CF_2=CF-OCF_3$	$CF_2=CF-Cl$	$CH_2=CH-CH_3$
$CH_2=CF_2$	X	X	X $(CF_2=CF_2)$	X	X $(CF_2=CF_2)$
$CF_2=CF_2$	X $(CH_2=CF_2)$	X $(CF_2=CF_2)$	X		X
$CH_2=CH_2$	X $(CH_2=CF_2)$		X $(CF_2=CF_2)$		

X Denotes commercially available fluoroelastomers.

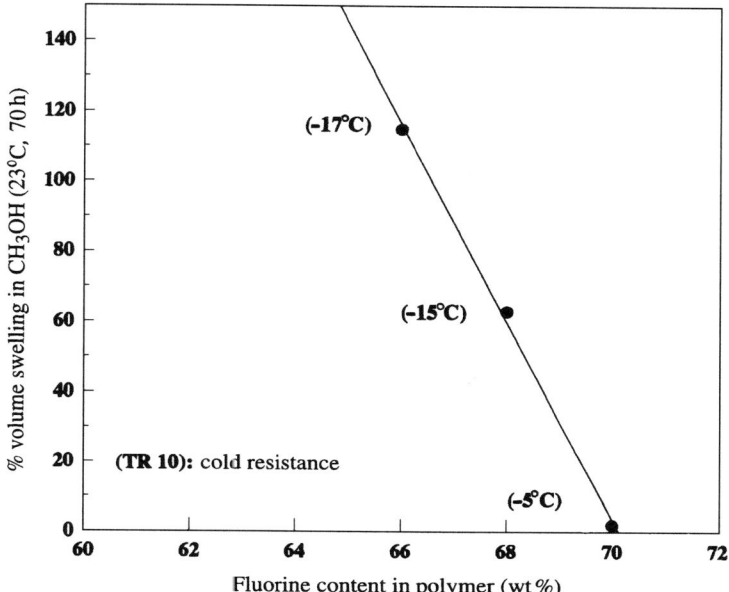

Figure 2.2. Cold and methanol resistance versus fluorine content

esters and ketones. Improvements can be obtained by reducing the amount of the VF2 monomer in the polymer, obtaining polymers with higher total fluorine content. However, lower VF2 monomer gives higher chain stiffness, so inferior low-temperature flexibility results (Figure 2.2). Fundamental properties of these materials can be related to their microstructure that can be described in terms of monomer sequences, i.e. monomer dyads triads and tetrads [9]. Glass transition temperatures of multimonomer systems, such as VF2/HFP/TFE or VF2/MVE/TFE systems, can be related to monomer dyads by the following equation [10]:

$$\frac{1}{T_g} = \frac{W_{11}}{T_{g1}} + \frac{W_{22}}{T_{g2}} + \frac{W_{33}}{T_{g3}} + \frac{W_{12}}{T_{g12}} + \frac{W_{13}}{T_{g13}} + \frac{W_{23}}{T_{g23}}$$

where W_{ij} is the weight fraction of dyads ij, T_{gi} the T_g of the homopolymer of monomer i and T_{gij} the T_g of the ideally alternating copolymer of monomers i and j.

To achieve useful high-temperature properties, as in the case of other thermoset elastomers, it is necessary to establish a stable and permanent network among the macromolecular chains by a crosslinking reaction.

VF2-based fluorocarbon elastomers can be crosslinked by ionic and radical systems, as discussed in a subsequent paragraph. Curing rate in ionic crosslinking reactions can be related to the molar fractions of the monomer triads N212 and

N213 (1 = VF2, 2 = HFP, 3 = TFE). The monomer tetrads N1111, N1113 and N1133 of VF2/TFE monomer blocks, in VF2/HFP/TFE elastomeric terpolymers, have been related to the storage shear modulus G'. The minimization of this parameter leads to improved low-temperature properties, measured as temperature retraction and compression set, although the glass transition temperature (T_g), measured by differential scanning calorimetry (DSC), did not change. Monomer dyads, triads and tetrads can be computed by a 'Markov chain model', once the polymer monomer composition and the monomer reactivity ratios are known.

3 MONOMER PREPARATION

3.1 CHLOROTRIFLUOROETHYLENE (CTFE)

The preparation of CTFE can be done by an hydrofluorination reaction starting from perchloroethane.

$$CCl_3-CCl_3 + HF \longrightarrow CCl_2F-CClF_2$$

The product 1,1-dichloro, 1-fluoro, 2-chloro, 2,2-difluoroethane (CFC113) may be dechlorinated by Zn to give CTFE [11].

$$CCl_2F-CClF_2 + Zn \longrightarrow CF_2=CFCl + ZnCl_2$$

3.2 VINYLIDENE FLUORIDE (VF2)

A process to produce VF2 has been described starting from acetylene. Two moles of HF are added in a reaction catalysed by a Lewis acid, such as BF3, to give 1,1-difluoroethane (CFC152) [12]. CFC152 is then chlorinated to 1-chloro-1,1-difluoroethane CFC142 [13] and subsequently dehydrochlorinated giving VF2 [14]. This process can be schematized by the following reactions:

$$CH{\equiv}CH + 2HF \xrightarrow{BF_3} CH_3CHF_2$$

$$CH_3CHF_2 + Cl_2 \longrightarrow CH_3CClF_2 + HCl$$

$$CH_3CClF_2 \longrightarrow CH_2=CF_2 + HCl$$

A somewhat different process starts from 1,1,1-trichloroethane which after the hydrofluorination reaction gives CFC142. CFC142 is subsequently dehydrochlorinated as before to VF2.

The reaction scheme is as follows:

$$CH_3CCl_3 + HF \longrightarrow CH_3CClF_2 + HCl$$

$$CH_3CClF_2 \longrightarrow CH_2=CF_2 + HCl$$

The dehydrochlorination reaction can be performed both thermally or catalytically. It is highly endothermic, therefore good heat transfer has to be provided.

The reaction environment is also highly corrosive requiring great care in the choice of construction materials.

A catalytic pyrolysis has been described to produce VF2 starting from 1,1,1-trifluoroethane [15,16]. The catalyst was prepared starting from commercial $CrF_3 \cdot 3H_2O$ heating it with hot air and subsequently passing HF gas through it. The so-prepared catalyst was packed in a Hastelloy-c pipe where 1,1,1-trifluoroethane was passed over it at about 400 °C obtaining VF2 at high conversion and purity.

4 POLYMERIZATION TECHNOLOGY

Fluorocarbon elastomers are generally prepared by high-pressure, free radical emulsion polymerization techniques [17]. Since the monomers can be in the gaseous form under the polymerization conditions, gas–liquid mass transfer can be a limiting step for the polymerization kinetics. In this case a proper reactor design has to be considered to perform useful reactions and obtain acceptable productivity. Organic or inorganic peroxy compounds, such as ammonium persulphate can be used as polymerization initiators. Inorganic initiators generally lead to ionic polymer chain ends, such as $-CH_2OH$ and $-CF_2COOH$, which contribute to the colloidal stability of the latex formed during the polymerization reaction. In this case, suitable emulsifying agents, such as ammonium perfluorooctanoate, are not strictly required [6]. Ionic chain ends deriving from the polymerization initiator also have important effects on polymer properties, such as rheology, mechanical properties and even sealing properties. For example, higher compound viscosities are obtained, increasing the amount of ionic chain ends, with the same molecular weight polymers. When inorganic peroxy compounds are used, ionic chain ends can be controlled by minimizing the amount of decomposed initiator and using a chain transfer agent to obtain the desired molecular weight. Appropriate molecular weight distribution is required to obtain good processing behaviour, such as easy milling, high flowability, and fast extrusion. Molecular weight control can be easily obtained in semi-batch process. In this case monomers and other chemicals are continuously added to the reactor, and after a proper time, the reaction is stopped and the polymer discharged. Mathematical models, simulating the polymerization reaction, can be used to determine the polymerization parameters, and to adjust them along the polymerization process to obtain the desired molecular weight distribution.

A mathematical model of the emulsion semi-batch copolymerization of fluorinated monomers has been recently presented. Experimental results related to VF2/HFP/TFE systems, showing the evolution of the reaction kinetics, monomer composition and molecular weight as a function of time are reported [18].

5 CURE CHEMISTRY

VF2-based-fluorocarbon elastomers can be crosslinked by ionic curing; however, radical curing can be also possible if the polymer has been prepared in the presence of a cure site monomer (CSM) bringing a functional group susceptible of radical attack, such as bromine (Br) or iodine (I).

5.1 IONIC CURING

VF2/HFP and VF2/HFP/TFE fluorocarbon elastomers can be vulcanized by bis-nucleophiles, such as bisphenols and diamines. Schmiegel in 1979 identified the triad monomer sequences N212 and N213 as 'base sensitive sites' by 19F-NMR spectroscopy. The formation of double bonds in the presence of a base from these sites is the starting step in the ionic crosslinking of these fluoroelastomers [19]. The following reaction mechanism has been recently proposed by Arcella *et al.* on the basis of a broad study involving solution investigations using 19F-NMR, 1H-NMR, FT–IR techniques, coupled with solid-state vulcanization experiments [20]:

1. Formation of $-C(CF_3)=CH-$ double bond by elimination of 'tertiary' fluorine.
2. Double bond shift catalysed by fluoride ion and formation of $-CH=CF-$ double bond.
3. Nucleophilic addition to the $-CH=CF-$ double bond with:
   (a) allylic displacement of fluoride affording the new $-C(CF_3)=CH-$ double bond;
   (b) addition/fluoride elimination from the same double bond.

The mechanism is schematized in Figure 2.3 where the bis-nucleophile $Nu-R-Nu$ represents a bisphenol or diamine crosslinker.

5.2 RADICAL CURING

Dehydrofluorination needed for the ionic crosslinking reaction produces double bonds beyond that are strictly required for the crosslinking itself. Smith and Perkins [21] found 4 moles of F- per mole of crosslink during vulcanization by diamines. Schmiegel predicted 6 moles of F- per mole of crosslink in vulcanization by bisphenol-AF in presence of a quaternary phosphonium salt [19]. Venkateswarlu *et al.* found even higher amounts in vulcanization by bisphenol-AF depending on the relative amount of a phosphonium salt [22]. Excess of unsaturations formed in ionically crosslinked elastomeric parts represents weak points susceptible of further nucleophilic attack by basic substances present in the contact fluid. This is, for example, the case of new oils and fuels containing basic additives [23,24]. Radical curing avoids the presence of unsaturations after the crosslinking reaction.

The reaction proceeds via a radical mechanism activated by the presence of a peroxide substance which thermally decomposes under moulding conditions. The fluoroelastomer has to be functionalized in order to perform a satisfactory crosslinking reaction giving a high density network. Bromine containing

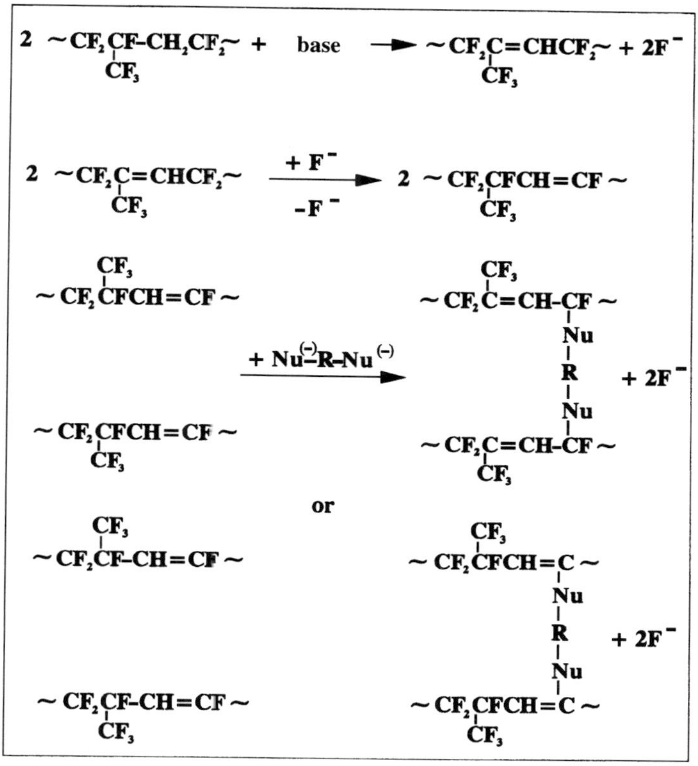

Figure 2.3. Reaction mechanism for ionic curing

fluoroelastomers have been prepared incorporating a Br-fluoro-olefine [25] or a Br-fluoro-vinylether [26], as the cure site monomer within the macromolecular chain, during the polymerization reaction. These fluoroelastomers are crosslinked in the presence of peroxides giving a very stable network. However, moulding problems, due to excessive mould fouling and consequently a high scrap rate, have limited the development of these polymers. The introduction of iodine-containing fluoroelastomers has represented a technological breakthrough, since these polymers show much fewer moulding problems, and allow the use of more sophisticated transformation techniques, such as injection moulding. In I-fluoroelastomers, the iodine functionality is generally introduced into the polymer, as terminal end group by performing the polymerization reaction in the presence of a iodine-containing fluorocarbon substance, acting as the polymerization chain-transfer agent [27]. The way of introducing the radical cure site by a chain-transfer reaction is fundamental in obtaining good moulding properties, since it allows the radical functionality to reach the low molecular weight polymer fraction. Low molecular weight chains are most responsible for mould fouling if they are not able to participate in the crosslinking reaction. In addition to excellent moulding, I-fluoroelastomers also show excellent sealing, although their thermal stability is lower than that of Br-fluoroelastomers. Br-fluoroelastomers have also been prepared with this concept, obtaining improved results [26].

The combination of I at chain ends and Br in the polymeric chain [28] has further improved, as expected, the Br-fluoroelastomers mentioned above [26]. In practice the iodine has been introduced following the leaching of I-fluoroelastomers [27], but replacing only the Br at the chain ends and leaving only the Br in the polymeric chain.

I/Br-fluoroelastomers combine the properties from iodine as chain end [27] and from Br in the polymeric chain [25]. As a result the thermal stability is decreased.

In a completely different composition the chain end iodine can be introduced via an iodine salt, in order to maximize the number of polymer chains containing I, and minimize the total amount of I required for processability improvement, without affecting the polymer's thermal stability. These I/Br-fluoroelastomers then have the same thermal stability of comparable Br-fluoroelastomers, but improved processability [29]. The formulation package includes a peroxide, such

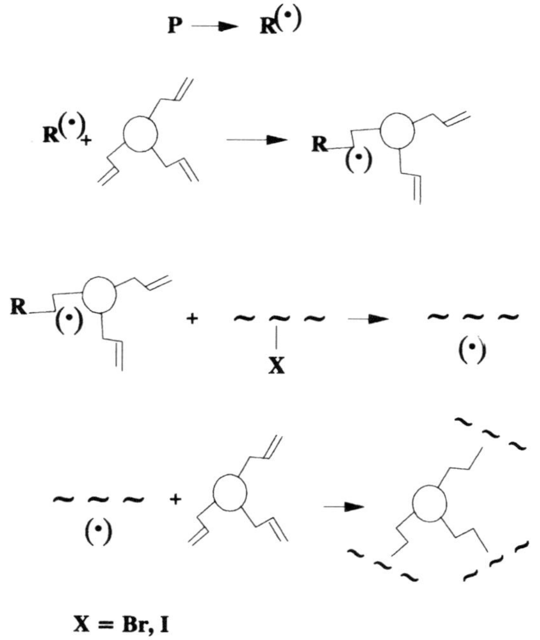

X = Br, I

Figure 2.4.   Reaction mechanism for radical curing

as 2,5-dimethyl-2,5-di-t-butylperoxyhexane and a crosslinking co-agent such as triallyl isocyanurate or triallyl cyanurate.

The reaction mechanism is schematized in Figure 2.4.

6 COMPOUNDING

Compounding of fluoroelastomers is relatively simpler when compared with other rubbers.

Ingredients are obviously the result of the particular cure chemistry, final properties to be achieved and processing technology to be used.

Usually the standard chemicals are: filler(s), processing aids, an acid scavenger (metal oxide), an activator (hydroxide), crosslinker and accelerator (cure system) as well as pigments in case of coloured items (Table 2.3).

Filler type and level are directly dependant on the desired final properties while combinations of fillers are sometimes the key to the optimization of processing behaviour.

For example, in black compounds small quantities of calcium silicate or barium sulphate are known to reduce fouling.

Medium thermal carbon black (N 990) is by far the most widely used grade for black compounds. In fact it is the best compromise to balance economic factors and physical properties.

More reinforcing carbon blacks (such as SRF) allow higher hardness and better physical properties at the expense of slightly higher compression set and cost.

Surface treated white fillers are sometimes used to improve flow, moisture resistance and tensile properties [30–32].

The proper combination of metal oxide and hydroxide is dictated by the chosen cure system (Table 2.4). The diaminic cure is nowadays restricted to applications where Food and Drug Administration (FDA) compliance is needed due to food contact. Some coating and extrusion applications also use the diaminic system.

The most popular cure system (about 80% of all applications) is the nucleophilic. It is based on the crosslinker (bisphenol AF) and accelerator (phase transfer catalyst such as phosphonium or amino-phosphonium salt).

Some grades are also available to be crosslinked by peroxidic cure systems: this technique requires special fluoroelastomers where cure site monomers can be activated to generate new stable bonds. Despite the superior resistance of peroxide cured FKMs, difficult processing has been for years the major obstacle to their steadily growing use. Recent improvements (in chemistry and polymerization) are now offering more opportunities for such class of polymers.

On the market-place, fluoroelastomers are sold either as pure gum polymers or as precompounded grades (incorporating a cure system) to meet the requirements of customers using only the nucleophilic cure system. Precompounded grades are adjusted by the supplier to give the best combination of accelerator and crosslinker. The final factory compounding is reduced to the addition of fillers and other chemicals necessary to obtain the application and processing performance.

Table 2.3. Typical Compounding Ingredients for Fluoroelastomers

Ingredient	Grade	phr	Notes
Crosslinker	Bisphenol AF	1–3	Nucleophilic cure
	Diamine	1–3	Diaminic cure
	TAIC	1–4	Peroxidic cure
Accelerator	Phosphonium		Phase transfer
	salt	0.2–1	for nucleophilic cure
Acid scavenger	Magnesium oxide		Diaminic cure
	*Low activity	10–15	improves bonding
	*High activity	3–6	
	Lead oxide		Steam and acid
			Resistance
			Also for peroxide cure
Activator	Calcium hydroxide	4–6	Activates
			dehydrofluorination
Black filler	MT N990	10–50	Best cost/performance
			compromise
	SRF N765	10–30	Better hot tear
White filler	Calcium silicate	10–40	General-purpose
	*Untreated		improved metal
	*Epoxy silane treated		bonding
	Barium sulphate	20–50	Steam and temperature
			Resistance
Pigments	Iron oxide	1–5	Red/brown compounds
	Chromium oxide	1–5	Green compounds
	Organic pigments	0.5–3	Large range
Processing aids	Carnauba wax	1–1.5	
	Low MW polyethylene	0.5–1	
	Organosilicone	0.5–1.5	
	Compound		
	Perfluoropolyethers	0.5–2	

Table 2.4. Typical cure systems for fluoroelastomers

Cure system	Crosslinker	Accelerator	Acid scavenger	Activator
Diaminic	Hexamethylene carbamide or NN'Dicinnamilidene 1,6-hexanediamine	—	MgO (10–15 phr) (low activity)	—
Bisphenolic	Bisphenol AF	Phosphonium or amino-phosphonium salt	MgO (10–15 phr) (low activity) or MgO (3–5 phr) (high activity)	Ca (OH)$_2$ (4–6 phr)
Peroxide	Triallyl isocyanurate or triallyl cyanurate	Peroxide	PbO (2–4 phr) or ZnO (2–4 phr)	—

'The art of compounding' is a wide area where chemists apply experience, expertise and knowledge, nevertheless some basic compounding rules are typical for fluoroelastomers.

The levels of the acid acceptor (MgO) and activator ($Ca(OH)_2$) in case of bisphenol cure strongly affect not only the cured network as reflected by the final properties but also the cure properties so that the cure system must be finely tuned to achieve the optimum balance of processing parameters (Figure 2.5).

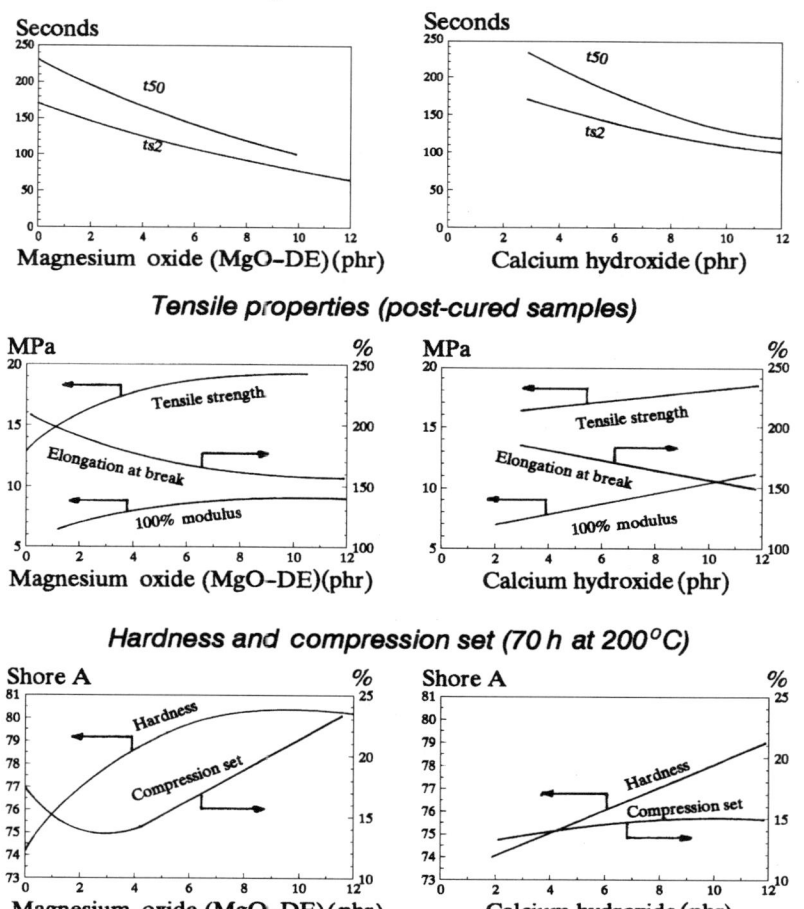

Figure 2.5. Effect of MgO and $Ca(OH)_2$ on curing and physical properties. Vinylidene fluoride–hexafluoropropylene copolymer. Standard test formulation: Tecnoflon FOR 65BI/R = 100: MgO–DE = 3 phr, $Ca(OH)_2$ = 6 phr, MT Black = 30 phr

7   MIXING

The manufacture of fluoroelastomeric items uses typical rubber technology equipment, i.e. for mixing, shaping and curing. Complete descriptions of such technologies is available in specialized literature [33–38].

7.1   OPEN MILLS

Open milling is quite popular with fluoroelastomers as for many specialty rubbers. When compared with internal mixers, open mills afford lower capital investment and are easier to clean (at lower manpower cost) to minimize cross-contamination risks. Several compounders feel more at ease in using open milling for expensive compounds where visual control by skilled operators is possible.

Several obstacles should be considered when using the method:

1.   Skilled people and highly manual operations are required.
2.   Productivity is limited by time vs volume of each batch.
3.   Environmental contamination (moisture, dust, etc.).
4.   Poor control of the main parameters (procedure, friction ratio, cooling, nip).
5.   Excessive operative cost once production volume reaches the break-even point.

These considerations probably explain why internal mixers are also becoming more popular for fluoroelastomers due to the increasing trend towards specialized large-scale compounding facilities.

7.2   INTERNAL MIXERS

When considering fluoroelastomers for internal mixing one should take into account some properties that are peculiar to this kind of rubber [39]:

1.   High heat buildup due to viscous shear.
2.   Low elongation at break ('green strength').

Intensive cooling and low starting temperature are necessary to avoid pre-scorched compounds. Generally the 'upside-down' procedure is preferred except for some recipes with white fillers where poor wetting of the filler by the unmasticated rubber can be a problem.

In most cases one-pass mixing is adopted except for some high hardness compounds requiring the addition of curatives in a second step. Tangential or intermeshed rotors are widely used.

After the fluoroelastomer exits the mixer, compounds are normally immediately sheeted off and cooled down by open milling and air-cooled batch-off.

8 MOULDING

Standard transforming technologies are also all applicable for fluoroelastomers eg. compression, transfer and injection moulding are widely used as well as extrusion and calendering. Below, some considerations are given with the object of highlighting some peculiarities typical of fluoroelastomers.

8.1 COMPRESSION MOULDING

About 60% of fluoroelastomers are shaped by compression moulding. The investments in presses and moulds are relatively modest. Mould design should take into account that fluoroelastomers may shrink at a different ratio (from 2.5 up to 3.0%) when compared with other rubbers (EPDM, NBR from 1.5 to 2.0%; VMQ from 3.0 up to 3.5%).

To minimize scrap, a preform is generally cut or extruded. The preform is placed in the mould cavity, pressed and cured under pressure and heat; if required some degassing is also applied.

Recently some compression moulding presses for fluoroelastomers have been equipped with vacuum devices to improve quality and minimize scrap.

Moulds are usually set at 170 up to 200 °C. Common practice is to leave the part in the mould until a suitable cure state has been reached, this time is usually near the oscillating disk rheometer t90 at moulding temperature. After press cure, an oven post cure is needed to improve and fix the physical properties of the final item.

8.2 INJECTION MOULDING

The first fluoroelastomers were difficult to injection mould, but recent developments in polymerization and cure chemistry (well designed and balanced molecular weight distribution; controlled ratio and distribution of the sequences of monomers along the backbone; improved emulsion and microemulsion technologies; chemistry of polymer chain terminals; special fluorinated processing aids; improved cure systems ...) enable excellent polymers to be produced by this technology.

The necessary investments are substantially larger than for compression moulding, either for both machines and for moulds, but faster cycles, higher automation, lower manpower, easier quality and process control [40] are the driving forces in the preference for injection moulding whenever a large number of parts have to be produced (eg. in the case of automotive applications).

In fact mass production for large consumption such as O-rings, seals and gaskets is steadily and rapidly moving towards injection moulding. Single station presses (vertical or horizontal) can be utilized depending on the items to be moulded. The barrel and nozzle are usually set at 70–100 °C with the mould at

180–220 °C. About 0.7–1 ton/cm^2 of moulded area is required for a 70–80 shore A hardness.

Processing aids are often necessary to improve flow and demoulding; unfortunately very small quantities (from 0.5 to 3 phr) of a few selected only chemicals may be added due to poor compatibility with the fluoroelastomer.

Flow lines and splice-knitting are the most common problems to be considered and avoided mostly by practical experience leading to the best compromise and balance of press, mould, fluoroelastomer type and compounding ingredients.

The best results are achieved by vacuum applied during the injection step in order to avoid air trapping, splitting and porosity. The mould must be designed to avoid as much as possible losses in flash, sprue and runners. Gates must be designed to avoid lack of flow into the cavities and minimize back-grinding during curing. Thick flashes not only represent additional cost but are also dangerous because of tear propagation in the deflashing operation. A carefully designed gate is the key to control this aspect: obviously the sharper the cutting edge the better the quality of the finished part but also the faster will be the wearing of the mould.

8.3 EXTRUSION

Extruded tubes, hoses and profiles represent a small percentage (less than 10%) of the total fluoroelastomer consumption. No practical continuous vulcanization method is used probably because pressure is needed to obtain usable items: for this reason the normal cure is achieved by steam or air autoclave. Extrusion is also useful to preform compounds into shapes to fill cavities for compression moulding.

The typical temperature pattern for fluoroelastomers is a gradual increase from the feeding zone to the die (maximum temperature at the die is about 100 °C). The screw is usually working at the same temperature as that of the feeding zone.

Processing aids are commonly added to improve surface and output. Here also strong constraints to the choice of the appropriate processing aid arise from the poor compatibility of most chemicals with fluoroelastomers [41–43].

8.4 CALENDERING AND COATING

Calendering is mostly used to produce sheets with or without fabric inserts. Usually low viscosity and high elongation grades are preferred for such applications.

A brief mention should be made of solution coating: the old technology based on fabrics coated by dipping or spraying with fluoroelastomeric solutions in organic solvents (methyl ethyl ketone, toluene . . .) is nowadays being replaced by the use of highly concentrated (about 70% w/w solids) waterborne emulsions of polymer (ex: TECNOFLON TN LATEX) to comply with lower solvent emission requirement.

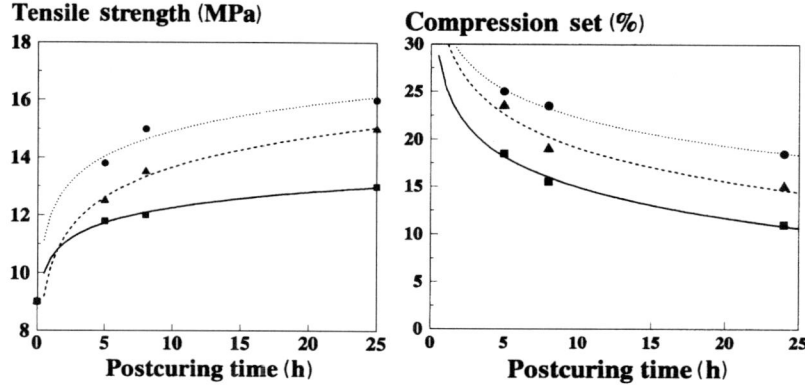

Figure 2.6. Effect of postcuring time and temperature on physical properties. Tecnoflon FOR 65BI/R. Standard black formulation. Postcuring temperature: ■ 200 °C; ▲ 225 °C; ● 250 °C

9 CURE AND POSTCURE

As already mentioned, a typical cure is performed from 170 up to 220 °C, while optimum properties are, however, achieved after postcure. Standard post cure conditions are 18–24 hours at 220–250 °C. Figure 2.6 gives the results of an experimental study on the effect of time/temperature on a typical copolymer. Obviously post cure conditions must in any case be chosen, also considering the specific item (mainly as far as thickness) in order to avoid splitting and cracks due to excessive thermal stress.

10 PROPERTIES AND APPLICATIONS

According to ASTM D2000 system fluoroelastomers are designated as HK polymers [44]. Up to 210 °C, fluoroelastomers remain serviceable indefinitely offering excellent resistance in most solvents and chemicals (Table 2.5).

Table 2.5. General chemical resistance of fluoroelastomers

Outstanding resistance	Good to excellent resistance	Poor resistance
Hydrocarbon solvents	Low-polarity solvents	Strong caustic
Automotive fuels[a]	Oxidative enviroments	(NaOH, KOH)
Engine oils[b]	Dilute alkaline solutions	Ammonia and amines
Apolar chlorinated solvents	Aqueous acids	Polar solvents
Hydraulic fuels	Highly aromatic solvents	(ketones)
Aircraft fuels and oils	Water and salt solutions	(methyl alcohol)

[a]Unleaded fuels give some problems due to the presence of methyl alcohol.
[b]Certain amine additives in engine oils can be detrimental.

As previously indicated, the maximum fluorine content for a VDF–HFP–TFE composition is about 70% w/w. Recent developments from producers have been able to offer easy processable and low compression set high-fluorine terpolymers, contradicting the previous statement that the higher the fluorine content the better the chemical resistance at the expense of processability and compression set. Heat and oil resistance are obviously to be considered together with other properties such as tensile, compression set, fluorine content, processability etc. that are the key parameters in the choice of the appropriate grade. General information can be found in the literature mainly available from producers.

Today VF2/HFP and VF2/HFP/TFE polymers are produced and marketed internationally by different companies such as Du Pont-Dow Elastomers (Viton$^{(TM)}$), Ausimont (Tecnoflon$^{(TM)}$), Dyneon (Fluorel$^{(TM)}$ and Daikin (Daiel$^{(TM)}$). Lower capabilities are attributed to Russia, China and Nippon Mektron (Japan). In 1994 the total worldwide market for fluoroelastomers was estimated at about 9000–10 000 tons of which about 60% was automotive, 10% chemical and petrochemical, 10% aerospace and 20% other markets. The yearly growth rate is placed between 5 and 8% due to new application and replacement of rubbers of inferior performance.

As far as items are concerned about 30–40% comprise O-rings and gaskets, 30% shaft seals and oil seals, 10–15% hoses and profiles [45–49].

In the automotive field the harsh environment in and near by the engine requires fluoroelastomers for shaft seals, valve stem seals, O-rings, engine head gaskets, filter casing gaskets, diaphragms for fuel pumps, water pump gaskets, fuel hoses, seals for exhaust gas and pollution control equipment, bellows for turbo-charger lubricating circuits, etc.

Aerospace and military uses are for shaft seals, hydraulic hoses, O-rings, electrical connectors, gaskets for firewalls, traps for hot engine lubricants, fuel tanks, heat-shrinkable tubing for wire insulation, flares ...

Chemical and petrochemical plants utilise O-rings, expansion joints, diaphragms, blow-out preventers, valve seats, gaskets, hoses, safety clothes and gloves, stack and duct coatings, tank lining, etc.

REFERENCES

1. O. R. Pierce, G. W. Holbrook, O. K. Johannson, J. C. Saylor and E. D. Brown, *Ind. Eng. Chem.*, 1960, **52**, 783.
2. S. H. Rose, *J. Polymer Sci.*, B, 1968, **6**, 837.
3. G. S. Kiker and T. A. Antkowiak, *Rubber Chem. Tech.*, 1974, **47**, 32.
4. M. E. Conroy, F. J. Honn, L. E. Robb and D. R. Wolf, *Rubber Age*, 1995, **76**, 543.
5. J. S. Rugg and A. C. Stevenson, *Rubber Age*, 1957, **82**, 102.
6. A. L. Logothetis, *Prog. Polym. Sc.*, 1989, **14**, 251.
7. G. Ajroldi, M. Pianca, M. Fumagalli and G. Moggi, *Polymer*, 1989, **30**, 2180.
8. R. E. Banks; *Fluorocarbons and Their Derivatives*, Macdonald London, 1970, 17.

9. G. Brinati and V. Arcella, *Kautschuk + Gummi − Kunststoffe*, 1992, **45**, 470.
10. N. W. Johnston, *J. Macromol. Sci. Rev. Macromol. Chem.*, 1976, C, **14**, 215.
11. H. S. Booth and P. E. Burchfield, *J. Am. Chem. Soc.*, 1933, **55**, 2231.
12. R. E. Burk, D. D. Coffman and G. H. Kalb, US Pat. 2 425 991 (1947).
13. J. D. Calfee and P. A. Florio, US Pat. 2 499 129 (1950).
14. C. F. Feasley and W. H. Stover, US Pat. 2 627 529 (1953).
15. Japan. Pat. Open., 130 507 (1979).
16. Japan. Pat. Open., 235 409 (1986).
17. W. M. Grootaert, C. H. Millet and A. T. Worm, *Kirk-Othmer: Encyclopedia of Chemical Technology*, 4th edn, Vol. 8, p 990 Wiley, New York (1995).
18. M. Apostolo, G. Brinati and V. Arcella, Paper presented at North American Research Conference on the Science and Technology of Emulsion Polymers/Polymer Colloids, Hilton Head (SC) (1993).
19. W. W. Schmiegel, *Angew. Makromol. Chem.*, 1979, **76/77**, 39.
20. V. Arcella, G. Chiodini, N. Del Fanti and M. Pianca, Paper 57 presented at the 140th ACS Rubber Division Meeting, Detroit, Mich., 1991.
21. J. F. Smith and G. T. Perkins; *J. Appl. Polym. Sci.*, 1961, **5**, 460.
22. P. Venkateswarlu, R. A. Guenthner, R. E. Kolb and T. A. Kestner, Paper 123 presented at the 136th ACS Rubber Division Meeting, Detroit, Mich., 1989.
23. V. Arcella, S. Geri, G. Tommasi and P. Dardani, *Kautschuk + Gummi − Kunststoffe*, 1986, **39**, 407.
24. V. Arcella, R. Ferro and M. Albano, *Kautschuk + Gummi − Kunststoffe*, 1991, **44**, 833.
25. D. Apotheker and P. J. Krustic, US Pat. 4 035 586 (1977).
26. V. Arcella, G. Brinati, P. Bonardelli and G. Tommasi, US Pat. 4 745 165 (1988).
27. M. Tatemoto, IX Intern Symp. on Fluorine Chem., Avignon, 1979.
28. A. L. Moore, US Pat. 4 948 852 (1990).
29. M. Albano, G. Brinati, V. Arcella, E. Giannetti, US Pat. 5 173 553 (1992).
30. D. Skudelny-Kunststoffe *German Plastics*, **77** (11), 17 (1987).
31. R. Ferro, G. Giunchi and C. Lagana, *Rubber and Plastic News*, 19 Feb. (1990).
32. H. Struckmeyer, *Kautschuk + Gummi − Kunststoffe*, **43**, Jahrgang 9, (1994).
33. N. P. Cheremisinoff. *Polymer mixing and extrusion technology*, Marcel, Dekker, New York and Basle (1987).
34. W. Hoffmann. *Rubber Technology Handbook*, Hanser, Munich (1988).
35. M. Morton. *Rubber Technology*, 3rd edn., Van Nostrand Reinhold, New York (1987).
36. Z. Tadmor and C. G. Cogos. *Principles of Polymer Processing*. SPE/Wiley/Interscience, New York (1979).
37. G. M. Blow and C. Hepburn, *Rubber Technology and Manufacture*, 2nd edn, Butterworths, London (1982).
38. K. Nagdi. *Gummi Werkstoffe — Ein Ratgeber fur Anwender*, Vogel-Verlag, Wurzburg (1981).
39. J. Leblanc and M. Chominiatycz, *Kautschuk + Gummi − Kunststoffe*, **42** (10) and **42** (11) (1989).
40. J. Leblanc, R. Polet and M. Vincent, *Kautschuk + Gummi − Kunststoffe*, **44**, Jahrgang 7 (1991).
41. M. J. Stevens. *Extruder Principles and Operation*, Elsevier, London and New York (1986).
42. R. Ferro, G. Fiorillo and G. Restelli, *Elastomerics*, March and April (1989).
43. J. Leblanc, *Elastomers and Plastics*, 22 (1990).

44. ASTM D200-80, *Standard Classification System for Rubber Products in Automotive Applications*.
45. G. Streit, Paper presented at the 3rd Brazilian Congress of Rubber Technology, San Paulo, Sept 1989.
46. J. Leblanc. Paper 900197, SAE meeting 26–2 Feb. Mar. Detroit, Mich. (1990).
47. S. Aloisio, Paper 90956, SAE meeting 28–3 Feb. Mar. Detroit, Mich. (1994).
48. R. Mastromatteo, Paper 900195, SAE meeting 26–2 Feb. Mar. Detroit, Mich. (1990).
49. G. Streit, *Kautschuk + Gummi – Kunststoffe*, **43**, Jahrgang 11 (1990).

3

Dielectric Properties of Fluoropolymers

PETER AVAKIAN and HOWARD W. STARKWEATHER, Jr.
DuPont Central Research and Development, Experimental Station,
Wilmington, DE USA

Nonpolar fluoropolymers, especially those based on tetrafluoroethylene (TFE), are widely used in electrical applications in spite of their high cost relative to other nonpolar polymers such as hydrocarbon polymers. Among the advantages of fluoropolymers are their resistance to chemical attack, especially oxidation, their high melting points, and their retention of useful properties over a very wide range of temperatures. PTFE, the homopolymer of TFE, has the smallest dielectric loss and the highest melting point. Because of its high molecular weight and melt viscosity, it cannot be fabricated by the usual melt processes, such as extrusion or injection molding, and compression molding or paste extrusion must be used. At a molecular weight corresponding to a useful melt viscosity, the level of crystallinity is excessive leading to a poor combination of properties. The introduction of comonomers gives a lower level of crystallinity combined with a melt viscosity which permits extrusion to produce film or coat wire. FEP resin is a copolymer of TFE with hexafluoropropylene (HFP). Another copolymer, PFA, contains perfluoro propyl vinyl ether. The larger side group permits control of crystallinity at a smaller mole fraction of comonomer. This results in a melting point closer to that of PTFE but at higher cost. Other fluoropolymers, such as polyvinylidene fluoride are quite polar and find utility because of that property. A largely alternating copolymer of ethylene and tetrafluoroethylene, ETFE, is slightly polar but finds utility as a wire coating especially in aircraft because of its mechanical properties and resistance to cut-through and creep. A number of other fluoropolymers are not generally used in electrical or electronic applications. Nevertheless, dielectric measurements on these materials are useful for the study of internal motions which determine the appropriate temperature ranges for various applications.

Modern Fluoropolymers. Edited by John Scheirs
© 1997 John Wiley & Sons Ltd

When a dielectric is placed between the plates of a capacitor, the capacitance is increased because of the displacement of positive and negative charges within the material due to the electric field. The ratio between the capacitances with and without the dielectric is called the dielectric constant or permittivity. Actually, it is not a constant but a complex variable which depends on the temperature and frequency.

$$\varepsilon^* = \varepsilon' - i\varepsilon'' \tag{1}$$

The imaginary part, ε'', is called the loss factor. The ratio,

$$\varepsilon''/\varepsilon' = \tan \delta \tag{2}$$

is the dissipation factor. The real part of the permittivity, ε', increases with increasing temperature or decreasing frequency. The loss factor and the dissipation factor exhibit maxima when the rate at which certain dipoles can reorient in an electric field is comparable to the experimental frequency. These phenomena are known as relaxations and are comparable to the viscoelastic relaxations which are observed in mechanical measurements except that only those which involved the reorientation of dipoles are dielectrically active [1].

1   POLYMERS OF TETRAFLUOROETHYLENE

Carbon–fluorine bonds are highly polar. However, in a perfluorocarbon molecule having tetrahedral bond angles, these dipole vectors should cancel out. Any dielectric activity is therefore evidence for nontetrahedral bond angles. This would not be surprising in view of the well-known helical conformation of the chains in crystalline polytetrafluoroethylene.

Polytetrafluoroethylene (PTFE) exhibits three dynamic mechanical relaxations which have been called α, β, and γ in the order of decreasing temperature [2]. In early work, all of these relaxations were said to have been dielectrically active [3,4]. However, subsequent studies have shown that only the low-temperature γ-relaxation which has been attributed to motions of relatively short chain segments in the amorphous regions exhibits a dielectric loss peak [5,6]. Presumably, the samples in the earlier work had been decorated with extraneous dipoles.

The dissipation factor measured at a frequency of 1 kHz is plotted against temperature in the region of the γ-relaxation in Figure 3.1 for PTFE and three copolymers of TFE [7]. FEP is a copolymer with hexafluoropropylene (HFP), and PFA is a copolymer with perfluoro propyl vinyl ether. Thus, they contain CF_3 and OC_3F_7 side groups, respectively.

At this frequency, the peak occurs at 194 K ($-79\,°C$) for PTFE, 220 K ($-53\,°C$) for FEP, 198 K ($-75\,°C$) for PFA, and 186 K ($-87\,°C$) for ETFE. The height of the loss peak is smallest for PTFE and much larger for ETFE than for the perfluorocarbon polymers. The Arrhenius activation energies for these relaxations defined in terms of the following expression, where f is the frequency in

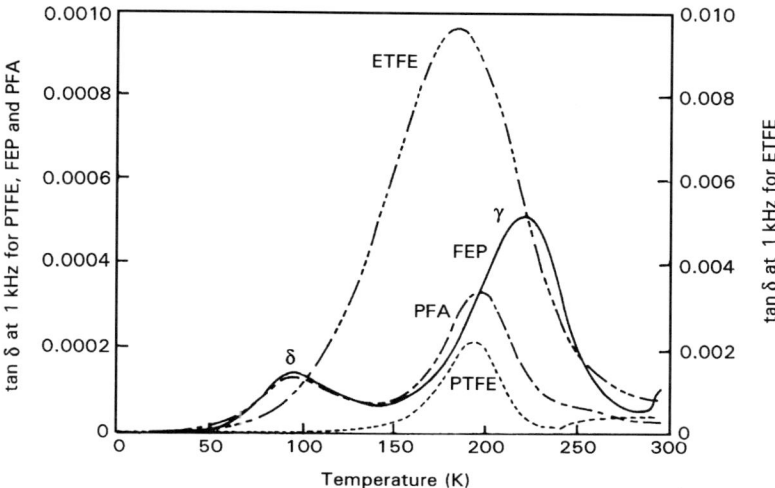

Figure 3.1. Dissipation factors at 1 kHz for polymers of TFE. (Adapted with permission from ref. 7. Copyright 1991 American Chemical Society)

hertz (cycles/second), R is the gas constant, and T is the temperature in kelvin,

$$E_a = -R \ d(\ln f)/d(1/T) \tag{3}$$

are about 67 kJ/mole for PTFE, FEP, and PFA and 45 kJ/mole for ETFE. It has been suggested that the fact that the temperature of the relaxation is higher in FEP than in the other polymers and that the peak height is greater than in PTFE and PFA may be due to an overlapping β-relaxation which has been attributed to internal motions in the crystalline regions.

The effect of pressures up to 3000 atm (304 MPa) on the γ-relaxation in PTFE and FEP has been investigated [8]. For PTFE, increasing the pressure causes the peak to shift to lower frequencies in isobaric frequency scans and to higher temperatures in scans at a defined frequency. With increasing pressure, the activation energy increases from 69 kJ/mole at 1 atm (0.10 MPa) to 88 kJ/mole at 2500 atm (253 MPa) for PTFE but remains essentially constant at 59 kJ/mole for FEP. The activation volume decreases with increasing temperature for PTFE but remains almost constant for FEP. The response of the γ-relaxation in PTFE to pressure is similar to that of the analogous relaxation in polyethylene. The different behavior in FEP is consistent with the suggestion that the observed relaxation in that polymer is actually a combination of the γ- and β-relaxations.

It is seen in Figure 3.1 that FEP and PFA exhibit an additional loss peak at 94 K ($-179\,^\circ$C), the δ-relaxation. It is attributed to motions of side groups which are not present in PTFE and ETFE. These relaxations have an activation energy of only 15.5 kJ/mole. This corresponds to an activation entropy close to zero meaning that the groups responsible move independently of each other.

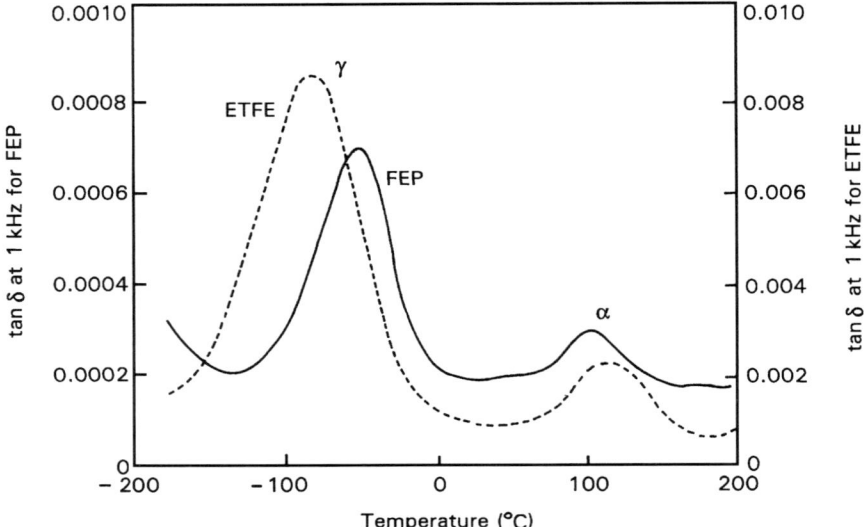

Figure 3.2. Dissipation factors of FEP and ETFE at higher temperatures at 1 kHz. (Reproduced with permission from ref. 9. Copyright 1994 American Chemical Society)

In Figure 3.2, the dissipation factor at 1 kHz in a higher temperature range is shown for FEP and ETFE [9]. The differences in tan δ_{max} in Figures 3.1 and 3.2 reflect measurements by different equipment on slightly different samples and are not significant for scientific understanding. In addition to the γ-relaxation, the higher temperature α-relaxation is seen at 103 °C (376 K) for FEP and 113 °C (386 K) for ETFE. This relaxation, which is associated with the glass transition, is attributed to the motions of longer chain segments in the amorphous regions.

2   THE EFFECT OF GUEST MOLECULES IN PTFE

While PTFE is resistant to many chemicals, it can absorb limited amounts of compounds which contain little or no hydrogen [10]. Dielectric studies were done on samples containing 1.4% chloroform, 2.4% carbon tetrachloride, and 5.6% chlorofluorocarbon-113 ($CCl_2F-CClF_2$), respectively [11]. The temperature dependence of the dissipation factor at 1 kHz for the samples saturated with chloroform and CFC-113 as well as a control are shown in Figure 3.3. The data for the sample saturated with carbon tetrachloride which has no dipole moment are similar to those for the control. Both of the polar additives increased the peak height for the γ-relaxation and introduced a new, large loss peak at a lower temperature. From the frequency–temperature relationships for the γ-relaxation,

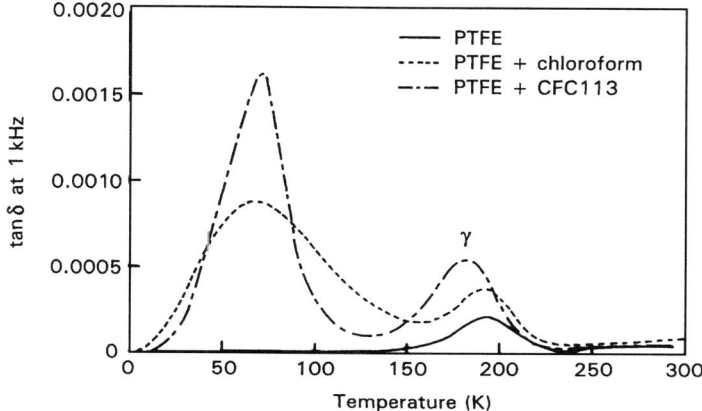

Figure 3.3. Effect of chloroform and CFC-113 on the dissipation factor of PTFE at 1 kHz. (Adapted with permission from ref. 11. Copyright 1992 American Chemical Society)

it was concluded that the presence of CFC-113 facilitates noncooperative motions of short chain segments in the amorphous regions of PTFE.

The low-temperature relaxations in Figure 3.3 occur at 67 and 73 K (-206 and $-200\,°C$) for the samples saturated with chloroform and CFC-113, respectively. In isothermal frequency scans, the maxima at 1 kHz are at 59 K ($-214\,°C$) with chloroform and 61 K ($-212\,°C$) with CFC-113. The activation energies are 8–13 kJ/mole, and the activation entropies are essentially zero. These relaxations are attributed to the reorientation of the absorbed molecules moving independently of each other. Experiments using hydrofluorocarbon additives have produced similar results [12].

3 POLY(PERFLUOROPROPYLENE OXIDE)

Krytox® oils are low molecular weight polymers of hexafluoropropylene epoxide which are used in lubrication and related applications. The chemical structure is $F(CF(CF_3)CF_2O)_n CF_2CF_3$. Dielectric measurements were made on samples of molecular weight 1850 and 8250, corresponding to values of n of about 10 and 49, respectively [13]. Figure 3.4 is a plot of the dissipation factor at 1 kHz vs temperature from 50 to 300 K (-223 to $+27\,°C$). A logarithmic scale is used in order to present both large and small peaks clearly.

In scans by differential scanning calorimetry (DSC) at $10\,°C/min$, glass transitions were observed at 195 and 219 K (-78 and $-54\,°C$) for the lower and higher molecular weight oils, respectively. At 1 kHz, the α-relaxations which correspond to the glass transition occurred at 213 and 232 K (-60 and $-41\,°C$). In work

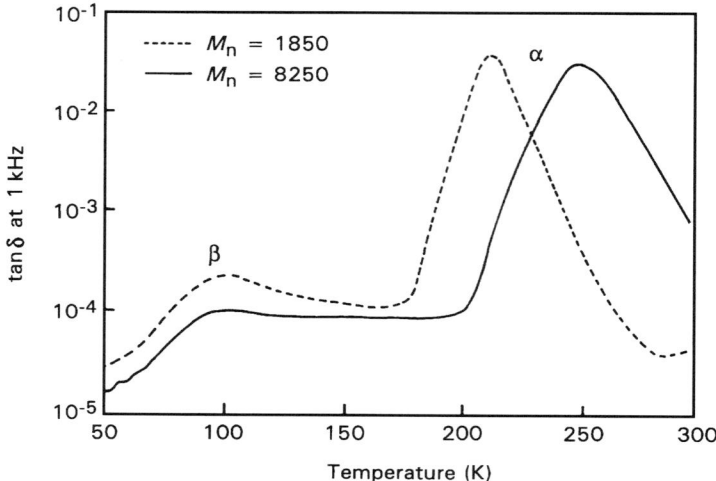

Figure 3.4. Dissipation factor at 1 kHz for poly(perfluoropropylene oxide). (Reproduced with permission from ref. 13. Copyright 1992 American Chemical Society)

by Alper and coworkers [14] on a sample of molecular weight 3700 ($n = 21$), the glass transition was observed at an intermediate temperature. Arrhenius plots ($\log f$ vs $1/T$) for the α-relaxations were curved, a familiar characteristic of glass transitions.

For both materials, there is a β-relaxation near 100 K ($-173\,°C$), about 7 K higher than the δ-relaxations which are associated with motions of the side groups on FEP and PFA. The activation energy for the β-relaxation in Krytox® is 18.4 kJ/mole, slightly higher than the value of 15.5 kJ/mole for the δ-relaxations in FEP and PFA. All of these relaxations have activations entropies close to zero and are attributed to noncooperative motions of fluorinated side groups.

It is noteworthy that while the peak heights for the α-relaxations are similar, that for the β-relaxation is about twice as large for the lower molecular weight sample. This is probably related to the fact that it contains more end groups which, like the side groups, are perfluoromethyls.

4   FLUORINATED ELASTOMERS

Elastomers should combine an absence of crystallinity, a low glass temperature (T_g), and a mechanism for forming crosslinks. A major class of fluorinated elastomers consists of copolymers of 20–70% vinylidene fluoride (VF_2), 20–60% HFP, and 0–40% TFE. The composition of greatest interest contains 60% VF_2 and 40% HFP with up to 30% TFE [15].

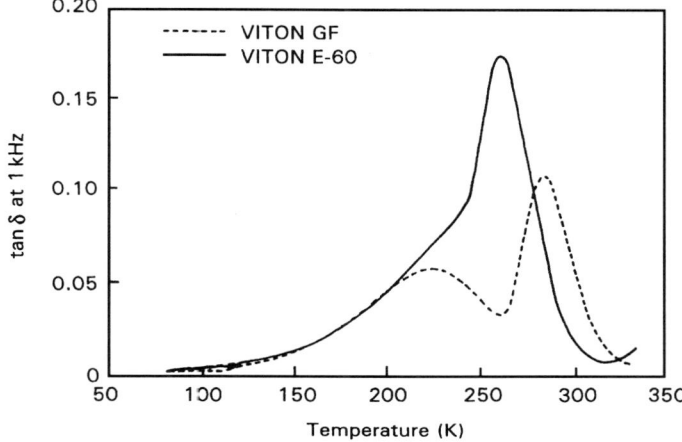

Figure 3.5. Dissipation factors at 1 kHz for Viton® fluoroelastomers. (Reproduced with permission from ref. 16. Copyright 1996 John Wiley & Sons, Limited, Chichester)

In Figure 3.5, the dissipation factors at 1 kHz are plotted for Viton® E-60 which is a binary copolymer of VF_2 and HFP and Viton® GF, a terpolymer with TFE, [16]. The values of T_g determined by DSC were -18 and $-8\,°C$, respectively. The dielectric α-relaxation was observed at $-7\,°C$ for the binary copolymer and $+13\,°C$ for the terpolymer. The peak height was much larger for the former.

The secondary β-relaxation appears as a maximum at 225 K ($-48\,°C$) in the temperature scan for the terpolymer, but only as a shoulder for VF_2/HFP. This is probably the result of overlap between the two relaxations. Both polymers exhibited maxima for the β-relaxation in isothermal frequency scans, at 223 K ($-50\,°C$) for VF_2/HFP and 205 K ($-68\,°C$) for VF_2/HFP/TFE. The activation energies were 50 and 40 kJ/mole, respectively.

Below 200 K ($-73\,°C$), the data are almost identical for the two materials. There is no evidence for the relaxation attributed to side groups which was observed at 94 K ($-179\,°C$) in FEP and PFA. That relaxation is very weak and may be obscured by the much stronger higher temperature relaxations in the Vitons®.

5 ALTERNATING COPOLYMERS OF METHYL VINYL ETHER WITH TFE OR HFP

These polymers can be viewed as ETFE, the largely alternating copolymer of ethylene and TFE, with side groups. That is, the ethylene has a methoxy substituent and, in the case of HFP/MVE, the TFE has a CF_3 side group. Whereas ETFE is a semicrystalline polymer, TFE/MVE and HFP/MVE are amorphous. The dissipation factors at 1 kHz are shown in Figure 3.6 [16]. For HFP/MVE, separate runs were done in the lower and higher temperature regions, and the data differ slightly in the overlap region.

The glass temperatures as determined by DSC were $-4\,^\circ$C for TFE/MVE and $22.5\,^\circ$C for HFP/MVE. The dissipation factors at 1 kHz exhibited well-defined maxima at $11\,^\circ$C for TFE/MVE and $39\,^\circ$C for HFP/MVE.

The most prominent secondary relaxations appear as maxima only in isothermal frequency scans. The next lower temperature relaxation below the α-relaxation, which we will designate the β-relaxation, was seen at 159 K ($-114\,^\circ$C) for TFE/MVE and 290 K ($+17\,^\circ$C) for HFP/MVE with activation energies of 26 and 69 kJ/mole, respectively. In the case of HFP/MVE, the α- and β-relaxations are very close. Extrapolations on an Arrhenius plot indicate that they would merge at a temperature of 300 K ($+27\,^\circ$C) and a frequency of about 30 kHz.

In HFP/MVE, there is a relaxation at 117 K ($-156\,^\circ$C) with an activation energy of 11.7 kJ/mole which appears to be analogous to the relaxations at 94 K ($-179\,^\circ$C) in FEP and PFA and is attributed to motions of the CF_3 side groups.

Finally, in both TFE/MVE and HFP/MVE, there is another relaxation at 9 K ($-264\,^\circ$C) and 13 K ($-260\,^\circ$C), respectively, in temperature scans and 7 K ($-266\,^\circ$C) in isothermal frequency scans. This probably reflects the reorientation of methoxy side groups in the MVE units.

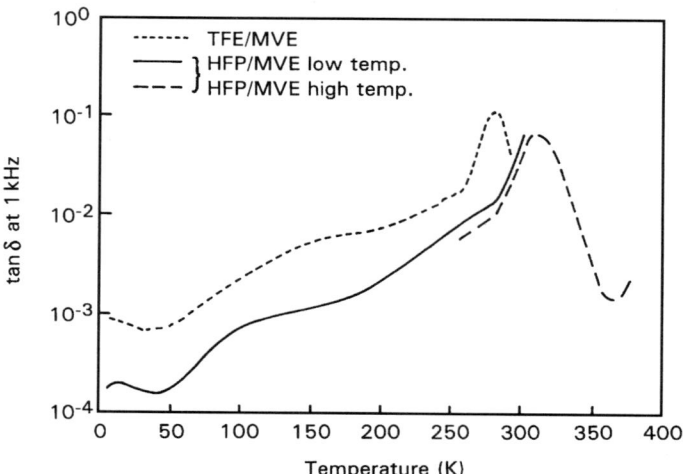

Figure 3.6. Dissipation factors at 1 kHz for copolymers of TFE and HFP with methyl vinyl ether (MVE). (Reproduced with permission from ref. 16. Copyright 1996 John Wiley & Sons, Limited, Chichester)

6 POLAR FLUOROPOLYMERS

6.1 POLYCHLOROTRIFLUOROETHYLENE

During World War II, polychlorotrifluoroethylene (PCTFE) was widely used in the Manhattan Project. It is considerably more polar than PTFE, but its dielectric properties are adequate for some applications. Dielectric studies have been done on samples over a wide range of crystallinity levels [1]. The polymer melts at about 220 °C, and the glass temperature by dilatometry is about 52 °C.

The α-relaxation was observed at about 150 °C at 10 Hz and 175 °C at 5 kHz with an activation energy of 335 kJ/mole. It has been assigned to motions in the crystalline regions. The data follow a straight line in an Arrhenius plot.

The β-relaxation which occurs in the range of 100 °C gives a curved relationship between log frequency and $1/T$ which is characteristic of glass transitions. It is strongest in samples of low crystallinity and is assigned to the amorphous regions. A mixed α- and β-relaxation is seen at intermediate levels of crystallinity. The possibility that the glass temperature may increase with increasing crystallinity has been discussed.

The γ-relaxation is the most prominent dielectric feature. With increasing crystallinity, it shifts to lower temperatures and higher frequencies. It has a frequency near 1 kHz at room temperature, and its temperature ranges from about -50 °C at 1 Hz to $+150$ °C at 10^7 Hz. It is thought that the γ-relaxation contains crystalline and amorphous components which at a frequency of 1 Hz occur at -40 and -10 °C, respectively. The activation energies have been estimated to be 56 kJ/mole for the crystalline relaxation and 73 kJ/mole for the amorphous relaxation. Both are attributed to motions of short chain segments.

6.2 POLYVINYL FLUORIDE

Polyvinyl fluoride (PVF) is used as a weather-resistant, dirt-repellent film for coating household siding [15]. The dielectric α-relaxation, corresponding to the T_g near 40 °C, is obscured by the effect of ionic conductivity which dominates the electrical properties at elevated temperatures [1]. This problem can be moderated by the use of a static field which presumably displaces or electrolyzes the ionic species [17]. For example, a loss peak at 75 °C and 100 Hz was observed in a sample of PVF to which a field of 30 kV/cm had been applied for 36 hours at 155 °C. The α-relaxation has also been observed near 55 °C in studies of thermally stimulated currents (TSC) [18].

At a frequency of 100 Hz, the γ- and β-relaxations are seen at -70 and $+3$ °C, respectively. With increasing frequency, they gradually merge to form a single peak. The activation energies are about 380 kJ/mole for the α-relaxation and 170 kJ/mole for the β-relaxation. For the γ-relaxation, it is about 33 kJ/mole which corresponds to an activation entropy close to zero, a characteristic of a noncooperative relaxation.

6.3 POLYVINYLIDENE FLUORIDE

ETFE and polyvinylidene fluoride (PVF_2) are isomeric polymers. The former is equivalent to a head-to-head polymer of vinylidene fluoride.

If both polymers are in a planar zig-zag conformation, ETFE has a center of symmetry every two chain atoms, whereas PVF_2 has a high degree of lateral polarity because the hydrogens are all on one side while the fluorines are all on the other. In PVF_2, this conformation is found in the so-called β crystal form which can be produced by stretching a film which can then be poled by heating in an electric field just below the melting point to form a ferroelectric [15]. The electrical field of this structure can be modulated by a mechanical stress. This is called the piezoelectric effect. This effect in PVF_2 is used in audio transducers in microphones and loudspeakers.

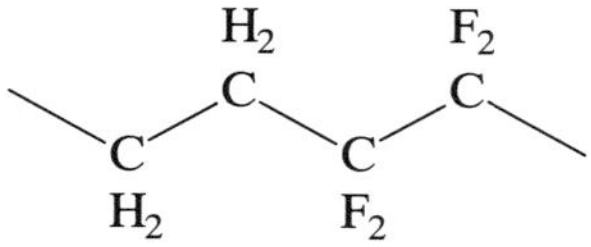

$$H_2 \quad F_2$$
$$C \qquad C$$
$$C \qquad C$$
$$H_2 \quad F_2$$

ETFE

$$F_2 \quad F_2$$
$$C \qquad C$$
$$C \qquad C$$
$$H_2 \quad H_2$$

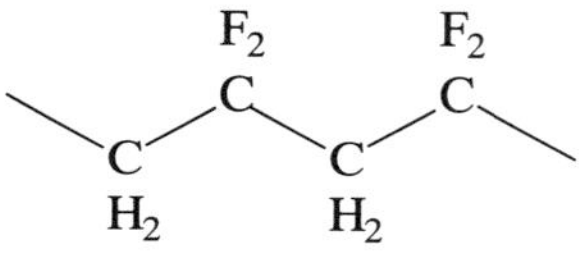

PVF$_2$

The T_g of PVF_2 at about $-35\,°C$ is reflected in the dielectric α-relaxation at about $-25\,°C$ for a frequency of 1 kHz [1]. There is a secondary relaxation at about $-75\,°C$ and 100 Hz.

Like PVF, PVF_2 is impacted by the effects of ionic conductivity at elevated temperatures. In this case also, the application of a static field is helpful in revealing an underlying dielectric relaxation [17]. Isothermal frequency scans were done on a sample to which a static field of 25 kV/cm had been applied at $160\,°C$ for 40 h. A loss peak appeared which varied from a frequency of 30 Hz at $61.5\,°C$ to 100 kHz at $160\,°C$, corresponding to an activation energy of about 96 kJ/mole. This relaxation has been attributed to internal motions in the crystalline phase.

REFERENCES

1. McCrum, N. G., Read, B. E., and Williams, G., *Anelastic and Dielectric Effects in Polymeric Solids*, (Wiley, New York, 1967; Dover Publications edition, New York, 1991).
2. McCrum, N. G., *J. Polym. Sci.*, **34**, 355 (1959).
3. Kabin, S. P., *Sov. Phys.-Tech. Phys.*, **1**, 2542 (1956).
4. Krum, F. and Muller, F. H., *Kolloid Z.*, **164**, 8 (1959).
5. Eby, R. K. and Wilson, F. C., *J. Appl. Phys.*, **33**, 2951 (1962).
6. Sacher, E., *J. Macromol. Sci., Phys.*, **B19**, 109 (1981).
7. Starkweather, H. W., Avakian, P., Matheson, R. R., Fontanella, J. J., and Wintersgill, M. C., *Macromolecules*, **24**, 3853 (1991).
8. Starkweather, H. W., Avakian, P., Fontanella, J. J., and Wintersgill, M. C., *Macromolecules*, **25**, 7145 (1992).
9. Avakian, P., Starkweather, H. W., Fontanella, J. J., and Wintersgill, M. C., *Proceedings Am. Chem. Soc., Div. Polym. Materials Sci. and Eng.*, **70**, 439 (Spring, 1994).
10. Starkweather, H. W., *Macromolecules*, **10**, 1161 (1977).
11. Starkweather, H. W., Avakian, P., Matheson, R. R., Fontanella, J. J., and Wintersgill, M. C., *Macromolecules*, **25**, 1475 (1992).
12. Starkweather, H. W., Avakian, P., Fontanella, J. J., and Wintersgill, M. C., *Macromolecules*, **27**, 610 (1994).
13. Starkweather, H. W., Avakian, P., Fontanella, J. J., and Wintersgill, M. C., *Macromolecules*, **25**, 3815 (1992).
14. Alper, T., Barlow, A. J., Gray, R. W., Kim, M. G., McLachlan, R. J., and Lamb, J., *J. Chem. Soc., Faraday Trans.*, **2**, 76, 206 (1980).
15. England, D. C., Uschold, R. E., Starkweather, H., and Pariser, R., *The Robert A. Welch Foundation Conferences on Chemical Research, XXVI Synthetic Polymers*, Houston, Texas, 1982.
16. Starkweather, H. W., Avakian, P., Fontanella, J. J., and Wintersgill, M. C., *J. Thermal Analysis*, **46**, 785 (1996).
17. Osaki, S., Uemura, S., and Ishida, Y., *J. Polym. Sci., Part A-2*, **9**, 585 (1971).
18. Sauer, B. B., Avakian, P., and Starkweather, H. W., *J. Polym. Sci., Part B: Polym. Phys.*, **34**, 517 (1996).

4

The Use of Fluoropolymers in Space Applications

BRUCE A. BANKS

National Aeronautics and Space Administration, Lewis Research Center, Cleveland, Ohio, USA

1 INTRODUCTION

Many of the applications of fluoropolymers in space parallel those found in terrestrial applications where a wide range in service temperature, low friction, resistance to chemical attack, or dielectric properties are important. As a result, fluoropolymers have found extensive application in space for bushings, lubricants, sleeves, tubing, seals, and as electrical insulation for wiring and in other forms for electronic packaging. The success of space applications of fluoropolymers greatly hinges upon the durability of their performance in the space environment. The intent of this chapter, however, is to present applications of fluoropolymers which are unique to space and durability issues relevant to their use in the space environment.

2 BACKGROUND — SPACE APPLICATIONS

2.1 THERMAL CONTROL AND RADIATOR SURFACES

The dominant use of fluoropolymers unique to space applications is for thermal control and radiator surfaces. Thermal control and radiator surfaces are frequently required on spacecraft to reject waste heat and maintain acceptable operating temperatures within a spacecraft. Typical characteristics of such surfaces include long-term survival over a wide operating temperature range, low solar absorptance, high thermal emittance, and freedom from contamination caused by outgassing. The use of fluoropolymers for spacecraft thermal control has been dominated by the use of FEP (fluorinated ethylene propylene) Teflon®, which has its second (unexposed) surface metalized with silver or aluminum. A cross-sectional view of the silvered FEP Teflon® thermal control blanket

Modern Fluoropolymers. Edited by John Scheirs
© 1997 John Wiley & Sons Ltd

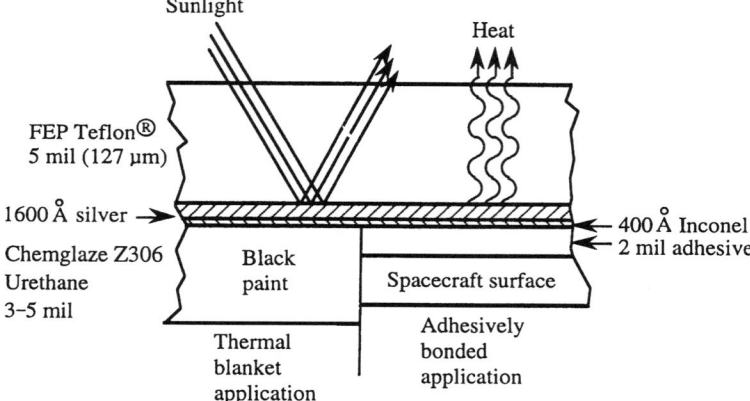

Figure 4.1. Cross-section view of two types of LDEF silvered FEP thermal control surfaces

used for the LDEF (long duration exposure facility) spacecraft [1,2] is shown in Figure 4.1. The spacecraft thermal control is achieved by means of high reflection of incident solar energy (which has an intensity peak near 0.55 μm) and high thermal emittance of FEP at spacecraft temperatures (which radiate near 10 μm) as shown in Figure 4.2 [2,3].

Other similar forms of thermal control surfaces or blankets consist of white Tedlar® (PVF (polyvinylidene fluoride)) and beta-cloth (PTFE (polytetrafluoroethylene) impregnated woven fiberglass). These two thermal control materials rely upon the low solar absorptance of their white color rather than second surface

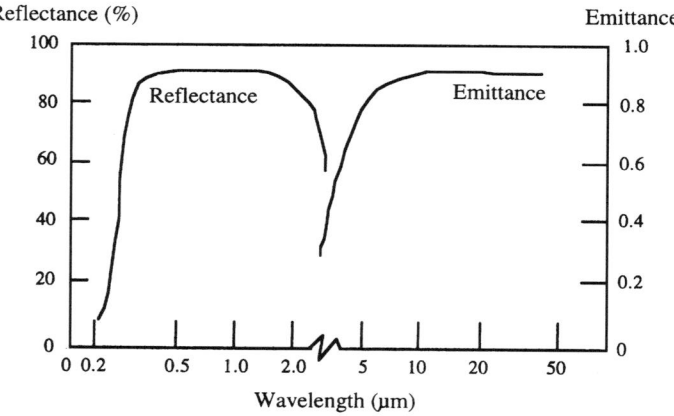

Figure 4.2. Radiation characteristics of silvered FEP thermal control surfaces

reflection from bright metals to reduce solar heating. Their emittance, similar to FEP, is based on the emittance of the bulk white Tedlar® or PTFE impregnated fiberglass.

Other advanced concept thermal control systems using fluoropolymer liquids have also been proposed. These include liquid droplet and liquid sheet radiators which utilize perfluoroethers which are sprayed out into space to radiatively cool and then collected and recirculated [4].

2.2 FLUOROPOLYMER-FILLED SiO_x ATOMIC OXYGEN PROTECTIVE COATINGS

PTFE filled SiO_x ($1.9 < x < 2.0$) has found unique application in space as high strain-to-failure coatings to prevent underlying polymers from atomic oxygen attack [5,6]. These molecularly mixed films are produced by ion beam sputter co-deposition from PTFE Teflon® and SiO_2 targets as shown in Figure 4.3. PTFE and FEP are somewhat unique in that sputter etching of these materials results in scission fragments consisting of about 85% C_2F_4 which are fully capable of repolymerizing in a more cross-linked fluoropolymer form as they deposit on the surfaces [7,8]. Thus, by simultaneous sputter etching of SiO_2 and PTFE, a molecular mixture of PTFE and SiO_x is produced which bears mixed properties. Such co-deposited films (up to 16% PTFE by volume) possess high transparency,

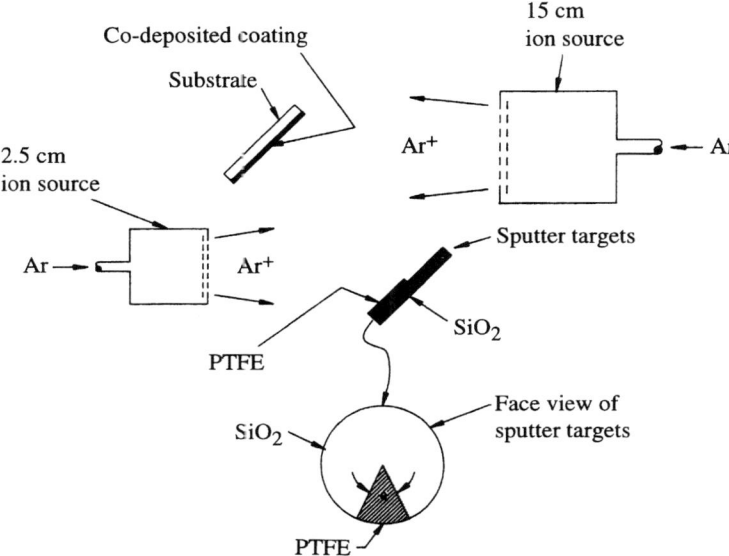

Figure 4.3. Ion beam sputter co-deposition process for deposition of fluoropolymer-filled SiO_x protective coatings

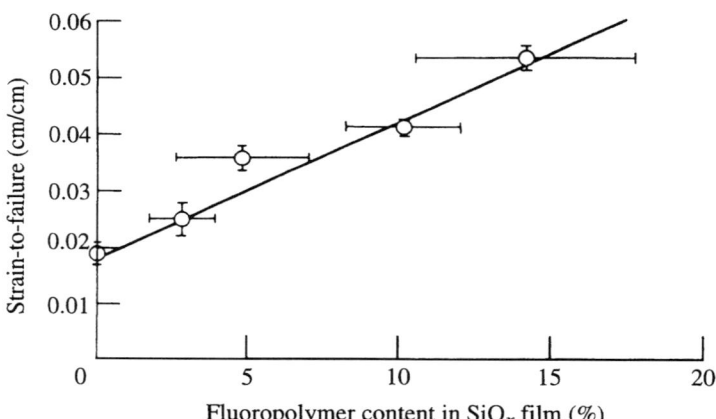

Figure 4.4. Strain-to-failure dependence upon fluoropolymer content in SiO_x (1.9 < x < 2.0) films

hydrophobicity, and atomic oxygen durability and increased strain-to-failure over pure SiO_2 films as shown in Figure 4.4 [9]. As Figure 4.4 illustrates, the addition of approximately 8% fluoropolymer to SiO_x films results in a doubling of a strain-to-failure. Such films have been used as atomic oxygen protective coatings on acrylic laser retroreflectors used for laser acquisition and range-finding of the EURECA spacecraft and the Japanese Space Flyer Unit [6].

3   IN-SPACE ENVIRONMENTAL INTERACTIONS, DURABILITY AND PERFORMANCE

3.1   OPERATING TEMPERATURE

Spacecraft in low Earth orbit (LEO) subject external fluoropolymer thermal control materials to a cycle of solar heating and radiative cooling every 90 minutes. These can result in temperatures potentially ranging from -80 to $+150\,°C$. The additional thermal load required to radiate waste heat can require significantly elevated temperatures which may challenge the maximum service temperature of FEP or its bonding to the substrates intended to be cooled. Because of the short orbital period in LEO, spacecraft which require a 15-year mission lifetime will cause such materials to undergo 87,660 thermal cycles. These must occur without cracking or debonding from the substrates intended to be cooled. Adhesives used to bond the silvered FEP Teflon® thermal control surfaces can diffuse through cracks in the metalized layers, allowing migration of the adhesive to the lower surface of the FEP Teflon® and thus subjecting it to solar radiation. This occurred on the LDEF mission which utilized a Y966 acrylic adhesive causing a gradual increase in the solar absorptance and a brown discoloration of the thermal control surface [10].

3.2 OUTGASSING AND CONTAMINATION

Frequently spacecraft surfaces and components are required to operate at low temperatures which may cause volatile materials to condense, thereby compromising the performance of a surface or device. The polymers used in space applications are required to produce minimal volatile materials which are condensable. Standardized testing (ASTM E-595-77/84/90) has been developed over the years to quantify the total mass loss (TML) and collected volatile condensable materials (CVCM) characteristics of materials used in space. Materials which are considered low-outgassing have a TML of 1% or less and a

Table 4.1. Outgassing properties of various fluoropolymers [11]

Material	Manu-facturer[a]	%TML[b]	%CVCM[c]	Application
TCK 10 Teflon® carbon coated Teflon®	Chemfab	0.53	0.01	Anti-static film
TCK 10 Teflon® coated Kevlar®	Chemfab	0.61	0.01	Anti-static film
TCK 1589 Teflon® coated Kevlar®	Chemfab	0.57	0.00	Anti-static film
TCK 1590 Teflon® coated Kevlar®	Chemfab	0.76	0.00	Anti-static film
TCK 6 Teflon® coated Kevlar®	Chemfab	0.36	0.00	Anti-static film
Tedlar® 150 BL 30 CC black film	DuPont	0.14	0.00	Film
Tedlar® coating on aluminum	Riegel Paper	0.14	0.05	Coating
Tedlar® E48678-155A three-layer black film composite	DuPont	0.58	0.02	Cond film
Tedlar® TZD15SH9 (M) black flame retard elec cond film	DuPont	0.27	0.00	Anti-static film
Teflon® FEP 2000L-BK black	DuPont	0.01	0.00	Film gasket
Teflon® FEP insulation TX22-731	High-Temp Wires	0.02	0.00	Insulation
Teflon® FEP Type A 5 mil film	DuPont	0.01	0.00	Film/sheet
Teflon® PFA film sheet TE-970	DuPont	0.00	0.00	Film
Teflon® TFE 0.5 mil film	Chemfab	0.01	0.00	Film
Tefzel® film 2 mil	DuPont	0.12	0.02	Blanket

[a]Chemfab = Chemfab/Chemical Fabrics Corp., Merrimac, NH; Dupont = E.I. DuPont de Nemours and Co., Inc., Wilmington, DE; Riegel Paper = Riegel paper Corp., New York, NY; High Temp Wires = High-Temp Wires Co., Westburg, Long Island, NY.
[b]Total mass loss.
[c]Collected volatile condensable materials.

CVCM of 0.1% or less [11]. Table 4.1 indicates the TML and CVC outgassing characteristics of various fluoropolymers [11].

3.3 ATOMIC OXYGEN

The most prevalent atmospheric specie in low Earth orbit between the altitude of 180 and 650 km is atomic oxygen. It is formed by photodissociation of upper atmosphere O_2 by short wavelength UV radiation from the sun. Spacecraft orbiting the Earth run into the atomic oxygen surrounding the Earth causing impact energies of 4.5 ± 1 eV with a sufficient flux to cause oxidation, resulting in gradual erosion of fluoropolymers [12].

Fixed direction arrival of atomic oxygen to fluoropolymer surfaces causes the development of a microscopic cone-like textured surface as shown in Figure 4.5. This textured surface development occurs for any material whose oxidation products are fully volatile. The textured surface produced by atomic oxygen causes light to reflect diffusely as opposed to specularly from a smooth surface. Although the change from a specular surface to a diffuse surface has little if any effect on the total reflectance or absorptance of a metalized thermal control blanket, the recession of the surface due to atomic oxygen attack does cause a reduction in thermal emittance. For example, on the LDEF mission which was 5.75 years

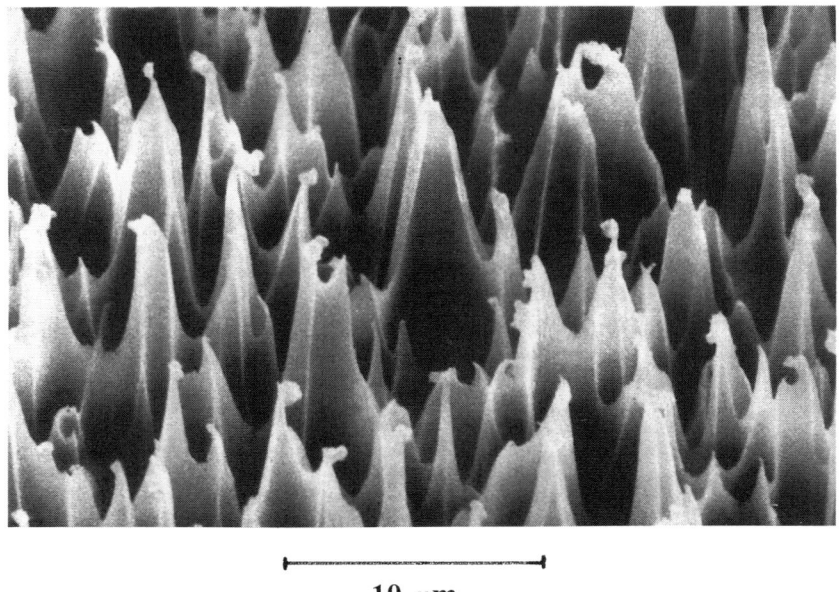

10 μm

Figure 4.5. Chlorotrifluoroethylene (Kel-F) after exposure on LDEF to an atomic oxygen fluence of 5.77×10^{21} atoms/cm^2

Thickness of FEP Teflon® (mm)

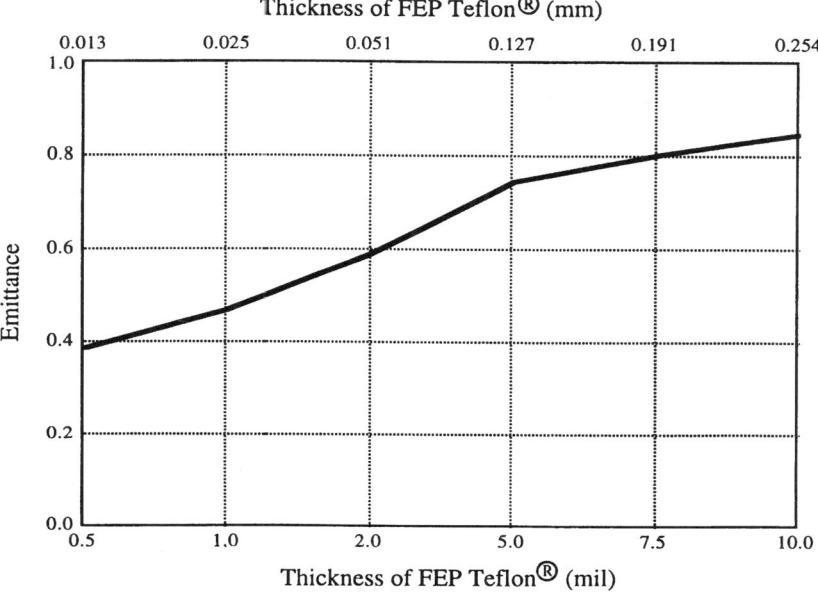

Figure 4.6. Dependence of emittance on FEP thickness for silvered FEP thermal control surfaces

in low Earth orbit, approximately 0.03 mm of FEP was eroded as a result of atomic oxygen attack at 8.1° from the ram direction. This caused a reduction in thermal emittance (because infrared radiation is emitted from a range of depths near the surface of the polymer) as can be seen in Figure 4.6, which shows the dependence of emittance on thickness of silvered FEP [13].

A measure of the susceptibility of materials to atomic oxygen attack is the atomic oxygen erosion yield, which is the volume of material removed through oxidation per incident oxygen atom. Table 4.2 lists the atomic oxygen erosion yield of various materials including fluoropolymers [13–23]. The thickness of material that will be lost through oxidation by atomic oxygen is simply given by the product of the atomic oxygen fluence for the mission and the erosion yield of the material. Potential synergistic effects on erosion yield of fluoropolymers from combined atomic oxygen and vacuum ultraviolet radiation exposure have been investigated, but no clear conclusions can be drawn as yet from in-space data [1,24,25].

3.4 MICROMETEOROID AND DEBRIS

Hypervelocity impact by micrometeoroid and debris particles can frequently penetrate fluoropolymer thermal control blankets because these blankets are

Table 4.2.   Atomic oxygen erosion yields of various materials in LEO

Material	Atomic oxygen erosion yield, cm^3/atom	Reference
FEP (in silvered Teflon®)	$0.337 \pm 0.005 \times 10^{-24}$	23 (corrected for LDEF ram fluence of 9.09×10^{21} atoms/cm^2)
	0.35×10^{-24}	21
PTFE (polytetrafluoroethylene Teflon®)	0.20×10^{-24}	21
	$0.37 \times 0.06 \times 10^{-24}$	20
White Tedlar®(crystalline polyvinylfluoride with white pigment)	0.29×10^{-24}	13, 21
CTFE (Kel-F® chlorotrifluoro-ethylene)	$1.97 \pm 0.12 \times 10^{-24}$	11 (corrected for LDEF ram fluence of 9.09×10^{21} atoms/cm^2)
ECTFE (Halar®ethylene-chlorotrifluoroethylene)	2.1×10^{-24}	13–15
	2.0×10^{-24}	21
Kapton®H (polyimide)	3.0×10^{-24}	13, 16, 17–19
	$2.89 \pm 0.6 \times 10^{-24}$	13, 20
Polyethylene	$3.97 \pm 0.23 \times 10^{-24}$	20
PMMA (polymethylmethacrylate)	$6.3 \pm 0.3 \times 10^{-24}$	20
PEEK (polyetheretherketone)	$3.7 \pm 1.0 \times 10^{-24}$	21
	2.3×10^{-24}	14
Kevlar®49	4.0×10^{-24}	21
Clear Tedlar® (crystalline polyvinylfluoride)	3.8×10^{-24}	13
Kevlar®29	$1.5 \pm 0.5 \times 10^{-24}$	21
Polysulfone	2.3×10^{-24}	21
Polystyrene	$4.17 \pm 0.17 \times 10^{-24}$	20
Nylon (polyamide)	$2.8 \pm 0.2 \times 10^{-24}$	20
Carbon (highly oriented pyrolytic graphite)	1.04×10^{-24}	20
	1.2×10^{-24}	16
Carbon (pyrolytic polycrystalline)	0.61×10^{-24}	20
	1.2×10^{-24}	16
Mylar A (polyethylene terephthalate)	$3.4 - 3.6 \times 10^{-24}$	13
Mylar D (polyethylene terephthalate)	3.0×10^{-24}	13
Diamond (single crystal natural Class IIA)	$0.0000\pm 0.000\,023 \times 10^{-24}$	22

typically less than 0.0254 cm thick. Penetration holes for thin polymer blankets and craters caused in thick fluoropolymers are typically sufficiently small in size not to cause significant changes in emittance or absorptance properties for missions of the order of a few years. However, on the LDEF mission micrometeoroid and debris impacts to silvered FEP thermal control blankets caused delamination of the FEP from an underlying silver/inconel/paint surface resulting in a decrease in thermal emittance of approximately 0.001/year [1]. In addition, the oxidation of the silver around each penetration site caused an increase in solar absorptance of approximately 0.0017/year [1].

3.5 ULTRAVIOLET AND SOFT X-RAY RADIATION

Although only 8% of the solar radiation has wavelength range between 0.1 and 0.4 μm, the photons in this range have sufficient energy to break organic bonds in polymers [26]. Vacuum ultraviolet (VUV) radiation (<0.2 μm) will cause crosslinking and embrittlement of the exposed surface of FEP Teflon®, leading to a reduction in tensile strength [1,26,27]. Soft x-ray radiation due to solar flares has been used to explain why FEP retrieved from the Hubble spacecraft during its servicing mission was far more embrittled than FEP retrieved from the LDEF mission, even though the Hubble exposure was of much shorter duration [28]. The Hubble mission allowed exposure of fluoropolymers to more solar flare soft x-rays than the LDEF mission, thus causing the observed increased embrittlement. In some cases on the Hubble retrieved hardware, FEP Teflon® under negligible stress had full thickness (0.0127 cm thick) cracks in it [29]. Soft x-rays penetrate farther into the fluoropolymers than VUV radiation and cause embrittlement that can extend much deeper than that caused by UV embrittlement.

3.6 CHARGED PARTICLE RADIATION

Electrons and protons trapped by the geomagnetic field in the Earth's Van Allen belts can cause impingement of tens of kilo-electron-volt to mega-electron-volt energy electrons and protons on fluoropolymer surfaces [1]. Degradation of FEP is dominated by electrons and does not appear to be great until fluences of $> 1 \times 10^{16}$ charged particles/cm^2 is reached [1,30]. Such fluences could be achieved after two years in geosynchronous orbit causing an increase in solar absorptance of silvered FEP of approximately 0.05 [30].

For LEO orbits the particle fluxes are reduced by orders of magnitude, thus causing minimal effects on fluoropolymer durability [1]. Galactic cosmic radiation consisting predominantly of nuclei of elements of hydrogen through iron, and solar particle events consisting of proton and other heavy nuclei, possess energies of $\geqslant 10$ MeV and do not possess fluxes sufficient to cause significant damage of fluoropolymers on spacecraft.

REFERENCES

1. Levadou F, Froggatt M, Rott M and Schneider E (1991), Preliminary investigation Into UHCRE thermal control coatings, LDEF first post-retrieval symposium, *NASA CP 3134*, pp. 875–898.
2. Silverman E (1995), Space environmental effects on spacecraft: LEO materials selection guide, *NASA CR 4661*, Part 2, 10, pp. 130–131.
3. Slemp W (1988), Ultraviolet radiation effects, NASA/SDIO Space Environmental Effects Workshop, *NASA CP 3035*, p. 425–445.
4. Gulino D and Coles C (1988), Oxygen plasma effects on several liquid droplet radiator fluids, *J of Spacecraft and Rockets*, **25** (2), 99–110.
5. Banks B, Mirtich M, Rutledge S, and Swec D (1984), Sputtered coatings for protection of spacecraft polymers, *NASA TM 83706*, p. 1–4.
6. Banks B and Rutledge S (1993), Performance characterization of EURECA retroreflectors with fluoropolymer-filled SiO_x Protective Coatings, *NASA CP 3275*, pp. 51–63.
7. Morrison D and Robertson T (1973), *J of Thin Solid Films*, **15**, 87.
8. Banks B, Sovey J, Miller M, and Crandall K (1978), *NASA TM 78888*, p. 3.
9. Banks B, Mirtich M, Rutledge S, Swec D, and Nahra H (1985), *NASA TM 87051*, p. 5.
10. Wilkes D, Brown M, Hummer L, and Zwiener J (1991), Initial materials evaluation of the thermal control surfaces experiment (S0069), *NASA CP 3134*, p. 905.
11. Campbell W N and Scialdone J (1993), Outgassing data for selected spacecraft materials, *NASA CP 1124*, Rev. 3, p. 387.
12. Banks B, de Groh K, and Rutledge S (1996), Prediction of in-space durability of protected polymers based on ground laboratory thermal energy atomic oxygen, *NASA TM 107209*, p. 1–2.
13. Silverman E (1995), Space environmental effects on spacecraft: LEO Materials Selection Guide, Part I, *NASA CR 4661*, 4, pp. 1–40.
14. Whitaker A (1993), Atomic oxygen effects on LDEF experiment AO171, *NASA CP 3194*, pp. 1125–1135.
15. Brower Jr W, Harish H, and Bauer R (1992), Effects of orbital exposure on halar during the LDEF mission, *NASA CP 3162*, pp. 391–415.
16. Banks B, Rutledge S, Paulsen P, and Stueber T (1989), Simulation of the low earth orbital atomic oxygen interaction with materials by means of an oxygen ion beam, *NASA TM 101971*, p. 8.
17. Visentine J, Leger L, Kuminecz J, and Spiker I (1985), STS-8 atomic oxygen effects experiment, AIAA paper 85-0415.
18. Zimcik D, and Magg C (1988), Results of apparent oxygen reactions with spacecraft materials during shuttle flight STS 41-G, *J. Spacecraft and Rockets*, **25** (2), 162–167.
19. Koontz S, King G, Dunnet A, Kirkendahl T, Linton R, and Vaughn J (1994), The ISAC atomic oxygen flight experiment, *J. Spacecraft and Rockets*, **31** (3), 475–488.
20. Gregory J (1992), On the linearity of fast atomic oxygen effects, *NASA CP 3257*, pp. 193–198.
21. Whitaker A, and Kamenetzky R (1992), Atomic oxygen erosion considerations for spacecraft materials selection, *NASA CP 3257*, pp. 117–123.
22. Banks B, Rutledge S, and Cales M (1993) NASA Lewis Research Center EOIM-III preliminary results, *Proceedings of the EOIM-III-3, BMDO Experiment Workshop*, p. 399–442.
23. Banks B, Dever J, Gebauer L and Hill C (1991), Atomic oxygen interactions with FEP Teflon and silicones on LDEF, *NASA CP 3134*, pp. 801–815.

24. Koontz S, Leger L, and Albyn K (1989), Vacuum ultraviolet radiation/atomic oxygen synergism in materials reactivity, *J. Spacecraft and Rockets*, **27** (3).
25. Rutledge S. and Banks B (1996), A technique for synergistic atomic oxygen and vacuum ultraviolet radiation durability evaluation of materials for use in low earth orbit, *NASA TM 107230*, p. 18.
26. Slemp W (1988), Ultraviolet radiation effects, *NASA CP 3035*, pp. 425–446.
27. Van Eesbeek M, Levadou F, Tupikov V, Cherniavsky A, Khatipov S, Milinchuk V, Stepanov V, and Milintchouk A (1994), Degradation of Teflon PTFE and FEP exposed to far ultraviolet radiation, *Proceedings of the Sixth International Symposium on Materials in a Space Environment*, p. 406–416.
28. Milintchouk A, van Eesbeek M, and Levadou F (1996), Soft X-ray radiation as a factor in the degradation of spacecraft materials, *Proceedings of the Third International Conference for Protection of Materials and Structures From the LEO Space Environment*, Toronto, Canada, April, 1996.
29. Zuby T, de Groh K, and Smith D (1995), Degradation of FEP thermal control materials returned from the Hubble Space Telescope, *NASA TM 104627*, p. 5.
30. Leet S, Fogdall L, and Wilkinson M (1995), Thermal–optical property degradation of irradiated spacecraft surfaces, *J. Spacecraft and Rockets*, **32** (5), 832–838.

5

Processing of Fluoroelastomers

STEPHEN BOWERS,
Du Pont Dow Elastomers SA, Geneva, Switzerland

1 FLUOROELASTOMERS: NATURE AND STATUS

Fluorohydrocarbon elastomers (ASTM D-1418 designation FKM, ISO-R1629 designation FPM) form by far the largest proportion of industrially used fluoroelastomers. Principal trade names are Viton* (from DuPont Dow Elastomers), Fluorel[†] (from 3M), Tecnoflon[‡] (from Ausimont) and Dai-El[§] (from Daikin). These elastomers offer the highest continuous heat resistance of any conventionally processed rubber, combined with excellent resistance to a wide range of chemical media including hydrocarbon fuels and oils, alcohols (depending on the base polymer composition) and lubricant additives.

Viton A, a copolymer cf vinylidene fluoride (VF_2) and hexafluoropropylene (HFP) containing 66% fluorine was originally introduced commercially to the market in 1958. Viton B, a terpolymer including tetrafluoroethylene (TFE), containing 68% fluorine, was introduced shortly after and provided a significant improvement in heat and fluid resistance. Today, about 40 years after the introduction of the first commercial grade, there exists a wide range of copolymers and terpolymers with fluorine levels as high as 70% as well as speciality grades incorporating perfluoromethylvinylether (PMVE), to enhance low-temperature flexibility. Peroxide curable tetra-polymers, incorporating low levels of peroxide specific cure site monomer provide enhanced resistance to aqueous chemical media and nucleophilic reagents. Copolymers of TFE and propylene offer a different property balance with emphasis on hydrolysis resistance.

---

* = Registered trademark of DuPont Dow Elastomers.
[†] Fluorel is a trademark of 3M.
[‡] Tecnoflon is a trademark of Ausimont.
[§] Dai-El is a trademark of Daikin.

Modern Fluoropolymers. Edited by John Scheirs
© 1997 John Wiley & Sons Ltd

Increasing demands for higher service performance of elastomeric parts, especially in automotive power train and fuel system applications, continue to promote volume growth in excess of 6% per year.

Historically, FKM required more care in processing than hydrocarbon elastomers, particularly when using the injection process, but today, due to innovations in polymer design, cure systems and process aid technology, this is no longer the case. A recent summary of fluoroelastomer technology is available [1]. If FKM is required to fulfill a specific service need then there are few processing-related restrictions which might limit its use. This chapter provides an overview of FKM processing for major end uses, primarily sealing devices such as O-rings, shaft seals and fuel system components. For a more complete discussion of the nature and properties of fluoroelastomers see the Chapter 32 by A. van Cleeff. Suppliers' literature and back-up advice are also accessible sources of more detailed information. Reference 2 is typical.

FKM is sold with gum Mooney viscosities (ML/121 °C) ranging from 10 to 160+ and is often available with a pre-incorporated bisphenolic cure system and/or process aid package. Developments in base polymers, typified by the IRP (improved rheology polymer) grades of Viton, have transformed flow characteristics and minimized polymer related mold fouling and sticking tendencies [3].

2 COMPOUNDING

2.1 GENERAL

End-use requirements usually dictate the type of FKM to be used, the vulcanization system and the fillers. However, compound recipes do have a significant effect on processability. Compared with hydrocarbon elastomers, FKM recipes tend to be simple, containing metal oxides for cure activation and thermal stability, a cure system, one or more fillers (eg low loadings of large particle size blacks), process aids and, in the case of mineral-filled stocks, pigments to provide coloration. Plasticizers are avoided since they evaporate during post cure, resulting in increased shrinkage and no reduction in hardness. Protective ingredients have no value since ozone, weathering and heat resistance are inherently excellent.

3 VULCANIZATION SYSTEMS

Technically important categories are:

1. Blocked polyfunctional amines.
2. Bisphenolic systems such as bisphenol AF [2,2-bis(4-hydroxyphenyl)hexafluoropropene] accelerated using a quaternary phosphonium salt such as triphenylbenzyl phosphonium chloride (BTPPC).
3. Selected peroxides with a co-agent to enhance cross-linking efficiency and ultimate properties, especially compression set/stress relaxation resistance.

3.1 POLYFUNCTIONAL AMINES

These are best known as DIAK™* No. 1 (hexamethylene diamine carbamate), DIAK No 3 (dicinnamylidene diamine carbamate) and DIAK No. 4 (4,4'-bis[aminocyclohexyl] methane carbamate). DIAK No. 1 and No. 4 are scorchy, all are sensitive to heat history prior to final processing and all are prone to mold foul. With inferior compression set/stress relaxation resistance they are most often used in compounds for roller covers, hose liners and jackets and, historically, DIAK Nos. 1 and 4 were used to meet US FDA food contact requirements.

3.2 BISPHENOLIC SYSTEMS

Introduced around 1970, these systems in their varied forms dominate the market today. Unlike amines or peroxides, if correctly used bisphenolic systems can provide excellent scorch safety at processing temperatures followed by a fast in-mold cure reaction at molding temperatures, typically 175 °C and above (Figure 5.1). They are inherently 'clean' and can be further assisted by correct selection of internal process aids. Curatives and accelerators are available as dispersions in low molecular weight FKM, but suppliers' precompounds predominate. The use of precompounds will ensure consistent dispersion of the cure system and may utilize proprietary curatives and process aids not otherwise available. Given the range of precompounds on offer, new mixes from individual components are rare, but they are sometimes used to provide specific curing characteristics or cross-link density. The latest breed of bisphenol cured

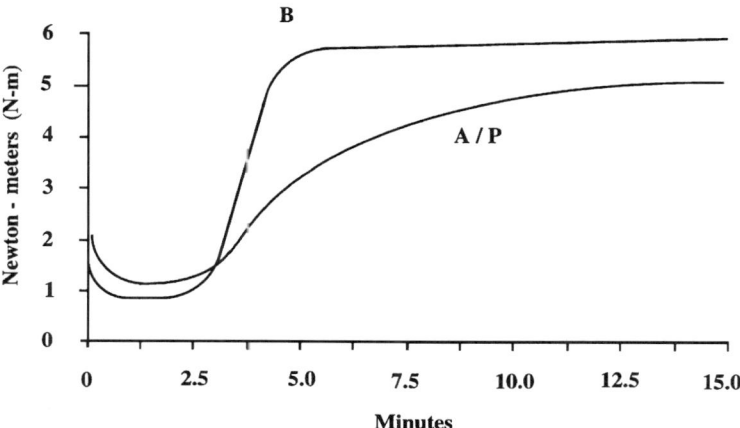

Figure 5.1. Indicative rheometer curves for amine (A), bisphenol (B) and peroxide (P) systems

---

*TM = trademark of DuPont Dow Elastomers.

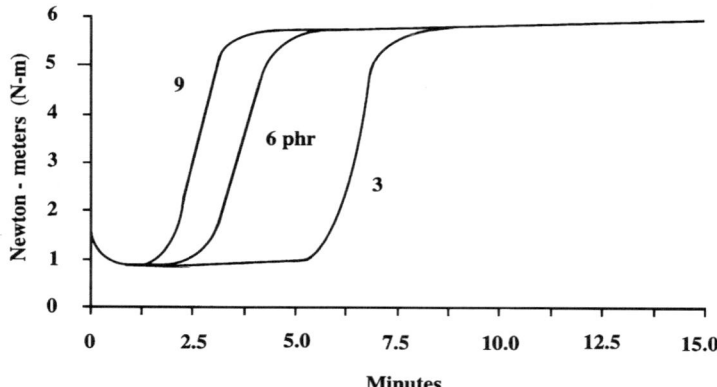

Figure 5.2. Effect of calcium hydroxide level on cure initiation

precompounds are low in viscosity and contain cure systems that are relatively inert at temperatures much below 190 °C, but become highly active above 200 °C. They are designed for injection molding at high temperature where the thermally depressed viscosity allows for excellent flow and is unrestricted by precure. Very short vulcanization times are practical and ease of release exceptional.

Bisphenol systems require activation by water. Rubber grade calcium hydroxide, Ca(OH)$_2$, included in the recipe releases water at curing temperatures. A typical level is 6 phr but cure initiation time may be varied by adjusting the level (Figure 5.2). Over the range 2 to 10 phr (parts per hundred) there is little effect on the cure rate as defined by the slope of the rheometer curve, or the state of cure as defined by the maximum torque values achieved. Bisphenol cure systems provide excellent compression set/stress relaxation resistance for periods comparable to the heat resistance of the polymer.

3.3   PEROXIDE VULCANIZATION

FKM modified for peroxide cure (1) requires a co-agent such as triallyl isocyanurate (DIAK No. 7) or trimethallyl isocyanurate (DIAK No. 8) to enhance cross-link density and hence vulcanizate properties. The safe processing peroxides most commonly used are 50% active 2,5-dimethyl,2,5-di(*t*-butyl-peroxy) hexane or less active hexyne-3, typically Varox DBPH or Luperco 130XL respectively. Peroxides tend to give less scorch safety than bisphenolic systems. Flow time may only be extended by the use of a less active peroxide or lower mold temperatures. Appropriate internal release agents are usually essential for demolding and to minimize mold fouling. Peroxide cure systems are used to enhance resistance to aggressive aqueous media and certain lubricant additives and are mandatory in polymers containing PMVE such as Viton GLT and GFLT.

4 AUXILIARY ACTIVE CHEMICALS

FKM mixes require metal oxides such as magnesium to enhance cure activity and vulcanizate thermal or chemical stability. When using an amine cure system 15 phr of a low-activity magnesia (Iodine No. 20) is typical. For bisphenol cure systems 3 to 6 phr of a high-activity grade (Iodine No. 120+) are preferred. Calcium oxide may be added to aid post cure of thick components. As noted, bisphenol cures also require activation by calcium hydroxide. Where resistance to aqueous media is needed the above three metal oxides are replaced by lead oxide (litharge), preferably fumed. Active magnesias and calcium hydroxide are difficult to disperse in FKM (see section 6), they are also hygroscopic and should be stored in hermetically sealed containers, preferably small sachets. For environmental health reasons it is essential that litharge is predispersed in low molecular weight FKM. Ideally, all metal oxides would be added in this way to minimize dispersion problems.

5 INTERNAL PROCESS AIDS

Compound flow is mainly determined by polymer viscosity, cure system and processing temperature. Selected internal process aids can assist flow and are often essential for easy mold release, to minimize mold fouling and to achieve a smooth surface finish on profiles and sheet. Proven process aids include low molecular weight polyethylene, hard waxes such as carnauba or VPA No. 2 and

Table 5.1. Representative formulations (75 to 80 Shore A)

	O-Ring	Colored, bonded seal	Best chemical/ alcohol/low- temp. resistance
Viton A402C[a]	100	—	—
Viton B651C[b]	—	100	—
Viton GFLT[c]	—	—	100
Magnesium oxide	3	6	—
Calcium hydroxide	6	3	—
Litharge	—	—	3
N990 black	25	—	30
Blanc fixe	—	35	—
Wollastonite	—	15	—
Inorganic pigment	—	5	—
Hard wax	0.5	1	1
Fatty acid amide	—	—	0.5
Stearic acid	—	—	0.25
TAIC	—	—	3
50% peroxide DBPH	—	—	3

[a] 40 ML/121 °C A-type IRP di-polymer with incorporated curatives and process aid.
[b] 60 ML/121 °C B-type IRP ter-polymer with incorporated curatives and bonding promoter.
[c] 75 ML/121 °C high fluorine PMVE-containing polymer.

sulfones such as VPA No. 3. Suppliers' precompounds often contain proprietary process aids, but it is often necessary to add more for optimum performance. Hard waxes in quantities of up to 1 phr especially assist in demolding, often with 0.5 to 1 phr of an organic sulfone. Up to 2 phr of hard wax will enhance the surface quality of extrusions or calendered sheet. A three-component system, specifically for peroxide cured recipes comprises 0.25 phr stearic acid, 0.2 to 0.5 phr fatty acid amide and 1 to 1.5 phr hard wax [2]. Excess of a single process aid may cause knitting problems in molding due to exudation at the flow front. An overall excess may increase shrinkage in post cure. Certain process aids may also affect cure rate and scorch safety. It should be noted that hard waxes are difficult to disperse in low-viscosity polymers and should be added in particulate form. Three example FKM recipes are given in Table 5.1.

6 MIXING

Basic principles in the preparation of consistent FKM compounds are freedom from contamination, energy input and minimum exposure to water. Contamination with lower performance elastomers during mixing can degrade the serviceability of products made from the mix. Contamination with unintended compounding ingredients or machine lubricants can influence vulcanization and cause various processing problems such as mold fouling and poor knitting. Absorption of water, e.g. from condensation on chilled rollers or rotors or in cooling baths will variably accelerate bisphenol-cured compounds.

Mixing of any elastomer compound is energy specific. The rheology of FKM is such that shear viscosity and energy input decrease rapidly with increasing temperature. Mixing procedures need to maximize energy input before the matrix becomes too soft. Starting cold, full cooling is required, preferably using chilled water. A formal rework of up to 5 minutes on cold equipment after standing 16 to 24 hours is virtually essential to completely disperse metal oxides, mineral fillers such as wollastonites or talcs and often hard waxes. Untoward variations in rheometer cure response, flow and vulcanizate properties are often due to omission of this step. Extended milling or internal mixing times or working on stock blenders is rarely effective. Cooling is best done in forced air or on a clean metal bench.

6.1 MILL MIXING

Mills should have hinged guides for ease of cleaning, set in to prevent edge contamination with oil, and a roll friction ratio of 1.1 : 1. Preblended ingredients less amines, peroxides or liquids are added as soon as a rolling bank has formed. Amines, peroxides and liquids should be mixed with a small amount of filler and added later. Cross blend and cigar roll several times. Typical mix time: 15 to 20 minutes.

6.2 INTERNAL MIXING

Mixers may have tangential or intermeshing rotors run at moderate speeds. Journals need to be well maintained to minimize lubricant seepage. Loading factors are typically 65–75%. A clean out batch, preferably scrap FKM should be run prior to the mix. Load the ingredients, less amines, peroxides and process aids 'upside down'. Add liquids with a small portion of filler at 2 to 2.5 minutes. Peroxides and amines may be added in the last minute of the internal mixer cycle, later on a mill or in a separate internal mixer pass. Dump onto a cool mill at 4 minutes (typical) or when the stock has reached a temperature of 120 °C.

Reworking to complete dispersion may be by mill or internal mixer. If a cooling bath is used at any stage, remove the stock as soon as possible and force-air dry. Only stack full compounds after complete cooling and drying.

Continuous mixing has been demonstrated but requires polymers or precompounds in particulate form. Subsequent reworking may still be necessary.

7 COMPOUND CONTROL

FKM compounds should only be sampled for rheometer (preferably MDR) and other tests after the reworking step. Bisphenol-containing precompounds in particular are capable of excellent reproducibility under production conditions. Data from extrusion rheometers are of more value in fundamental polymer development and injection molding simulation than product quality control.

8 MOLDING

Given the high volume cost of FKM the choice of molding process, where it exists, will be heavily influenced by the ratio of salable product to scrap, in addition to the usual factors of volume, tool and equipment amortization, etc. Molds should be made from a high-quality Cr/Ni tool steel, properly finished and hardened and should be closely fitting since FKM flows easily into parting lines. Good temperature control over the mold surface is essential.

8.1 COMPRESSION MOLDING

This process allows flexible, economic production of relatively small volumes of components. Using precisely extruded or cut preforms, molding scrap can be 10% or less. High-viscosity base polymers can be specified where technically desirable and high hardness compounds satisfactorily utilized. Depending on curatives and part dimensions, mold temperatures may be 160 to 195 °C with vulcanization times from 30 to 2 minutes. Bumping should occur soon after closing the press before the stock fully plasticizes.

8.2 TRANSFER MOLDING

This technique gives better dimensional control of finished parts than compression molding, but usually a higher proportion of molding scrap. Thanks to the rheological characteristics of FKM, compounds with seemingly high Mooney viscosities (ML/100 °C) can be molded where press tonnage allows. Cycle times are similar to those for compression molding.

8.3 INJECTION MOLDING

Availability of optimized polymer/bisphenol cure combinations has encouraged the injection molding of FKM, especially for high-volume precision parts such as O-rings and shaft seals. Dimensional tolerances can be much closer than with compression molding and vulcanizate properties essentially equal. Automatic operation with rotating brush or other automatic demolding methods is established. Where cavity numbers and configuration permit, runner scrap may be minimized using hot runner molds or, preferably, multiple nozzle injection. The new breed of bisphenol-cured precompounds, inactive at temperatures much below 200 °C, is adaptable to either technique. Depending on part volume, vulcanization times of 30 seconds or less with mold temperatures in the region of 200 °C are achievable with outstanding ease of release. Computer-based injection molding simulation programs can be used to determine an optimum combination of machine, mold configuration, runner design, material characteristics and process conditions for a given material and thus reduce development time and start-up costs.

9 EXTERNAL MOLD RELEASE AGENTS

Semi-permanent types baked onto new, or chemically cleaned, mold surfaces prior to start-up assist breaking in, especially with high cure state, low demolding strength compounds. Otherwise, release agents should not be necessary with modern bisphenol-cured compounds and should be avoided. If required, proprietary dispersions based on polyethylene (up to 180 °C) or modified silicones are effective with amine cures. Semi-permanent release agents work best with peroxide cures but require intermittent touch up.

10 SHRINKAGE

FKM exhibits higher shrinkage from the mold than hydrocarbon rubbers, further increased during post cure by the loss of process aids and curative by products (Figure 5.3). There is some variation between different cure systems, fillers and compound hardness but, in general, 3 to 3.5% linear shrinkage (or diametral for O-rings) must be allowed for in mold design. Note that shrinkage is temperature dependent. It is rarely possible to reduce shrinkage isotropically through compounding without compromising vulcanizate properties. Given the

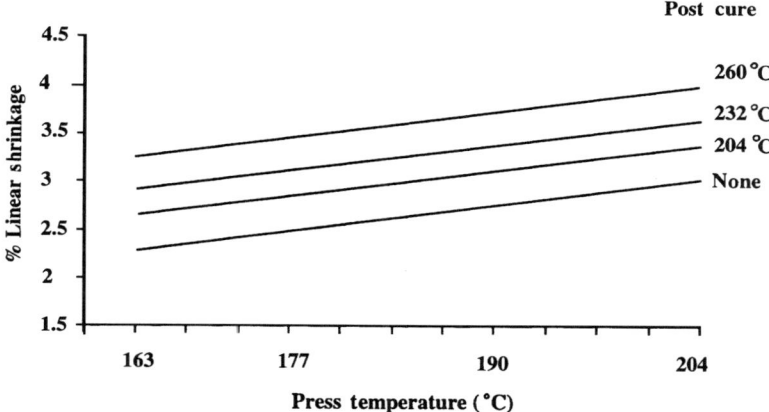

Figure 5.3. Shrinkage of FKM versus temperature

higher shrinkage of FKM, a normal variation around the mean leads to greater variation in actual part dimensions and this should be considered when specifying part tolerances.

11 BONDING

Every major supplier of bonding agents offers one- or two-part systems for the invulcanization bonding of FKM. The former are best adapted for mass production, being applied to chemically cleaned metal inserts by dipping, either diluted or as-received. Two-part systems may be required for situations where internal cavity pressure is low or for roller covers. Pre-baking the primed inserts as specified by the bonding system supplier (10 minutes at 120 °C is typical) minimizes wiping due to compound flow.

Compounds should contain the minimum levels of process aid necessary for mold release. With bisphenol systems the usual ratio of magnesium oxide to calcium hydroxide of 3 : 6 phr should be reversed to 6 : 3. Precompounds containing incorporated bonding promoters are available from FKM suppliers [3]. Post-cure temperatures for bonded parts should not exceed 200 °C, with a slow rise from near cold to prevent thermal shock at the bond interface.

12 OTHER VULCANIZATION PROCESSES

FKM is amenable to vulcarization in steam or pressurized hot air with support to prevent collapse (in the case of tubing and hollow profiles), also continuous curing of sheets and radiation curing are possible. Steam curing tends to activate most cure systems, leading to 'tighter' vulcanizates. With the exception of thin solution-derived coatings, pressureless vulcanization leads to sponging and profile collapse.

13 CALENDERING

Compounds for calendering should be of medium viscosity and contain up to 2 phr of hard wax. IRP-type polymers give smooth surfaces, but may benefit from blending with higher molecular weight grades to improve green strength. Typical process temperatures are 50–70 °C for the top roll and 45 to 65 °C for the center roll. Prewarmed compound should be fed uniformly across the nip to minimize feed temperature variations.

14 EXTRUSION

With FKM the imperative is to minimize frictional adhesion of hot compound to the die, so preventing surface fracture and a consequent poor extrudate surface finish. Influencing factors include process aids, polymer green strength and machine/die characteristics [2,4]. Roller feed is required.

A typical starting setup could be:

Machine	roller feed, from 10 to 16 : 1 L/D, breaker fitted;
Screw	2 start, compression ratio 1.5 to 2 : 1, cooled;
Die	tapered, moderate land length;
Temperature profile	cold feed, 40 to 65 °C, head 75 to 85 °C, die 90 to 110 °C.

Ram extrusion of preforms is relatively insensitive to compound or temperature, 60 to 80 °C being typical.

15 POST CURING

All FKM compounds require post curing, usually in air, to develop optimum properties. Properties which especially improve during post cure are compression set and stress relaxation resistance. Temperatures are typically 200 °C (amine cures and bonded parts), 230 to 260 °C (bisphenol and peroxide cures) for 16 to 24 hours. A programmed rise from cold or 100 °C is advisable, especially for thick or bonded parts and those containing litharge. Thick articles may also require inclusion of 1 to 3 phr of calcium oxide to prevent fissuring. Ovens must be positively vented to atmosphere and should achieve 10 air changes per hour. Oven loading should not exceed 10% of its volume. Overloading and/or poor air circulation may lead to oven fires [5].

16 HEALTH AND SAFETY

Polymer and ingredient suppliers' safety in use information should always be studied prior to processing FKM for the first time or before introducing a new grade. With accepted good practice, FKM presents no hazards unique in the rubber industry. Depending on national and local regulations, disposal may be by landfill or combustion in approved incinerators with provision to neutralize acidic exhaust gases.

REFERENCES

1. A. L. Logothetis, *Fluorelastomers. Organofluorine Chemistry: Principles and Commercial Applications*, ed. R. E. Banks *et al.*, Plenum Press, New York, 1994.
2. Processing Viton fluoroelastomer, *Du Pont Bulletin VDS 451*.
3. Viton selection guide *Du Pont Bulletin H-53647*.
4. R. Ferro, G. Fiorello, G. Restelli *Extruding fluoroelastomers to meet higher performance* needs, Parts 1 and 2, *Elastomerics*, March/April 1989.
5. R. E. Tarney and E. W. Thomas An investigation of fluoroelastomer post cure oven fires. ACS Rubber Division Meeting, Fall, 1991, Detroit.

6

Fluorinated Polyurethanes

ROBERT F. BRADY, Jr.

Naval Research Laboratory, Chemistry Division, Washington, DC, USA

1 INTRODUCTION

Polyurethanes are perhaps the most versatile polymers. Materials with a wide variety of physical and chemical properties can be formulated from the many commercially available and relatively inexpensive polyisocyanates and polyols. Polymers with block, comb, or random morphologies are produced by careful selection of reaction conditions, and the materials may take any form from brittle glasses to elastomers.

Introducing fluorine into polyurethane resins brings about changes in properties similar to those seen when other polymers are fluorinated. Chemical, thermal, hydrolytic, and oxidative stability are enhanced, and the polymer becomes more permeable to oxygen. Surfaces become more biocompatible and less able to bond to substances in contact with them.

Fluorourethanes are widely used in modern chemical technology. They are used in products ranging from hard, heat-resistant electrical components to biologically compatible surgical adhesives. Perhaps the largest use of fluorourethanes is in surface coatings for industrial and residential structures, automobiles, ships and aircraft. Fluorourethanes are also used widely in medical products and as surface-enhancing treatments for textiles, leather and carpets. The literature of fluorourethanes is abundant; the bulk of citations are found in the patent literature and relatively little information is found in journals and books. Properties of a particular fluorourethane are determined by the raw materials and manufacturing processes used, and useful generalizations about properties cannot be made without considering the use for which the material is designed. The performance properties most valuable for particular applications are summarized in section 5.

Modern Fluoropolymers. Edited by John Scheirs
© 1997 John Wiley & Sons Ltd

1.1 SURFACE ENERGY

Fluorourethanes are one of a very small number of polymers with surface energies less than 25 mJ/m^2 [1]. Low surface energy imparts an antiadhesive character to a polymer and is usually the primary or an important secondary reason for choosing a fluorourethane for a particular application. Surface coatings, medical and prosthetic devices, and treatments for carpets, leather, and textiles are particularly valuable when they fail to bond to oil and water, soils and stains, and biological matter. Fluorourethanes are not yet used in self-lubricating surfaces, bearings or valves, but developments in these areas can be expected in the near future.

The surface energy of a polymer is related to the functional groups which gather at its surface [2], and fluorinated groups are known to produce polymers with low surface energy [3]. Substitution of fluorine in hydrocarbons causes the surface energy to decrease in the order $-CH_2- > -CH_3 > -CF_2- > -CF_3$. This is readily observed in the surface energies of polymers with surfaces composed of ordered single groups such as poly(ethylene) (33.7 mJ/m^2), poly(dimethylsiloxane) (21.2), poly(tetrafluoroethylene) (18.6) and poly[methyl (3,3,3-trifluoropropyl) siloxane] (6.0).

Completely fluorinating a polymer produces a low surface energy, of course, but gives little latitude in tailoring other desired properties such as glass transition temperature, processability, the absence of porosity, and low cost. Low surface energy is preferably attained by deploying a small number of perfluoroalkyl groups on the surface, where they will be most effective. Producing polymers with an effective number of fluorinated groups is relatively easy, but causing them to form ordered aggregates with the fluorinated groups aligned and closely packed on the surface is far more difficult. Creating an ordered surface is augmented by a backbone which has sufficient mobility to present the fluorinated group to the air interface, and is thermodynamically favored because it reduces the interfacial free energy between the polymer and air. Fluorourethanes excel at this behavior and possess very hydrophobic and oleophobic surfaces.

In polymers with mobile backbones and sidechains the surface rearranges in response to the medium in which it is immersed. For instance, when a fluorourethane is placed in contact with water fluorinated groups migrate into the interior of the polymer and more polar groups move to the surface, thereby minimizing the free energy at the interface with water. This rearrangement can be rapid [4-6]; the surface of a fluorourethane cycled between water and air rearranges in about four minutes reversibly and repeatedly [7]. Understanding the factors which control the reorganization of the surface is fundamental to the use of fluorourethanes and is a focus of research activities in laboratories around the world.

2 MANUFACTURE

2.1 ADDITION REACTIONS

The most versatile and frequently employed method of manufacturing fluorourethanes is the addition reaction of polyisocyanates with polyols. In the first steps, the oxygen of the alcohol forms a new bond with the isocyanate carbon and the hydrogen of the alcohol bonds to the isocyanate nitrogen, forming a urethane linkage. The hydrogen remains active and can react with another isocyanate group to produce an allophanate linkage. The reaction ends when all isocyanate groups are consumed, when steric hindrance prevents further reaction, or when the polymer gels and functional groups lack the mobility to react. In the absence of water, no volatile side products are formed in these reactions.

Fluorine is most frequently introduced through the polyol component. Although fluorinated polyisocyanates are commercially available, they are more difficult to obtain and are a more expensive route to introduce a fixed amount of fluorine into the polymer. A wide variety of fluorinated alcohols and polyols, and fluorinated and nonfluorinated polyisocyanates are commercially available; those used most frequently are listed in Tables 6.1–6.6.

The manufacturing process usually involves more than one polyisocyanate and a mixture of fluorinated and unfluorinated alcohols and polyols. The order of addition of reactants is critical to achieve the desired polymer structure and properties. Vastly different polymers can be obtained from the same starting materials by varying the order of addition.

Fluorourethanes are also prepared as resins dispersed in water. A typical production method begins with the addition polymerization of several monomers, one of which contains a perfluoroalkyl moiety and others which contain hydroxyl groups. This polymer is then reacted with a polyisocyanate and sometimes other polyols to produce an aqueous emulsion of a hydroxy-functional resin. This can subsequently be cured with an emulsified polyisocyanate.

For example, toluene diisocyanate is reacted with one-half equivalent of 2-hydroxyethyl methacrylate and one-half equivalent of 2,2,3,4,4,4-hexafluorobutan-1-ol. The resulting material is polymerized with butyl acrylate, methyl methacrylate, methacrylic acid and acrylic acid in water to give [8] an emulsion with a minimum film-forming temperature of $10\,^\circ$C.

Acrylic latexes are readily modified by including the acrylate or methacrylate ester of a $1H$, $1H$, $2H$, $2H$-perfluoroalkyl alcohol in the polymerization reaction. Alkyl chains of 6 to 16 carbons are used [9]. For example, an acrylate emulsion polymer containing $1H$, $1H$, $2H$, $2H$-perfluorooctyl methacrylate has been prepared [10].

2.2 PHOTOCURABLE RESINS

Acrylic-modified fluorinated polyurethane resins are cured by irradiation with ultraviolet (UV) light. These resins are used most frequently as contact lenses and as cladding on optical fibers. Reactive fluorinated oligomers for radiation-cured coatings have been reviewed [11].

For example, a photocurable fluorinated urethane resin is prepared by condensing 1,2-dihydroxy-3-perfluorohexylpropane, 2-hydroxyethyl acrylate and isophorone diisocyanate. The resulting polymer is then mixed with

2-perfluorooctylethyl acrylate, trimethylolpropane triacrylate and catalyst and irradiated with UV light to give a transparent cured resin [12]. Resins are prepared by adding acrylated urethane oligomers made from dipropylene glycol and hexamethylene diisocyanate to a perfluoroalkylethyl acrylate and photocured at 200–450 nm [13].

2.3 FLUORINATION OF POLYURETHANES

Fluorinated polyurethanes are also prepared by treating the surface of an unfluorinated material with a cold plasma of elemental fluorine [14] or carbon tetrafluoride [15]. Because fluorinated polyurethanes are frequently employed for their surface properties, this technique provides an effective and relatively inexpensive way to place fluorine at the surface. Polyurethane membranes treated with elemental fluorine are used to separate gases [14].

Table 6.1. Fluorinated alcohols used to prepare polyurethane resins

Alcohol	Application	Reference
CF_3CH_2OH	Hard contact lenses	143
$CF_3CH_2CH_2OH$	Anti-fogging coating for glass	167
$(CF_3)_2CHOH$	Solvent	191
$CF_3CHFCF_2CH_2OH$	Water-based coatings	8
$F(CF_2)_nOH$		
[$n = 6$–12]	Oil-, water- and soil-resistant textile finishes	118
[$n = 7$]	Leather substitutes	134
$(CF_3)_2CF(CF_2CF_2)_nCH_2CH_2OH$		
[$n = 3$–5]	Oil- and water-resistant textile finishes	192
$F(CF_2CF_2)_nCH_2CH_2OH$		
[$n = 3$–6]	Oil- and water-resistant textile finishes	117, 118
[$n = 3$–7]	Emulsion polymers	10
$F(CF_2CF_2)_nCH_2CH_2SH$		
[n unspecified]	Oil- and water-resistant textile finishes	125, 126
$C_7F_{15}CH_2OH$	Rigid insulating foams	193
$HOCH_2CF_2CF_2OCF(CF_3)CF_2OCF{=}CF_2$	Elastomers	194
$C_6F_{13}(CH_2)_2S(CH_2)_3OH$	Cladding for optical fibers	154
$H(CF_2CF_2)_5CH_2OH$	Coating for magnetic recording tape	173
$C_8F_{17}CH_2CH_2OH$	Coatings for textiles and leather	195
$C_8F_{17}CH_2CH_2OCH_2CH_2OH$	Housings for office machines	195

3 RAW MATERIALS

3.1 FLUORINATED ALCOHOLS

Dramatic changes in the properties of polyurethanes can be achieved simply by adding a fluorinated alcohol to the formulation. The low surface energy of the fluorinated compound causes it to migrate to the air interface, and small molecules migrate quickly. The compound then reacts with isocyanate and is bound at the surface. Fluorinated alcohols used for this purpose are listed in Table 6.1.

Straight-chain alcohols with all but the α and β carbons fluorinated are most often used. These are named as fluorinated alcohols, protonated perfluoroalcohols or, somewhat improperly but more conveniently, as 2-perfluoroalkyl-substituted ethanol. Thus, $CF_3CF_2CF_2CF_2CH_2CH_2OH$ is found in the literature as 3,3,4,4,5,5,6,6,6-nonafluorohexan-1-ol, $1H, 1H, 2H, 2H$-perfluorohexan-1-ol, or 2-perfluorobutylethyl alcohol.

3.2 FLUORINATED DIOLS

Fluorinated diols are used directly in resin formulations or are reacted with fluorinated oxiranes to produce polyol intermediates. Diols are named using the same conventions as alcohols. Diols used in fluorourethanes are given in Table 6.2.

3.3 FLUORINATED POLYOLS

Polyhydroxylated fluorinated resins with molecular weights between 500 and 10 000 are preferred for many applications. Many hydroxyl groups provide adhesion to the substrate, increase reactivity with isocyanate resins, and produce a resin with higher crosslink density, improved physical properties and superior chemical resistance.

3.3.1 Fluorinated Ethylene Vinyl Ether (FEVE) Polyol Resins

The fluorinated polyol used most frequently in polyurethane coatings is a copolymer of chlorotrifluoroethylene and various functionalized vinyl ethers, as shown in Figure 6.1 [16]. Chlorotrifluoroethylene imparts weatherability to the

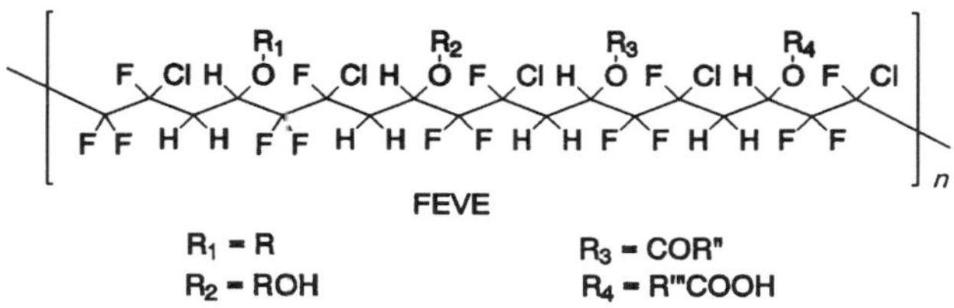

$R_1 = R$ $R_3 = COR''$
$R_2 = ROH$ $R_4 = R'''COOH$

Figure 6.1. The structure of fluorinated ethylene–vinyl ether (FEVE) polyols

Table 6.2. Fluorinated Diols Used to Prepare Polyurethane Resins

Diol	Application	Reference
$HOCH_2(CF_2)_2CH_2OH$	Segmented polyether polyurethanes	196, 197
$HOCH_2(CF_2)_3CH_2OH$	Elastomers	194
	Segmented polyether polyurethanes	191, 196, 197
$HO(CH_2)_2(CF_2)_4(CH_2)_2OH$	Cladding for optical fibers	155
	Colored resin powders	109
$HO(CF_2)_2O(CF_2)_4O(CF_2)_2OH$	Antithrombogenic elastomers	152
$H(CF_2)_4CH_2OCH_2CH(OH)CH_2OH$	Photocurable polyurethane acrylate coatings	198
$CF_3(CF_2)_5CH_2CH(OH)CH_2OH$	Photocurable polyurethane acrylic resins	12
$C_6F_{13}SO_2N(CH_3)CH_2CH(OH)CH_2OH$	Oil- and water-repellent resins	199
$C_6F_{13}CH_2CH_2OCH_2CH(OH)CH_2OH$	Oil- and water-repellent resins	200
$HOCH_2CH_2CFCF_2(CF_2CF_2)_nCH_2CH_2OH$ $\overset{|}{CF_3}$ $[n = 2]$	Coatings	201, 202
$[n = 0\text{-}7]$	Transparent elastomers	163
	Coatings	203
	Electrical insulation	176
	Marine coatings	91

Structure	Application	Reference
OH H₃C—C—CH₃ (with fluorinated benzene ring, F substituents) H₃C—C—CH₃ OH	Polyurethane resins	204
$C_8F_{17}SO_2N(CH_2CH_2OH)_2$	Nonthrombogenic medical implants	205
	Coatings	28
	Lubricants for recording media	38
	Printed circuit boards	169
	Thermal printer components	178
	Elastomers	206, 207
	Toluene diisocyanate polymers	137
(structure: HO—C₆H₄—C(CF₃)₂—C₆H₄—OH) $(CF_2)_4[OCF_2CF_2SO_2N(CH_2CH_3)CH_2CH_2OH]_2$		
(structure with $n\text{-}C_8F_{17}$, CF₃, F₃C—C—CF₃, OH groups on benzene ring)	Modified epoxy resins	208
$HO(CH_2CH_2CH_2CH_2O)_{6\text{-}1}(CH_2CHO)_{4\text{-}0}(CH_2CH_2O)_{0\text{-}9}H$ with $CF_3(CF_2)_{4\text{-}5}CH_2$	Poly(fluoroalkylether)urethanes	7

Table 6.3. Properties of FEVE fluoropolyol solutions

	Zaflon®	Lumiflon®	Fluorobase®	Fluonate®
Fluorine content (w%)	21	25–35		
OH value (mg KOH/g)	44–53	40–150	29	42–54
COOH value (mg KOH/g)	3	0–50	4	
M_n (by GPC) $\times 10^4$	1.3	0.2–10		
M_w (by GPC) $\times 10^4$		0.4–20		
Glass transition temperature T_g ($°C$)	35	20–70		
Decomposition temperature T_{dec} ($°C$)		240–250		
Solids, weight (%)	60	60	50	50
Viscosity (Cps at 25 °C)	1500	4000	590	
Weight per gallon (lb)		9.41	9.19	9.08
Specific gravity		1.13	1.11	1.09
Flash point (Seta) (°F)	77	77	45	77
Solvent	Xylene	Xylene	Toluene/ n-BuOAc	n-BuOAc

copolymer. The ethers may include alkyl vinyl ethers to provide solubility in organic solvents, transparency, hardness and gloss; hydroxyalkyl vinyl ethers to give adhesion and a site for reaction with isocyanates; and carboxylated alkyl vinyl ethers to provide pigment compatibility and adhesion [17]. A wide variety of substituted vinyl ethers, including the cyclic vinyl ether 2,3-dihydrofuran [18], is used in these resins, and their proportions vary across a wide range. Manufacturing processes lead to random or block [19] copolymers. FEVE resins are manufactured under different brand names by four major producers: Asahi Glass (Lumiflon®), Ausimont (Fluorobase®), Dainippon (Fluonate®), and Toa Gosei (Zaflon®). Information from manufacturers' data sheets is compiled in Table 6.3 where it can be seen that the resins, although not identical, are very similar in composition. These resins are used principally in organic surface coatings.

3.3.2 Polyol Resins Derived from Hexafluoroacetone

A surface composed of closely packed trifluoromethyl groups exhibits [2] the lowest surface energy known, 6 mJ/m^2. Desiring to produce organic surface coatings which possess a high proportion of these groups, Griffith and co-workers synthesized a polyol with numerous trifluoromethyl groups evenly arranged along the polymer backbone [20]. To do so they first reacted [21] hexafluoroacetone with 0.5 equivalent of benzene to give a 9 : 1 mixture of 1,3-bis(2-hydroxyhexafluoro-2-propyl)benzene (Scheme 1) (**1**) and 1,4-bis(2-hydroxyhexafluoro-2-propyl)benzene (**2**). They also reacted [22] hexafluoroacetone with 0.5 equivalent of propene to give an 82 : 4 mixture

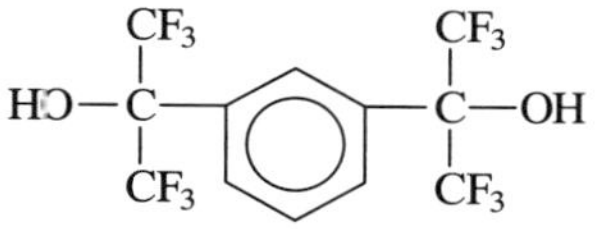

1

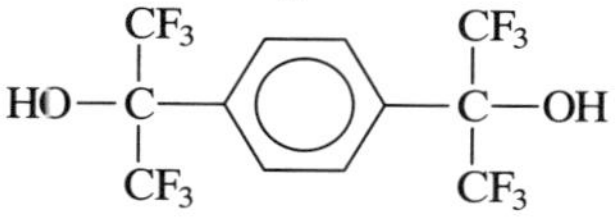

2

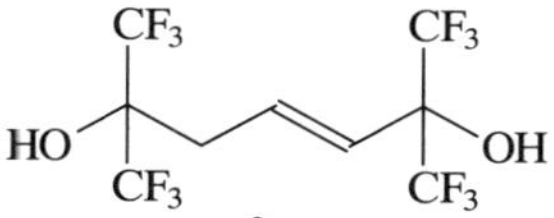

3

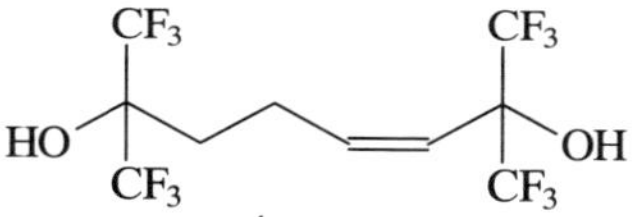

4

$HOCH_2CF_2CF_2CF_2CH_2OH$

6

$HOCH_2CH_2CH_2CH_2OH$

7

Scheme 1

of *trans*-1,1,5,5-tetrakis(trifluoromethyl)-1,5-dihydroxy-pent-3-ene (**3**) and *cis*-1,1,5,5-tetrakis(trifluoromethyl)-1,5-dihydroxy-pent-3-ene (**4**). The reaction of equal molar amounts of these two mixtures with one equivalent of epichlorohydrin and 1.1 equivalents of sodium hydroxide produces [23] the highly fluorinated polyol resin (**5**) shown in Figure 6.2. Similar polyol resins

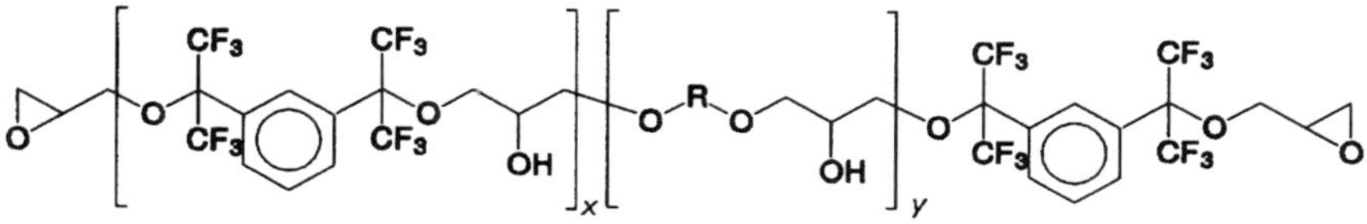

5

$R = C(CF_3)_2CH_2CH{=}CHC(CF_3)_2$

$R = CH_2CF_2CF_2CF_2CH_2$

$R = CH_2CH_2CH_2CH_2$

Figure 6.2. The structure of fluorinated polyols based on hexafluoroacetone

have been formed by reacting the mixture of (1) and (2) with 2,2,3,3,4,4-hexafluoropentane-1,5-diol (6) (obtained by reduction of perfluoroglutaric acid [24]) and with 1,4-butanediol (7).

These resins possess excellent resistance to water, chemicals and corrosive agents. The trifluoromethyl groups protect the backbone from attack and impart low surface energy. Hydroxyl groups provide adhesion and reactivity, and aromatic rings provide rigidity and temperature stability. The epoxy rings on the chain ends are too few to permit effective curing reactions, and these polyol resins cannot be used in epoxy chemistry. However, they are readily cured by polyisocyanates to form durable fluorinated polyurethane coatings. The surface energy of poly(tetrafluoroethylene) (PTFE) is close to that of these polyols and PTFE can be dispersed in the resin as if it were a conventional paint pigment. Polymers derived from hexafluoroacetone have been reviewed [25–27].

3.3.3 Fomblins®

Fomblin® polyols are composed of a perfluoropolyoxyalkyl central block containing the structure

$$-[-CF_2O\ (CF_2\ CF_2\ O)_a\ (CF_2\ O)_b\ CF_2-]-$$

The central block may also contain structural units such as

$$-(-CF_2-)-,\ \ -(-CF_2\ CF_2O-)-,\ \ -(-CF_2CF_2CF_2O-)-$$

and

$$-(-CF_2CF\ (CF_3)\ CF_2O-)-$$

These polyols have a molecular weight between 400 and 2200 and are widely used in the preparation of elastomeric fluorourethanes. Various end groups are attached to impart reactivity with a variety of curing agents.

The central block has been capped with hydroxymethyl groups to produce a diol which reacts with Bisphenol A6F to produce coatings [28]. The diol also reacts with ethylene oxide to produce diols used in elastomers [29], and with glycidol to produce polyols endcapped with the $-CH_2OCH_2CHOHCH_2OH$ group and used for surface treatments for graphite, silicates, and leather [30] as well as in protective coatings for electronics and optics [31].

The perfluoropolyether core has been endcapped with toluene diisocyanate to give a diisocyanate intermediate [30] with isophorone diisocyanate to produce ophthalmic devices [32], and with dicyclohexylmethane diisocyanate to produce elastomers [33] and gaskets, adhesives, and sealants [34].

Table 6.4. Nonfluorinated isocyanates used to prepare polyurethane resins

Isocyanate	Name	Reference
	Emulsion resins	8
	Lubricants for	
	recording media	38
	Leather treatments	133, 137
	Optical fiber cladding	153, 154
	Thermal printer	
	components	178
	Leather treatments	133
	Pigmented coatings	209
	Thermal recording	
	materials	177
	Elastomers	29, 164
	Thermoplastic resin	37
	Elastomers	33
	Gaskets, adhesives,	
	sealants	34
	Coatings	88, 90, 108
	Carpet treatments	138
	Coatings	91–95, 103
	Leather treatments	132

(*continued overleaf*)

Table 6.4. (*continued*)

Isocyanate	Name	Reference
	Coatings for optical components and mirrors	167
	Ophthalmic devices Lubricants for recording media Powder coatings Optical fiber cladding Masonry treatments Heat exchanger linings	32, 140 38 115, 116 155 166 185
	Undersea optical cables	157

Table 6.5. Fluorinated isocyanates used to prepare polyurethane resins

Isocyanate	Application	Reference
$F(CF_2)_n CH_2 CH_2 NCO$ $[n = 3-22]$	Coatings	203
$OCNCH_2(CF_2)_4 CH_2 NCO$	Surgical adhesives Chromatographic liquid phase Antithrombogenic elastomers	147 148 152
$OCNCH_2CH_2(CF_2)_n CH_2 CH_2 NCO$ $[n = 4]$ $[n = 2-16]$	 Transparent elastomers Coatings	 163 203
	Dental materials	144

3.3.4 Fluorinated Acrylic Polyols

These materials are valuable for their hardness, high glass transition temperature, and clarity. Many are photocured and used for fiber cladding and contact lenses [35–37].

3.3.4.1 Isocyanates and Polyisocyanates

Nonfluorinated isocyanates are usually used to produce fluorourethanes because they are less expensive than their fluorinated counterparts. Those most frequently used are shown in Table 6.4.

The Fomblin® perfluoropolyether may be endcapped with toluene diisocyanate [30] and with isophorone diisocyanate [31]. These perfluoropoly-oxyalkylene diisocyanates contain primary isocyanate groups and are used to prepare lubricants for recording media [38]. Other fluorinated isocyanates are shown in Table 6.5.

3.3.4.2 Miscellaneous Fluorinated Precursors

Various amines, anhydrides, oxiranes, alkenes, and carboxylic acid chlorides are used to impart distinctive properties to fluorourethanes. Some of these compounds are listed in Table 6.6.

4 CHARACTERIZATION

Many modern analytical techniques have been employed in the determination of physical and chemical properties of fluorourethanes. Leading references to modern instrumental methods are given in Table 6.7.

5 APPLICATIONS AND PROPERTIES

5.1 SURFACE COATINGS

5.1.1 Solvent-borne Coatings

Fluorourethane coatings [39,40] are particularly valuable for their low surface energy and resistance to chemicals, corrosion, and weather. They may be deposited in a desired location and thickness, and do not require heat to cure; other fluoropolymers usually cannot be processed in this way. Fluorourethane coatings are now being used in niche markets where their exceptional properties justify their high cost. Work to improve the solubility and processability of these polymers is ongoing in many industrial and academic laboratories and is expected to produce improved and less costly fluorourethane coatings which will become more widely used.

5.1.1.1 FEVE Technology

The majority of solvent-borne fluorinated polyurethane coatings are based on FEVE polyol resins. These resins are produced commercially by several manu-facturers, available in a range of compositions and molecular weights, and easy to formulate into coatings. The copolymers are soluble in conventional organic solvents [41] such as toluene, xylene, or butyl acetate and are cured with a variety of isocyanates to produce resins with excellent adhesion and resistance to chemical and weather.

Two considerations must be kept in mind when using FEVE-based polyols. First, joining chlorotrifluoroethylene to a vinyl ether in the polymer backbone

Table 6.6. Miscellaneous fluorinated starting materials used to prepare polyurethane resins

Precursor	Application	Reference
CF_3COCl	Poly(amide urethane) block polymers	5
$CF_2{=}CF_2$	Coatings	92, 99, 100, 102, 103
$CH_2{=}CF_2$	Coatings	99, 100, 101, 102, 103
$CF_3CF{=}CF_2$	Coatings	107
CF_3CF_2COCl	Poly(amide urethane) block polymers	4, 5
$CF_3(CF_2)_2COCl$	Poly(amide urethane) block polymers	5
tetrafluorosuccinic anhydride (five-membered ring with O, two C=O, and CF_2CF_2)	Fluorinated polyurethanes	210
$ClCO(CF_2)_3COCl$	Poly(amide urethane) block polymers	4, 5
$CF_3CF_2CH_2OCOCH{=}CH_2$	Thermoplastic resins	37
$HCF_2CF_2CH_2OCOCH{=}CH_2$	Acrylic resins	35
	Elastomers	162
$HO_2C(CF_2)_4CO_2H$	Poly(amide urethane) block polymers	4, 5

Structure	Application	Reference
$H_3C-CH-CH-C_4F_9$ (epoxide, O)	Fluorinated polyol intermediates	211
$CF_3(CF_2)_6COCl$	Poly(amide urethane) block polymers	5
(fluoro-substituted phthalic anhydride)	Poly(ester urethanes)	212
$CF_3(CF_2)_5CH_2$ (epoxide)	Fiber cladding	158
$HOOC\;\;F_3C\;\;CF_3\;\;CF_3\;\;COOH$	Fiber cladding	158
$H_2C-CH-O-(CF_2CF_2)_3CF(CF_3)_2$	Polyol intermediate	213

(continued overleaf)

Table 6.6. (continued)

Precursor	Application	Reference
$C_8F_{17}CH=CH_2$	Acrylic resins	36
	Elastomers	206
$CF_3(CF_2)_7CH_2CHOHCH_2OCOCH=CH_2$	Fiber cladding	158
$CF_3(CF_2)_7CH_2CHOHCH_2OCOC(CH_3)=CH_2$	Fiber cladding	158
	Elastomers	206
	Oligomeric polyols	214
	Diol intermediate	210

Table 6.7. Instrumental methods used to characterize fluorinated polyurethanes

Method	References
Biocompatibility	152, 205
Compression set	164
Differential scanning calorimetry	3, 7, 110, 164, 191, 196, 202, 206, 212, 215, 216, 217, 218, 219
Dynamic mechanical analysis	7, 110, 164, 202, 216, 217
Electrochemical impedance spectroscopy	220
Electron spectroscopy for chemical analysis (ESCA)	3, 11, 215
Ferroelectric behavior	221
Gel permeation chromatography	7, 202, 204, 215, 222
Hydrolytic stability	110, 206, 216
Infrared spectroscopy	3, 5, 7, 164, 191, 196, 197, 202, 204, 210, 212, 215, 219, 223
Mass spectrometry	5, 210, 222
Mechanical properties	110, 206, 216
Moisture vapor transmission rate (MVTR)	11
Nuclear magnetic resonance spectroscopy	3, 5, 202, 204, 210, 212, 217, 222
Polarizing optical microscopy	217, 222
Secondary ion mass spectrometry (SIMS)	215, 224
Surface energy	4, 5, 7, 11, 219, 223
Surface imaging techniques	223
Tensile strength	7, 110, 152, 164, 202, 205
Thermal stability	206, 212
Thermogravimetric analysis	212
UV-visible spectroscopy	204
Viscometry	5, 202, 212, 217, 219
Water resistance	110
X-ray diffraction spectrometry	
(small angle)	222
(wide angle)	218
X-ray photoelectron spectroscopy (XPES)	7, 196, 197, 224

places a chlorine atom and a hydrogen atom on adjacent carbons and makes possible the loss of hydrogen chloride. This process creates in the polymer backbone a double bond which facilitates the subsequent loss of additional hydrogen chloride and eventual degradation of the polymer. To avoid this, a ultraviolet stabilizer should be used in the coating, and the presence of an acid scavenger such as calcium carbonate is sometimes useful. Secondly, the FEVE polyol does not wet pigments well, and wetting and dispersing aids are needed in the formulation.

A large number of FEVE-based polyurethane coatings intended for industrial, architectural, automotive, and marine use are disclosed in the patent literature. Formulations contain aliphatic isocyanates (usually based on hexamethylene diisocyanate), pigments, UV stabilizers, rheology modifiers and organic solvents. The coatings are valuable for their resistance to abrasion, acid [42–44], corrosion [45–48], ice, impact, soil [49], scratching [50–52], staining [53,54], low temperature [55], thermal shock [56], water [57], and weather [58–61]. They demonstrate excellent gloss [62], color and gloss retention [63], durability, hardness, and strong adhesion to metals [64], to nonfluorinated coatings [65], and to plastics such as poly(ethylene terephthalate) [66] and polycarbonate [67]. The coatings cure at ambient temperature and are used most frequently for exterior applications on steel [68], aluminum [69–75], concrete [76], glass, [77] and wood [78]. They are valuable as coatings for luxury automobiles [79–84] and are also useful as flooring materials [85]. Coatings prepared from FEVE polyols and polymers containing cationic groups have been applied to phosphated steel by electrodeposition [86].

5.1.1.2 FEVE Polyol Resins Modified with Acrylates or Methacrylates

Hydroxy-functional acrylic resins are frequently blended with FEVE polyols to improve surface reflectance [87], clarity, and hardness and reduce cost. For example, a FEVE polyol and a methacrylic resin react with a polyisocyanate to produce a stain-resistant high-gloss coating [88]. A fluorinated acrylic polyol made from adipic acid, 1,4-butanediol, fumaric acid and 2,2,3,3,3-pentafluoropropyl methacrylate reacts with methylene diphenyl diisocyanate (MDI) at $120\,^\circ$C for 1 hour to produce a thermoplastic resin [37].

5.1.1.3 Other Resins based on FEVE Technology

Many modifications and variations of FEVE technology have been reported. A 19 : 23 : 6 : 52 copolymer of 2,5-norbornadiene, vinyl caproate, 2-hydroxyethyl crotonate and chlorotrifluoroethylene has a glass transition temperature of $55\,^\circ$C. When cured with hexamethylene diisocyanate a coating with excellent adhesion to steel is produced [89]. Silicone oligomers (MW 2500) containing terminal vinyl ester bonds can also be incorporated into fluorinated polyols. A 50 : 0.5 : 42 : 7.5 copolymer of chlorotrifluoroethylene, unsaturated silicone oligomer, vinyl pivalate and 2-hydroxyethyl crotonate reacts with hexamethylene diisocyanate to produce a coating with good weatherability, water repellency, low dynamic coefficient of friction, low and stable surface energy, and high gloss [90].

5.1.1.4 Polyol Resins Derived from Hexafluoroacetone

Coatings formulated with HFA-based polyol resins and powdered poly(tetrafluoroethylene) (PTFE) exhibit not only the hardness and toughness of conventional polyurethane coatings but also the low surface energy and easy cleanability of PTFE. When the biuret of hexamethylene diisocyanate is used as the curing agent, the coatings demonstrate excellent color and gloss stability and

Figure 6.3. Fluorourethane coatings resist strong bonds to marine organisms and are cleaned easily if cleaned frequently. The 65-foot patrol boat shown here has a nontoxic coating containing 24 vol% of PTFE on its hull (courtesy of US Navy)

resistance to weather and corrosion. Clear coatings in flat, semigloss, or gloss finishes are formulated with these resins, and conventional pigments are used without difficulty [91]. Useful coatings may contain as much as 38% by volume of PTFE, but are somewhat soft and easily marred. Formulations containing 24 vol% of PTFE represent the best compromise between good hardness and low surface energy [92] and perform successfully in military and industrial applications. These coatings are being used as anticorrosion coatings for ships' bilges [93] and tanks [94] and as nontoxic fouling-release coatings on small boats such as the one shown in Figure 6.3; fouling accumulates but can be removed with ease if cleaning is frequent [95].

These linings are also specified by the US Army and Navy for the interior of bulk fuel storage tanks (Figure 6.4) [96]. Large tanks always contain water at the bottom which accumulates naturally from condensation. Fluorourethane linings resist attack by both fuel and water, which is more corrosive than the fuel. The low-energy surface of the lining also permits cleaning with high-pressure water only; no detergent is required. Oily washings are processed in an oil–water separator and clear water is deposited in a municipal sewer, thus avoiding the high cost of disposing of large volumes of oily water.

Unpigmented coatings are used as clear, abrasion-resistant, water-shedding finish coats on submarine radomes (Figure 6.5) [97]. A radome is made more

Figure 6.4. Tanks for bulk storage of petroleum products are lined with a fluorourethane coating which is not attacked by condensed water or by fuel. Service lifetimes of greater than 20 years are expected for these linings (courtesy of US Navy)

Figure 6.5. A submarine radome (the tallest structure, rising from the spotted mast) must remain free water to achieve stable operations. A clear fluorinated coating causes water to bead and drain rapidly (courtesy of US Navy)

effective by a coating which allows water to drain rapidly from its surface, for films of water interfere with transmission and reception of signals. Radomes must be retuned as water drains from their surface, and a strong and steady signal is obtained only when water is absent.

5.1.1.5 Fluorinated Acrylic Resins

A resin prepared from perfluorooctylethylene, 2-hydroxyethyl methacrylate, methyl methacrylate, dimethyl maleate, butyl methacrylate, cyclohexyl vinyl ether, methacrylic acid and 1,2-(dimethylamino)ethyl methacrylate reacts with a polyisocyanate to produce a stain-, water- and weather-resistant coating with high gloss and good gloss retention [36]. An hydroxylated unsaturated polyester resin reacts with 2,2,3,3,3-pentafluoropropyl methacrylate to give a polyol which reacts with a polyisocyanate to give a hard, weather-resistant high-gloss coating [35]. A water-repellent coating is prepared from an unspecified fluorinated acrylic resin and a polyisocyanate [98].

5.1.1.6 Other Polyol Resins

5.1.1.6.1 Resins containing tetrafluoroethylene Coatings with good weather resistance are prepared by reacting a polyisocyanate with a copolymer of hydroxybutyl vinyl ether and tetrafluoroethylene [99], or with a 45 : 18 : 37 terpolymer of vinyl benzoate, hydroxybutyl vinyl ether and tetrafluoroethylene [100], or with a polyol prepared by saponification of a vinyl ether — tetrafluoroethylene copolymer [101]. A high gloss coating with good solvent, stain and boiling water resistance is prepared by reacting a polyisocyanate with a 40 : 10 : 40 : 10 terpolymer of tetrafluoroethylene, hexafluoropropylene, vinylidene fluoride, and 2-hydroxyethyl allyl ether [102]. A 14 : 8 : 10 : 68 copolymer of 4-hydroxybutyl vinyl ether, 2-acetoxyethyl vinyl ether, tetrafluoroethylene and vinylidene fluoride reacts with the biuret trimer of hexamethylene diisocyanate to form a coating with good stain resistance, strength and adhesion to aluminum and poly(ethylene terephthalate) [103].

5.1.1.6.2 Resins containing vinylidene fluoride Vinylidene fluoride copolymers are compatible with acrylic resins and resist weather, staining, and loss of gloss [104]. A coating with high gloss and good resistance to impact and weather is prepared by reacting a polyisocyanate with a 60 : 31 : 9 terpolymer of vinylidene fluoride, vinyl pivalate, and 2-hydroxyethyl crotonate [105]. Reaction of a copolymer containing vinylidene fluoride, ethyl vinyl ether, cyclohexyl vinyl ether, 4-hydroxybutyl vinyl ether, and butyl maleate with a polyisocyanate produces a weather-resistant high gloss finish which retains 94% of its gloss after two years of outdoor weathering [106].

5.1.1.6.3 Resins containing hexafluoropropylene Polymerization of hexafluoropropylene, 4-hydroxybutyl vinyl ether, ethyl vinyl ether, vinyl acetate and ethoxylated vinyl acetate gives a copolymer which reacts with a polyisocyanate to produce a coating with high gloss and excellent gloss retention [107].

5.1.1.6.4 Marine coatings A coating based on a 52 : 30 : 18 copolymer of chlorotrifluoroethylene, vinyl propionate and 2-hydroxyethyl crotonate and hexamethylene diisocyanate has shown good resistance to marine fouling [108].

5.1.1.6.5 Color Pigments Water-resistant ultrafine colored resin powders containing 1,4-bis(2-acryloyloxyethyl)-tetrafluorobutane, benzoyl peroxide and C. I. Solvent Red 49 are blended into a fluorinated polyurethane methacrylate binder and cured to produce a coating with excellent color stability [109].

5.1.2 Water-borne Coatings

The water resistance and mechanical properties of water-borne polyurethane coatings are significantly improved when small amounts of fluorinated polyols are incorporated into the formulation [110]. FEVE copolymers may be applied as aqueous dispersions and cured at 160 °C for 25 minutes to produce coatings with high gloss and good water and weather resistance [111]. They may also be cured with water-based hardeners [112].

A water-borne perfluoropolyether polyol and a hydroxy-functional polyester are cured with a water-borne polyisocyanate to produce a coating with a low surface energy and low coefficient of friction [113].

5.1.3 Powder Coatings

Fluorinated polyurethanes are effective powder coatings due to their good water repellency, lubricity, adhesion, and resistance to acid and weather. The glass transition temperature of the resin should be between 35 and 120 °C to optimize flow out and cure at the annealing temperature. Blocked isocyanates, which form free isocyanates when heated, are frequently used [114]. For example, a powder coating is made from a polyol containing 30% fluorine, caprolactone-blocked isophorone diisocyanate, wax and titanium dioxide pigment [115]. Another useful powder coating contains a hydroxyl- and carboxyl-functional FEVE copolymer and a blocked isophorone diisocyanate [116].

5.2 TEXTILE TREATMENTS

The second most important use of fluorinated polyurethanes is in surface treatments for textiles, carpets, glass, wood, paper and leather. Fluorourethanes are applied in a one-step treatment. They resist soil, stains, oil and water, impart smoothness and a soft hand to leather and fabrics, and resist removal by many cycles of laundering or dry cleaning.

5.2.1 Fabrics

The oil and water resistance of conventional urethane-based fabric treatments is significantly improved when hydroxylated fluorinated compounds are added to the process. Fabrics are usually treated with an emulsion of the fluorourethane, dried, and sometimes given a mild heat cure.

Materials used for this purpose contain long perfluoroalkyl segments. Examples include a mixture of n-perfluoroalkylethyl alcohols containing 8 to 14 carbons (average 11) [117], n-perfluoroalkanols containing 6 to 12 carbons [118], and

$$C_8F_{17}CH_2CH_2SCH_2CON(CH_2CH_2OH)_2 \ [119].$$

Compounds having the structure [120]

$$HOCH_2(CH_2SCH_2CH(R_F))_2CH_2OH$$

or the structure [121]

$$(HOCH_2)_2C(CH_2SCH_2CH_2R_F)_2$$

in which R_F is a perfluoroalkyl group containing 6 to 16 carbons, are also used for this purpose.

Effective fabric treatments are made by using a n-perfluoroalkylethyl acrylate alone [122,123] or with an organosilsesquioxane oligomer [124]. A mixture of 2-perfluoroalkylethane thiols in which the straight chains average 12 carbons in length is also used to produce fabric treatments [125,126].

Other effective fluorinated compounds are based on the adduct of 1,2-dihydroxy-3-(2-perfluorohexyl)ethyloxypropane with ethylene oxide [127], a hydroxyl-terminated vinyl fluoride polymer, [128,129] a terpolymer of a propylene oxide adduct of 4-hydroxybutyl vinyl ether, chlorotrifluoroethylene and ethyl vinyl ether [130], and a copolymer of perfluorooctylethyl methacrylate and 2-hydroxypropyl methacrylate [131].

5.2.2 Leather

A fluorourethane polymer, prepared by reacting a perfluoroalkylethyl alcohol and an acrylated polyol with either hexamethylene diisocyanate biuret trimer [132] or toluene diisocyanate [133], is copolymerized with an acrylic resin in water to produce an emulsion which imparts oil and water resistance to leather.

Perfluoroheptan-1-ol, aliphatic diols and polyisocyanates react to form a polymer which is emulsified in polyoxyalkylene ethers, organic solvents and water. This emulsion imparts good water repellency, surface smoothness and soil resistance to fabrics used as leather substitutes [134]. Siloxane modifications of similar formulations are also employed for this purpose [135,136].

Dioxaperfluorodecyl-bis [*N*-(ethyl)-*N*-(2-hydroxyethyl) sulfonamide] is poly-merized with toluene diisocyanate to give a coating for leather which resists water and chemicals [137].

5.2.3 Carpets

A water-based graft copolymer for oil- and waterproofing carpets is prepared in two steps. Glyceryl monostearate, *N*-perfluorooctylsulfonyl diethanol amine, hexamethylene diisocyanate and methyldiethanolamine are polymerized in water to obtain an aqueous dispersion which reacts further with perfluorooctylsulfonyl methacryloylethyl methyl amine, stearyl methacrylate, and vinyl acetate to give a dispersion containing 21.4% fluorine [138]. Dispersions of fluorinated acrylic resins in water are also used for this purpose [139].

5.3 MEDICAL AND DENTAL APPLICATIONS
5.3.1 Contact Lenses

Soft hydrogels with good strength, wettability, oxygen permeability, optical clarity, flexibility, biocompatibility, and boiling water resistance are valuable for use as contact lens materials. A hydroxymethyl-terminated perfluoropolyoxyalkyl ether reacts with isophorone diisocyanate, then with polyethylene glycol, then with 2-isocyanatoethyl methacrylate and is then photocured to give a polymer for ophthalmic devices [32,140]. Silicone-containing monomers are also added to this synthetic process [141,142]. Hard contact lenses with good oxygen permeability use 2,2,2-trifluoroethanol in an acrylic-modified polyurethane resin [143].

5.3.2 Dental Materials

A urethane prepared from 2-hydroxypropyl methacrylate and 1,2-bis (3-isocyanatophenyl)-tetrafluoroethane is cured with a methacrylate resin to produce a dental material with good mechanical properties [144]. A resin prepared from 2-isocyanatoethyl methacrylate and a polyol derived from hexafluoroacetone is filled with quartz and cured with other methacrylates to produce dental composites with good strength and hydrophobicity and relatively low shrinkage [145]. Fluorourethanes are also used in linings for dentures [146].

5.3.3 Medical Products

A polyol based on ethylene oxide — propylene oxide reacts with 2,2,3,3,4,4,5,5-octafluorohexyl-1,6-diisocyanate to give a resin with a residual isocyanate content of 2.5%. Surgical adhesives prepared from this polymer show excellent adhesion to tissue, flexibility, and low toxicity [147]. The same diisocyanate, ethylenediamine and polyoxytetramethylene diol are copolymerized and coated on glass beads to create a packing for column chromatography. B-lymphocytes adsorb on the resin, allowing T-lymphocytes to elute [148]. Fluorourethanes are also used to construct an intra-aortic balloon catheter [149], and as hydrophobic microporous membranes used in breathable, sterile fabrics [150,151]. Antithrombogenic elastomers are prepared from 2,2,3,3,4,4,5,5-octafluorohexane-1,6-diisocyanate, hydroxyl-terminated polyoxyalkanes, and fluorinated dioxa-1,10-decanediols [152].

5.4 OPTICAL FIBER CLADDING COMPOUNDS

Fluorinated polyurethanes have a low index of refraction (less than 1.43), low permeability to water liquid and vapor, good adhesion, and very low water adsorption. These features make these polymers attractive materials for applications in wet environments where preventing water from gathering at optical interfaces is critical.

Quartz and optical-quality glass must have organic claddings to prevent the growth of microcracks which weaken and ultimately sever the fiber. Fluorourethane cladding serves this purpose and also prevents the ingress of water which is destructive to these vitreous materials. Its low index of refraction increases the degree of internal reflection and reduces optical losses.

A fluorinated epoxy-polyurethane coating is prepared by reacting $1H, 1H, 2H, 2H$-heptadecafluorodecyl acrylate and glycidyl methacrylate with 2-mercaptoethanol, then with toluene diisocyanate, and finally esterifying with acrylic acid to produce an acrylic resin. This resin is blended with $1H, 1H, 3H$-tetrafluoropropylacrylate and a photocatalyst to give a resin which may be UV-cured to give a cladded optical fiber stable between -30 and $80\,°C$ [153]. Photocurable fiber cladding resins are also produced by adding the reaction product of toluene diisocyanate, 2-hydroxyethyl acrylate, and 3-[2-perfluorohexylethyl)thio]-propanol to an unspecified urethane — acrylic polymer [154].

The reaction of 1,4-bis(2-hydroxyethyl) tetrafluorobutane, 1,1,2,2-tetrafluoro-glycidyl ether, isophorone diisocyanate and 2-hydroxyethyl acrylate gives a fluorinated acrylated polyurethane. Reaction of this polymer with 1,4-bis(2-acryloyloxyhydroxyethyl)tetrafluorobutane produces a fiber cladding resin with n_D^{25} 1.420 and good elasticity, hardness, and adhesion [155].

A cladding for quartz is prepared by reacting a polyurethane acrylate, 2-perfluorooctylethyl acrylate, 2,2-dimethylpropane-1,3-diacrylate, trimethylol-propane triacrylate, 1-hydroxycyclohexyl phenyl ketone and N-glycidyloxy allyl 3-trimethoxysilylpropyl amine [156]. A cladding for undersea optical cable is formed in the reaction of $(R_F(CH_2)_2SCH_2)_2C(CH_2OH)_2$ (in which R_F is a perfluorinated alkyl group containing $1-10$ carbons) with 1,6-diisocyanato-2,2,4-trimethylhexane, followed by reaction with 1,4-butanediol monoacrylate; lastly a reactive diluent and a photocatalyst are added to give a photocurable resin [157].

A cladding resin is formed in the reaction of perfluorohexylmethyleneoxirane and perfluoro-2,2,4-trimethylhexane-1,6-dioic acid; this product reacts with 1,6-diisocyanato-2,2,4-trimethylhexane at $80\,°C$ for 15 hours, and that product reacts with 3-perfluorooctyl-2-hydroxypropyl acrylate. The resulting resin is mixed with a photocatalyst and irradiated with UV light to give a cured resin with n_D^{25} 1.379 and useful hardness and elongation [158].

5.5 ELASTOMERS

Fluorinated polyurethane elastomers are valuable for their good mechanical properties and their resistance to solvents, chemicals, heat and cold [159]. Frequently, formulations are modified with siloxanes to optimize certain properties [160]. FEVE-based polyols are used to manufacture elastomeric automobile bumpers and interior trim components [161]. Tertiary amines are fugitive catalysts compatible with perfluoropolyoxyalkylene diols and are used to produce elastomers with no catalyst residue [33].

Elastomers with good low-temperature properties suitable for gaskets, adhesives, and sealants are prepared from a hydroxymethyl-terminated perfluoropolyoxyalkylene (MW 2103), 4,4'-dicyclohexylmethane diisocyanate, and 1,4-bis(hydroxymethyl)cyclohexane [34]. An elastomer with good flexibility at low temperature and good mechanical properties at high temperature is prepared from a similar perfluoropolyoxyalkylene (MW 2825) and diphenyl methane diisocyanate [29].

A fluorourethane elastomer is prepared in three steps. First, a copolymer is prepared from 2,2,3,3-tetrafluoropropyl acrylate, 2-hydroxyethyl mercaptan, 2-hydroxyethyl acrylate, and catalyst. A second copolymer is prepared by substituting 2-isocyanatoethyl methacrylate for 2-hydroxyethyl acrylate in the reaction above. Lastly an equimolar mixture of these two copolymers reacts to produce an elastomer with good mechanical strength, compatibility, and transparency [162].

A heavily fluorinated transparent elastomer is made by the reaction of 3,3,4,4,5,5,6,6-octafluorooctane-1,8-diisocyanate and 3-trifluoromethyl-3,4,4,5,5,-6,6,7,7,8,8-undecafluorodecane-1,10-diol [163]. Highly fluorinated thermoprocessable elastomers are prepared from perfluoropolyoxyalkylene diols, nonfluorinated diols and diphenylmethane diisocyanate [164].

5.6 TREATMENTS FOR VITREOUS MATERIALS

Soil-, water- and weather-resistant ceramic articles are prepared by priming inorganic substrates with a white epoxy resin, topcoating with a FEVE-based polyurethane resin, and baking [165]. Other FEVE-based coatings for concrete are available [84, 161].

Buildings, stone, and fibrous materials are protected from weather by treatment with a fluorourethane. A stable aqueous dispersion of an elastomeric fluorinated polyurethane suitable for this purpose is prepared from a hydroxymethyl-terminated perfluoropolyoxyalkylene (MW 2103), isophorone diisocyanate and dimethylolpropionic acid [166].

Antifogging coatings for mirrors and optical components are prepared by reacting the isocyanurate trimer of hexamethylene diisocyanate with 3,3,3-trifluoropropyl alcohol. The polymer is coated on glass and cured at 100 °C to give a coating with excellent light transmission and high water contact angle [167].

5.7 ELECTRICAL APPLICATIONS

Fluorine imparts superior electrical insulating ability to polymers. Placement of fluorine-containing groups along or within the polymer backbone improves the processability and thermal stability of the polymer. Fluorination may also impart a low dielectric constant to resins. The C−F bond is a strong dipole, so when fluorine atoms are few they must be placed so that their dipoles are opposed [168]. Alternatively, the resin must be so heavily fluorinated that there is a statistical cancellation of dipoles.

5.7.1 Materials for Printed Circuit Boards

The density of electrical components on printed circuit boards is inversely proportional to the dielectric constant of the resin from which the board is made. Resins with low dielectric constants are better insulators and conductive paths on the boards can be placed closer together, thus reducing the size and weight of components. Urethanes containing Bisphenol A6F give printed circuit boards with good flexural and peel strength, high resistance, and a low dielectric loss tangent [169].

An unspecified perfluoropolyether polyurethane is useful in the production of printed circuit boards. A 10 μm film of this resin on aluminum is etched with laser radiation (248 nm, 45 mJ pulse in 15 ns, peak power 3 MW), completely ablating the film under the beam without affecting the adjacent resin [170]. Fluorourethanes are also employed as antistatic agents on printed circuit boards [171].

5.7.2 Recording Media

FEVE-based coatings are useful as abrasion- and weather-resistant coatings for magnetic recording materials [172]. Lubricants for recording media are prepared by reacting a fluorinated diol with a perfluoropolyoxyalkylene isocyanate. For example, Bisphenol A6F and a toluene diisocyanate-endcapped perfluoropolyoxyalkylene (MW 1800) react to produce a fluorinated polyurethane which is applied to the surface of a magnetic disk, giving the disk a coefficient of friction of 0.38 [38]. The alcohol $1H$, $1H$, $11H$-perfluoroundecan-1-ol is used in a coating for magnetic recording tape [173]. Perfluoropolyether diols are used to manufacture magnetic recording tape having a coefficient of friction of 0.28 [174,175].

5.7.3 Insulation

A thermoplastic insulating polymer is prepared by extruding onto copper wire a mixture of poly(tetramethylene) glycol (MW 2000), 1,4-butanediol, 1,3-bis(2-hydroxyhexafluoro-2-propyl)benzene and methylene diphenyl diisocyanate.

Tensile strength, solvent resistance, and elongation are significantly improved compared to the insulation made without the fluorinated diol [176].

5.8 PRINTING TECHNOLOGIES

Fluorinated polyurethanes are used widely in printing technologies where low surface energy and stability to high heat, at least for a short period, is required. The polymers are useful lubricants on printing heads and internal parts of copying machines, and on polyester printing ribbons and transparency sheets.

Thermal recording media are treated with heat-resistant fluorourethane release layers to prevent sticking and staining of the recording heads. A resin for a polyester thermal recording layer is prepared from diphenyl methane diisocyanate, 1,4-butanediol, a poly(butylene adipate) polyol, and a polyethylene glycol terminated with a $5H$-perfluoropentyl radical [177].

Toner fusing elements in thermal printers are treated with fluorourethane resins to promote easy release of toners to polyester or paper sheets. A coating for this purpose is prepared from a silazane-terminated polydimethylsiloxane, Bisphenol A6F, and 2,4-toluene diisocyanate [178].

Toners contain carbon black and ferrite particles coated with a resin which imparts high chargeability and abrasion resistance. A resin for ferrite particles is prepared from perfluorooctylethyl methacrylate, styrene, methyl methacrylate and 3-methacryloxypropyltrimethoxysilane [179]. Charge-generating layers on electrophotographic photoreceptors are manufactured from FEVE-based polyurethane resins [180].

Thermal transfer printing ribbons are prepared by coating an ink onto one side of a film and treating the other side with a smooth lubricating and heat-resisting polyurethane layer [181]. This layer is made from an isocyanate-functional perfluoropolyoxyalkane and a silicone compound containing active hydrogen [182], or alternatively from a perfluoroalkylacrylate, a saturated polyester and a polyisocyanate [183].

5.9 MISCELLANEOUS USES

5.9.1 Heat Exchangers

Surfaces of heat exchangers must be kept free from films of liquid water as well as deposits and corrosion products, for all of these obstruct heat flow and diminish efficiency. A polyol based on hexafluoroacetone is used in a urethane coating which successfully promotes water-beading and draining from the interior surfaces of copper–nickel tubes in a shipboard heat exchanger (Griffith, JR, private communication). A fluorinated coating on the interior surfaces of copper–nickel tubes in a steam turbine condenser prevents scale formation for more than 14 months [184]. A second resin made from an unspecified fluoropolyol and blocked isophorone diisocyanate is used to coat the aluminum fins of heat exchangers [185].

5.9.2 Binders for Explosives

Urethane resins made from polyols and polyisocyanates, one or both of which is fluorinated, are used as binders in high-energy explosives. Because the carbon–fluorine bond energy (540 kJ/mol) is 24% greater than the carbon–hydrogen bond energy (435 kJ/mol) [186], the presence of fluorine raises the decomposition energy of the binder. The binder is not vulnerable to accidental detonation from bullet or shell impact and is less sensitive than its unfluorinated counterpart to shock [187].

5.9.3 Sealants

Fluorinated acrylic–polyurethane–silicone sealants are used on liquid-crystal display panels to prevent moisture ingress leading to corrosion and deformation of the substrate [188]. Urethane-cured FEVE-based polyols provide sealants with good breaking strength, elongation, and retention of elongation [189]. Sealants are also made from hydroxy-functional fluorinated acrylic resins, polyols and polyisocyanates [190].

REFERENCES

1. Brady, Jr, RF *Nature* 368 (6466), 16–17 (1994).
2. Zisman, WA in *Contact Angle, Wettability, and Adhesion, Adv. Chem. Ser.* **43**, pp. 1–51 (Amer Chem Soc, Washington, DC, 1964).
3. Edelman, PG, Castner, DG and Ratner, BD *Polym. Prepr.* **31**(1), 314–15 (1990).
4. Chapman, TM and Marra, KG *Macromolecules* **28**, 2081–5 (1995).
5. Chapman, TM, Benrashid, R, Marra, G and Keener, JP *Macromolecules* **28**, 331–5 (1995).
6. Katano, Y, Tomono, H and Nakajima, T *Macromolecules* **27**, 2342–4 (1994).
7. Yu, XH, Okkema, AZ and Cooper, SL *J. Appl. Polym. Sci.* **41**, 1777–95 (1990).
8. Huth, HU and Angelmayer, KH German Patent DE 4 006 098, August 29, 1991; *Chem. Abstr.* **115**, 280315n (1991).
9. Wu, HS and Kaler, EW PCT Intl. Appl. WO 94 22 928, October 13, 1994; *Chem. Abstr.* **123**, 144917k (1995).
10. Kuriyama, S Japanese Patent 05 25,354, February 2, 1993; *Chem. Abstr.* **119**, 162482m (1993).
11. Head, RA, Powell, RL and Fitchett, M *Polym. Mater. Sci. Eng.* **60**, 238–42 (1989).
12. Onishi, H, Matsumoto, S and Aoki, K Japanese Patent 03 106 915, May 7, 1991; *Chem. Abstr.* **115**, 209462q (1991).
13. Tokuda, H, Rainaa, BF Japanese Patent 06 279 555, October 4, 1994; *Chem. Abstr.* **122**, 291784x (1995).
14. Ozerin, AN, Rebrov, AV, Feldman, VI, Krykin, MA, Storojuk, IP, Kotenko, AA and Tul'skii, MN *React. Funct. Polym.* **26**, 167–75 (1995).
15. Benoist, P and Legeay, G *Europ. Polym. J.* **30**, 1283–7 (1994).
16. Munekata, S *Prog. Org. Coatings* **16**, 113–34 (1988).
17. Handforth, V *J. Oil Colour Chem. Assoc.* **73**(4), 145–48 (1990).
18. Nakabayashi, A and Kamiyanagi, K Japanese Patent 04 01,216, January 6, 1992; *Chem. Abstr.* **116**, 237506w (1992).

19. Kodama, S, Ishida, T and Myake, H Japanese Patent 07 26,204, February 3, 1995; *Chem. Abstr.* **123**, 172881n (1995).
20. Field, DE and Griffith, JR US Patent 4,157,358, June 5, 1979.
21. Farah, BS, Gilbert, EE and Sibilia, JP *J. Org. Chem.* **30**, 998 (1965).
22. Urry, WH, Niu, JHY and Lundsted, LG *J. Org. Chem.* **33**, 2302 (1968).
23. Field, DE *J. Coatings Technology* **48**(615), 43–7 (1976).
24. McBee, ET, Marzluff, WF and Pierce, OR *J. Am. Chem. Soc.* **74**, 444 (1952).
25. Cassidy, PE, Aminabhavi, TM, Reddy, VS and Fitch JW *Europ. Polym. J.* **31**, 353–61 (1995).
26. Cassidy, PE *J. Macromol. Sci., Rev. Macromol. Chem. Phys.* **C34**, 1–24 (1994).
27. Cassidy, PE, Aminabhavi, TM and Farley, JM *J. Macromol. Sci., Rev. Macromol. Chem. Phys.* **C29**, 365–429 (1989).
28. Re, A and Terenghi, T Europ. Pat. Appl. EP 86 110 195, March 4, 1987; *Chem. Abstr.*, **106**, 214920f (1987).
29. Turri, S, Gianotti, G, Levi, M and Tonelli, C Europ. Pat. Appl. EP 621 298, October 26, 1994; *Chem. Abstr.*, **122**, 242066f (1995).
30. Tonelli, C and Simeone, G Europ. Pat. Appl. EP 525 795, February 3, 1993; *Chem. Abstr.*, **118**, 256736d (1993).
31. Re, A and DeGiorgi, M Europ. Pat. Appl. EP 294 829, December 14, 1988; *Chem. Abstr.*, **110**, 194773n (1989).
32. Goldenberg, M US Patent 4 933 408, June 12, 1990; *Chem. Abstr.*, **113**, 232716t (1990).
33. Re, A and DeGiorgi, M Europ. Pat. Appl. EP 291 855, November 23, 1988; *Chem. Abstr.*, **110**, 136760v (1989).
34. Re, A and Giavarini, F Europ. Pat. Appl. EP 359 273, March 21, 1990; *Chem. Abstr.*, **113**, 61078e (1990).
35. Kanetani, K Japanese Patent 95 26,204, January 27, 1995; *Chem. Abstr.*, **123**, 172871j (1995).
36. Nakajima, S and Akyama, H Japanese Patent 94 345 823, December 20, 1994; *Chem. Abstr.*, **123**, 113131v (1995).
37. Kanetani, K Japanese Patent 06 93 075, April 5, 1994; *Chem. Abstr.*, **121**, 84809y (1994).
38. Anon. (to Ausimont SpA) Japanese Patent 87 187 798, August 17, 1987; *Chem. Abstr.*, **107**, 201844e (1987).
39. Scheirs, J, Burks, S and Locaspi, A *Trends in Polymer Science*, **3** (3), 74–82 (1995).
40. Khan, AK, Saxena, MS and Chandra, S *Pigment and Resin Technology*, **7** 4–11 (April 1992).
41. Izumi, T, Murakami, S, Inagaki, H and Hirakuri, Y Japanese Patent 03 281 611, December 12, 1991; *Chem. Abstr.*, **116**, 216468v (1992).
42. Senda, A, Shimizu, Y, Kaneko, H and Endo, K Japanese Patent 05 287 230, November 2, 1993; *Chem. Abstr.*, **120**, 273227x (1994).
43. Nagaso, T and Sukejima, H, Japanese Patent 04 371 266, December 24, 1992; *Chem. Abstr.*, **120**, 194154t (1994).
44. Satomoto, Y and Nakasuji H Japanese Patent 04 311 779, November 4, 1992; *Chem. Abstr.*, **118**, 171165s (1993).
45. Koyama, M, Nishio, T, Marumoto, E and Iida, A Japanese Patent 04 341 377, November 27, 1992; *Chem. Abstr.*, **119**, 51460h (1993).
46. Komazaki, S, Oka, M, Yoshida, S, Kawai, I and Toyoda, M Japanese Patent 01 197 510, August 9, 1989; *Chem. Abstr.*, **112**, 139998f (1990).
47. Deflorian, F, Fedrizzi, L, Lenti, D and Bonora PL *Prog. Org. Coatings*, **27**, 39–53 (1993).

48. Kano, M, Umetsu, Y and Kono M Japanese Patent 61 118 466, June 5, 1986; *Chem. Abstr.*, **106**, 34789m (1987).

49. Nagaso, T and Sukejma, H Japanese Patent 04 363 172, December 16, 1992; *Chem. Abstr.*, **119**, 74706g (1993).

50. Watanabe, M, Mototani, S and Hirohata, H Japanese Patent 06 25 592, February 1, 1994; *Chem. Abstr.*, **121**, 59823r (1994).

51. Kawakami, S, Okamoto, S and Hanami, K Japanese Patent 05 112 752, May 7, 1993; *Chem. Abstr.*, **120**, 10466s (1994).

52. Kawakami, S, Okamoto, S and Hanami, K Japanese Patent 05 111 675, May 7, 1993; *Chem. Abstr.*, **119**, 252266y (1993).

53. Maeda, K, Nakamura, T and Tsutsumi, K Japanese Patent 07 233 343, September 5, 1995; *Chem. Abstr.*, **124**, 32227z (1996).

54. Maeda, K, Hirashima, Y and Tsutsumi, K Japanese Patent 07 233 344, September 5, 1995; *Chem. Abstr.*, **124**, 32228a (1996).

55. Maruyama, T and Watanabe, S Eur. Pat. Appl. EP 416 501, March 13, 1991; *Chem. Abstr.*, **115**, 116498r (1991).

56. Nishio, T, Koyama, M, Marumoto, E and Iida, A Japanese Patent 05 111 673, May 7, 1993; *Chem. Abstr.*, **120**, 10464q (1994).

57. Nakano, Y, Miyazaki, H and Watanabe, K Japanese Patent 62 73 944, April 4, 1987; *Chem. Abstr.*, **107**, 98345s (1987).

58. Sugyama, M, Ishikawa, N and Okada, K Japanese Patent 05 65 455, March 19, 1993; *Chem. Abstr.*, **119**, 98117s (1993).

59. Iida, A, Nishio, T, Maruki, E and Inukai, H Japanese Patent 06 248 224, September 6, 1994; *Chem. Abstr.*, **122**, 136313z (1995).

60. Mogami, M Japanese Patent 05 125 146, May 21, 1993; *Chem. Abstr.*, **119**, 228177z (1993).

61. Sugiyama, M, Ishikawa, N and Okada, K Japanese Patent 05 236 642, February 2, 1992; *Chem. Abstr.*, **119**, 141328a (1993).

62. Nakabayashi, A, Shimizu, A and Sasahara, H Japanese Patent 04 366 114, December 18, 1992; *Chem. Abstr.*, **119**, 141312r (1993).

63. Iida, A, Nishio, T and Marumoto, E Japanese Patent 06 100 834, April 12, 1994; *Chem. Abstr.*, **122**, 12206x (1995).

64. Marumoto, E, Nishio, T, Iida, A and Oodera, A Japanese Patent 05 247 140, September 24, 1993; *Chem. Abstr.*, **120**, 79633m (1994).

65. Marumoto, E, Nishio, T, Koyama, M and Iida, A Japanese Patent 04 346 871, December 2, 1992; *Chem. Abstr.*, **119**, 74694b (1993).

66. Katsuragawa, S, Nanba, S and Katsuhara, Y Japanese Patent 02 173 128, July 4, 1990; *Chem. Abstr.*, **114**, 25871v (1991).

67. Nanba, S and Kobayashi, S Japanese Patent 06 271 807, September 27, 1994; *Chem. Abstr.*, **122**, 163694f (1995).

68. Watanabe, K, Fuka, T, Osawa, K, Kotani, T, Asakawa, H and Ishimura, H Japanese Patent 03 76 641, April 2, 1991; *Chem. Abstr.*, **116**, 61681h (1992).

69. Kodama, S and Washida, H Japanese Patent 07 196 973, August 1, 1995; *Chem. Abstr.*, **123**, 343696n (1995).

70. Myazaki, N, Takayanagi, T and Uchino, B Japanese Patent 07 82 520, March 28, 1995; *Chem. Abstr.*, **123**, 317062h (1995).

71. Hirashima, Y, Maeda, K and Tsutsumi, K Japanese Patent 07 62 290, March 7, 1995; *Chem. Abstr.* **123**, 259983w (1995).

72. Maeda, K, Nakamura, T and Tsutsumi, K Japanese Patent 06 248 222, September 6, 1994; *Chem. Abstr.* **122**, 268267x (1995).

73. Maeda, K, Nakamura, T and Tsutsumi, K Japanese Patent 06 264 020, September 20, 1994; *Chem. Abstr.* **122**, 190566x (1995).

74. Nishio, T, Iida, A, Maruki, E and Inukai, H Japanese Patent 06 248 225, September 6, 1994; *Chem. Abstr.* **122**, 190508e (1995).

75. Moyle, RT and Soltwedel, JN US Patent 5 178 915, January 12, 1993; *Chem. Abstr.* **118**, 215050g (1993).

76. Myazaki, N, Takayanagi, T and Uchino, B Japanese Patent 07 179 808, July 18, 1995; *Chem. Abstr.* **123**, 317119g (1995).

77. Nishio, T, Koyama, M, Marumoto, E and Iida, A Japanese Patent 04 346 873, December 2, 1992; *Chem. Abstr.* **119**, 74695c (1993).

78. Tsukito, H, Iwatsuka, J and Takahashi, H Japanese Patent 06 270 107, September 27, 1994; *Chem. Abstr.* **122**, 84049w (1995).

79. Shibamoto, K and Nakano, M Japanese Patent 07 133 448, May 23, 1995; *Chem. Abstr.* **123**, 317068q (1995).

80. Miyazaki, N, Kodama, S-I, Takayanagi, T and Uchino, B PCT Intl. Appl. WO 95 02 645, January 26, 1995; *Chem. Abstr.* **122**, 293587j (1995).

81. Nakama, S, Sakurai, M, Kuwabara, I and Nakao, K Japanese Patent 05 331 407, December 14, 1993; *Chem. Abstr.* **121**, 182115p (1994).

82. Takizuka, N and Enokida, Y, Japanese Patent 05 51 556, April 6, 1993; *Chem. Abstr.* **119**, 98122q (1993).

83. Takemoto, A and Sukejima, H Japanese Patent 04 15 276, January 20, 1992; *Chem. Abstr.* **116**, 237462d (1992).

84. Maruyama, T and Watanabe, S Europ. Pat. Appl. EP 416 501, March 13, 1991; *Chem. Abstr.* **115**, 116498r (1991).

85. Yokota, M, Myazaki, N, Miura, R, Okuno, H, Gomi, H and Abe, J Japanese Patent 07 118 597, May 9, 1995; *Chem. Abstr.* **123**, 342060p (1995).

86. Washida, H, Miura, R, and Kodama, S Japanese Patent 07 118 598, May 9, 1995; *Chem. Abstr.* **123**, 260025s (1995).

87. Nakao, I Japanese Patent 06 248 231, September 6, 1994; *Chem. Abstr.* **122**, 163697j (1995).

88. Hirashima, Y, Maeda, K and Tsutsumi, K Japanese Patent 07 76 667, March 20, 1995; *Chem. Abstr.* **123**, 289869f (1995).

89. Marumoto, E, Iida, A and Nishio, T Japanese Patent 06 122 730, May 6, 1994; *Chem. Abstr.* **121**, 159498g (1994).

90. Inukai, H, Nishio, T, Marumoto, E and Iida, A Japanese Patent 06 322 053, November 22, 1994; *Chem. Abstr.* **122**, 315425j (1995).

91. Brady, Jr, R F *Polym. Mater. Sci. Eng.* **74**, 118–19 (1996).

92. Brady, Jr, R F, Griffith, J R, Love, K S and Field, D E *J. Coatings Technology* **59** (755), 113–19 (1987).

93. Brady, Jr, R F *Pitture e Vernici Europe* **1995** (6), 13–18 (1995).

94. Brady, Jr, R F *Europ. Coatings J.* **31**, 267–9 (1995).

95. Brady, Jr, R F, Griffith, J R, Love, K S and Field, D E in *Polymers in a Marine Environment*, D. Goring, Ed, pp. 191–5 (The Institute of Marine Engineers, London, 1989).

96. Brady, Jr, R F, Griffith, J R and Thomas, R V *Navy Civil Engineer* **32** (2), 23–5 (1993).

97. Griffith, J R and Brady, Jr, R F *Chemtech* **19** (6), 370–73 (1989).

98. Kashiwada, S, Okamoto, N, Wakimoto, M, Tanura, K and Haruta N Japanese Patent 06 93 225, April 5, 1994; *Chem. Abstr.* **121**, 136282t (1994).

99. Takagi, T and Kitamura, T Japanese Patent 06 145 585, May 24, 1994; *Chem. Abstr.* **122**, 12217b (1995).

100. Mori, H, Mitsuhata, H, Tano, K and Shimizu, Y Japanese Patent 06 184 243, July 5, 1994; *Chem. Abstr.* **122**, 12243g (1995).

101. Schlipf, M and Merten, G *Vortr. Poster — Symp. Materialforsch.* **3**, 2465-7 (1991); *Chem. Abstr.* **120**, 137173z (1994).
102. Horibatake, T, Ishikawa, S, and Kasai, K Japanese Patent 07 18 214, January 20, 1995; *Chem. Abstr.* **123**, 146862n (1995).
103. Shimizu, Y, Mohri, H, Wada, S and Saito, H Europ. Pat. Appl. EP 483 750, May 6, 1992; *Chem. Abstr.* **117**, 133010j (1992).
104. Saitoh, H, Shimizu, Y, Oka, M, Chida, A and Sakakura, A Eur. Pat. Appl. EP 458 290, November 27, 1991; *Chem. Abstr.* **116**, 153942x (1992).
105. Koyama, M and Oodera, A Japanese Patent 05 86 321, April 6, 1993; *Chem. Abstr.* **119**, 98135w (1993).
106. Tanaka, H, Toyoshima, S, Hamada, K and Ishikawa, N Japanese Patent 04 317 776, November 9, 1992; *Chem. Abstr.* **118**, 236051y (1993).
107. Okada, K, Tanaka, H, Ishikawa, N and Sugiyama, M Japanese Patent 04 258 613, September 14, 1992; *Chem. Abstr.* **118**, 193724a (1993).
108. Nishio, T, Marumoto, E, Iida, A and Inukai, H Japanese Patent 07 102 211, April 18, 1995; *Chem. Abstr.* **123**, 260027u (1995).
109. Yokoshima, M Japanese Patent 04 36 326, February 6, 1992; *Chem. Abstr.* **117**, 50851h (1992).
110. Yang, S, Xiao, H X, Chen, W P, Kresta, J, Frisch, K C and Highley, D P *Prog. Rubber Plast. Technol.* **7**, 163-73 (1991).
111. Okazaki, H, Fujii, S and Tonomura, S Japanese Patent 04 131 165, May 1, 1992; *Chem. Abstr.* **117**, 214655a (1992).
112. Kodama, S, Yamauchi, M, Hirono, T and Kitahata, H Japanese Patent 07 179 809, July 18, 1995; *Chem. Abstr.* **123**, 316350p (1995).
113. Tamaki, Y and Murakawa, A Japanese Patent 06 145 598, May 24, 1994; *Chem. Abstr.* **122**, 12215z (1995).
114. Sugimoto, K and Saka, K Japanese Patent 05 186 565, July 27, 1993; *Chem. Abstr.* **119**, 205614a (1993).
115. Sagawa, C, Katagiri, M and Kurashina, T US Patent 5 223 562, June 29, 1993; *Chem. Abstr.* **120**, 137272f (1994).
116. Yasumura, T, Kobayashi, S and Komoriya, H Europ. Pat. Appl. EP 556 729, August 25, 1993; *Chem. Abstr.* **120**, 109596w (1994).
117. Matsuo, H, Tamura, M and Ito, K Japanese Patent 62 64 883, March 23, 1987; *Chem. Abstr.* **107**, 156314u (1987).
118. Wehowsky, F, Kleber, R and Jaeckel, L German Patent DE 3 540 147, May 14, 1987; *Chem. Abstr.* **107**, 177996h (1987).
119. Tatsu, H, Kumamoto, S and Enokida, T Japanese Patent 62 184 087; August 12, 1987; *Chem. Abstr.* **108**, 114168q (1988).
120. Luedemann, S, Bernheim, M, Roessler, E and Mosch, F Europ. Pat. Appl. EP 467 083, January 22, 1992; *Chem. Abstr.* **116**, 153791 (1992).
121. Luedemann, S, Bernheim, M and Roessler, E German Patent DE 4 022 443, January 16, 1992; *Chem. Abstr.* **116**, 153087 (1992).
122. Coppens, D M and Allewaert, K E A, Europ. Pat. Appl. EP 648 890, April 19, 1995; *Chem. Abstr.* **123**, 172559p (1995).
123. Suzuki, Y Japanese Patent 04 53 842, February 21, 1992; *Chem. Abstr.* **117**, 50287d (1992).
124. Shinjo, M, Okamoto, S, Katakura, Y and Takubo, S Europ. Pat. Appl. EP 271 054, June 15, 1988; *Chem. Abstr.* **109**, 172089w (1988).
125. Luedemann, S, Bernheim, M, Sander, B, Roessler, E and Vogel, H B Europ. Pat. Appl. EP 325 918, August 2, 1989; *Chem. Abstr.* **112**, 79351c (1990).
126. Luedemann, S, Bernheim, M and Sander, B German Patent DE 3 802 633, August 3, 1989; *Chem. Abstr.* **112**, 38084z (1990).

127. Namaki, R and Enomoto M Japanese Patent 04 163 373, June 8, 1992; *Chem. Abstr.* **117**, 214457n (1992).
128. Furuta, T, Kamemaru, K and Nakagawa, K Japanese Patent 04 194 082, July 14, 1992; *Chem. Abstr.* **118**, 23709g (1993).
129. Furuta, T, Kamemaru, K and Nakagawa, K Japanese Patent 04 146 275, May 20, 1992; *Chem. Abstr.* **117**, 214493w (1992).
130. Yokota, M, Myazaki, N and Kamata, T Japanese Patent 05 321 151, December 7, 1993; *Chem. Abstr.* **121**, 11685v (1994).
131. Enomoto, M and Ogasawara, K Japanese Patent 06 02 278, January 11, 1994; *Chem. Abstr.* **121**, 37607w (1994).
132. Shioya, G, Ito, K and Kamata, T Japanese Patent 03 265 700, November 26, 1991; *Chem. Abstr.* **116**, 237780f (1992).
133. Akazawa, T Japanese Patent 06 146 175, May 27, 1994; *Chem. Abstr.* **121**, 257492v (1994).
134. Hanada, K, Misaizu, I, Kashiwamura, M, Goto, T and Kuriyama, K Japanese Patent 01 24 823, January 26, 1989; *Chem. Abstr.* **111**, 155405v (1989).
135. Karydas, A and Rodgers, J US Patent 5 057 377, October 15, 1991; *Chem. Abstr.* **116**, 60257n (1992).
136. Hanada, K, Misaizu, I, Kashiwamura, M, Goto, T and Kuriyama, K Japanese Patent 01 43 520, February 15, 1989; *Chem. Abstr.* **111**, 116498h (1989).
137. Lan, Y, Xu, Y, Xu, X and Ge, W *Pige Keji* **6**, 20–2 (1988); *Chem. Abstr.* **110**, 175179 (1989).
138. Roettger, J, Passon, K H, Schroer, W D and Kortmann, W Europ. Pat. Appl. EP 452 774, October 23, 1991; *Chem. Abstr.* **116**, 42239d (1992).
139. Shioya, G and Kamata, T Japanese Patent 06 41 520, February 15, 1994; *Chem. Abstr.* **121**, 85831m (1994).
140. Goldenberg, M US Patent 4 929 692, May 29, 1990; *Chem. Abstr.* **113**, 232273w (1990).
141. Hogi, T and Morita, N Japanese Patent 04 50 814, February 19, 1992; *Chem. Abstr.* **117**, 14460w (1992).
142. Mueller, K F and Plankl, W L Europ. Pat. Appl. EP 406 161, January 2, 1991; *Chem. Abstr.* **115**, 214919y (1991).
143. Sakagami, T, Machida, K and Kokubu, K Japanese Patent 63 253 918, October 20, 1988; *Chem. Abstr.* **111**, 180766w (1989).
144. Reiners, J, Winkel, J, Klauke, E Sueling, C and Podszun, W German Patent DE 3 516 257, November 13, 1986; *Chem. Abstr.* **106**, 90233n (1987).
145. Antonucci, J M, Stansbury, J W and Venz, S *Polym. Mater. Sci. Eng.* **59**, 388–96 (1988).
146. Fock, J, Hahn, G and Wagenknecht, G German Patent DE 3 841 617, May 10, 1990; *Chem. Abstr.* **114**, 69106d (1991).
147. Matsuda, T, Takakura, T and Itoh, T Europ. Pat. Appl. EP 332 405, September 13, 1989; *Chem. Abstr.* **112**, 240560g (1990).
148. Tokunaga, E, Sasagawa, S, Miyamoto, M, Takakura, T and Kojima, G Japanese Patent 61 221 123, October 1, 1986; *Chem. Abstr.* **106**, 81228n (1987).
149. Yoshioka, Y, Koyanagi, H and Tsutsui, N *Jinko Zoki* **17**, 500–3 (1988); *Chem. Abstr.* **110**, 44895q (1989).
150. Tanny, G B, Keningsberg, Y and Shchori, E US Patent 5 126 189, June 30, 1992; *Chem. Abstr.* **117**, 193311t (1992).
151. Tanny, G B, Keningsberg, Y and Shchori, E Europ. Pat. Appl. EP 216 622, April 1, 1987; *Chem. Abstr.* **107**, 238151c (1987).
152. Takakura, T, Kato, M and Yamabe, M *Makromol. Chem.* **191**, 625–32 (1990).

153. Kinaga, Y, Seko, K and Murofushi, S Japanese Patent 01 11 116, January 13, 1989; *Chem. Abstr.* **111**, 99069z (1989).
154. Barraud, J Y, Gervat, S, Ratovelomanana, V, Boutevin, B, Parisi, J P, Cahuzac, A and Jocteur, R Europ. Pat. Appl EP 565 425, October 13, 1993; *Chem. Abstr.* **120**, 114606d (1994).
155. Yokoshima, M Japanese Patent 03 277 618, December 9, 1991; *Chem. Abstr.* **116**, 175712z (1992).
156. Mishima, T, Nishimoto, H and Yamanaka, H Japanese Patent 05 222 136, August 31, 1993; *Chem. Abstr* **120**, 136581u (1994).
157. Barraud, J Y, Gervat, S, Ratovelomanana, V, Boutevin, B and Jocteur, R French Patent FR 2 712 291, May 19, 1995; *Chem. Abstr.* **123**, 288213a (1995).
158. Taniguchi, N and Yokoshima, M Japanese Patent 07 138 342, May 30, 1995; *Chem. Abstr.* **123**, 289910n (1995).
159. Menough, J *Rubber World*, 9–20 (January 1989).
160. Koike, N and Sato, S Japanese Patent 06 234 923, August 23, 1994; *Chem. Abstr.* **122**, 163229h (1995).
161. Maruyama, T and Nakamoto, M Japanese Patent 03 167 276, July 19, 1991; *Chem. Abstr.* **116**, 23023n (1992).
162. Nakagome, S and Sasaki, Y Japanese Patent 02 182 712, July 17, 1990; *Chem. Abstr.* **114**, 165106z (1991).
163. Baum, K, Archibald, T G and Malik, A A US Patent 5 204 441, April 20, 1993; *Chem. Abstr.* **119**, 140994c (1993).
164. Tonelli, C, Trombetta, T, Scicchitano, M, Simeone, G and Ajroldi, G *J. Appl. Polym. Sci.* **59**, 311–27 (1996).
165. Tanaka, Y Japanese Patent 05 329 439, December 14, 1993; *Chem. Abstr.* **122**, 268238p (1995).
166. Cozzi, E, Guidetti, V and Palazzi, S, Europ. Pat. Appl. EP 533 159, March 24, 1993; *Chem. Abstr.* **119**, 162521y (1993).
167. Honda, T and Kaetsu, I Japanese Patent 06 172 675, June 21, 1994; *Chem. Abstr.* **122**, 33660f (1995).
168. Hougham, G G, Shaw, J M and Viehbeck, A Europ. Pat. Appl. EP 594 947, May 4, 1994; *Chem. Abstr.* **121**, 301653z (1994).
169. Iwaya, Y and Ikeda, T Japanese Patent 04 198 217, July 17, 1992; *Chem. Abstr.* **118**, 104303x (1993).
170. Occhiello, E, Re, A, Malatesta, V and Garbassi, F Europ. Pat. Appl. EP 315 152, May 10, 1989; *Chem. Abstr.* **112**, 58283s (1990).
171. Davletbaeva, I M, Rakhmatullina, A S, Kuzaev, A I, Tselikova, E P, Mikhalkin, VI and Berdnikov, VI Russian Patent RU 2 028 317, February 9, 1995; *Chem. Abstr.* **124**, 9712v (1996).
172. Yonemura, A, Sekiya, M and Chiba, K Japanese Patent 02 301 040, December 13, 1990; *Chem. Abstr.* **114**, 230815w (1991).
173. Hanada, K Japanese Patent 62 252 414, November 4, 1987; *Chem. Abstr.* **108**, 222607r (1988).
174. Anon. (to BASF A.-G.) Japanese Patent 03 152 115, June 28, 1991; *Chem. Abstr.* **115**, 137928w (1991).
175. Anon. (to BASF A.-G.) Japanese Patent 03 152 116, June 28, 1991; *Chem. Abstr.* **115**, 137927v (1991).
176. Nishiyama, H and Yanagida, T Japanese Patent 03 157 413, July 5, 1991; *Chem. Abstr.* **115**, 234871e (1991).
177. Hanada, K, Misaizu, I, Kashiwamura, M, Goto, T and Kuriyama, K Japanese Patent 01 11 887, January 17, 1989; *Chem. Abstr.* **111**, 48228t (1989).

178. Chen, J H, Chen, T J and Demejo, L P US Patent 5 233 008, August 3, 1993; *Chem. Abstr.* **119**, 251331d (1993).
179. Anon (to Sanyo Chem. Indus. Ltd), Japanese Patent 02 16 573, January 19, 1990; *Chem. Abstr.* **113**, 14795b (1990).
180. Tsunemi, K Japanese Patent 03 229 261, October 11, 1991; *Chem. Abstr.* **116**, 184593p (1992).
181. Yaegashi, H, Kushida, N, Toma, K Suzuki, T and Hasegawa, T Japanese Patent 63 74 687, April 5, 1988; *Chem. Abstr.* **109**, 83665p (1988).
182. Miyamoto, M Japanese Patent 03 254 978, November 13, 1991; *Chem. Abstr.* **116**, 108510u (1992).
183. Shiraishi, S Japanese Patent 63 221 091, September 14, 1988; *Chem. Abstr.* **110**, 156261m (1989).
184. Zhits, V M and Ratner F Z *Tyazh. Mashinostr.* **1993**, 19–24; *Chem. Abstr.* **121**, 111510c (1994).
185. Kashiwada, S Japanese Patent 04 268 384, September 24, 1992; *Chem. Abstr.* **118**, 215017b (1993).
186. Brady, Jr, R F *Chem. Britain* **26**(5), 427–430 (1990).
187. Hoeller, R and Rudolf, K Europ. Pat. Appl. EP 316 891, May 24, 1989; *Chem. Abstr.* **111**, 137029w (1989).
188. Ogata, K and Ooshima, N Japanese Patent 04 180 027, June 26, 1992; *Chem. Abstr.* **117**, 242896y (1992).
189. Miura, R, Moriwaki, K, Takeyasu, H, Washita, H and Miyazaki, N Europ. Pat. Appl. EP 343 527, November 29, 1989; *Chem. Abstr.* **113**, 25058q (1990).
190. Murata, T and Tanaka, H Japanese Patent 63 202 611, August 22, 1988; *Chem. Abstr.* **110**, 24810e (1989).
191. Yoon, S C, Sung, Y K and Ratner, B D *Macromolecules* **23**, 4351–6 (1990).
192. Amimoto, Y, Yoshida, T and Enomoto, K Japanese Patent 02 92 985, April 3, 1990; *Chem. Abstr.* **113**, 174047n (1990).
193. De Vos, R, Thorpe, D and Van Essche, G Europ. Pat. Appl. EP 605 105, July 6, 1994; *Chem. Abstr.* **122**, 134853b (1995).
194. Cohen, G M, Farnham, W B and Feiring, A E US Patent 5 185 421, February 9, 1993; *Chem. Abstr.* **119**, 9324s (1993).
195. Anon. (to Hokushin Kogyo, Inc.) Japanese Patent 03 181 564, August 7, 1991; *Chem. Abstr.* **115**, 257921k (1991).
196. Yoon, S C and Ratner, B D *Macromolecules* **21**, 2392–400 (1988).
197. Yoon, S C and Ratner, B D *Macromolecules* **21**, 2401–4 (1988).
198. Takamatsu, Y, Niimoto, M and Sato, M Japanese Patent 04 93 317, March 26, 1992; *Chem. Abstr.* **117**, 193685t (1992).
199. Yoshizumi, M and Yamashita, Y Japanese Patent 02 258 821, October 19, 1990; *Chem. Abstr.* **114**, 165158t (1991).
200. Yoshizumi, M and Yamashita Y Japanese Patent 02 258 822, October 19, 1990; *Chem. Abstr.* **114**, 165159u (1991).
201. Tortelli, V and Tonelli, C *J. Fluorine Chem.* **47**, 199 (1990).
202. Ho, T and Wynne, K J *Macromolecules* **25**, 3521–7 (1992).
203. Baum, K, Malik, A A and Tzeng, D D US Patent 4 942 164, July 17, 1990; *Chem. Abstr.* **114**, 7461g (1991).
204. Guo, X A and Hunter, A D *J. Polym. Sci., Polym. Chem. Ed.* **31**, 1431–9 (1993).
205. Kashiwagi, T, Ito, Y and Imanishi, Y *J. Biomater. Sci., Polym. Ed.* **5**, 157–66 (1993).
206. Furukawa, M *J. Appl. Polym. Sci., Appl. Polym. Symp.* **53**, 61–76 (1994).
207. Takaishi, K and Samura, T Japanese Patent 01 245 012, September 29, 1989; *Chem. Abstr.* **112**, 218544h (1990).

208. Harnish, D F, Pickens, D and Zweig A M US Patent 5 004 790, April 2, 1991; *Chem. Abstr.* **115**, 51063d (1991).
209. Hamada, K, Tanabe, K and Matsuo, K Japanese Patent 04 11 670, January 16, 1992; *Chem. Abstr.* **117**, 50894z (1992).
210. Hamel, N N and Gard, G L *J. Fluorine Chem.* **68**, 253-9 (1994).
211. Collet, A, Commeyras, A, Viguier, M and Hirn, B PCT Intl. Appl. WO 9 410 222, May 11, 1994; *Chem. Abstr.* **122**, 161745z (1995).
212. Tamareselvy, K, Venkatarao, K and Kothandaraman, H *J. Polym. Sci., Polym. Chem.* **28**, 2679-93 (1990).
213. Shinjo, M and Hayashi, K Japanese Patent 63 248 827, October 17, 1988; *Chem. Abstr.* **111**, 79198n (1989).
214. Pechhold, E Europ. Pat. Appl. EP 331 307, September 6, 1989; *Chem. Abstr.* **112**, 57028a (1990).
215. Yoon, S C, Ratner, B D, Ivan, B and Kennedy, J P *Macromolecules* **27**, 1548-54 (1994).
216. Yang, S, Xiao, H X, Higley, D P, Kresta, J, Frisch, K C, Farnham, W B and Hung, M H *J. Macromol. Sci., Pure Appl. Chem.* **A30**, 241-52 (1992).
217. Wilson, L M and Griffin, A C *Macromolecules* **27**, 1928-31 (1994).
218. Papadimitrakopoulos, F, Sawa, E and MacKnight, W J *Macromolecules* **25**, 4682-91 (1992).
219. Honeychuck, R V, Ho, T and Wynne, K J *Polym. Mater. Sci. Eng.* **66**, 521-2 (1992).
220. Deflorian, F, Fedrizzi, L, Locaspi, A and Bonora P L *Electrochim. Acta* **38**, 1945-50 (1993).
221. Jayasuriya, A C, Tasaka, S, Shouko, T and Inagaki, N *Polym. J. (Tokyo)* **27**, 122-6 (1995)
222. Wilson, L M and Griffin, A C *Macromolecules* **27**, 4611-14 (1994).
223. Tingey, K G and Andrade, J D *Langmuir* **7**, 2471-8 (1991).
224. Hearn, M J, Briggs, D, Yoon, S C and Ratner, B D *Surf. Interface Anal.* **10**, 384-91 (1987).

7

Adhesion Properties of Fluoropolymers

D. M. BREWIS and I. MATHIESON
Institute of Surface Science and Technology, Loughborough University,
Leicestershire, UK.

1 INTRODUCTION

Adhesion properties of substrates are determined by the chemical nature of the surface, the topography and the cohesive strength of the surface regions. The chemical nature of a surface determines two critical factors in adhesion, namely: the wetting by a liquid component, and the forces of interaction across an interface where wetting has occurred.

The relative surface energies of the solid and liquid determine the degree of wetting. In general, a liquid will exhibit a zero contact angle, i.e. will wet a substrate, when the surface energy of the solid is greater than the surface tension of the liquid. Surface energies are determined by the nature of chemical groups (see Table 7.1). The chemical groups also determine the type of interactions

Table 7.1. Surface energies and polar and dispersion components of surface energy of some polymers[2]

Polymer	Chemical structure of monomer compared with ethylene	γ_s^d (mN/m)	γ_s^p (mN/m)	γ_s (mN/m)
Polytetrafluoroethylene	4H replaced by F	18.6	0.5	19.1
Polytrifluoroethylene	3H replaced by F	19.9	4.0	23.9
Poly(vinylidene fluoride)	2H replaced by F	23.2	7.1	30.3
Poly(vinyl fluoride)	1H replaced by F	31.3	5.4	36.7
Low-density polyethylene	—	33.2	—	33.2

Key: γ_s^d Dispersion component of surface energy; γ_s^p polar component of surface energy; γ_s total surface energy.

Modern Fluoropolymers. Edited by John Scheirs
© 1997 John Wiley & Sons Ltd

across an interface. These range from chemical bonds to various types of van der Waals' interactions and hydrogen bonding [1].

Topography can also influence the bondability of a substrate. In certain cases increasing the roughness can increase the area of contact of an adhesive and thereby increase the bond strength; contact will be favoured if the viscosity of the adhesive is low and it hardens slowly. In addition, with the correct topography there is the potential for mechanical anchoring. However, if the topography involves deep narrow crevices, then the degree of wetting achieved may be low; this will be especially true if the viscosity of the adhesive is high.

Finally, the cohesive strength of surface regions can have a large influence on bond strength. Regions of low-molecular-weight material and/or contamination can give rise to weak boundary layers (WBL) when joined [3].

Materials lacking any of the requirements for good adhesion described above will often need a pretreatment prior to bonding, painting or printing.

2   ADHESION OF FLUOROPOLYMERS

The 'non-stick' nature of polytetrafluoroethylene (PTFE) can be explained by the factors described above; namely, lack of chemical functionality resulting in poor wetting and weak interfacial interactions and regions of low cohesive strength at the surface [3]. On removal of weak layers, and without any chemical modification, a moderate improvement in adhesion can be achieved [4]. For good adhesion, however, chemical modification of the PTFE appears necessary.

Partially fluorinated materials like poly(vinyl fluoride) (PVF) and poly(vinylidene fluoride) (PVDF) have higher surface energies than PTFE. They also possess a relatively high polar contribution to surface energy compared with, say, polyethylene or PTFE (see Table 7.1). As such, PVF and PVDF might be expected to have good adhesion with certain adhesives where there is the potential for polar interactions across the interface. This has been found to be the case with some batches of PVF [5]. In other cases poor adhesion to PVF has been reported with the same adhesive[4]. This has been shown to be due to a region of low cohesive strength at the surface [4].

3   PRETREATMENTS

Effective pretreatments for PTFE were developed in the 1950s. These were sodium in liquid ammonia [6] and sodium naphthalenide in tetrahydrofuran (THF) [7]. Many other treatments for PTFE have since been investigated including plasma treatment [8], direct electrochemical reduction [9], treatment with an alkali metal amalgam [10] and reduction with benzoin dianion [11]. However, none of these has proved as effective as the original two methods which remain the main commercial pretreatments.

The sodium complex methods are also effective with other fluoropolymers. For example, large increases in adhesion have been observed with ethylene–chlorotrifluoroethylene copolymer (ECTFE), PVF and PVDF [12]. It was found, however, that the treatment rate with these polymers was much slower than with PTFE.

Although the sodium in liquid ammonia and sodium naphthalenide treatments are very effective, especially with PTFE, they have undesirable environmental features. There is clearly scope for more environmentally friendly alternatives to the two methods used commercially. In the last few years interest has centred on plasma treatment, but other work has included the treatment of partially fluorinated polymers with flames and Group(I) hydroxides. The various pretreatments for fluoropolymers are now discussed; only those which are commercially reasonable will be dealt with.

3.1 SODIUM COMPLEXES

The treatment of fully fluorinated polymers such as PTFE with sodium in liquid ammonia or sodium naphthalenide in THF is very rapid. In the 1970s the changes in surface chemistry caused by these pretreatments were studied using X-ray photoelectron spectroscopy (XPS). Brecht, Mayer and Binder[13] showed that treatment of PTFE with sodium naphthalenide for 30 s reduced the F/C ratio in the surface from 2 to 0.17 and introduced substantial quantities of oxygen (O/C ratio = 0.2). Dwight and Riggs [14] treated fluorinated ethylene–propylene copolymer (FEP) with sodium in liquid ammonia. Complete defluorination took place and large quantities of oxygen were introduced into the surface including carbonyl and carboxylic acid groups. Extensive roughening of the PTFE surface was reported.

The changes in surface chemistry with partially fluorinated polymers are less pronounced after a given time, as can be seen from the results in Table 7.2.

The details of the bond test are described elsewhere [12].

3.2 USE OF PLASMAS

Although flames and corona discharges consist of plasma the term 'plasma treatment' usually refers to a process carried out at reduced pressure, typically 1 Torr. Much research was carried out in the 1960s and 1970s on the use of plasma to pretreat fluoropolymers and other polymers [15–19] and has continued to the present date. Some recent studies are described in references 20–32. The joint strengths obtained after plasma treatment of PTFE are usually much lower than obtained with the sodium complex treatments; typical results are given in Table 7.3. However, plasma treatment of partially fluorinated polymers is much more effective in terms of bondability and the chemical changes are much more pronounced than with PTFE (Table 7.3). This is of course the opposite to the changes caused by the sodium naphthalenide treatment (Table 7.2).

Table 7.2.   Effect of 'Tetra Etch'[a] treatment on the surface compositions of PTFE, PVF, PVdF and ECTFE[b] and the failure loads of composite lap shear joints involving these polymers with an epoxide adhesive[12]

Polymer		Colour	Surface compositions by XPS (atom %)				Failure load (N)
			C	Cl	F	O	
PTFE	none	White	38.4	—	61.6	—	420
	10 s	Brown	87.6	—	0.8	11.6	4280
	1 min	Black	82.2	—	0.9	16.9	4260
PVF	none	Colourless	70.4	—	28.8	0.8	360
	10 s	Colourless	72.4	—	26.7	0.9	800
	1 min	Colourless	75.4	—	23.0	1.6	2080
	60 min	Colourless	87.3	—	11.4	1.3	3020
PVDF	none	Colourless	51.4	—	47.9	0.7	1580[c]
	1 min	Faint brown	77.4	—	12.9	9.7	2450[c]
	60 min	Faint brown	79.5	—	9.2	11.3	2940[c]
ECTFE	none	Cream	53.2	14.3	32.5	—	240
	1 min	Cream	72.5	3.7	17.7	6.0	3300

[a]'Tetra-Etch' is a product of W Gore & Associates. It is a sodium complex in organic ether; it produces similar results to sodium naphthalenide in THF.
[b]ECTFE is a copolymer of ethylene and chlorotrifluoroethylene.
[c]A cyanoacrylate adhesive was used for these tests.

Table 7.3.   Effects of plasma treatment on the surface composition and failure loads of composite lap shear tests using PVF and PTFE with an epoxide adhesive[12]

Polymer	Plasma and treatment time	Surface composition by XPS (atom %)			Failure load (N)
		C	O	F	
PVF	None	70.4	0.8	28.8	360
	Ar 1 m	71.3	11.0	17.7	4060
	Ar 30 m	78.0	11.9	10.1	4540
	O_2 1 m	66.3	12.4	21.3	3420
	Air 1 m	66.8	8.0	25.2	3080
	N_2 1 m	66.5	5.8	27.7	2720
PTFE	None	34.6	—	65.4	420
	Ar 1 m	42.1	4.5	53.4	1340
	Ar 5 m	33.1	1.6	65.3	1400
	Ar 30 m	33.5	0.9	65.6	1860
	Ar 60 m	32.8	0.2	67.0	1660
	O_2 1 m	34.4	—	65.6	1080
	O_2 10 m	33.6	0.2	66.2	1440
	O_2 30 m	34.0	0.5	65.5	640

Table 7.4. Effect of flame treatment of PVF, PTFE and ECTFE on surface composition and adhesion in a composite lap shear test with epoxide adhesive[12]

Treatment (ss)	Surface composition by XPS (atom%)				Failure load (N)
	C	Cl	F	O	
PVF none	70.4	—	28.8	0.8	360
PVF 0.06	67.6	—	28.0	4.4	3240
PTFE none	38.4	—	61.6	—	420
PTFE 0.04	34.0	—	66.0	—	80
ECTFE none	53.2	14.3	32.5	—	240
ECTFE 0.06	68.8	8.0	17.2	6.0	2980

3.3 FLAME

The flame treatment has been widely used commercially for many years to enhance adhesion to polyolefins [33–35]. It is ineffective with PTFE but with partially fluorinated polymers such as PVF and ECTFE it is very effective (see Table 7.4).

3.4 USE OF GROUP (I) HYDROXIDES

Brewis et al. [36] carried out a detailed study on the effect of Group (I) hydroxides, especially potassium hydroxide, on the bondability and surface chemistry of PVF and PVDF. Some of the results for PVF are summarised in Tables 7.5 and 7.6 and Figure 7.1 and those for PVdF in Table 7.7. It can be seen that alcoholic

Table 7.5. Effect of treatment time on surface compositions and adhesion levels of PVF treated with aqueous and alcoholic KOH solutions (5M, 80°C)[36]

Time	Colour	Failure load(N)	Surface composition by XPS (atom%)			
			C	F	O	K
Aqueous						
0	None	350	69.5	29.4	1.0	—
1 min	None	440	71.2	27.9	0.9	—
10 min	None	450	69.6	28.0	2.4	—
30 min	None	430	71.6	23.7	4.7	—
1 h	None	580	73.1	20.2	6.7	—
Alcoholic						
0	None	350	69.5	29.4	1.0	—
10 s	None	500	74.5	24.9	0.6	—
30 s	None	530	68.6	29.6	1.8	—
1 min	Faint brown	890	69.2	28.7	1.1	1.2
10 min	Faint brown	3020	69.1	28.0	2.1	0.8
30 min	Light brown	3600	69.2	23.8	4.8	2.2
1 h	Brown	3490	70.6	21.8	6.2	1.4

Table 7.6. Effect of temperature on the treatment of PVF using 5M aqueous KOH + 0.15 g TBAB per 50 ml of solution[36]

Temp (°C)	Time	Failure load (N)
50	10 s	410
	1 min	670
	1 h	450
80	10 s	430
	30 s	310
	1 min	320
	1 h	4220 MF[a]
100	30 s	890
	1 min	2060
	1 h	4470 MF
110	10 s	650
(boiling)	30 s	3280
	1 min	3980 MF
	1 h	4350 MF

[a]MF = material failure in PVF.

Table 7.7. The effect on adhesion level and surface composition for various KOH treatments of PVDF[36]

Treatment	Colour	Failure load (N)	Surface composition by XPS (atom%)		
			C	F	O
None	None	1300	51.0	49.0	—
Aq KOH (5M 80°C)					
10 s	None	4120			
30 secs	None	4040	52.7	45.7	1.6
1 min	faint br	4010	53.9	43.1	3.0
10 min	faint br	4560	58.0	37.0	5.0
1 hr	light br	4320	62.0	28.9	9.1
Aq KOH + TBAB 0.15 g[a] *(5M 80°C)*					
10 secs	Light br[b]	4250	62.0	28.4	9.6
30 secs	Light br	3940	63.1	24.4	12.5
1 min	Brown	4860	69.7	16.7	13.6
10 min	Brown	4430	74.6	9.9	15.5
1 hr	Dark br	4740	77.7	4.3	16.8[c]
Alc KOH (5M 80°C)					
10 secs	Light br	4150	59.3	34.7	6.0
1 min	Brown	4630	62.2	27.5	10.3
5 mins	Dark br	4340	65.7	21.7	12.6
10 min	Black	4150	67.0	18.7	14.3
1 hour	Black	3230	74.0	8.9	17.1

[a]TBAB per 50 ml of solution.
[b]br = Brown.
[c]0.7 atom% of N and 0.5 atom% of Na were also present.

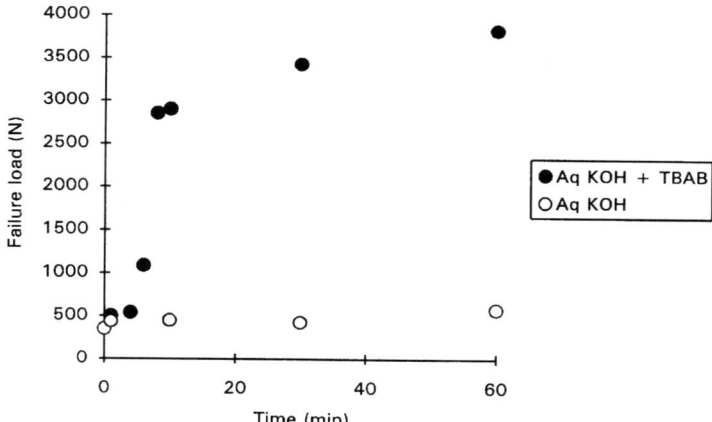

Figure 7.1. Bond strength vs treatment time for PVF treated with 5M aqueous KOH and 5M aqueous KOH + 0.03 g of tetrabutylammonium bromide (TBAB) per 50 ml of KOH solution at 80 °C [Reprinted from International Journal of Adhesion & Adhesives *16*, Brewis *et al.* 87, copyright (1996) with kind permission from Elsevier Science Ltd, The Boulevard, Langford Lane, Kidlington OXS IGB, UK 36]

KOH is much more reactive than aqueous KOH towards PVF and PVDF, and of these two polymers the latter is much more reactive. The addition of a small amount of TBAB to the aqueous KOH solution greatly enhances its reactivity. Group (I) hydroxides are ineffective with PTFE

4 CONCLUSIONS

To utilise the special properties of fluoropolymers it is often necessary to bond them to another substrate. To achieve satisfactory adhesion with fully fluorinated polymers such as PTFE it is necessary to pretreat them. Provided partially fluorinated polymers do not have a cohesively weak surface layer they should be easy to bond. However, they often possess such a layer and pretreatment will be necessary. There is a marked contrast in the effectiveness of different pretreatments between fully fluorinated polymers on the one hand and partially fluorinated polymers on the other. The sodium complex methods are particularly effective with fully fluorinated polymers, whereas partially fluorinated polymers may be effectively treated with plasmas, flames and Group (I) hydroxides. Plasmas are reasonably effective with PTFE, but flames and Group (I) hydroxides are ineffective with PTFE.

REFERENCES

1. R J Good, *J Adh. Sci. Technol.*, **6**(12) (1992) 1269.
2. D G Rance in *Industrial Adhesion Problems*, ed. D M Brewis and D Briggs, Orbital Press (1985) p. 63.

3. D M Brewis, *Int. J. Adhesion & Adhesives*, **13**(4) (1993) 251.
4. D M Brewis, I Mathieson, I Sutherland and R A Cayless, *J. Adhesion*, **41**(1–4) (1993) 113.
5. D M Brewis, *Progress in Rubber and Plastic Technology*, **1**(4) (1985) 1.
6. R J Purvis and W R Beck US Patent 2, 789, 063 (16 April 1957) to Minnesota Mining and Manufacturing Co.
7. A A Benderly, *J. Appl. Polym. Sci.*, **6**(20) (1962) 221.
8. H Schonhorn. and R H Hansen., *J. Appl. Polym. Sci.*, **11**, (1967) 1461.
9. R H Dahm, D J Barker, D M Brewis. and L R H Hoy, *Adhesion* 4, ed. K W Allen, Applied Science, London (1980) p. 215.
10. J Jansta, F P Dousek. and J Riha, *J. Appl. Polym. Sci.*, **34**, (1975) 3201.
11. C A Costello and R J McCarthy, *Macromolecules*, **17** (1984) 2941.
12. I Mathieson, D M Brewis, I Sutherland and R A Cayless, *J. Adhesion* **46** (1994) 49.
13. V H Brecht, F Mayer and M Binder, *Die Angewandt Makromol. Chem.*, **33** (1973) 89.
14. D W Dwight and W M Riggs, *J. Coll. Int. Sci.*, **47**(3) (1974) 650.
15. J R Hall, C A L Westerdahl, A T Devine and M J Bodnar, *J. App. Polym. Sci.*, **13** (1969) 2085.
16. H Schonhorn and R H Hansen, *J Appl. Polym. Sci.*, **11** (1967) 1461.
17. H Yasuda, *J Macromol. Sci. Chem.*, **A10**(3) (1976) 383.
18. R R Sowell, N J Delollis, H J Gregory and O Montoya, *J. Adhesion*, **4** (1972) 15.
19. B W Malpass and K Bright in *Aspects of Adhesion 5*, R J Alner (ed.) University of London Press (1969) p. 224.
20. T Kasemura, S Ozawa and K Hattori, *J. Adhesion*, **33** (1990) 33.
21. M Morra, E Occhiello and F Garbassi, *Surf. Int. Anal.*, **16** (1990) 412.
22. G P Hansen, R A Rushing, R W Warren, S L Kaplan and O S Kolluri, *Int. J. Adhesion & Adhesives*, **11**(4) (1991) 247.
23. H J Griesser, D Youxian, A E Hughes, T R Gengenbach and A W M Mau, *Langmuir*, **7** (1991) 2484.
24. D Youxian, H J Greisser, A W H Mau, R Schmidt and J Liesegang, *Polymer*, **32**(6) (1991) 1126.
25. T G Vargo, J A Gardella, A E Meyer and R E Baier, *J. Poly. Sci. Part A*, **29** (1991) 555.
26. G L Anderson, D A Dillard and J P Wightman, *J. Adhesion*, **36**(4) (1992) 213.
27. A Kinbara, A Kikuchi, S Baba and T Abe, *J. Adhesion Sci Technol.*, **7**(5) (1993) 457.
28. S L Kaplan, E S Lopata and J Smith, *Surf. Int. Anal*, **20** (1993) 331.
29. M A Golub, E S Lopata and L S Finney, *Langmuir*, **10** (1994) 3629.
30. M K Shi, A Selmani, L Martinu, E Sacher, M R Wertheimer and A Yelon, *J. Adhesion Sci. Technol*, **8**(10) (1994) 1129.
31. T R Gengenbach, X Xue, R C Chatelier and H J Griesser, *J. Adhesion Sci. Technol.*, **8**(4) (1994) 305.
32. M E Ryan and J P S Badyal, *Macromolecules*, **28** (1995) 1377.
33. R L Ayres and D L Shofner, *SPE Journal*, **28**(12) (1972) 51.
34. D Briggs, D M Brewis and M B Konieczko, *J. Mat. Sci.*, **14** (1979) 1344.
35. I Sutherland, D M Brewis, R J Heath and E Sheng, *Surf. Int. Anal.*, **17** (1991) 507.
36. D M Brewis, I Mathieson, I Sutherland, R A Cayless and R H Dahm, *Int. J. Adhesion & Adhesives*, **16** (1996) 87.

8

Hexafluoroisopropylidene-containing Polymers

PATRICK E. CASSIDY and JOHN W. FITCH III
Department of Chemistry, Southwest Texas State University, San Marcos, TX, USA

1 INTRODUCTION

Monomers containing the hexafluoroisopropylidene (**6F**) group have found wide application in the synthesis of high-performance polymers. Many of these monomers are readily available commercial materials, and those which are not can usually be prepared by Friedel-Crafts or Grignard condensation reactions with hexafluoroacetone. The **6F** group functions as a highly stable, bulky, yet flexible, non-conjugating linking group which increases the free volume and reduces the crystallinity in polymers containing it. These polymers show dramatic improvement of properties when compared to non-fluorinated analogues. In general, the presence of the **6F** group in a polymer increases solubility, oxidative and thermal stability, optical transparency, flame resistance and resistance to UV radiation, while decreasing crystallinity, dielectric constant, water absorption and surface energy [1,2].

Numerous applications for polymers containing hexafluoroisopropylidene groups have been suggested, including water and heat-resistant coatings, fibers, adhesives and even dental prostheses. Most of the work has involved polymers derived from several commercially available **6F** monomers. The more common and readily available **6F** monomers are listed in Table 8.1 along with two **6F-OH** monomers which are also commercially available and which contain the 2-hydroxyhexafluoro-2-propyl group. Condensation polymerizations involving the **6F-OH** monomer series have not been widely studied, presumably because of the low reactivity of the somewhat acidic and sterically hindered hydroxyl group in these monomers. However, monomers derived by functionalizing the hydroxyl

Modern Fluoropolymers. Edited by John Scheirs
© 1997 John Wiley & Sons Ltd

Table 8.1. Commercially available 6F monomers

Structure	Name (acronym[a])	Mp or bp (mm-Hg) °C
	2,2-bis(4-aminophenyl)hexafluoro-propane (bis-A-AF)	200–202
	2,2-bis(3-amino-4-methylphenyl)hexafluoro-propane (bis-AT-AF)	105–106
	2,2-bis[4-(4-aminophenoxy)phenyl]hexafluoropropane (bis-AF-A)	163–164
	2,2-bis(3-amino-4-hydroxyphenyl)hexafluoro-propane (bis-AP-AF)	240–243
	2,2-bis(4-hydroxyphenyl)hexa-fluoropropane (bis-AF)	162

Structure	Name	
	2,2-bis(4-carboxyphenyl)hexa-fluoropropane (bis-B-AF)	268–275
	4,4'-(hexafluoroisopropylidene) diphthalic anhydride (6FDA)	241–243
	1,3-bis(2-hydroxyhexafluoro-2-propyl)benzene (1,3-HFAB)	99–100 (20 mm-Hg)
	1,4-bis(2-hydroxyhexafluoro-2-propyl)benzene (1,4-HFAB)	83–85

[a] Acronyms are those adopted by Central Glass Co., Ltd. Tokyo, Japan.

function with more reactive groups (acrylates and epoxides) have received some study [3].

2 POLYETHERS AND POLYETHER KETONES

Aromatic polymers containing the **6F** group, which have been known since 1965, possess good mechanical, thermal and electrical properties. Compound **1**, prepared from 2-(4-phenoxyphenyl)hexafluoro-2-propanol under Friedel-Crafts conditions, was patented in 1965 as a thermally stable polymer which can be compression molded at 330–350 °C and which exhibits excellent mechanical properties [4].

Compound **2**, a thermoplastic also exhibiting excellent thermal and oxidative stability, was prepared by coupling the sodium salt of bisphenol AF with 4,4'-dibromobiphenyl in the presence of a copper(I) salt [5]. The most common preparative route to **6F**-containing aromatic polyethers and polyether ketones involves nucleophilic aromatic substitution of activated dihaloarenes with diphenols. In this way, **1** can also be prepared by the reaction of bisphenol AF with 2,2-bis(4-fluorophenyl)hexafluoropropane in tetramethylene sulfone containing K_2CO_3 [6].

The reaction of bisphenol AF with decafluorobiphenyl to produce **3** is also representative of this synthetic route (Equation 1):

(1)

Compound **3** has excellent solubility and thermal stability (10% weight loss in air at 500 °C), a dielectric constant of 2.17, low water absorption, a tensile strength of 5270 psi and a tensile modulus of 221 600 psi [7–9]. Crosslinkable analogues of **3** are obtained when propargyl bromide is used as an endcap. These polymers crosslink at 200–350 °C to give a low moisture-absorbing (0.15%) material [10]. Bisphenol AF-based polyethers bearing o-nitrobenzyl groups in the main chain are also known and undergo smooth decomposition upon UV irradiation, making them useful as positive photoresists [11].

Sulphone- and 6F-containing polyethers have been prepared by reacting bisphenol AF with dichlorodiphenylsulfone in high boiling, polar, aprotic solvents. The resulting poly(ether-sulfone) is an amorphous thermoplastic with $T_g = 205$ °C [12]. Membranes prepared from the fluorinated polysulfone were found to be both more permeable and more selective than the **6H** analogue toward He, CO_2 and CH_4 [13–15]. Similar results have been obtained in comparisons of **12F** to **12H**–**PEK**s (see below) and polycarbonates derived from bisphenol A and AF [13].

A series of poly(ether ketone)s (**PEK**s) has been prepared in which linking isopropylidene groups were systematically replaced by **6F** groups (Table 8.2) [16]. The most highly fluorinated polymer, **12F**–**PEK**, is a highly soluble, colorless film former. Its dielectric constant is 2.40 (measured at 10 GHz) and it shows only 20% weight loss after 200 h at 350 °C in air. By comparison, Victrex PEEK is insoluble in most solvents, has a dielectric constant of 2.85 and suffers a 96% weight loss after 200 h at 350 °C [16].

A comparison of the properties of **12F**–**PEK** to **12H**–**PEK** shows that both the oxidative stability and tensile strength are considerably reduced in the non-fluorinated analogue (Table 8.2). Of greater interest is the observation that replacement of the **6F** group in the bisphenol AF-derived portion of the polymer (R' in Table 8.2) by a **6H** group causes a dramatic reduction in both oxidative

Table 8.2. PEK Series [16]

Name	Structure	Inherent viscosity (dL/g)	T_g (°C)	TGA (°C) (10% weight loss)		Tensile strength (psi)
				Air	Nitrogen	
12F-PEK	R=R′=C(CF₃)₂	0.71	180	537	552	10600
6H6F-PEK	R=C(CH₃)₂ R′=C(CF₃)₂	0.89	174	553	550	9150
6F6H-PEK	R=C(CF₃)₂ R′=C(CH₃)₂	1.09	172	485	542	7600
12H-PEK	R=R′=C(CH₃)₂	0.73	169	485	536	8200

stability and tensile strength whereas replacement of the other **6F** group (R in Table 8.2) by the **6H** group has very little effect on properties. Thus, the 12F and **6H6F–PEK's** have properties similar to each other and are both superior to the **6F6H** and **12H** analogues. Placement of the hexafluoroisopropylidene group in a polymer backbone is thus concluded to be of importance in determining both the properties and, in this case, the cost of the derived polymer.

A series of fluorinated poly(ether ketone)s was prepared from decafluo-robenzophenone and various bisphenols by solution polycondensation [17]. The resulting polymers are soluble in aprotic solvents and were cast into flexible, creasable films. The T_g's of the polymers ranged from 155 to 223 °C. These polymers showed low to moderate dielectric constants ((2.68–2.98 (measured at 10 KHz and 0% relative humidity (RH)) and 2.79–3.19 (measured at 10 KHz and 60% RH)) and thermal stability in air of 359–442 °C.

3 POLYESTERS

A considerable number of polyesters containing the **6F** group are known, and the location of the group in the polymer backbone can also affect properties in polymers such as **4a–d**.

4a: R = R′ =⸺ C(CF₃)₂ ⸺
4b: R =⸺C(CH₃)₂⸺; R′ =⸺ C(CF₃)₂⸺
4c: R =⸺ C(CF₃)₂ ⸺ ; R′ =⸺C(CH₃)₂⸺
4d: R = R′ =⸺ C(CH₃)₂⸺

Polymers **4a–d** form clear, colorless films from solution in common organic solvents. Thermal stabilities in air (dynamic — 10% weight loss at 440–490 °C; isothermal — 100 h at 350 °C, 7–86% weight loss) were again highest when the **6F** group replaced the **6H** group in the bisphenol rather than in the acid. Polymer **4a**, which has a tensile strength of 6890 psi, had a dielectric constant of 2.34 (measured at 10 GHz) [18]. Polymers containing fluorine in both monomer segments are quite water repellent, having a water contact angle of 95° compared to 85° for a related polyester having only one **6F** group in the repeat unit [19].

The terephthalate and isophthalate polyesters of bisphenol AF, which have been prepared and found to be more soluble than the 6H analogues, are stable in air to 450–470 °C [20,21]. Related bisphenol AF polyesters, **5a–c** have T_g = 400, 120 and 80 °C respectively [22–25].

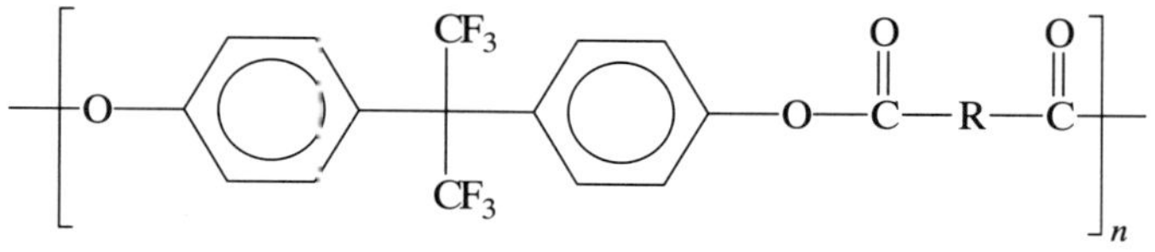

5a: R = biphenylene, **5b**: R = $(CF_2)_4$, **5c**: R = $(CF_2)_8$

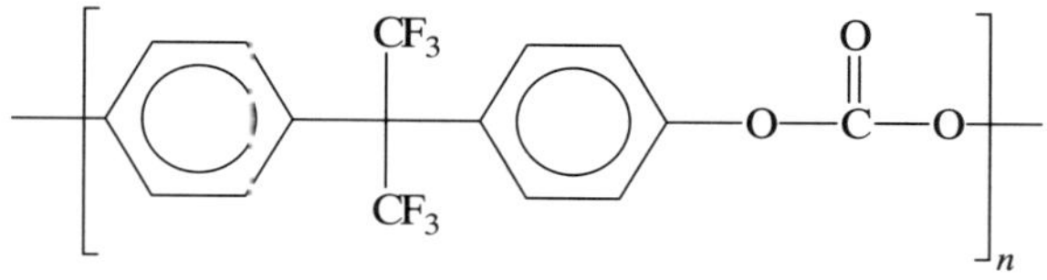

6

The T_g of polycarbonate, **6**, is 154 °C [26] which is slightly higher than that of the **6H** analogue (140 °C), and it has been proposed for use in electrical apparatus for household applications [27].

There have also been a few reports of polyesters derived from 1,3-and 1,4-bis(hexafluoro-2-hydroxy-2-propyl)benzene (**1,3-** and **1,4-HFAB**).

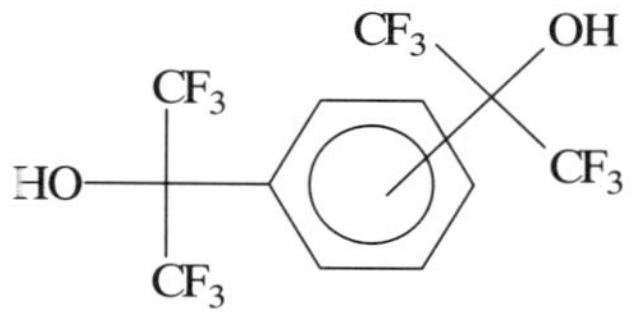

1,3- or 1,4-HFAB

Keller reported in 1984 that **1,3-HFAB** reacts with either adipyl or glutaryl chloride to give, by combining the diol and diacid chloride in non-stoichiometric ratios, oligomeric alcohol-terminated or acid chloride-terminated prepolymers (DP = 4–6). The acid chloride-terminated oligomers could then be converted to low surface energy polyesters which wet the surface and adhered strongly to poly(tetrafluoroethylene) [28]. A series of rather low molecular weight polyarylates derived from **1,3-** and **1,4-HFAB** has been reported [29]. In contrast to hexafluoroisopropylidene-linked polymers, **1,3-** and **1,4-HFAB** polyarylates were found to be very poorly soluble in most organic solvents.

4 POLYACRYLATES

Poly(fluoroacrylate)s can be combined with poly(tetrafluoroethylene) to give tough, hybrid materials with low water absorption which have been suggested for biomedical devices such as artificial human prostheses and dental materials [30]. For example, **7** can be homopolymerized or copolymerized with non-fluorinated acrylic monomers to produce polymers showing promise in dental applications [31,32].

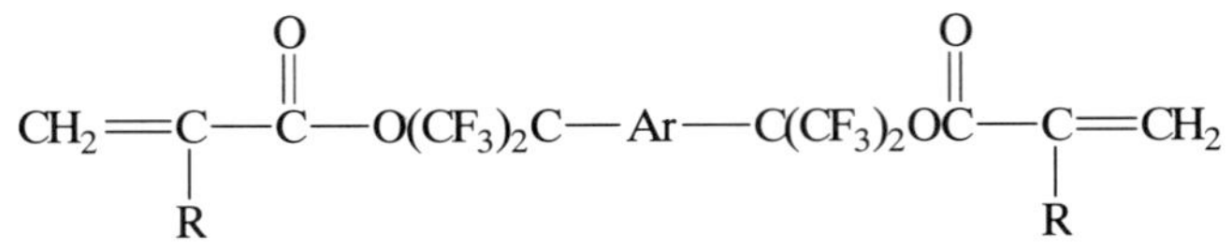

$$CH_2=C-C-O(CF_3)_2C-Ar-C(CF_3)_2OC-C=CH_2$$

R = H, CH₃
Ar = 1,3- or 1,4-R'C₆H₃ (R' = H, F)

7

Additional hydrophobic polymers (water contact angles of 90–91°) have been obtained from monomers **8**.

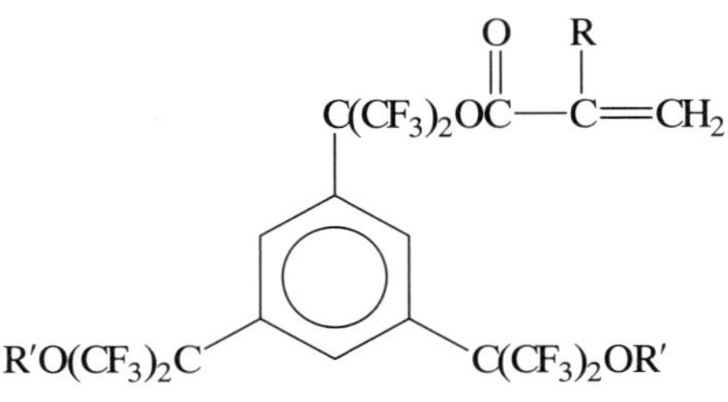

R = H, CH₃
R' = C₁₋₃ alkyl

8

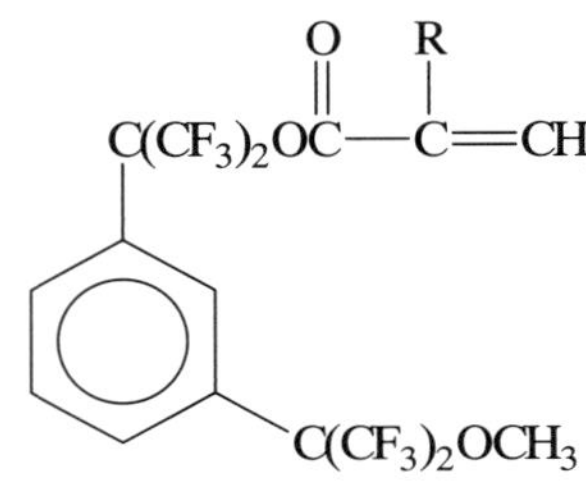

R = H, CH₃

9

Very similar polyacrylates derived from **9** also formed brittle films and had water contact angles as high as 160° [33]. Poly(tetrafluoroethylene), for comparison, has a water contact angle of 108° [34]. It has been suggested by an Electron Spectroscopy for Chemical Analysis (ESCA) study of poly(fluoroalkyl methacrylate)s that fluoroacrylates orient at the interface of a coated surface so that the backbone aligns with the surface and the fluorocarbon tails project into the air so as to minimize the interfacial energy [35].

Extensive additional work on complex poly(fluoroacrylate)s and epoxies derived from **1,3-** and **1,4-HFAB** has been conducted primarily by the research group at the US Naval Research Laboratory Group. The reader is referred to a recent review for additional material [3].

5 POLYIMIDES

Polyimides are, perhaps, the oldest and most developed of the thermally stable polymers. Yet, the introduction of fluorine-containing moieties can improve the properties, such as decreasing color and increasing processability and solubility (tractability).

The usual synthesis of polyimides, condensation of dianhydrides with diamines to produce a polyamic acid which is subsequently cyclodehydrated, is not trivial. The use of 6F-containing monomers allows the final imidization step to be conducted in solution and, therefore, more completely to provide a more perfect backbone. As will be seen in the following text, there are several ways to obviate the imide formation step.

The uses of polyimides covers a wide range; the major application is for aircraft composites, but they are also found in the microelectronics industry due to favourable processing, mechanical and dielectric properties and low water absorption. Therefore, the addition of fluorine to these polymers serves to improve these important properties. Other applications are optical fiber sheaths, optical waveguides, carbon fiber composites [36], gas separation membranes [37] and surface acoustic wave (SAW) sensors [38].

The simplest procedure for polyimide synthesis is to condense 6FDA (hexafluoroisopropylidene diphthalic anhydride) with several diamines to yield high T_g (350 °C), tough, flexible, transparent films [39–46] (see **10** below). Here, T_g's are reported as high as 380 °C and thermal decomposition nearly 600 °C.

10

Of course, the 6F function could be only in the diamine portion of the backbone, the dianhydride being pyromellitic dianhydride, 1,4,5,8-naphthalenetetracarboxylic dianhydride (benzophenonetetracarboxylic dianhydride), or other similar monomers [47–53].

Highly transparent (90%) low dielectric (2.56–2.75 at high frequency, 8–12 GH_2) nearly colorless, film-forming polyimides, **11**, have been reported which contain a dimethyl–silyl link between the aromatic imide rings [54].

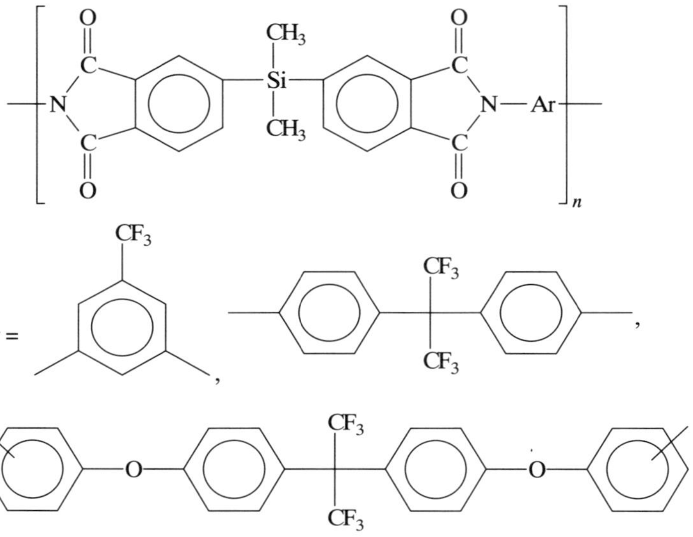

11

Yet another approach to incorporating fluorine into polyimides is to add pendant fluorinated groups. This was accomplished by the synthesis of the diamine, **12**, and reaction of it with several common dianhydrides [55].

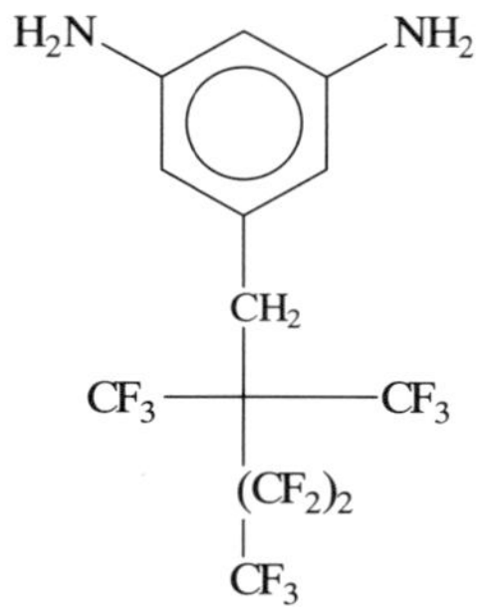

12

The result was film-forming polymers with low dielectric constants (2.3–2.7 at 1 mHz), low water absorption (0.5–1.1% at 85% RH) and quite high (for polyimides) thermal expansion coefficient (67–98 ppm/°C).

The miscibility of fluorine-containing polyimides continues to be of interest in an effort to improve their application properties. Fully miscible alloys are possible if the dianhydride portions are identical, even if the **6F**-containing diamines are of different catenation [56]. Even if the dianhydrides are different, the polyimides are partially miscible with *meta* and *para* **6F** diamines.

A significant effort has been made to incorporate ester, amide and sulfone functions with imides to determine effects of these cofunctions and possible improvements in properties. Ester-imides were synthesized by condensation of diacyl chlorides with dihydroxy-terminated imide monomers, thereby obviating the need to use imidization as the chain-extension step [57,58]. Colorless, solution-cast films were produced with low to moderate dielectric constants (2.8–3.2), thermal stabilities over 400°C and T_g's in the range of 215–272°C for the fully aromatic systems. Comparison of these materials to those non-6F-containing analogues demonstrate again the value of incorporating fluorine.

Poly(imide-amide-sulfone)s were prepared by the condensation of sulfone-containing diamines with imide- and **6F**-containing diacid chlorides [59]. These polymers are soluble in polar, aprotic solvents from which colorless, flexible films could be cast and showed moderate dielectric constants (3.5–3.7), T_g's of 280–360°C and thermal stability in air of 460–480°C.

Finally, a crosslinkable poly(imide-amide) is made possible by providing pendant cyano groups [60]. The cyano functions are contained in the diamine monomer which is condensed with diacid chlorides which contain **6F** and preformed imide moieties. Heating these soluble (in polar aprotic solvents) polymers, after being cast into films, up to 370°C for five hours and finally at 400°C for one hour afforded insoluble products. After crosslinking, the dielectric constants decreased (from 3.5–3.9 to 3.1–3.7) and decomposition temperatures were maintained at 480–490°C. Of course T_g's were not detectable after crosslinking whereas before they were 260–300°C.

6 POLYAMIDES

The common condensation of diamines with diacid chlorides (with **6F** groups in either or both) is the anticipated and successful approach to fluorinated polyamides [61–63]. The **6F** group decreases the rigidity of the backbone and, therefore, the crystallinity usually observed in these types of systems and, thereby, increases solubility. Solution-cast, flexible films were stable to above 450°C in air and give T_g's above 200°C. Films were superior in mechanical properties and had lower water absorption than non-fluorinated analogues [64,65].

The oxygen bridge is introduced to the fluorinated polyamide by the use of an ether- and 6F-containing diamine, thereby adding increased flexibility to the backbone. Diamines of this type were reacted with several diacid chlorides to give fluorinated poly(ether amides) with T_g's from 200 to 256 °C and stabilities in nitrogen from 450 to 470 °C. Solubilities, however, were limited to polar, aprotic solvents [65,66].

The effect of the presence of a single trifluoromethyl (**3F**) group has also been investigated. Three new monomers, 1,1-bis(p-carboxyphenyl)-2,2,2-trifluoroethanol (**3FOH**), 1-methoxy-1,1-bis(p-carboxyphenyl)-2,2,2-trifluoroethane (**3FM**) and 1-chloro-1,1-bis(p-chloroformylphenyl)-2,2,2-trifluoroethane (**3FCl**) were prepared and characterized [67].

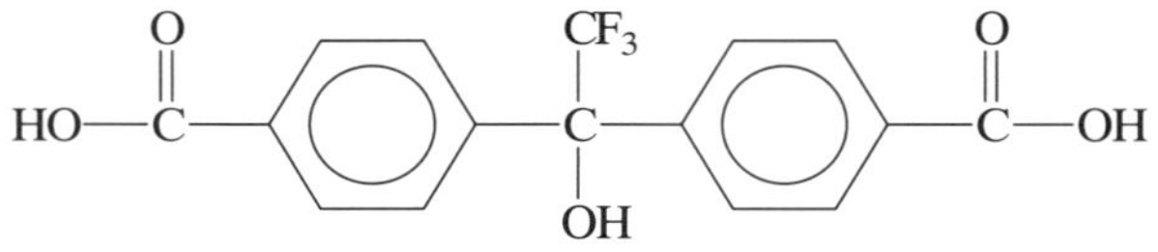

3FOH

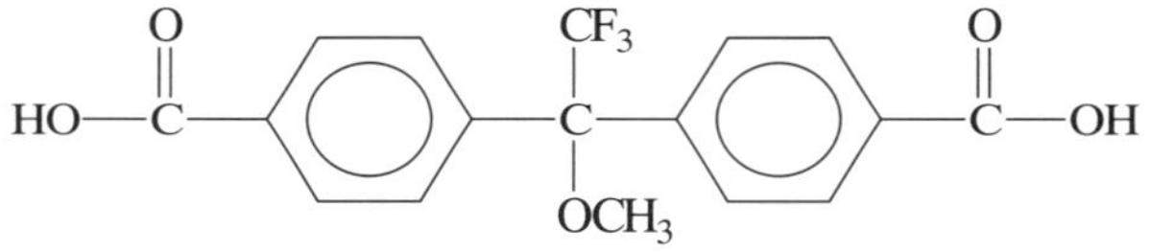

3FM

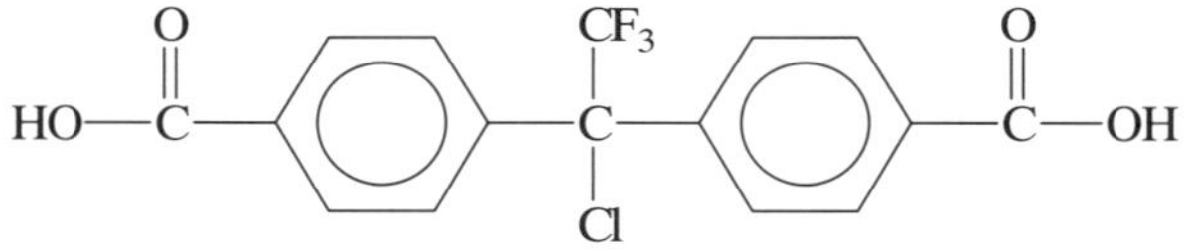

3FCl

A 3FM polyamide series was synthesized by reacting equal molar amounts of the 3FM diacid chloride with various diamines in an NMP/pyridine solvent. A representative polymer of the 3FCl series was obtained by similarly polymerizing the 3FCl diacid chloride with the 4-BDAF diamine. All polymers formed tough, creasable films and were soluble in DMAc, NMP, pyridine and THF.

In order to evaluate the effect of the 3F-methoxy linking group on polyamide properties, the properties of the **3FM** series were compared to an analogous **6F** series. The **3FM** analogues gave a low solution viscosity similar to the **3FCl** polymer. The T_g's of the **3FM** series ranged from 238 to 311 °C and those of the 6F series ranged from 262 to 337 °C. The **3FM** T_g's were lower on average by 29 °C compared to the **6F** series. The 10% weight loss temperatures ranged from 462 to 491 °C and 500 to 518 °C for the **3FM** and **6F** series, respectively. Again, the **3FM** series' weight-loss values were about 32 °C lower, on average, than the **6F** series' values.

Although the weight loss and T_g temperatures of the **3FM** series are consistently lower than weight loss and T_g temperatures for the **6F** series, the thermal stability of the **3FM** series still compares favorably to the **6F** series and is well within the definition of thermally stable polymers.

In an effort to improve the thermal characteristics of polyamides and the processability of polyimides, a fluorine-containing copolymer was synthesized [68]. This was accomplished by the preformation of the imide function by condensing **6FDA** with *p*-aminobenzoic acid giving a diacid-terminated diimide. This diacid was then converted to the diacyl chloride and condensed with numerous aromatic diamines which contained ether, **6F** and sulfone groups between phenylene rings. Thermal stabilities were quite high, above 500 °C in air.

The next step in the investigation of fluorinated imide–amide polymers was to produce *N*-methylated versions to improve thermal stability (up to 470 °C) and decrease color, water absorption (less than 0.2%) and T_g's (around 240 °C) [69].

Ester functions have also been added to amide fluoropolymers to attempt further property improvements [57]. This series, synthesized by the condensation of ester-containing diacid chlorides with diamines, gave T_g's of 214–266 °C, decomposition temperatures in air above 400 °C and dielectric constants of 3.6 to 4.0.

7 MISCELLANEOUS POLYHETEROCYCLES

7.1 POLYOXADIAZOLES

Although the oxadiazole function has shown great promise in polymer backbones, in terms of both ease of synthesis and mechanical and thermal properties, little has been done to take advantage of it. The incorporation of the **6F** moiety, of course, only enhances these properties and very few papers have appeared in this vein [70]. Nearly all of this work, in fact, incorporated other groups also with the oxadiazole, such as imide and ether-ketone. This means that the synthesis can occur through formation of hydrazide or by imide- or ether-forming condensations, all excellent routes to high polymers.

The 1,3,4-oxadiazole moiety

By reaction of the dihydrazide of hexafluoroisopropylidene bis(benzoic acid) with various diacid chlorides, which contain imide functions preformed, and subsequent cyclodehydration, polymers such as **13** are produced [71]. Polymers

13

of this type were stable to dynamic heating above 400 °C and isothermally above 300 °C in air.

The imide group's synthesis can also be used to form analogous systems by condensing the 6F diacid (**6FDA**) with various tetraamines which contain ether and preformed oxadiazole functions [72]. These tough, film-forming materials could be solution-cast to yield high T_g's (*ca.* 250 °C), and thermal stabilities ($\approx$500 °C) and moderately low dielectric constants ($\approx$2.90). Similarly imide-amide-oxadiazole polymers with 6F were derived from the reaction of 6F and imide-containing diacid chlorides with oxadiazole-containing diamines [73].

Finally, ether formation by nucleophilic aromatic substitution is quite useful to condense bisphenol AF with *p*-fluorophenyl-terminated oxadiazole monomers [74]. Although the thermal stability was maintained ($\approx$500 °C), the T_g dropped from earlier examples (176 °C) and the dielectric constant increased (3.09).

7.2 POLYBENZAZOLES

Polybenzoxazoles, polybenzimidazoles and polybenzthiazoles have been used as a platform for the introduction of the **6F** group to introduce tractability to these difficulty processed materials, albeit with limited success. The simplest of these, **14**, was synthesized by the usual condensation of tetraamine with dicarboxylic acid [75].

Thermo-oxidative stability was excellent, but solubility improved only slightly. Due to the difficulty of benzimidazole formation, better polymers were obtained by condensing fluorophenyl-terminated benzimidazoles with BisAF to give poly(ether benzimidazoles) which were soluble in common solvents, formed films and were stable to well above 500 °C [76].

Fluorinated polybenzoxazoles which contain imide functions, **15**, are possible by either imide or oxadiazole chain extension reactions [57,77,78].

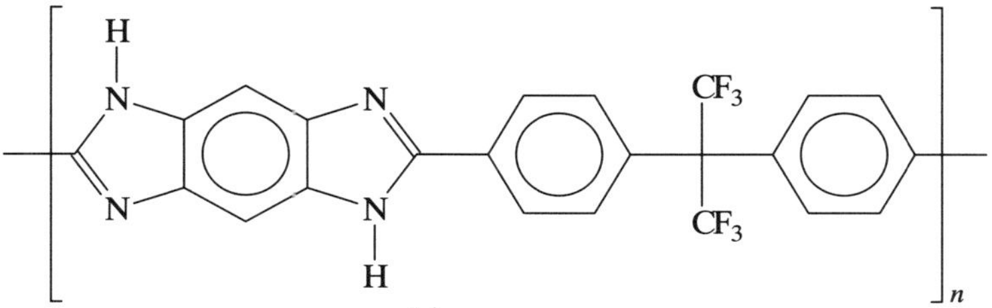

14

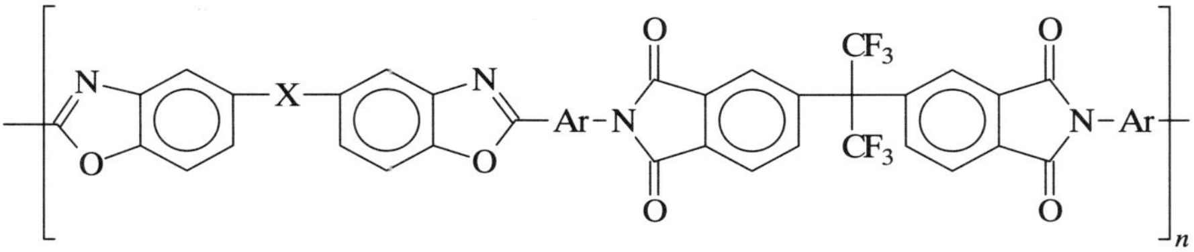

$X = C(CF_3)_2$; $Ar = p$- or m-C_6H_4

15

These materials were used in glass fiber composites and demonstrated the usual good thermal stability (over 500 °C) and dielectric constants (2.80–3.10).

Finally, fluorinated polybenzthiazoles, **16**, show some solubility, no T_g and stability above 500 °C.

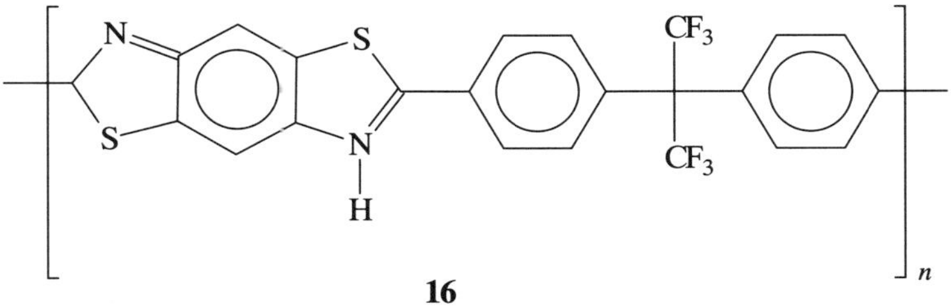

16

8 CONCLUSION

Successful incorporation of fluorine into numerous types of high-temperature polymer backbones has shown that it provides several advantages in processing and properties. This inclusion has occurred mostly in the form of a hexafluoroisopropylidene group in or pendant to the backbone, but also can be achieved by introduction of a trifluoromethyl group and by aryl fluoride compounds.

Polymers of imides, esters, ethers, ether ketones, amides and several heterocycles (all aromatic), as well as polyacrylates, have all benefited by the addition of fluorine moieties in terms of increased solubility and stability and decreased color, dielectric constant and water absorption. The high cost of these materials, however, limits their use to small-scale and specialty applications such as microelectronics, aerospace and medical devices.

REFERENCES

1. Cassidy, P. E., Aminabhavi, T. M. and Farley, J. M., *J. Macromol. Sci.-Rev. Macromol. Chem. Phys.*, **C29(2&3)**, 365 (1989).
2. Bruma, M., Fitch, J. W. and Cassidy, P. E., *J. Macromol. Sci.-Rev. Macromol. Chem. Phys.*, **C36(1)**, 119 (1996).
3. Cassidy, P. E., Aminabhavi, T. M., Reddy, V. S. and Fitch III, J. W., *Eur. Polym. J.* **31(4)**, 353 (1995).
4. Stamatoff, G. S. and Wittmann, J. W., French Patent 1 394 897 (April 9, 1965); *Chem. Abstr.*, **63**, 18297c (1965).
5. Farnham, A. G. and Johnson, R. N, US Patent 3 332 909 (1967); *Chem. Abstr.*, **68**, 69543a (1968).
6. Lau, K. S. Y. and Dougherty, T. K., US 4 827 054 (1989); *Chem. Abstr.*, **111**, 174813t (1989).
7. Irvin, J. A., Neef, C. J., Kane, K. M., Cassidy, P. E, Tullos, G., and St Clair, A. K., *J. Polym. Sci., Part A: Polym. Chem.*, **30**, 1675 (1992).
8. Mercer, F. W., Goodman, T. D., Lau, A. N. K., Vo, L. P., and Sovish, R. C., US Patent 5 114 780 (1992).
9. Mercer, F. W., Duff, D., Wojtowicz, J., and Goodman, T. D., *Polym. Mater. Sci. Eng.*, **66**, 198 (1992).
10. Mercer, F. W., Goodman, T. D., Lau, A. N. K., and Vo, L. P., PCT Int. Appl. WO 91 16 370 (1991); *Chem. Abstr.*, **116**, 236362j (1992).
11. Iizawa, T., Kodou, H., and Nishikubo, T., *J. Polym. Sci., Part A: Polym. Chem.*, **29**, 1875 (1991).
12. Johnson, R. N., Farnham, A. C., Clendinning, R. A., Hale, W. F., and Merriam, C. N., *J. Polym. Sci., Part A-1*, **5**, 2375 (1967).
13. Mohr, J. H., Paul, D. R., Tullos, G. L., and Cassidy, P. E., *Polymer*, **32**, 2387 (1991).
14. Kawakami, J. H., Bikson, B., Gotz, G., and Ozcair, Y., US Patent 4 971 695 (1990); *Chem. Abstr.*, **114**, 124183d (1991).
15. Wang, Z., Chen, T, and Xu, J., *J. Appl. Polym. Sci.*, **51**, 1533 (1994).
16. Tullos, G. L. Cassidy, P. E. and St. Clair, A. K., *Macromolecules*, **24**, 6059 (1991).
17. Mercer, F. W., Goodwin, A. A., Fone, M. M., and Reddy, V. N., *Polymer*, **38**, 1987 (1997).
18. Kane, K. M., Wells, L. A. and Cassidy, P. E., *High Perform. Polym.*, **3**, 191 (1991).
19. (a) Yoshimura, T. and Fukuda, T., Japanese Kokai Tokkyo Koho JP 02 167 337 (1990); *Chem. Abstr.*, **114**, 63052m (1991); (b) Tomata, H. and Ito, K., Japanese Kokai Tokkyo Koho JP 02 182 722 (1990); *Chem Abstr.*, **114**, 82778u (1991).
20. Kakimoto, M. A. and Imai, Y., *J. Polym. Sci., Polym. Chem. Ed.*, **24**, 3555 (1986).
21. Korshak, V. V., Vinogradova, S. V. and Pankratov, V. A., British Patent 1 122 201 (July 1968).
22. Korshak, V. V., Vinogradova, S. V. and Pankrativ, V. A., *Vysokomol. Soedin.*, **7**, 1689 (1965).
23. Korshak, V. V., Vinogradova, S. V. and Pankratov, V. A., *Izv. Akad. Nauk SSSR, Ser. Khim.*, 1649 (1965).

24. Korshak, V. V., Manucharova, I. F., Vinogradova, S. V. and Pankratov, V. A., *Vysokomol. Soedin.*, **7**, 1813 (1965).
25. McCune, L. K., *Text. Res. J.*, **32**, 762 (1962).
26. Mark, V. and Hedges, C. V., PCT Int. Appl. WO 8 202 402 (July 1982); *Chem. Abstr.*, **97**, 217318g (1982).
27. Tokuda, T. and Furukawa, K, Japanese Kokai Tokkyo Koho JP 62 141 061 (June 1987); *Chem. Abstr.*, **107**, 218858k (1987).
28. Keller, T. M., *J. Polym. Sci., Polym. Chem. Ed.*, **22**, 2719 (1984).
29. Cassidy, P. E., Fitch, J. W., Reddy, V. S., Lunceford, B. and Person, D., Society of Plastics Engineers, ANTEC, Vol. 40, 2162 (1994).
30. Keller, T. M., *J. Polym. Sci., Polym. Chem. Ed.*, **23**, 2557 (1985).
31. Hirisawa, T. and Hirabayashi, S. Japanese Kokai Tokkyo Koho, JP 02 129 145 (1990).
32. Hirisawa, T. and Hirabayashi, S. Japanese Kokai Tokkyo Koho, JP 02 129 146 (1990).
33. Griffith, J. R., *Am. Chem. Soc., Polym. Mat.*, **50**, 422 (1984).
34. Griffith, J. R. and O'Rear, J. G., US Patent 4 578 508 (1986).
35. Ramharack, R., *Polymer Preprints*, **29**(1), 146 (1988).
36. Hartman, C. R., Eur. Pat. Appl. EP 330 821 (1989); *Chem. Abstr.*, **112**, 120555n (1990).
37. Matsumoto, K., Minamizaki, Y. and Xu, P., *Maku, 17*, 395 (1992); *Chem. Abstr.*, **118**, 16896n (1993).
38. Hoyt, A. E., Ricco, A. J., Yang, H. C. and Crooks, R. M., *J. Am. Chem. Soc.*, **117**, 8672 (1995).
39. McGrath, J. E., Rogers, M. E., Arnold, C. A., Kim, Y. J. and Hedrich, J. C., *Macromol. Chem., Macromol. Symp.*, **51**, 103 (1991).
40. Sonnett, J. M, McCullough, R. L., Beeler, A. J. and Gannett, T. P., *Int. SAMPE Tech Conf.*, **24**, T735 (1992); *Chem. Abstr.*, **118**, 192363v (1993).
41. Laius, L. A., Zhukova. T. I., Kuznetsov, N. P., Kudryavtsev, V. V., Svetlichnyi, V. M., Simakov, B. V., Ostrovskii, V. I., Rastorgueva, N. M. and Nikiforova, G. N., *Vysokomol. Soedin. Ser. B*, **33**, 851 (1991).
42. Hougham, G., Tesoro, G. and Shaw, J., *Macromolecules*, **27**, 3642 (1994).
43. Misra, A. K., Tesoro, G., Hougham, G. and Pendharkar, S. M., *Polymer*, **33**, 1078 (1992).
44. Yusa, M., Takeda, S. and Miyaders, Y., Eur. Pat. Appl. EP 450 926 (1991); *Chem. Abstr.*, **116**, 84434u (1992).
45. Sasaki, S., Matsuura, T., Ando, S. and Nishi, S., *NTT R&D*, **40**, 967 (1991); *Chem. Abstr.*, **116**, 130626h (1992).
46. Nishi, S., Masuura, T., Ando, S. and Sasaki, S., *J. Photopolym. Sci. Technol.*, **5**, 359 (1992); *Chem. Abstr.*, **118**, 170159f (1993).
47. Hayes, R. A., Eur. Pat. Appl. EP 336 998 (1989); *Chem. Abstr.*, **112**, 159207a (1990).
48. Park, J. W., Lee, M., Lee, M. H., Liu, J. W., Kim, S. D., Chang, J. Y., and Rhee, S. B., *Macromolecules*, **27**, 3459 (1994).
49. Ghasemi, H. and Hay, A. S., *Macromolecules*, **27**, 3116 (1994).
50. Jones, R. J. and Silverman, E. M., *Int. SAMPE Tech. Conf.*, **20**, 542 (1988); *Chem. Abstr.*, **111**, 11641n (1989).
51. Jones, R. J. and Silverman, E. M., *SAMPE J.*, **25**, 41 (1989); *Chem. Abstr.*, **111**, 79104d (1989).
52. Klobucar, W. D., Eisenbraun, A. L. and Zumstein, R. C., Eur. Pat. Appl. EP 373 647 (1990); *Chem. Abstr.*, **114**, 24819x (1991).
53. Lee, B. and Li, H. M., US 4 973 661 (1990); *Chem. Abstr.*, **114**, 165153n (1991).

54. St Clair, A. K., St Clair, T. and Pratt, J. R., *NASA Tech Briefs*, p. 55 (Dec. 1994). [US Patent 5 093 453]
55. Auman, B. C., Higley, D. P., Scherer, K. V., McCord, E. F. and Shaw, W. H., *Polymer*, **36**, 651 (1995).
56. Chung, T-S and Kafchinski, E. R., *Polymer*, **37**(9), 1635 (1996).
57. Bruma, M., Mercer, F., Fitch, J. W. and Cassidy, P., Chapter in *Fluoropolymers: Synthesis and Properties*, eds Hougham, G., Davidson, T. and Cassidy, P., Plenum (in press).
58. Bruma, M., Sava, I., Mercer, F., Negulescu, I., Daly, W., Fitch, J. and Cassidy, P., *High Perform. Polym.*, **7**, 411 (1995).
59. Bruma, M., Mercer, F., Fitch, J. and Cassidy, P., *J. Appl. Polym. Sci.*, **56**, 527 (1995).
60. Bruma, M., Mercer, F., Schulz, B., Dietel, R., Fitch, J. and Cassidy, P., *High Perform. Polym.*, **6**, 183 (1994).
61. Vora, R. H., US 4 914 180 (1990); *Chem. Abstr.*, **114**, 7455h (1991).
62. Cassidy, P. E., Thaemlitz, C. J. and Brewer, K. W., *Polym. Prep. (Am. Chem. Soc., Div. Polym. Chem.)*, **31**, 582 (1990).
63. Mueller, W. H., Khanna, D. N., Vora, R. H. and Erckel, J. R., Eur. Pat. Appl. EP 317 944 (1989); *Chem. Abstr.*, **112**, 57018x (1990).
64. Vora, R. H., US 4 962 181 (1990); *Chem. Abstr.*, **114**, 63467a (1991).
65. Negi, Y. S., Suzuki, Y-I, Kawamura, I., Kakimoto, M-A and Imai, Y., *J. Polym. Sci.: Part A: Polym. Chem.*, **34**, 1663 (1996).
66. Liaw, Der-Jang and Wang, Kun-Li, *J. Polym. Sci.: Part A: Polym. Chem.*, **34**, 1209 (1996).
67. Boston, H. G., Reddy, V. S., Cassidy, P. E. and Fitch, J. W., No. 654, *Proceedings of the Joint Regional Meeting of the ACS*, Memphis, Tennessee, November 1995.
68. Irvin, D. J., Cassidy, P. E., Fitch, J. W. and Cameron, M., *J. Polym. Mater.*, **12**, 83 (1995).
69. Irvin, D. J., Cassidy, P. E., Meurer, D. L., Fitch, J. W., Taylor, D. A., St Clair, A. and Stoakley, D., *Polymer*, **37**(11), 2227 (1996).
70. Fitch, J. W., Cassidy, P. E., Weikel, W. J., Lewis, T. M., Trial, T., Burgess, L., March, J. L., Glowe, D. E., and Rolls, G. C., *Polymer*, **34**, 4796 (1993).
71. Thaemlitz, C. J., Weikel, W. J. and Cassidy, P. E., *Polymer*, **33**, 3278 (1992).
72. Mercer, F., *High Perform. Polym.*, **4**, 2 (1992).
73. Bruma, M., Schulz, B. and Mercer, F. *Proceedings 3rd European Technical Symp. Polyimides & High Temp. Polym.*, Montpellier, France, June 1993.
74. Mercer, F. W., *High Perform. Polym.*, **5**, 69 (1993).
75. Hutzler, R. F., Meurer, D. L., Kimura, K. and Cassidy, P. E., *High Perform. Polym.*, **4**, 161 (1992).
76. Kane, J. J. and Gao, F., *Polym. Prep. (Am. Chem. Soc., Div. Polym. Chem.)*, **35**, 631 (1994).
77. Khanna, D. N. and Lee, W. R., Eur. Pat. Appl. EP 387 060 (1990); *Chem. Abstr.*, **114**, 82719 (1991).
78. Bruma, M., Schulz, B., Mercer, F., Dietel, R. and Neumann, W., *Polymers for Advanced Technologies* **5**, 535 (1994).

9

Polyfluoroacrylates

N. N. CHUVATKIN and I. YU. PANTELEEVA
Orgsteklo Company, Research and Engineering Centre, Dzherzhinsk, Russia

The Orgsteklo Company located in the Nizhny Novgorod region of Russia produces high heat and impact resistance polymer 'glass' based on phenyl 2-fluoroacrylate for the canopies of supersonic aircraft. This chapter is dedicated to the memory of Prof. L. S. Boguslavskaya who made a valuable contribution to the development of fluoroacrylate chemistry.

1 INTRODUCTION

The progress in supersonic aviation has necessitated the need to develop 'organic' glasses which exceed the current capabilities of acrylics such as poly(methyl methacrylate) in terms of heat and impact resistance. In the USA and some other countries, a polycarbonate glass has been used for glazing supersonic aircraft, while in the former Soviet Union poly (phenyl α-fluoroacrylate), having a higher heat resistance, was used. In the middle of the 1980s there arose an renewed interest in fluoroalkyl α-fluoroacrylate polymers as a potential material for light-guides. These polymers are characterized by a higher glass transition temperature, enhanced heat resistance, good mechanical strength and flexibility in comparison with the widely used fluoroalkyl methacrylate polymers. Today, however, this interest has waned probably due to the difficulty of synthesizing the monomer.

Under ordinary conditions the derivatives of perfluoroacrylic, α-(trifluoro-methyl)acrylic and perfluoromethacrylic acids do not homopolymerize according to a free-radical mechanism in practice. However, due to their ability to copolymerize with different olefines including those containing fluorine, these acrylates can be successfully used in the production of various copolymers with high potential.

At present, among polyfluoroacrylates, the polymers and the copolymers of fluoroalkyl acrylates and fluoroalkyl methacrylates have the most practical use.

They are used in the production of plastic lightguides, resists, water-, oil- and dirt-repellent coatings and in other advanced applications.

2 SYNTHESIS OF FLUOROACRYLATE MONOMERS

At present three review articles [1–3] have been published in which different methods of α-fluoroacrylates synthesis were considered. The most attractive method used readily available 2,2,3,3-tetrafluoropropanol and methyl lithium and could be performed under mild conditions achievable in the laboratory [4] (equation 1).

$$HCF_2CF_2CH_2OH \xrightarrow{MeLi} CF_2{=}CFCH_2OLi \xrightarrow{H_2O} CF_2{=}CFCH_2OH$$

$$\xrightarrow{H^+} CH_2{=}CFCOF \xrightarrow[-HF]{ROH} CH_2{=}CFCOOR \qquad (1)$$

Fluoroalkyl esters of α-fluoroacrylic acid can be easily synthesized with high yield from the corresponding acrylic esters and either of BrF_3 or ClF [2,5] (equation 2). It should be remembered nevertheless, that highly reactive fluorinating agents such as BrF_3 and ClF are extremely hazardous and proper handling precautions must be used.

$$CH_2 = CHCOOR_F \xrightarrow{Br_2} CH_2BrCHBrCOOR_F \xrightarrow[\text{or ClF}]{BrF_3} CH_2FCHFCOOR_F$$

$$\xrightarrow{R_3N} CF_2 = CFCOOR_F \qquad (2)$$

The first commercial production of α-fluoroacrylates (in particular phenyl α-fluoroacrylate) occurred in the 1960s, initially at the Kargin Polymer Research Institute pilot plant and subsequently at the Orgsteklo Company (Dzerzhinsk, Russia) [2,3]. A five-stage reaction was used as shown in equation (3).

$$CH_2{=}CHCH_2OH \xrightarrow[HF]{NCl} CH_2ClCHFCH_2OH \xrightarrow{HNO_3} CH_2ClCHFCOOH$$

$$(3)$$

$$\downarrow SOCl_2$$

$$CH_2{=}CFCOOPh \xleftarrow{PhONa} CH_2{=}CFCOCl \xleftarrow{R_3N} CH_2ClCHFCOCl$$

The resulting monomer yield was low ($\sim$20–25%) and this difficult synthesis also suffered from technological limitations. Thus, it was only viable in the former USSR for special military purposes. Until now, none of the proposed methods of α-fluoroacrylate synthesis have been suitable for commercial production.

A similar situation exists for the commercial production of perfluoroacrylic acid derivatives [6–9]. The old methods which use hexafluoropropylene (equation 4) continue to be the simplest; however, the fluoroolefin precursor conversion rate is only about 50% [7,9].

$$CF_3CF{=}CF_2 \begin{cases} \xrightarrow{NE_3} CF_3CHFCN \xrightarrow{KF \text{ or } KCl} CF_2{=}CFCN \\ \xrightarrow{MeOH} CF_3CHFCF_2OMe \xrightarrow{H_2SO_4} CF_3CHFCOOMe \end{cases} \quad (4)$$

$$CF_3CHFCOOMe \xrightarrow[\text{KCl}]{600-800\,°C} CF_2{=}CFCOOMe$$

Perfluoromethacrylic acid derivatives can also be easily produced from perfluoroisobutylene [9,10]. In a recent review [1], some methods for the synthesis of α-(trifluoromethyl)acrylic acid and its esters are given. According to one of these methods the synthesis proceeds in a one-step reaction [11] (equation 5).

$$CH_2{=}CBrCF_3 + CO + H_2O \xrightarrow[\text{Et}_3N]{PdCl_2 \, (PPh_3)_2} CH_2{=}\underset{\underset{CF_3}{|}}{C}COOH \quad (5)$$

The synthesis of some fluoroalkyl acrylates and fluoroalkyl methacrylates by way of esterification was realized in commercial production a long time ago. Due to the acidic nature of fluorine-containing alcohols, esterification proceeds slowly. In order to synthesize esters of strongly acidic and sterically hindered secondary and tertiary fluorinated alcohols, a well-known method of synthesis based on the corresponding acylchlorides in the presence of HCl acceptors can be used (equation 6).

$$CH_2{=}\underset{\underset{R}{|}}{C}{-}\underset{\underset{O}{\|}}{C}Cl + R'{-}\underset{\underset{R_F}{|}}{\overset{\overset{R_F}{|}}{C}}{-}OH \xrightarrow{-\,HCl} CH_2{=}\underset{\underset{R}{|}}{C}{-}\underset{\underset{O}{\|}}{C}{-}O\underset{\underset{R_F}{|}}{\overset{\overset{R_F}{|}}{C}}{-}R' \quad (6)$$

$$R = H, CH_3. \qquad\qquad R' = H, Alk, Alk_F$$

3 POLYMERIZATION OF FLUOROACRYLATES

α-Fluoroacrylates possess higher polymerization activity than the corresponding acrylic and methacrylic acid derivatives [3,12–14]. For example under the same conditions 2,2,3,3-tetrafluoropropyl α-fluoroacrylate (TFPFA) polymerizes

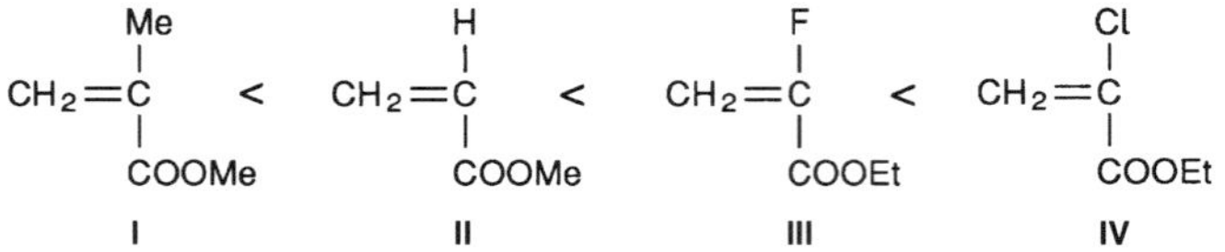

$$
\begin{array}{ccccccc}
\text{Me} & & \text{H} & & \text{F} & & \text{Cl} \\
| & & | & & | & & | \\
\text{CH}_2\!=\!\text{C} & < & \text{CH}_2\!=\!\text{C} & < & \text{CH}_2\!=\!\text{C} & < & \text{CH}_2\!=\!\text{C} \\
| & & | & & | & & | \\
\text{COOMe} & & \text{COOMe} & & \text{COOEt} & & \text{COOEt} \\
\text{I} & & \text{II} & & \text{III} & & \text{IV}
\end{array}
$$

Figure 9.1. Series of α-substituted acrylates in order of increasing rate of poly-merization

more rapidly than the corresponding methacrylate. The ratio constants of the chain propagation and chain termination reactions $(K_p/K_t)^{0.5}$ at 60 °C is 1.15 $(l/mol.s)^{0.5}$ for the α-fluoroacrylate and 0.22 for the methacrylate [12].

The polymerization of ethyl α-fluoroacrylate has been described previously [13]. It is shown that a substituent in the α-position of an acrylate can affect the rate of its polymerization according to the order shown in Figure 9.1. The absolute values of the chain propagation constants at 30 °C in the case of low conversion of the acrylates shown in Figure 9.1 are as follows: 450 (I), 720 (II), 1120 (III) and 1660 (IV) l/mol.s [13]. In bulk polymerization of α-fluoroacrylates when the degree of monomer conversion exceeds 4–5%, the polymerization rate increases rapidly [12] by a factor of 5–8%.

Figure 9.2 shows the dependence of TFPFA bulk polymerization rate on the degree of conversion at 60 °C [12]. The heat of polymerization of α-fluoroacrylate is 1.3 times higher than that of the methacrylates. For TFPFA the heat of polymerization is 73 kJ/mol [12]. The α-fluoroacrylate polymerization is accompanied by a significant volume contraction. In the case of 100% conversion of ethyl α-fluoroacrylate, the shrinkage is 32% [13].

Due to the relatively high rate of α-fluoroacrylate polymerization, its exothermic nature and accompanying shrinkage, the production of a defect-free sheet of 'organic glass' is rather difficult. In order to produce a high-quality organic glass of poly(phenyl α-fluoroacrylate), the use of azobisisobutyronitrile as an initiator is recommended for the phenyl α-fluoroacrylate polymerization [15]. In addition, better results can be achieved in the production of a 'glass' with thickness above 10 mm, when this monomer is polymerized at 18–20 °C in the presence of 0.01–0.05% mass dicyclohexyl percarbonate and 0.0005–0.001% mass benzoquinone. After a gradual increase of temperature to 150–170 °C, an organic glass of poly(phenyl α-fluoroacrylate) having molecular weight of 5–8×10^6 is obtained. The example of TFPFA polymerization illustrates that the addition of trace levels of benzoquinone controls an otherwise dangerous increase in the α-fluoroacrylate polymerization rate (lower curve in Figure 9.2).

The radical copolymerization of methyl- and ethyl α-fluoroacrylate with methylacrylate, methylmethacrylate, styrene and some other vinyl monomers has been studied [13,14]. From the constants of copolymerization with styrene, the authors of this work obtained the Alfrey–Price values Q and e. The Q and e values are extremely useful for the prediction of whether or not copolymerization

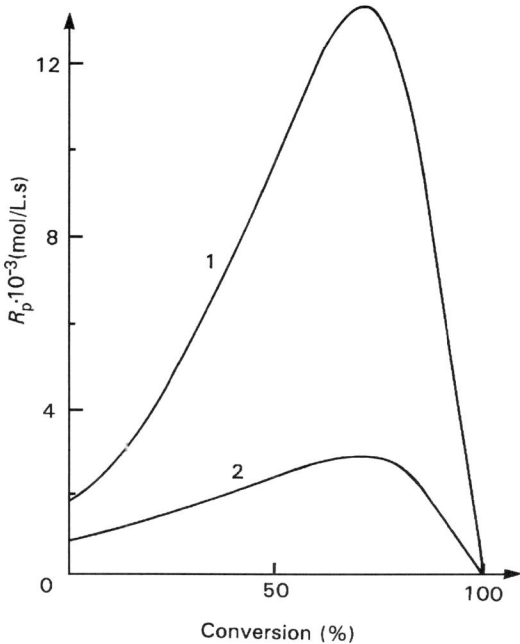

Figure 9.2. Dependence of the rate of bulk polymerization of TFPFA on the degree of conversion at 60 °C with a benzoyl peroxide initiator (0.1% mass). Curve 1 – free of inhibitor; curve 2 – with 0.005 wt% 1,4-benzoquinone [12]

will occur. For methyl α-fluoroacrylate, Q was 0.47 and e was 0.73; and for ethyl α-fluoroacrylate Q was 0.49 and e was 0.68. i.e. these values did not differ significantly from the corresponding values for the similar acrylates [16]. The small difference between the e values for α-fluoroacrylates that are non-substituted in the α-position provides evidence that a fluorine atom in α-fluoroacrylates has little effect on the polar factor of the copolymerization.

α-Fluoroacrylates are easily oxidized by atmospheric oxygen according to a free-radical mechanism to give peroxides that consist of alternating low molecular copolymers of monomers and oxygen [3] (equation 7).

$$n\text{CH}_2{=}\overset{\text{F}}{\underset{\text{O}={}\text{COR}}{\text{C}}} + n\text{O}_2 \longrightarrow -\text{O}-\text{CH}_2\overset{\text{F}}{\underset{\text{O}={}\text{COR}}{\text{C}}}-\text{O}-\left[\text{OCH}_2\overset{\text{F}}{\underset{\text{O}={}\text{COR}}{\text{C}}}-\text{O}\right]_{n-1} \qquad (7)$$

Other conditions being equal, α-fluoroacrylates are oxidized by oxygen far more rapidly (by a factor of 100 and even more) than their acrylate and methacrylate counterparts. That is why it is advisable to stabilize the α-fluoroacrylates for polymerization with the addition of a small amount of an antioxidant such as *para*-methoxyphenol. Otherwise the poly(α-fluoroacrylate) produced from monomer containing polyperoxide contaminants markedly loses its light transmission after heating, due to polyperoxide decomposition and HF formation.

α-Fluoroacrylate polymerization is greatly inhibited by tertiary amines [17], which are normally used in the final stage of the monomer synthesis (see section 2). Besides, even a small amount of tertiary amine addition to poly (α-fluoroacrylate) reduces light transmission of the polymer after heating to an even greater extent than do polyperoxides.

Various free-radical polymerization inhibitors can be used to prevent α-fluoroacrylate polymerization during their purification by distillation. Good results are obtained with the use of nonvolatile derivatives of paranaphthylene diamine, i.e. di-β-naphthyl-paraphenylene diamine, or their mixtures with benzoquinone and *para*-methoxyphenol [18].

Further substitution of a hydrogen atom by a fluorine atom in the acidic part of α-fluoroacrylate derivative molecule, as well as substitution of a fluorine atom by CF_3 group in the α-position, leads to almost complete loss of ability for free-radical homopolymerization. Only some highly fluorinated acrylates can form low molecular polymers in the process of intensive γ-irradiation or when heated under very high pressure in the presence of a peroxide initiator [19–23]. However, all these fluoroacrylates are copolymerized with various vinyl monomers and tetrafluoroethylene [19–31].

Because of the limited length of this chapter we cannot consider copolymerization of trifluoroacrylic, α-(trifluoromethyl)acrylic and perfluoromethacrylic acid derivatives with various copolymers in detail. Instead, more consideration is given to the overall chemistry of these reactions.

The low electron density on β-carbon atom of highly fluorinated acrylates, as well as electrophilic properties of the free radicals formed from them enables alternating copolymerization of these compounds with monomers. The monomers possess a comparatively high electron density of C=C bonds (equation 8).

$$\underset{\underset{R}{|}}{\overset{\overset{H}{|}}{\text{-----CH}_2\text{C}\cdot}} \quad \xrightarrow{\overset{\delta+}{\text{CF}_2=\text{CX}}\overset{\text{C(O)Y}}{|}} \quad \underset{\underset{R}{|}\;\underset{X}{|}}{\overset{\overset{\text{C(O)Y}}{|}}{\text{-----CH}_2\text{CHCF}_2\text{C}\cdot}}\overset{\delta-}{\underset{\delta+}{|}} \quad \xrightarrow{\text{CH}_2=\text{CHR}} \quad \begin{array}{c}\text{alternating}\\\text{copolymer}\end{array} \quad (8)$$

R = H, Alk, OAlk, Ph.   X = F, CF_3.   Y = Alk, F.

The marked tendency of the highly fluorinated acrylates towards alternating copolymerization is reflected by the high positive **e** Alfrey–Price values for these compounds, given in Table 9.1. It is known that in most cases the greater the difference in the *e* values of the two copolymers, the greater the tendency to form alternating copolymers [24]. Unfortunately the values of Q and e for some fluoromonomers, in particular for methyl trifluoroacrylate, given in various works differs considerably (see footnote[a] of Table 9.1).

Copolymerization of fluoroacrylates has been carried out in bulk, solution or emulsion with initiation by γ-irradiation, peroxide initiators or azobisisobutyronitrile. For example, methyl trifluoroacrylate which is not polymerizable by itself is copolymerized readily with ethylene, propylene and isobutylene. This reaction can be carried out in an autoclave or in a glass ampule under γ-irradiation with a dose of 6×10^6 rad. This results in a copolymerization rate of 1.3–1.9 mol/l.hour, monomer conversion in the range of 66–99% and copolymer molecular mass (generally not greater than 1×10^5) being observed at an equimolar ratio of the reagents [21]. The molecular weight of copolymers increases with an increase in the electron donor characteristics of olefins. Similar results are obtained during copolymerization of the same monomers in benzene solution when heated in the presence of a peroxide [25]. The alternating nature of these copolymers is in part responsible for their considerably high heat resistance [24,25].

The rate of copolymerization of trifluoroacrylic or perfluoromethacrylic acid derivatives with tetrafluoroethylene always rises with an increase in the tetrafluoroethylene concentration. As a result copolymers are formed in which separate chains of the corresponding fluoroacrylates are included in

Table 9.1. Q and e values of common fluoroacrylates

No.	Fluoroacrylate	Q	e	Refs
1	$CF_2{=}CFCOOCH_3^{a}$	0.048	+1.20	16
2	$CF_2{=}CFCN$	0.04	+1.92	23
3	$CF_2{=\!=}CCOF$ $\|$ CF_3	0.061	+2.81	28
4	$CH_2{=\!=}CCOOCH_3$ $\|$ CF_3		+2.90	22
5	$CH_2{=\!=}CCOOCH(CF_3)_2$ $\|$ CH_3	1.38	+1.30	33
6	$CH_2{=}CHCOOCH_2CF_3$	0.97	+1.13	33
7	$CH_2{=}CHCOOCH_2CH_3^{b}$	0.41	+0.55	16
8	$CH_2{=}CFCOOCH_2CH_3$	0.49	+0.68	13

[a] There are published other Q and e values as well, such as Q from 0.012 to 0.196 and e from 1.50 to 2.6524.
[b] Is given for comparison.

the poly(tetrafluoroethylene) polymer chain in a random order [26–28]. The copolymerization with hexafluoropropylene does not proceed [26].

α-(Trifluoromethyl)acrylates and α-(trifluoromethyl)acrylonitrile are able not only to copolymerize with other monomers according to radical mechanisms [22,29,30] but also to form homopolymers according to anionic mechanisms [30,31].

Radical homopolymerization of fluoroalkyl acrylates and methacrylates does not differ significantly compared to polymerization of similar alkyl acrylates [32]. However, during copolymerization of fluoroalkyl acrylates and methacrylates with the monomers having negative e values, a significant difference related to the electron-acceptor influence of fluoroalkyl group, appears. For example, during copolymerization of hexafluoroisopropyl acrylate [CH_2=CHC(O)OCH(CF_3)$_2$] with styrene an alternating copolymer is formed [33]. The e values for fluoroalkyl acrylates and methacrylates are considerably higher than for nonfluorinated ethers (Table 9.1, nos. 5–7).

Fluoroalkyl methacrylates and acrylates also polymerize according to anionic mechanism. However, their polymerization rate, yield and molecular weights are considerably lower, than in the case of radical polymerization [34].

4 PROPERTIES AND USES OF POLYFLUOROACRYLATES

The presence of a fluorine atom in the α-position to the ester group in a polyacrylate ester imparts to the corresponding polymer a unique combination of good optical and strength properties with high heat resistance and thermal stability [2,3,35–40]. As can be seen from Table 9.2, glass transition temperature and mechanical strength of poly(α-fluoroacrylate)s (Table 9.2, nos 1–12) are considerably higher, compared with the polymers of the similar methacrylates [41] (Table 9.2, nos 14–21).

Among the polymers of this class, poly(phenyl α-fluoroacrylate) (PPhFA) (Table 9.2, nos 11 and 12) has the best properties and it is recommended for use as a construction material for glazing cockpits of supersonic aircraft [15,39,40]. This material was described for the first time in a French patent [15]. However, long before this it was produced in the former Soviet Union in the Orgsteklo Company under the name 'E-2'*. The main advantages of this glass were high glass transition temperature (180 °C) and high heat resistance (above 270 °C). These properties are of great importance during high-speed climbing when a supersonic aircraft passes through the lower atmospheric layer. Poly (*para*-chlorophenyl α-fluoroacrylate) has an even higher glass transition temperature value, but it has low impact strength (Table 9.2, no. 10), while the copolymer of phenyl ester and parachlorophenyl ester have intermediate properties.

---

* This name was given by A. Ya. Jakubovich who pioneered the synthesis of PPhFA in 1958

Table 9.2. Physical properties of α-fluoroacrylic and methacrylic polymers

$$-(-CH_2-\underset{\underset{O=COR'}{|}}{\overset{\overset{R}{|}}{C}}-)_{\overline{n}}$$

No.	R'	T_g (°C)	n_D	a_n (kJ/m^2)	σ_b (mPa)	ε_b (%)	Refs
			Poly(α-fluoroacrylate)s, R=F				
1	CH_3	142	1.459	22	96	4.6	13,35
2	CH_2CF_3	123	1.385	17	63	12.5	35,36
3	$CH_2CF_2CF_2H$	95	1.389	21	54	66	35,36
4	$CH_2CF_2CF_3$	105	1.370	—	—	—	36
5	$C(CH_3)_2CF_2CF_2H$	133	1.400	—	—	—	35,36
6	$CH_2(CF_2)_3CF_2H$	77	1.379	—	18	24	35
7	$CH(CF_3)_2$	109	1.355	—	—	—	37
8	$CH_2CF(CF_3)_2$	108	1.360	—	—	—	35
9	C_6F_5	174	1.457	4	56	3.1	35,38
10	$p\text{-}C_6H_5Cl$	226	—	13	—	—	—
11	C_6H_5	180	1.552	19	96	5.5	15,39
12	C_6H_5[a,b]	180	—	48	110	12	39
13	Lexan LS3[b,c]	153	1.586	>100	70	120	[d],39
			Poly(methacrylate)s, R=CH$_3$				
14	CH_3[c]	113	1.491	14	78	4	35
15	CH_2CF_3	76	1.416	7	48	2	35,36
16	$CH_2CF_2CF_2H$	71	1.422	6	47	2.7	35,36
17	$CH_2CF_2CF_3$	72	1.395	—	—	—	36
18	$C(CH_3)_2CF_2CF_2H$	101	1.42	—	—	—	41
19	$CH_2(CF_2)_3CF_2H$	50	1.400	—	—	—	35
20	$CH(CF_3)_2$	91	1.38	—	—	—	41
21	C_6F_5	115	1.487	—	—	—	38

[a] An oriented glass.

[b] Light transmission at 10 mm sheet thickness is 88–89%. The light transmission of the rest of the glasses is 90–92%.

[c] There are given for comparison.

[d] The properties of 2.2-bis-(4-oxyphenyl) propane optical polycarbonate are taken from the prospectus of the General Electric Plastics Co. published in 1991.

Unfortunately, PPhFA possess inadequate resistance to UV-rays, humid air as well as periodical one-sided heating and cooling which occurs during aircraft take-off and landing. After two years' exposure in subtropical conditions, the polymeric 'glass' turns yellow, its light transmission reduces to 10–15% and on the outside surface of the 'glass' a large number of small cracks appear. The cracks dramatically reduce the materials impact strength. After planar orientation to 40%, which is performed at 220 °C, this organic glass has very high cracking resistance and its strength properties increase significantly (Table 9.2, no. 12). Using a proprietary method at a specified temperature, an oriented glass of PPhFA can be moulded into a desired shape. Such glass is an excellent material for glazing cockpits of supersonic aircraft. Nevertheless, the supersonic aircraft of the US Air Force fly successfully, having cockpits glazed by optical-grade polyaryl carbonate. This material is similar to that produced by General Electric Plastics and Bayer under the trade marks Lexan® and Makrolon® respectively [39]. From

(a)

(b)

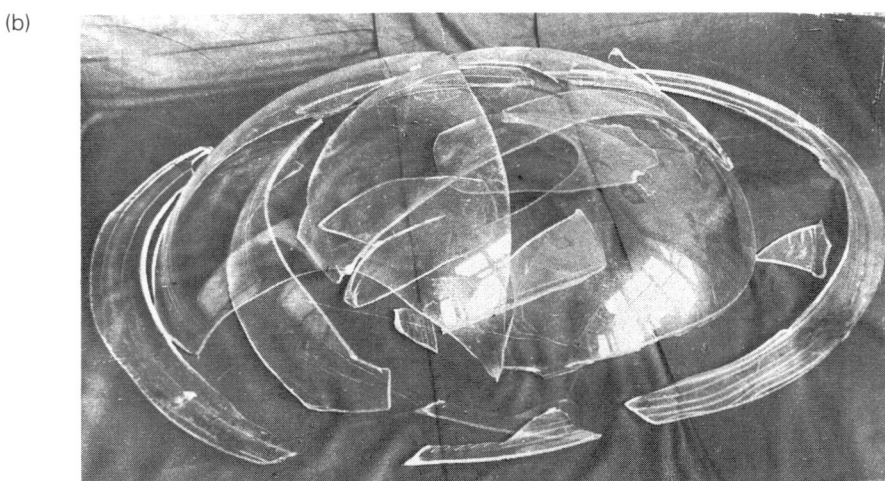

Figure 9.3. Result of large-calibre machine-gun fire against of aircraft canopy of (a) orientated poly(phenyl α-fluoroacrylate) (PPhFA), and (b) PMMA

Table 9.2 it can be seen that the organic glass made from PPhFA is superior to the glass under the Lexan® trademark, according to the glass transition temperature (T_g) and ultimate tensile stress values (σ_b). However, it is inferior to LEXAN in rupture elongation value (ε_b) and impact strength (a_n). The oriented PPhFA does, however, possess an impact strength sufficient for its main application.

In Figure 9.3(a) test results are shown of large-calibre machine-gun fire against a semi-hemispheric canopy of oriented PPhFA. The test material was placed in a chamber under a pressure of 0.5 atm to simulate the pressure drop experienced in flight conditions. For comparative purposes Figure 9.3(b) shows the results of similar testing on an article of poly(methyl methacrylate) (PMMA). It should be noted, however, that due to its complex production technology, this organic glass is much more expensive than other organic glasses for similar applications.

Block poly(α-fluoroacrylate)s and especially poly(methyl α-fluoroacrylate) and PPhFA are very resistant to crazing by organic solvents. For example, PPhFA is mildly soluble only in dimethylformamide. It is known that many polymer properties such as solubility and glass transition temperature depend on steric regularity. In the case of poly(methyl α-fluoroacrylate) it is shown, that this class of polymer is characterized by a higher content of isotactic dyads than that of polymeric methacrylates and α-chloroacrylates [13,42]. Since the glass transition temperature also depends on the volume and polarity of substituting agents, α-chloro- and α-fluoroacrylate polymers have almost the same glass transition temperature [2,13]. Irrespective of this, poly(α-chloroacrylate)s cannot be used as heat-resistant polymers due to their low thermal stability [13]. Figure 9.4 shows thermogravimetric (TG) weight-loss curves for poly(ethyl α-chloroacrylate) (PECA), poly(ethyl α-fluoroacrylate) (PEFA) and poly(methyl α-fluoroacrylate) (PMFA). The maximum degradation rate temperatures of PMFA and PEFA were 402 and 370 °C respectively (Fig. 9.4) [13].

The TG kinetic data for poly(fluoroalkyl-α-fluoroacrylate) (PFAFA) degradation demonstrates that on the basis of thermostability they are comparable with poly(fluoroalkyl acrylate)s (PFAA) that are nonsubstituted in α-position. In addition, they are superior to poly(fluoroalkyl methacrylate)s, which in turn are more thermostable than their nonfluorinated analogues [43,44] (Figure 9.5).

In general, poly(fluoroalkyl α-fluoroacrylate) degradation starts around 280–300 °C, while poly(fluoroalkyl methacrylate)s (PFAMA)s begin to degrade at temperatures approximately 30–50 degrees lower. After extended periods of heating optical defects start to form on the surface of the polymer. The onset temperature at which these defects occur is around 220–240 °C for PFAFA and 170–180 °C in the case of PFAMA [2,35].

When considering heat resistance of fluoroalkyl acrylate and methacrylate polymers, it should be noted, that owing to the presence of fluoroalkyl groups in a side chain, these polymers are more heat resistant than their nonfluorinated counterparts and the fluoropolymers do not oxidize during thermal destruction in air [44].

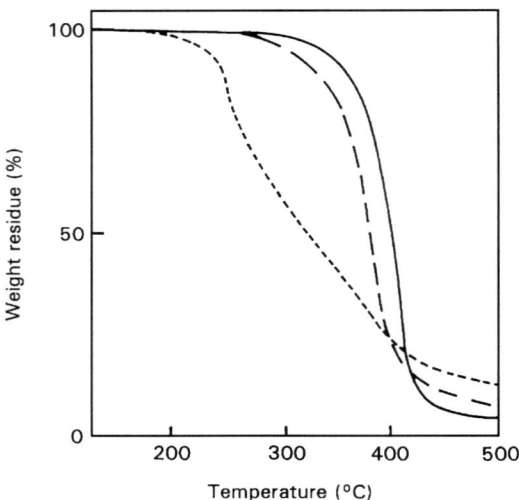

Figure 9.4.   Thermogravimetric curves for PECA (...), PEFA (...) and PMFA (...) heated at 10 °C min^{-1} in nitrogen. Reproduced by permission of John Wiley and Sons Ltd

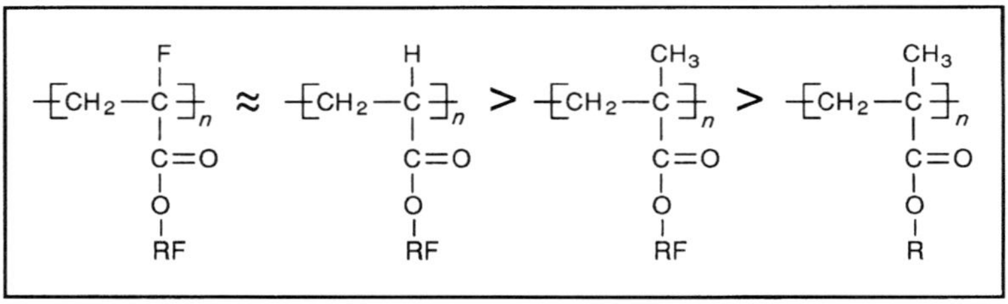

$$\left[CH_2-\underset{\underset{\underset{RF}{O}}{\overset{\overset{F}{|}}{\underset{|}{C=O}}}{C}\right]_n \approx \left[CH_2-\underset{\underset{\underset{RF}{O}}{\overset{\overset{H}{|}}{\underset{|}{C=O}}}{C}\right]_n > \left[CH_2-\underset{\underset{\underset{RF}{O}}{\overset{\overset{CH_3}{|}}{\underset{|}{C=O}}}{C}\right]_n > \left[CH_2-\underset{\underset{\underset{R}{O}}{\overset{\overset{CH_3}{|}}{\underset{|}{C=O}}}{C}\right]_n$$

Figure 9.5.   Series of α-substituted acrylates in order of decreasing thermal stability as measured by thermogravimetry. Note: R represents an alkyl group while RF denotes a perfluoroalkyl group

5   LIGHTGUIDES

Good optical properties combined with improved thermal stability and flexibility give the poly(fluoroalkyl α-fluoroacrylate)s a considerable advantage over poly(fluoroalkyl methacrylate)s (PFAMA) which have been widely studied and used in fibre optics. Table 9.2 gives properties of some PFAFAs which are proposed as a material for light-reflecting coatings of plastic lightguides [2,3,35–37*].

---

* The list of patents concerning this application is given in a review publication [3]

It can also be seen that these polymers have considerably higher values of glass transition temperatures (by about 20 °C) and deformation strength, with lower refractive index than the similar PFAMAs have (Table 9.2, nos. 1–8 and nos. 14–20). The value of critical angle at which the lightguide operates depends on the difference in refractive indices of the light-reflecting coating and lightguide core [41].

Generally the molecular weight of block PFAFAs is in the range 1–5 $\times 10^6$. The polymers with molecular weights of 1–3 $\times 10^5$, used in the production of lightguides, are obtained by polymerization in bulk in the presence of molecular weight regulators, e.g. BuSH [36,37]. Such polymers can be processed by extrusion at 210–230 °C.

The only known attempt of pilot production of lightguides having a coating of poly(α-fluoroacrylate) was made in Russia during the end of the 1980s. Lightguide cores of PMMA were covered with a light-reflecting coating of poly(1H,1H,3H-perfluoropropyl α-fluoroacrylate) (Table 9.2, no. 3) or copolymer of α-fluoroacrylate and 1H,1H,5H-perfluoroamyl acrylate [45]. These lightguides possess excellent flexibility, even though their fibre diameters were only 0.5–2 mm [45]. This feature made them superior to similar lightguides with coatings of PFAMA. However, these α-fluoroacrylates were very expensive and when low-cost plastic lightguides from Japan and the USA appeared on the Russian market, the domestic lightguides lost their competitiveness.

The majority of lightguides produced these days in many countries have a coating of PFAMA. The characteristics of some of these polymers are given in Table 9.2, nos. 15–20. Lightguides of this type are described in detail in the book edited by Ishikawa *et al.* [41]. In the last decade there were a great number of patents published which considered this application. Improvements in the coating material's refractive index, glass transition temperature, bending strength and adhesion to the lightguide core is generally achieved by variation of fluoroalkyl radical structure in the initial monomer or by copolymerization with other monomers. As this technology matured, certain companies developed fibre orientation procedures to obtain flexible lightguides.

6 MICROLITHOGRAPHY

Some poly(fluoroalkyl methacrylate)s due to their ability to degrade under the influence of UV, X-rays and electron beam irradiation have found application in microlithography such as the production of integrated circuits and electron boards. They can be used as positive resists, i.e. resists where the irradiated area is easily removed under action of a solvent [41,46]. The firm Daikin Kogyo (Japan) produces the resist FBM-110, which is the polymer shown in Figure 9.6(a).

The resist FBM-120 is a refinement of FBM-110 and is characterized by higher adhesion to the substrate material. Until recently, these materials were among the best resists available with excellent sensitivity and resolution capabilities.

However, progress in microlithography has brought new resists with even higher performance. For further reading on poly(fluoroalkyl acrylate) resists see Chapter 26 in this book.

7 TEXTILE TREATMENT

The low surface energy of fluoroalkyl acrylate and methacrylate polymers makes it possible to use them in the production of water- and oil-repellent compounds for fibre and textile treatment [41]. The limited scope of this chapter does not allow an extensive discussion on this subject. It should be noted only that at present the greatest amount of fluoroacrylates and methacrylates produced in the world are used in this field of application. At present aqueous dispersions of perfluoroalkyl acrylates and perfluoroalkyl methacrylates copolymers are widely used. These compounds compete successfully with other fluorine-containing compounds for the same application [41]. To achieve the required water- and oil-repellent effect it is necessary to use the copolymers having a perfluoroalkyl pendant group with at least seven carbon atoms. For example the acrylic polymer shown in Figure 9.6(b) has an ultimate surface tension value (γ_c) of 10.6 dyn/cm and is characterized by an excellent water- and oil-repellent ability [47]. In comparison, the γ_c of PTFE is 18.5 dyn/cm. The surface characteristics of the polymers of fluoroalkyl acrylates and methacrylates were well known long ago and are considered in detail by Pittman in the book edited by Wall [47].

Poly(fluoroalkyl acrylate)s exhibit some properties which are typical elastomers. However, as mentioned in the literature [47], they are not used

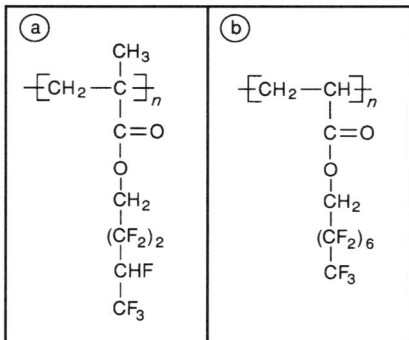

Figure 9.6. Structures of commercial poly[fluoroalkyl(meth)acrylate]s: (a) FBM-110 (Daikin Kogyo, Japan) used as a positive resist in microlithographic processing to produce integrated circuits; (b) a poly(perfluoroalkyl acrylate) with a free surface energy almost half that of PTFE and which is used in water- and dirt-repellant textile coatings (the largest market for fluoroacrylates at present)

commercially as elastomers because there are a number of other fluoroelastomers that have far superior characteristics.

The derivatives of perfluoroacrylic and perfluoromethacrylic acid may be used in the production of tetrafluoroethylene copolymers, which are melt processable at elevated temperatures. For instance, a TFE copolymer with 26 mol% of methyl perfluoromethacrylate melts at 296 °C and starts to decompose only at 350 °C [48]. This polymer no longer possesses the high crystallinity which is characteristic of PTFE, but retains its exceptionally high resistance to concentrated nitric and sulphuric acids as well as organic solvents [48]. Due to their low refractive index copolymers of this type may be used as light-reflecting coatings for specialty lightguides. The copolymer of TFE with perfluoromethacryloyl fluoride [CF_2=C(CF_3)C(O)F] may be used in the production of special membranes containing carboxylic groups [28].

REFERENCES

1. Wakselman, C., *Macromol. Symp.*, **82**, 77 (1994).
2. Boguslavskaya, L. S. and Chuvatkin N. N., *Macromol. Symp.*, **82**, 51 (1994).
3. Boguslavskaya, L. S., Panteleeva, I. Yu., Morozova, T. V., Kartashov, A. V. and Chuvatkin N. N., *Russ. Chem. Rev.*, **59**, 906 (1990).
4. Nguyen, T. and Wakselman, C., *J. Org. Chem.*, **54**, 5640 (1989).
5. Kartashov, A. V., Chuvatkin N. N. and Boguslavskaya, L. S., *Zh. Org. Khim.*, **27**, 2522 (1991).
6. Dedek, V., Paleta, O., Pánek, P. and Posta, A., *Sb. VSCHT Praze*, **C28**, 89 (1983).
7. Rendall, L. and Pearison, W. H., US Patent., 2 795 601 (1957).
8. Norman, R., US Patent, 4 404 398 (1983).
9. England, D. C., Solomon, L. and Krespan, C. G., *J. Fluor. Chem.*, **3**, 63 (1973).
10. Knunyants, I. L., Abduganiev, Yo. G., Rohhlin, E. M., Okulevich, P. O. and Karpushina, N. I., *Tetrahedron*, **29**, 595 (1973).
11. Fuchikami, T., Yamanouchi, A. and Ojima, I., *Synthesis*, 766 (1984).
12. Cherep, E. I., Samarina, A. V., Panteleeva, I. Yu. *et al.*, *1st All-Union Conference 'Radical Polymerization'*, Gor'kii, 1989, Russia, 138 (1989).
13. Yamada, B., Kontani, T., Yoshioka, M. and Otsu, T., *J. Polym. Sci. Polym. Chem. Ed.*, **22**, 2381 (1984).
14. Pittman, C. U., Ueda, M., Iri, K. and Imai, Y., *Macromolecules.*, **13**, 1031 (1980).
15. Bloch, B., Cavalli, C. and Charier, D., French Patent, 24 444 052 (1980).
16. Greenley, R. Z., *J. Macromol. Sci. Chem.*, **A14**, 427 (1980).
17. Suefuji, M., Japanese Patent Application, 60 158 136 (1984).
18. Boguslavskaya, L. S., Panteleeva, I. Yu., Kartashov, A. V. and Morozova, T. V., *Khim. Prom.*, 588 (1991).
19. Weise, J. K., *Polym. Prepr. Am. Chem. Soc., Div. Polym. Chem.*, **12** (1), 512 (1971).
20. Laita, Z., Paleta, O., Posta, O. and Liska, F., *Coll. Czech. Chem. Comun*, **40**, 2059 (1975).
21. Matsuda, O., Watanabe, T., Tabata, Y. *et al.*, *J. Polym. Sci., Polym. Lett. Ed.*, **16**, 283 (1987).
22. Iwatsuki, S., Kondo, A. and Harashina, H., *Macromolecules*, **17**, 2473 (1984).
23. Matsuda, O., Kostov, G., Tabata, Y. and Maert, S. M., *J. Appl. Polym. Sci.*, **24**, 1053 (1979).

24. Boutevin, B. and Ameduri, B., *Macromol. Symp.*, **82**, 1 (1994).
25. Zúrcova, E., Bouchal, K., Vacik, J., Kállar, J., Paleta, O. and Dedek, V., *IUPAC Macro 83, Bucharest, Abst. Sec. I*, 57 (1983).
26. Watanabe, T., Momose, T., Ishigaki, I., Tabata, Y. and Okato, I., *J. Polym. Sci.: Polym. Lett. Ed.*, **19**, 599 (1981).
27. Yul'chibaev, A. A. and Agzamova, R. T., *Izv. Vyssh. Uchebn. Zaved., Khim., khim. Tekhnol.*, **26**, 257 (1983).
28. Yul'chibaev, A. A., Agzamova, R. T. and Kuzieva, H. Yu., *5th All-Union Conference on Fluoro-Organic Chemistry, Mosow, 1986*, Mosow, 179 (1986).
29. Kleiner, E. K., US Patent, 3 386 977 (1968).
30. Ito, H., Miller, D. and Wilson, G., *Macromolecules*, **15**, 915 (1982).
31. Narita, T., Hagiwara, Hamana, H. and Maesaka, S., *Polym. J.*, **20**, 519 (1988).
32. Rostovsky, E. N. and Rubinovich, L.D., *Vysokomol. Soedin.: Karbocep. vysokomol. soed.*, 140 (1963).
33. Narita, T., Hagiwara, T. and Hamana, H., *Macromol. Chem. Rapid Commun.*, **6**, 5 (1985).
34. Narita, T., Hagiwara, T., Hamana, H. and Goto, M., *Macromol. Chem.*, **187**, 731 (1986).
35. Boguslavskaya, L. S., Samarina, A. V., Lebedeva, V. I., Panteleeva, I. Yu. and Chuvatkin, N. N., *Plast. Massy*, 15 (1988).
36. Ohmori, A., Tomihashi, N. and Kitahara, T., European Patent., 128 516 (1985).
37. Wichers, G., Heümuller, R., Groh, W. and Herbechtsmeier, P., BRD Patent Application, 3 602 275 (1987).
38. Bose, D., Boutevin, B., Petrasanta, Y. and Roussau, A., French Patent Application, 2 623 510 (1989).
39. Gudimov, M. M. and Perov, B. V., *Organicheskoe Steklo (Organic Glass)*, Khimiya, Mosow, 215 (1981).
40. Wakselman, C., Lampin, J. P., Molines, H. *et al.*, *Ann. Chim. (Paris)*, **9**, 719 (1984).
41. Ishikawa, N., *Fluorine Compounds, Modern Technology and Application*, Mir, Moscow (1984), translated into Russian from Japanese (edited in Japan, 1981).
42. Majumdar, R. and Harwood, H., *Polym. Bull.*, **4**, 391 (1981).
43. Gorelov, Yu. P., Boguslavskaya, L. S., Bulovyatova, A. B. and Tsareva, L. A., *Vysokomol. Soedin. Ser. B*, **25**, 514 (1983).
44. Sazanov, Y. P., Budovskaya, L. D., Ivanova, V. I., Fedorova, G. N. and Rostovsky, E. N., *Vysokomol. Soedin. Ser. B*, **19**, 366 (1977).
45. Boguslavskaya, L. S., Panteleeva, I. Yu., Morozova, T. V., Monich, I. M. and Chuvatkin, N. N., *1st International Conference 'Chemistry, Technology and Application of Fluorocompounds in Industry', St. Petersburg, Russia*, St Petersburg, 87 (1994).
46. Moreau, W. M., *Microdevices*, Plenum Press, New York and London (1988).
47. Wall, L. A., *Fluoropolymers*, Wiley-Interscience, New York (1972).
48. Yul'chibaev, A. A., Hodzhaev, S. G., Usmanov, H. U. and Yupusov, T. K., *Uzb. Khim. Zh.*, 47 (1978).

10

Perlast® Perfluoroelastomer

PETER CUMMINGS

Precision Polymer Engineering Limited, Clarendon Road, Blackburn, UK

1 ELASTOMERIC MODIFICATION OF PTFE

Perfluoroelastomers (ASTM designation FFKM) represent the highest point in fluoroelastomer development. An elastomer which is resistant to swelling and degradation when exposed to almost all solvents and chemicals is of such outstanding practical value that it has found applications in a huge variety of industrial sectors. Perlast® is a typical perfluoroelastomer, being based on patents dating back approximately 10 years, and is representative of the second generation of the genre. The first commercially available perfluoroelastomer was DuPont's Kalrez®, a copolymer of tetrafluoroethylene (TFE) and perfluoromethylvinylether (PMVE), developed and patented in the mid-1970s.

The reason for the development of the perfluoroelastomer family was the need for elastomers capable of use at high temperatures and able to withstand hostile chemical environments. It was thought by fluorine chemists that if PTFE, which already had these properties, could be modified chemically so that elastomeric behaviour could be induced then this might result in suitable material for a high-performance rubber. The fluorocarbon chain in PTFE which is produced by the polymerisation of TFE and which is the main constituent of perfluoroelastomers is so inert that the choice of suitable cure site monomers (CSM) and the design of cure systems capable of maximising the properties of the elastomer almost proved to be the fluorine chemists' equivalent of scaling the north face of the Eiger.

PTFE has no major shortcomings in its resistance to chemical attack, but the designer has had to cope with such effects as creep (a permanent deformation resulting from continuously applied loads) and other disadvantages of PTFE including the tendency to fret contact surfaces, pressure sensitivity and stiffness which is not generally a problem with fluoroelastomers. The occasional requirement for an elastomer to survive at temperatures greater than 300 °C has also been met with some grades [1].

Modern Fluoropolymers. Edited by John Scheirs
© 1997 John Wiley & Sons Ltd

Figure 10.1. Perlast® is specified in chemical transport applications for its virtu-ally universal chemical resistance. Reproduced with permission

2 HISTORY AND PROPERTIES

The story of FFKM development has been fully reviewed elsewhere [2], but some detail is necessary to place Perlast® in its particular niche. The first perflu-oroelastomer grades were notable for setting new standards of thermal stability with hitherto unheard-of resistance to almost all chemicals. Even today, Kalrez® 4079 is the only elastomer capable of withstanding temperatures of 316 °C in extended service. This resistance to very high temperatures with which some Kalrez® grades are associated is due to the bisphenol AF and triazine systems employed in their curing. The *sine qua non* of perfluoroelastomers in the vast majority of applications is their outstanding resistance to swelling and chem-ical attack over a significant temperature range, and in this respect Kalrez® and Perlast® are broadly similar, albeit over different temperature ranges. It is in the retention of flexibility at low temperatures that Perlast® excels, combining this critical property with its resistance to chemical attack.

The high-temperature limit of Perlast® is similar to that of vinylidene fluoride-based FKM elastomers in that it is capable of extended service at above 200 °C and even higher in the absence of oxygen. Hot air ageing tests on Perlast® G70B show limited degradation in comparison with typical fluoroelastomers at 250 °C. The performance of a perfluoroelastomer seal depends on the particular application: static or dynamic, true temperature of the seal and duration of the temperature peaks, soakback effects, system pressures and fluctuations and the influence of solvents or chemicals in the system. It is therefore important to treat quoted performance data as a guide only and in every case for the user to ascertain the suitability of the elastomer for the environment.

Uncommonly for elastomers, perfluoroelastomers tend to soften slightly with the influence of time and high temperature and this actually improves their sealing performance especially in the case of dynamic seals [3].

3 INERTNESS OF THE POLYMER

The key to the performance of Perlast® and indeed all commercialized FFKM's is the fact that the polymer backbone contains only carbon and fluorine atoms. The prefix 'per' in perflucroelastomer signifies 'fully'. The polymer chain being fully fluorinated contains no hydrogen atoms. This feature is remarkably useful. Firstly, the C–F bond is very strong; the bond energy, i.e. the energy that would be required to break the bond being relatively high.

	C–H	C–F
Bond energy (kJ/mol)	413	485

Secondly, the covalent radius of the fluorine atom is much larger than that of hydrogen (F $0.72\,\text{Å}$, H $0.37\,\text{Å}$) and the C–F bond is short. The fluorine atoms are forced to nestle together, twisting the chain into a spiral structure that repeats every 13 carbon atoms for a PTFE segment. It is this fluorine sheath that protects the carbon backbone from chemical attack.

4 LOW-TEMPERATURE FLEXIBILITY

The temperature range over which a typical fluoroelastomer will function without loss of sealing force or degradation is adequate for many design requirements. The capability of fluoroelastomers to function above $200\,^\circ\text{C}$ allows sufficient range for demanding applications; this temperature level can be comfortably exceeded by Perlast®, especially in non-oxidising environments. It is often at the lower end of the temperature scale, below $0\,^\circ\text{C}$ for example, where retained flexibility is crucial in maintaining a seal that the limitations of an elastomer are more obvious. Only at temperatures above the glass transition temperature (T_g) is sufficient thermal energy available to allow motion of the segments in the main backbone chain of the polymer. Thus for low-temperature flexibility, a low T_g is required. The key to the low T_g $(-19\,^\circ\text{C})$ and low-temperature capability of Perlast© is the inherent flexibility of the chain backbone which results from the irregularly placed pendant perfluorocarbon ether groups. In contrast, PTFE has a highly regular chain structure which affords high crystallinity and an inflexible chain backbone. Perlast® has greater chain flexibility than other perfluoroelastomers and, therefore, is significantly more capable of sealing at low temperatures. However, gaining improved low-temperature flexibility properties for perfluoroelastomers is fraught with complications and remains the subject of extensive research in this branch of fluorine chemistry [4].

The advantageous low-temperature properties of Perlast® have important implications for sealing performance because the vast majority of applications for perfluoroelastomers exist below 200 °C. Uschold [5] observed that fluorinated elastomers normally have their high-temperature stability at the expense of low-temperature stiffness; in other words the overall temperature range of each type of polymer is fairly similar. Whilst discussing the preparation of perfluorinated elastomers with glass temperatures as low as −76 °C, he conceded that the search for lower temperature fluoroelastomers has been a continuing challenge.

Tests performed by RAPRA (Rubber and Plastics Research Association) compared the low-temperature performances of Perlast® and three other seal materials. These consisted of: tetrafluoroethylene propylene (TFE/P), nitrile rubber (NBR) and another perfluoroelastomer (FFKM2). The four materials divided into three distinct classes in terms of low-temperature flexibility measured using a Gehman apparatus conforming to the requirements of BS 903, Part A13 (Figure 10.2).

As expected, the nitrile rubber exhibited the best properties with the next best performance being given by Perlast®. The other two materials tested were substantially inferior.

From the measurements taken during these tests, the relative modulus values shown in Table 10.1 were calculated. They are stated in °C. Modulus is a measure of 'stiffness'; the higher the modulus the greater the stiffness. The figures in the table for a particular material are the temperatures at which the relative

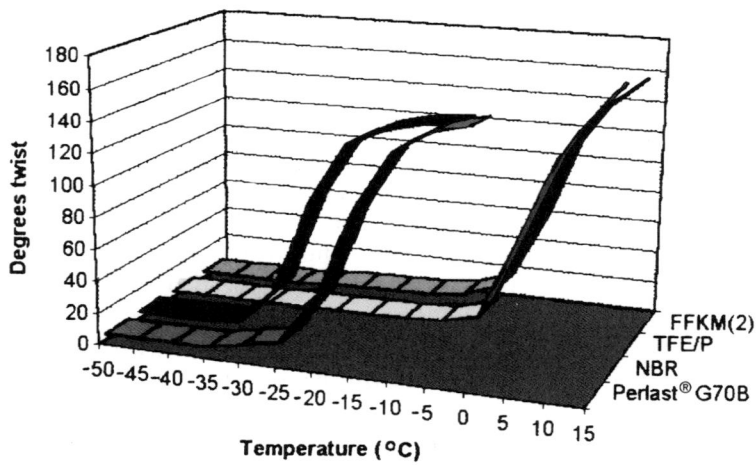

Figure 10.2. Low-temperature performance comparison graph for four elastomers

Table 10.1. Relative modulus of elastomers at low temperatures

	T2	T5	T10	T100
Perlast® (G70B)	−8	−14	−16	−26
FFKM2	8	3	1	−10
TFE/P	7	3	1	−5
NBR	−15	−23	−26	−33

modulus of the material is twice, 5 times, 10 times and 100 times respectively the room temperature value of its modulus. For example, if we assume that a given rubber has a modulus at room temperature of 0.20 MPa, then the T2 value is the temperature at which the modulus has increased to 0.40 MPa; T5 is the temperature at which the modulus is 1.00 Pa; T10, 2.00 MPa and T100, 20.00 MPa.

Since some rubbers start to stiffen at lower temperatures than others (for example the NBR in Table 10.1 compared to the TFE/P) their corresponding T2, etc. values are lower. Some rubbers increase in stiffness quite quickly once the critical temperature is reached and as the temperature falls so their T2, T5, etc. values are closer together. Other rubbers change more slowly with temperature and their corresponding T2, T5, etc. values are further apart. In the table, the range of the values for TFE/P (12 °C) is much smaller than that (18 °C) for the other three materials. Thus we can use the 'T' values to judge the low-temperature characteristics of one elastomer relative to another.

5 VULCANIZATION

The iodine-based cure system used with Perlast® is unique and, whilst not affording the levels of extreme high-temperature resistance associated with the triazine-cured Kalrez® 4079 (the only commercially available vulcanizate capable of continuous use at 316 °C), the Perlast® system has the important advantage of containing no metal oxide or other metal-based catalyst. Cure and filler systems exert the biggest influence on the performance of different perfluoroelastomers. In aqueous or acidic environments any metal oxides present in the curing system are prone to attack; the subsequent release of metal ions is highly undesirable, particularly in the semiconductor industry where purity and freedom from contamination are paramount.

The bisphenol AF and especially the triazine cure systems have particular advantages at elevated temperatures. The bisphenol AF system allows vulcanizates to withstand temperatures of above 270 °C and the more recent triazine-based system affords vulcanizates which can withstand prolonged use around 316 °C depending on other conditions. In both curing systems nucleophilic reactions are involved with the result that the cured structures have poor resistance to other nucleophilic reagents at high temperatures; thus the nitrile (triazine) cured types have less resistance to sodium hydroxide based media, amines and water/steam than cured Perlast® which has been crosslinked with a peroxide system and therefore is not capable of being hydrolytically split by these media [6].

6 CHEMICAL RESISTANCE

Perfluoroelastomers as a class are similar as far as chemical attack is concerned. Such attack as occurs can cause swelling (usually a reversible event), which if limited can be welcome. Swell of up to approximately 10% can beneficially increase the sealing force produced by an elastomeric gasket, but too much swell can cause the seal to become damaged resulting in premature failure. Chemical attack can degrade the material and this should be avoided at all costs. The chemical resistance of perfluoroelastomers is such that it seems more logical to provide a list of chemicals that should be avoided than one showing chemicals that have little or no effect.

Perlast® is affected by strong oxidising and reducing agents. Boiling tertiary amines and molten alkali metals (lithium, potassium, sodium, etc.) should be avoided. This is so for all fluorinated polymers.

Extremely strong oxidising agents such as fluorine gas and related compounds can produce volatile mixtures after reacting with fluoropolymers. Whilst chemicals which would cause a large volume change should be avoided (these include HCFC, CFC, fluorinated oils, CTFE oils), such a change is not necessarily fatal to the product: the effect is one of swell rather than degradation.

Ethylene and propylene oxide have very little swelling effect on Perlast® but consultation is advisable where these monomers are present because of their potential effect on the elastomer backbone.

Strong oxidizing agents such as fuming nitric acid will attack carbon black: accordingly Perlast® G70W incorporating a non-black filler should be used in such environments.

7 MIXED PROCESS STREAMS

There are many advantages in having a single elastomer that is resistant to most chemicals. Table 10.2 lists some chemicals that, although they do not affect Perlast®, do attack just about every other common elastomer. Thus problems of material selection for critical seals are alleviated by the ability to specify an elastomer with such broad resistance.

The chemicals in Table 10.2 are used in one plant by a pharmaceutical manufacturer and the choice of perfluoroelastomer as a standard seal material has a beautiful simplicity. In the plant in question Perlast® is used for pipeline, pump and valve seals at temperatures up to 150 °C. The solvents involved are used in the pure form as solutions and slurries and the use of Perlast® seals has given extended life, security and lower maintenance costs whilst replacing FKM, PTFE

Table 10.2. Solvent resistance of a range of elastomers

	Perlast©	NBR	EPDM	Silicone	Fluorosilicone	Fluorocarbon
Acetone	1	4	1	3	4	4
Benzene	1	4	4	4	3	1
Ethyl acetate	1	4	2	2	4	4
Formaldehyde	1	3	1	2	4	4
Hexane	1	1	4	4	1	1
Methanol	1	1	1	1	1	4
Methylene chloride	1	4	3	4	2	2
Methyl formate	1	4	2	No data	No data	No data
Methyl ethyl ketone	1	4	1	4	4	4
Methyl isobutyl ketone	1	4	2	4	4	4
Pyridine	1	4	2	4	4	4
Tetrahydrofuran	1	4	3	4	4	4
Thionyl chloride	1	4	3	No data	No data	2
Toluene	1	4	4	4	2	1

Classification system:
1 = little or no swelling (<10%) and the elastomer should not be adversely affected by the media.
2 = there may be greater swelling (10–20%) and the elastomer may suffer some loss of physical properties. These may not be fatal to the application and performance may be satisfactory, but seal life will probably be affected.
3 = the elastomer may have some functionality, but is clearly affected by the media. It is of even greater importance to test in field conditions.
4 = the elastomer is unsuitable for use with this substance.

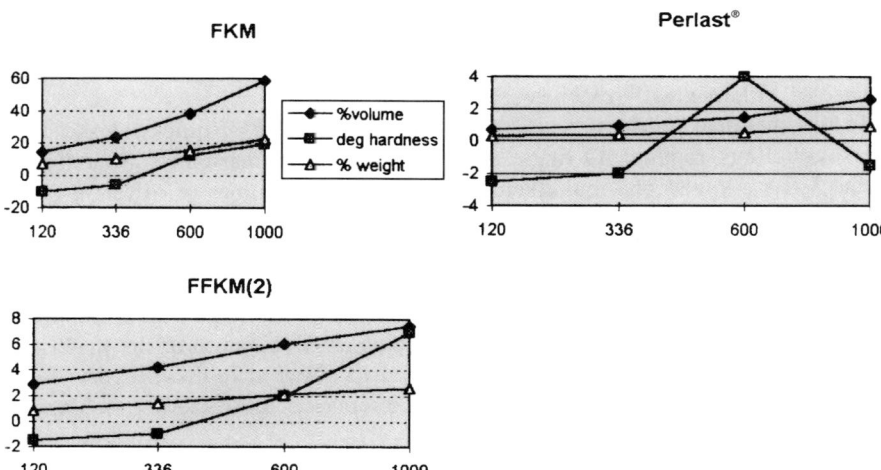

Figure 10.3. Comparison of the swell characteristics of three fluorinated elastomers. Tests were run for 120, 336, 600 and 1000 hours each (x-axis). Temperatures: FFKMs 250 °C, FKM 200 °C. Fluid: DOD-L-85734 lubricant. Note that the effects on the Perlast® were within the boundaries of +/−4 units on the scale. This compares with +8/−2 for the other perfluoroelastomer and +60/−20 for the fluoroelastomer

and non-elastomers often with much simpler solutions. Mixed process streams, cocktails of chemicals and solvents do not necessarily behave like an imaginary single fluid with unique properties [7], so the presence of such mixtures can greatly complicate elastomer selection.

The swell characteristics of Perlast® are compared with those of a typical fluoroelastomer copolymer (FKM) and another common perfluoroelastomer (FFKM2) in Figure 10.3.

Both the Perlast® and the other perfluoroelastomer performed quite satisfactorily over this extended test and the values of three physical measurements taken remained within narrow bands. In contrast the FKM swelled at a catastrophic rate well before the 336 hour interval and its hardness had dropped by approximately 10 points after only 120 hours. Its hardness rapidly increased throughout the rest of the test as the FKM became brittle due to the impact of temperature and chemical attack.

8 APPLICATIONS

Anywhere gases or liquids have to be managed presents a possible application for elastomers as seals. The advantages of using elastomers over other materials include the freedom to use imperfectly finished mating surfaces with no risk of leakage, avoidance of shaft fretting or damage. Their resilience removes the need for constant adjustment and there is no need to employ a separate energiser.

Fire-resistant lubricants containing phosphate esters and lubricating oils containing amines (to reduce 'coking') are both interesting applications where the use of FFKM greatly extends seal life.

In the chemical transport and storage industry, the use of quick-release dry-break couplings requires O-rings with extensive fluid resistance. Specifiying Perlast® has allowed one manufacturer to reduce the number of different elastomers used (from ethylene propylene (EPR), nitrile rubber (NBR), silicone (VMQ), fluorosilicone (FMQ) and fluoroelastomer (FKM)) whilst extending the range of materials carried.

A pump manufacturer uses Perlast® seals in three very large pumps supplied to a major UK chemical producer. Due to the aggressive nature of the materials carried the pumps must be continually cycled to allow strip-down every three months. To date there has been no unscheduled replacement of the Perlast© seals.

9 SEMICONDUCTOR PRODUCTION

It is in manufacturing systems based in the production of semiconductors that the properties of FFKM elastomer are stretched to the full. In this rapidly growing area just about every advantage of FFKM is utilised: chemical resistance, thermal stability, low extractables, low permeability and low outgassing are all important. The sheer reliability of the elastomer is crucial when it is extremely costly to stop a production process for seal replacement.

Figure 10.4. Semiconductor manufacture places some of the highest demands on perfluoroelastomers. Purity and resistance to the most aggressive chemical cocktails are paramount. Reproduced with permission

For example, the dry etching process requires door and window seals and exhaust valves resistant to highly aggressive chemicals (Figure 10.4). A similar requirement exists for seals used in filters, couplers and valves in the photo-masking process. Plasma CVD (chemical vapour deposition) processes involve seals in vacuum ports, chamber lids, etc. which must be resistant to plasma. The sputtering process calls for seals which have the lowest possible outgassing rates.

One semiconductor equipment manufacturer has conducted extensive tests on several potential materials for use as a wafer seal that would withstand downstream chemical plasma attack and would have a long and reliable life. Downstream exposure tests have shown that Perlast® is equivalent to or better than other perfluoroelastomers tested in dry etch process conditions.

The process in question is one of the most aggressive in the industry. It will strip photoresist at over 2.5 μ/minute on a 150 mm wafer. Some FFKM materials are suitable and others show degradation, becoming covered, for example, in a thin layer of white powder. The test conditions were as follows.

A variety of O-rings were mounted on a silicon wafer and placed in a vacuum process module with a downstream inductively coupled plasma photoresist strip chamber. Process conditions were as follows (sccm = standard cubic centimetres per minute):

$$400 \text{ sccm } O_2 + 40 \text{ sccm } CF_4 + 40 \text{ sccm } N_2.$$

$$750 \text{ m Torr, } 400 \text{ W RF, } 15\% \text{ heating } (130\,°\text{C}), 17 \text{ hours}$$

Table 10.3. Effect of chemical plasma attack on a range of elastomers

Type	Colour	Hardness IRHD	Max. temp (°C)	Result of exposure	Rating	Cost
FKM	Black	65–95	200	Fully degraded to powder	Very poor	1
FFKM1	White	80	230+	Thin surface, layer of white powder	Quite good	21
FFKM2	White	80	230+	Thin surface, layer of white powder	Quite good	21
Silicone 1	White	70	220	Degraded to powder, thin core left	Poor	1
Silicone 2	Red	70	220	Degraded to powder, thin core left	Poor	1
FFKM3	Black	80–90	300+	Intact, no visible degradation	Good	23
Perlast® G70B	Black	66	230+	Intact, no visible degradation	Good	14

The 'cost' column shows the relative costs of each elastomer type; '1' being the cost of a typical fluoroelastomer seal.
Note: IRHD stands for international rubber hardness degrees.

On visual inspection the Perlast® sample retained a smooth and moulded surface condition. No pitting of the surface and a cross-section indicated no obvious degradation of the material occurring to any significant depth. In fact, none of the perfluoroelastomers tested showed any significant damage, whilst the fluorocarbon (FKM) and silicone (VMQ) types completely degraded (see Table 10.3).

10 MTBE AS AN OCTANE BOOSTER

Environmental considerations now prevent the use of tetraethyl lead compounds as an octane booster for automotive applications. Methyl-*t*-butyl ether (MTBE) is a useful replacement and world output is estimated to exceed £20 billion annually during the 1990s. At the beginning of the decade nearly 30 new plants were scheduled for completion. MTBE is particularly aggressive and FKM seals are subject to swelling and show poor property retention when exposed to the concentrated or neat additive. Perfluoroelastomers are relatively unaffected by the chemical, although FKM may be acceptable for most blended MTBE–gasoline applications [8].

11 FUGITIVE EMISSIONS

Leaks of volatile organic compounds (VOCs) account for over 33% of all the air toxic volatiles emanating from valves, pumps, flanges and rotating

Figure 10.5. Perlast® elements in mechanical seals and valve stem seals reduce fugitive emissions, keeping manufacturing plants within the law. Reproduced with permission

machinery [9] (Figure 10.5). Regulations concerning the release of emissions into the atmosphere are understandably becoming more widespread and stringent. In the USA the Clean Air Act's Equipment Leak Rule took effect in 1994 giving CPI (chemical process industry) plants until 1997 to eliminate leaks. It has been suggested that, when the 500 ppmv (parts per million volatiles) emissions threshold takes effect in 1997, monitoring costs could exceed $600 000/year for a typical chemical plant containing roughly 10 000 valves [10]. Tests by one valve manufacturer, however, have shown that the greatest potential for reducing valve leaks is by improving stem seals [11].

The use of Perlast® seals in valve stems can give significantly better sealing performance than other elastomers with all the associated benefits of greater reliability, longer life and lower cost of maintenance. A valve stem packing consisting of chevron-shaped Perlast® seals sandwiched between graphite or acetal packers can greatly reduce the costs of specially designed systems, particularly when an expensive bellows-type seal would be the alternative.

12 GROOVE DESIGN

There are several factors contributing to the need for caution when designing grooves to contain FFKM seals. Firstly, the material has a high inherent

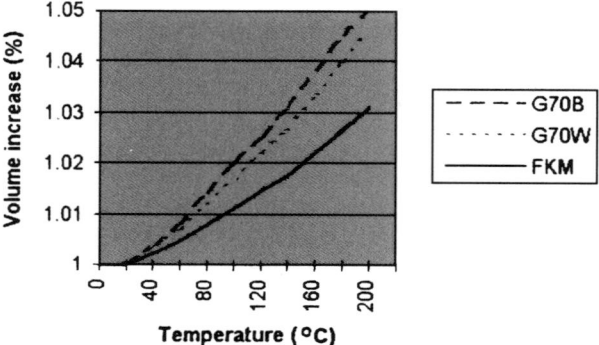

Figure 10.6. Thermal expansion comparison for fluorocarbon elastomer and per-fluoroelastomers

coefficient of linear expansion. Figure 10.6 compares the behaviours of samples of Perlast® G70B, G70W and a typical FKM when subjected to a temperature rise of about 180 °C.

Perfluoroelastomers are capable of being used over a wider temperature range than other elastomers and Perlast® is no exception to this. Swell will also occur with certain media and the result is that the total volume expansion can be as much as 50% higher than that for other elastomers. If the grooves are not designed with this in mind the material can extrude or the stresses built up can become so high that the seal is destroyed. It is important to consult the manufacturer for advice in this area. Also significant is the change in stress ratios that this thermal expansion can cause [12].

FFKMs have a tendency to become softer with increasing temperature. This can increase the risk of extrusion and if problems are forseen a harder compound should be used, whilst in extreme cases the use of a concave back-up ring may be necessary [13].

13 COMPRESSION SET

Where an elastomer is to be utilised for sealing over a period of time, the measurement of compression set has been the traditional indicator of sealing performance. Although an unsophisticated method of predicting long-term effectiveness (see section 14 below), it is nevertheless a crucial property and appears in most elastomer specifications. Figure 10.7 shows the compression set of Perlast© compared with two other FFKMs and a typical FKM at high temperatures.

The Perlast® sample performed better than the other three candidates at 200 °C and displayed properties close to the FKM at the higher temperature. This is consistent with the use of Perlast® at lower service temperatures rather than, for example, DuPont's Kalrez® 4079 and emphasises the importance of selecting the right elastomer for the relevant environment.

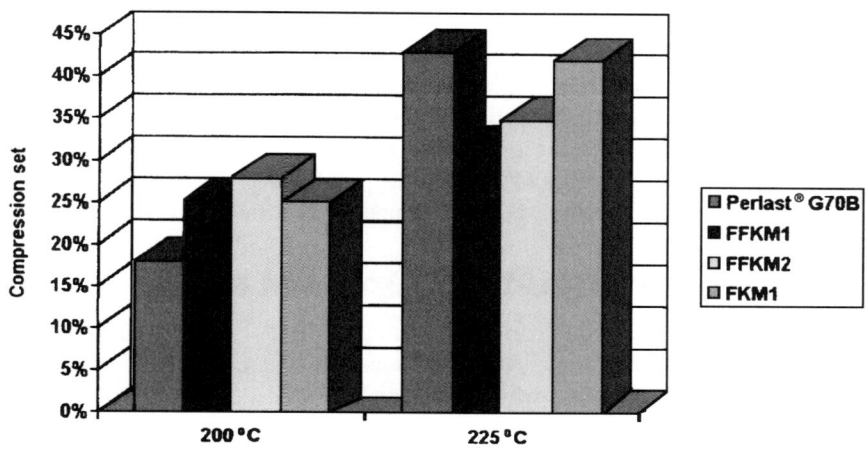

Figure 10.7. Comparison of compression set performance for various perfluoroelastomers and a fluoroelastomer at elevated temperatures Test conditions: BS 113
O-ring, 72 hours, 25% compression

14 STRESS RELAXATION

Whilst simple compression set data are useful in predicting sealing performance,
they can be deceptive. The retained sealing force over a given period can be
measured using a development of the 'Lucas test' (conducted on an instrument

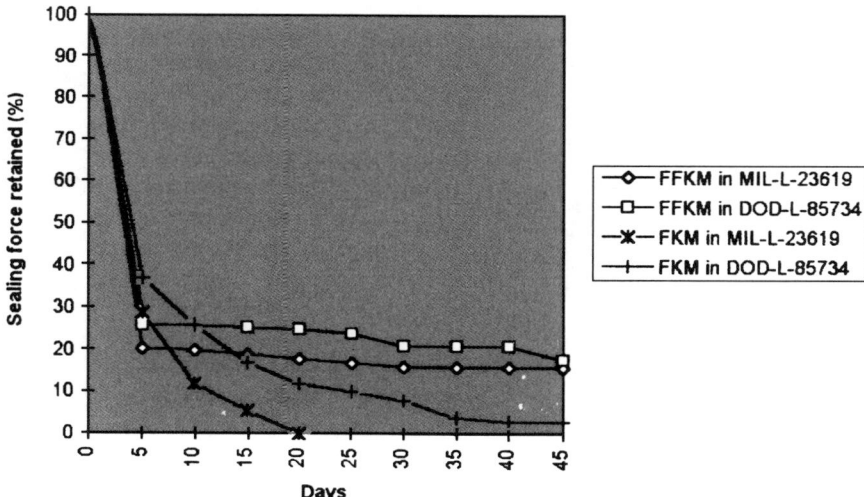

Figure 10.8. Stress relaxation curves of FFKM and FKM in two lubricants at
200 °C

produced by H. W. Wallace Ltd). The capability of FFKM in this respect is known to be excellent and can be seen in Figure 10.8. For comparison, data for a typical fluoroelastomer are also given.

For the first 100 hours both types of elastomers exhibit similar behaviour with the FKM seals performing marginally better. As the time period extends, however, the ability of the FFKM material to provide a greater sealing force is evident; this retention of sealing ability extends beyond the end of the test.

15 PERLAST® PERFLUOROELASTOMER PARTS

O-rings are by far the most common item produced in perfluoroelastomers (Figure 10.9). The elastomer is also available as rod, sheet, reinforced sheet and as custom-moulded parts to a particular design.

Encapsulated O-rings are often used where the requirement is a seal that has the chemical resistance and surface properties of PTFE in the form of FEP (copolymer of tetrafluoroethylene and hexafluoropropylene), or PFA (perfluoroalkoxy), but with an elastomeric insert to add resilience. Typically these

Figure 10.9. Perlast® parts include precision O-rings, sheet, reinforced sheet, rod, special mouldings, bonded parts and FEP/PFA encapsulated O-rings

Table 10.4.   Specific advantages of perfluoroelastomers in various industries

Industry type	Advantageous properties of perfluoroelastomers
Semiconductor manufacture	Resistance to elution, extraction and aggressive chemicals
Chemical processing	Chemical and steam resistance. Control of fugitive emissions. Mechanical seals
Food processing	Chemical resistance, extraction resistance. Tolerance of aggressive cleaning materials
Pharmaceuticals	Resistance to elution, extraction and aggressive chemicals
Gas turbine manufacture	Tolerance of lubricants and additives
Transportation	Virtually universal chemical resistance and tolerance of cleansing media
Paints, dyes, lacquers	Solvent resistance
Agricultural chemicals	Chemical resistance

elastomeric inserts are based on silicone or fluorocarbon rubber. The core material is chosen to have the nearest properties in terms of chemical resistance that would be required by the medium being sealed. Should the outer sheath of FEP or PFA become damaged, however, the resultant chemical attack on the elastomeric core could, in rare circumstances, have catastrophic repercussions for the process. To fill this need, Perlast® FEP and Perlast® PFA O-rings can be used so that the core will not suffer damage in this way.

16   PERFLUOROELASTOMER MARKETS

When absolute costs are compared, perfluoroelastomers may appear astonishingly expensive. The fact is that where they are specified, the savings in terms of replacement, maintenance and reliability can be spectacular. Consequently the elastomer has, in recent years, found niches in some areas where there is simply no comparable alternative. It is not improvident to pay a few hundred pounds for a seal when it can be left in place for months, replacing less expensive options that lasted only days or weeks. Downtime and the cost of stripping and rebuilding equipment are costly. The sheer reliability of a perfluoroelastomer is often reason enough to justify its use, irrespective of the longer service life which it will normally give.

Table 10.4 lists some of the application areas in which perfluoroelastomers are increasingly used in the form of O-rings, diaphragms, lip seals, sheet (including reinforced sheets), tubes, cord and specially moulded custom items.

REFERENCES

1.   Kalrez® 4079 is capable of service at 316° C. Kalrez® is the registered trademark of Du Pont.
2.   S. Smith, *Preparation, Properties and Industrial Applications of Organofluorine Compounds*, R. E. Barks, ed., Ellis Horwood, UK, 1982. The reader is also

referred to the insider's view provided by Herman Schroeder in *Advances in Fluoroelastomers*, Elsevier Science Publishing, 1986.

3. R. J. Weston, Very high temperature perfluoroelastomer parts., 1989 Eur. Conference on High Temperature Materials, p. 7.
4. R. H. Mobbs, F. Heatley, C. Price and C. Booth, *Progress in Rubber and Plastics Technology*, **11**, 94 (1995).
5. R. E. Uschold, *Polymer Journal*, **17**, No. 1, 253 (1985).
6. G. Streit, *Gummi Fasern Kunstoffe*, **47**, 598 (1994).
7. A. Beerbower *et al.*, Predicting elastomer fluid compatibility for hydraulic systems, *Transactions of American Society of Lubrication Engineers*, **6**, 246 (1963).
8. Michael Beron, *Perfluoroelastomer Seals for Higher-Octane Performance*, DuPont Company, ME May 1989 p. 60.
9. US Environmental Protection Agency figure (EPA, Wahington DC).
10. *Environmental Engineering*. EE-8 (Sept. 1994).
11. *Chemical Engineering*, 41 (Jan. 1995).
12. *Gummi Fasern Kunstoffe*, No. 9, 598 (1994).
13. Consigli per utilizzare perfluoroelastomeri negli O-Ring, *L'Industria della Gomma*, **36**, No. 11, 23 (Nov. 1992).

11

Melt Processable Tetrafluoroethylene – Perfluoropropylvinyl Ether Copolymers (PFA)

KLAUS HINTZER and GERNOT LÖHR
Dyneon GmbH D-84508 Burgkirchen, Germany

1 INTRODUCTION

Perfluoroalkoxy (PFA)-perfluoropolymers are produced by radically copolymerizing tetrafluoroethylene (TFE) with perfluoroalkylvinylethers, preferably perfluoropropyl vinylether (PPVE). The typical vinylether content is in the range 1–5 mol% and the molecular weight is about $1-5 \times 10^5$ g/mol. These copolymers are in contrast to polytetrafluoroethylene (PTFE), melt processable as conventional thermoplastics. They are semicrystalline with a melting point in the range 300–315 °C.

PFA consists of a fluorocarbon backbone with randomly distributed perfluoro propoxy side chains. Since these side chains cannot be incorporated into the crystalline phase, the degree of crystallinity is greatly reduced to about 60 wt% for typically processed PFA. Thus, excellent mechanical properties are achieved combined with the unique properties of the nonmelt processable PTFE such as chemical inertness, heat resistance, non-flammability, antistick characteristics and exceptional dielectric properties.

PFA was first commercialized by E. I. du Pont de Nemours & Co. in the late 1960s as a new class of perfluoropolymers which can be processed by established thermoplastic processing technologies and whose performance profile rivals that of PTFE, in contrast to the perfluoropolymer FEP which is also melt processable. FEP, a copolymer of TFE with hexafluoropropylene (HFP), has a much lower temperature rating due to its lower thermal stability.

Modern Fluoropolymers. Edited by John Scheirs
© 1997 John Wiley & Sons Ltd

The PFA polymerization processes in aqueous and non-aqueous reaction systems are detailed in pioneering patents [1–3]. The economically feasible synthesis of perfluoroalkyl vinylethers, in particular PPVE, starting from hexafluoropropylene oxide is described in several patents [4–6]. An economic route for manufacturing perfluoromethylvinylether (PMVE) has been recently disclosed [7].

A very interesting route for manufacturing PFA is the copolymerization of TFE with 1,1,1,2,2 pentafluoropropyl-trifluoro vinylether in an aqueous or non-aqueous reaction system followed by postfluorination [8]. The partially fluorinated vinylether copolymerizes with higher yields. The necessary fluorination of the incorporated comonomer provides, in one step stable, perfluorinated end groups.

The perfluoroalkoxy side chains have an exceptionally high thermal stability and survive virtually unaffected at processing temperatures up to 450 °C. However, perfluoro vinylethers still are very expensive comonomers. Therefore to reduce manufacturing costs and to lower processing temperatures, further comonomers were introduced such as PMVE and notably HFP [9,10]. The terpolymer with PMVE is called MFA [11,12]. To improve certain properties of FEP like stress cracking, FEP is modified with perfluoro vinylethers [9], in particular with perfluoro butylvinylether [13]. Blends of PFA with FEP are described [14] with better molding and mechanical properties.

These melt-processable perfluoropolymers including PFA and FEP are extensively covered in Chapter 21, this volume [12], specifically as to their performance with different applications. PFA is detailed in an excellent review article [15]. This article, only dealing with PFA, focusses on the control and influence of the molecular weight distribution (MWD) and the end group chemistry. In contrast to PTFE where polymer particle morphology plays a major role to the end properties due to the different processing technology, the end-use properties of PFA are largely determined by the MWD and the type and amount of end groups apart from the comonomer content.

2 MANUFACTURING OF PFA

2.1 POLYMERIZATION

2.1.1 Nonaqueous Polymerization [21]

Suitable organic solvents are chlorofluoroalkanes or fluoroalkanes which may contain non chain transferring hydrogen atoms. The preferred solvent was 1,1,2-trichloro-1,2,2-trifluoroethane (R 113) which is being phased out as a chlorofluorocarbon (CFC) an ozone-depleting agent. Initiators have to be soluble in the chosen solvent and essentially free of chain-transferring hydrogen atoms, like perfluoro- or ω-hydroperfluoroalkanoic peroxide such as bis-perfluoropropionyl peroxide or perfluoro acyl hypofluorite [16]. Chain transfer agents can be used in both the liquid and gaseous states. Useful are the lower alcohols, in particular methanol [2], lower alkanes like methane, ethane, and chloro- or fluoroalkanes, such as 1,1,1-trifluoro-2-fluoroethane (R 134a) [17].

The polymerization is carried out under moderate TFE pressure, preferably 2–10 bar, due to the relatively high solubility of TFE in the above-mentioned solvents, and at lower temperatures, mostly below 50 °C, to suppress chain transfer of the perfluoropropoxyradical to the monomer TFE; the so-called rearrangement reaction [2] is

$$\sim\!\sim\!\sim\!CF_2\!-\!\underset{\underset{O-C_3F_7}{|}}{CF^*} \longrightarrow \sim\!\sim\!\sim\!CF_2\!-\!CFO + C_3\!-\!F_7{}^* \tag{1}$$

To control chemical homogeneity, TFE pressure is kept constant and PPVE is replenished according to conventional copolymerization kinetics over the polymerization time. The *apparent* copolymerization parameter $r_1 = k_{11}/k_{12}$ (TFE suffix 1 and PPVE suffix 2) is approximately 3. For simplicity r_2 is assumed to be zero, although homopolyperfluoroalkylvinyl ethers have been prepared [1].

Under adequate stirring conditions the reaction is not mass transfer controlled. High space time yields depending on TFE pressures of more than 600 g/l h are readily achieved. The resultant copolymer is highly swollen with the solvent. At a solid content of 5% by weight the reaction medium is practically a gel which cannot be efficiently agitated to keep the MWD under control. Therefore copolymerization has to be stopped.

The onset of this 'gelling' appears to occur at the very beginning of the reaction since an acceleration of the polymerization rate (Figure 11.1) or a sudden increase of the molecular weight is not observed as it is for the gel effect at the polymerization of for example, styrene and methacrylates. From this point of view the non-aqueous polymerization is best suited to be run as a continuous process.

2.1.2 Aqueous Polymerization [3,10]

The copolymerization is carried out via an aqueous emulsion polymerization under very similar reaction conditions as practiced and described with the emulsion for PTFE fine resin powders[18]. The ammonium salt of perfluorooctanoic acid is the preferred emulsifier. Water-soluble initiators like ammonium and potassium persulfate [3] or potassium permanganate [10] are used in the presence of suitable buffer substances like NH_3, $(NH_4)_2CO_3$ and ammonium oxalate respectively.

The polymerization is run at TFE pressures of 10–25 bar and at temperatures from 40 to 90 °C. The TFE pressure is kept constant and the PPVE has to be properly replenished. The *apparent* copolymerization parameter (based on the TFE pressure) $r_1 = k_{11}/k_{12}$ has a value of 5 at 70 °C and decreases slightly with decreasing reaction temperature.

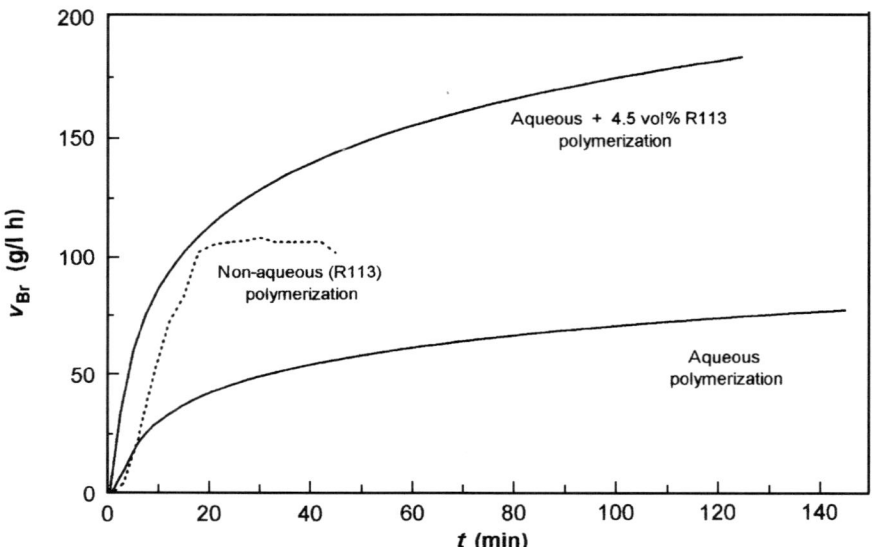

Figure 11.1. TFE uptake rate for aqueous and nonaqueous polymerization of PFA of comparable molecular weight and PPVE content. Aqueous reaction conditions: TFE pressure = 13 bar, $T = 60\,°C$, Ammonium peroxodisulfate = 3.6×10^{-4} mol/l. Nonaqueous reaction conditions: TFE pressure = 0.6 bar, $T = 47\,°C$, perfluoro propionyl peroxide = 1.0×10^{-3} mol/l. Note: v_{Br} is the polymerization rate

The same chain transfer agents as used for the nonaqueous copolymerization can be used. Hydrogen is particularly preferred [3] despite the hazards connected with it, because it leads to more thermally stable $-CF_2H$ end groups as also does R 134a [17]. Alcohols as chain transfer agents may cause colloidal instabilities of the dispersions. In general, PFA dispersions are colloid chemically more stable than PTFE dispersions. The latex particle size typically is around 200 nm at a practically achievable solid content of up to 35% by weight. At higher solid contents the dispersion tends to partially coagulate. Under such conditions the reaction system can no longer be efficiently agitated to keep the MWD and the chemical homogeneity under control.

The emulsion polymerization appears to be heavily mass transfer controlled under the practicable agitation conditions due to both the poor solubility of TFE in water and the large surface of the latex particles which is the locus of poly-merization. In consequence the space–time yield is only about 100 g/l h at a TFE pressure of 15 bar; that is, it is a magnitude lower than that of the nonaqueous reaction at comparable pressures (Figure 11.1).

The mass transfer rate can be considerably enhanced by adding organic solvents such as those used for the nonaqueous polymerization, for example

R113 [3]. Another way to considerably enhance the reaction rate is micro-emulsion polymerization [11].

Figure 11.1 shows polymerization rate/time curves for the aqueous process with and without addition of R 113 and for the nonaqueous polymerization process. Some reaction conditions are given in the legend of Figure 11.1. With the aqueous process the polymerization rate is steadily increasing with time, thus indicating that the locus of polymerization is the surface of the growing latex particles.

The presence of R 113 does not alter the shape of the polymerization rate curves. R 113 obviously plays only the role of an efficient gas carrier. As in the nonaqueous process the polymerization rate does not change basically with increasing solid content, an indication that the locus of polymerization is the organic solvent.

2.2 WORK UP

2.2.1 Recovery and Granulation

For the most important applications, PFA has to be pelletized by melt extrusion. For that purpose the polymer has to be recovered from the reaction systems in the form of fine free agglomerates which readily can be fed into and conveyed through the extruder. The granules have to be free from polymerization ingredients since PFA is very sensitive to brownish or greyish discolorations at temperatures above the melting point.

In the case of the aqueous polymerization the dispersion is coagulated by applying high shearing forces and/or adding inorganic acids like HCl or HNO₃. The granulation of the coagulated dispersion is achieved by adding organic solvents like acetone [10] or water-immiscible solvents like low boiling gasoline [19] under vigorous agitation. The granulation process is very much the same as that for manufacturing free flowing granular (suspension-polymerized) PTFE [20]. The agglomerated polymer is carefully washed with demineralized water and dried at elevated temperatures up to 280 °C [10] to remove the emulsifier and other polymerization ingredients.

In the nonaqueous process the organic solvent and the unconverted initiator are removed by drying the polymer at elevated temperatures up to 160 °C [2]. The polymer then is compacted for melt extrusion. Granulation of the polymer can also be accomplished by vigorously stirring the polymer suspension in presence of larger amounts of water.

2.2.2 Melt Pelletizing

Melt extrusion is carried out at temperatures up to 420 °C essentially without chain degradation. Degradation, however, occurs via the thermally unstable end groups in a consecutive reaction sequence which may be tentatively described as follows:

$$\sim\!\sim\!\sim CF_2-CF_2-COOH \longrightarrow \sim\!\sim\!\sim CF=CF_2 + CO_2 + HF$$
$$\sim\!\sim\!\sim CF=CF_2 + O_2 \longrightarrow \sim\!\sim\!\sim CFO + COF_2 \qquad\qquad (2)$$
$$\sim\!\sim\!\sim CF_2-CF_2-CFO \longrightarrow \sim\!\sim\!\sim CF=CF_2 + COF_2$$

In absence of atmospheric oxygen, the formed double bond splits off *difluorocarben* which further reacts to lower perfluoro olefins:

$$\sim\!\sim\!\sim CF=CF_2 \longrightarrow \sim\!\sim\!\sim CF: + :CF_2 \qquad\qquad (3)$$
$$n:CF_2 \longrightarrow TFE, HFP, PFIB$$

(PFIB is the very toxic perfluoro isobutylene). Difluorocarben is also supposed to react with the generated TFE according to the familiar reaction

$$:CF_2 + CF_2=CF_2 \longrightarrow CF_4 + C + :CF_2 \qquad\qquad (4)$$

thus producing carbon which is assumed to be the origin of grayish discoloration. Brownish discoloration is supposed to originate via dehydrofluorination by hydrogen containing components.

Practically all kinds of non-perfluorinated end groups like $-CONH_2$ [22,23], $-CF_2H$ [3] or $-CH_2OH$ [2] are more or less split off during melt extrusion, albeit to a lesser extent than the carboxyl end groups.

As a consequence of the decomposition reaction of the end groups the melt extruder and the processing equipments have to be constructed with corrosion-resistant nickel-based alloys.

2.2.3 Degassing

Because of the decomposition of the end groups the polymer pellets have to be degassed and the carbonylfluoride end groups eliminated either by treating the pellets with ammonia containing air [22] or nitrogen [19] or water [21] at elevated temperatures. The fluoride ion content is significantly reduced by both processes, the $-CFO$ end groups are transformed to carboxyl or amide groups.

2.2.4 Postfluorination

For high-purity applications mainly associated with the semiconductor industry the end groups have to be perfluorinated to reduce leaching and release of cations and anions into equipment.

Fluorination is performed with elemental fluorine mostly diluted with nitrogen as described in refs. [8,19,22,23] at elevated temperatures up to 250 °C. Melt pellets and polymer granules can be fluorinated without discernible chain degradation. Fluorination occurs by splitting off the end groups, e.g.

$$\sim\sim\sim CF_2-CFO + F_2 \longrightarrow \sim\sim\sim CF_3 + COF_2 \qquad (5)$$

Fluorine is removed by degassing [19,22]. The $-CFO$ end group is the most resistant to conversion to stable $-CF_3$ groups [19] in comparison to the other mentioned end groups.

3 CHARACTERIZATION

3.1 CHEMICAL COMPOSITION, HOMOGENEITY AND AMORPHOUS CONTENT

The perfluoroalkylvinylether content can be conveniently determined via IR-spectroscopy by comparing the absorbance at 995 cm^{-1} for PPVE, 1090 cm^{-1} for perfluoroethylvinylether and at 889 cm^{-1} for PMVE, respectively with the absorbance at 2365 cm^{-1} as described in refs. [9] and [13]. Corrections for the overlapping of the HFP band also are given in these patents. Chemical homogeneity can be estimated from the narrowness of DSC melt thermograms. It can be controlled as described in ref 24. The amorphous content also can be conveniently estimated by the absorbance at 770 cm^{-1}.

3.2 END GROUPS

Possible end groups resulting from the used initiator and chain transfer agent are $-CFO$, $-CO_2H$ (associated and non-associated), $-CONH_2$, $-CH_2OH$, $-CF_2H$ among others. The $-CF=CF_2$ end group is produced at the melt processing stage via equation (2). All these endgroups can be easily determined by IR-spectroscopy as described in refs. [2,19,25,26]. In favorable cases the number average molecular weight can be calculated from such measurements. End groups influence significantly the thermal stability and the crystallization kinetics from the melt, and hence the processing and end-use properties.

3.3 MOLECULAR WEIGHT, MOLECULAR WEIGHT DISTRIBUTION

In contrast to PTFE, the number average molecular weight M_n can be deduced by quantification of the end groups and the weight average M_w from the (extrapolated) plateau modulus of the complex viscosity via dynamic melt viscoelasticity. The dynamic rheometry additionally allows the determination of the MWD [27,28]. The potential of this method is demonstrated for PFA and other fluoropolymers in refs. [29,30].

As with any other melt-processable polymer proper tailoring of the MWD is required to obtain improved processing properties such as high extrusion rates. The control of the MWD for aqueous emulsion polymerization systems is very similar to that for suspension-polymerized modified PTFE [24]. To suppress the rearrangement reaction (equation 1) lower polymerization temperatures are advantageous. A rough estimate of the extent of this reaction and its influence of the molecular weight is given in ref. 24.

In general, nonaqueous polymerization yields a narrower MWD ($M_w/M_n \sim$ 1.8) than aqueous emulsion systems ($M_w/M_n \sim 2.5$). This fact can be deduced from the polymerization rate/time curves of Figure 11.1. At any instant of polymerization the (number average) degree of polymerization P_n is given by the familiar relation

$$P_n = v_{Br}/(v_I + v_{tr} + v_{re})$$ (6)

whereby v_{Br} is the polymerization rate, v_I the rate of initiation, v_{tr} the rate of chain transfer and v_{re} the rate of the rearrangement reaction (equation 1). This relation holds for both aqueous and non-aqueous reaction systems regardless of a possible mass transfer control as is the case for aqueous emulsion polymerization. For this system the polymerization rate is steadily increasing (Figure 11.1) due to increasing particle surface. As a consequence the molecular weight also steadily increases even if the concentration of the initiator and of the transfer agent and the polymerization temperature are kept constant.

The nature of the chain transfer agent, however, appears to play an important role for controlling the narrowness of the MWD as, for example, shown for methanol [2]. With this agent lower swelling ratios and higher flexlife could be achieved, thus demonstrating the tailoring of MWD that can be achieved.

4 PROPERTIES

4.1 PHYSICAL PROPERTIES

Commercial PFA grades have a melting point of 300–315 °C depending on the PPVE content. The melting point is lowered by about 13 K/PPVE mol%. The bulky perfluoropropoxy side groups do not enter the crystal lattice and thus reduce considerably the crystallinity to about 60% for typically processed products. The degree of crystallization would exceed 95% in the absence of the side groups at the considered useful molecular weight range of 3–5 $\times 10^5$ g/mol, as is the case with the so-called PTFE micropowders.

The side chains change the first-order transitions of PTFE at 19 °C and 30 °C respectively to but one transition at -5 °C and shift the second-order transition at 125 °C to 85 °C, as can be seen from torsion pendulum measurements (Figure 11.2). The second-order transition at -90 °C remains unchanged. Similar behaviour is observed with FEP [15] where the $-CF_3$ side groups partially enter the lattice [31]. With PFA, the side groups seems to only perturb the PTFE helix of the crystalline phase [32,33].

Figure 11.3 shows the specific volume as a function of temperature at various pressures. This PVT diagram was obtained by isobaric cooling. Similar to FEP the melting point increases by 0.09 K/bar deduced from the onset temperature

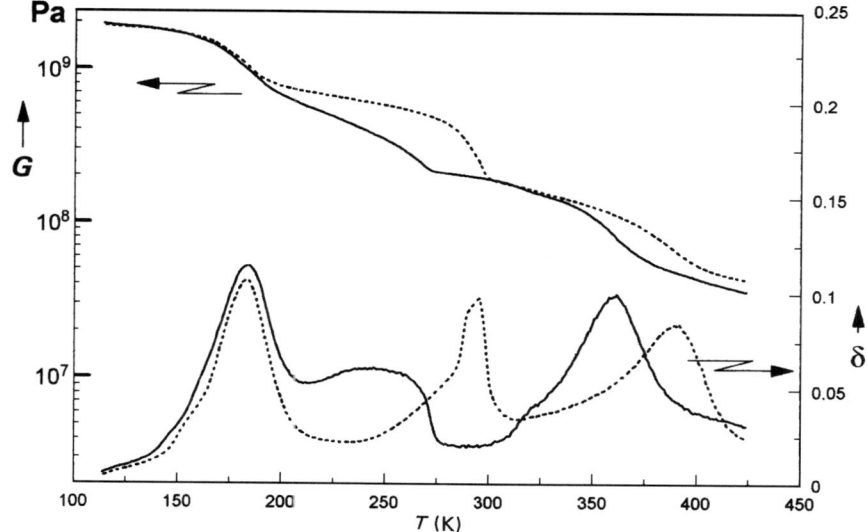

Figure 11.2. The torsion modulus G and the internal friction δ of PFA (————), 3.5 w% PPVE and PTFE ($\cdots\cdots$). Note: the first-order transition of PTFE at room temperature is absent in the case of PFA

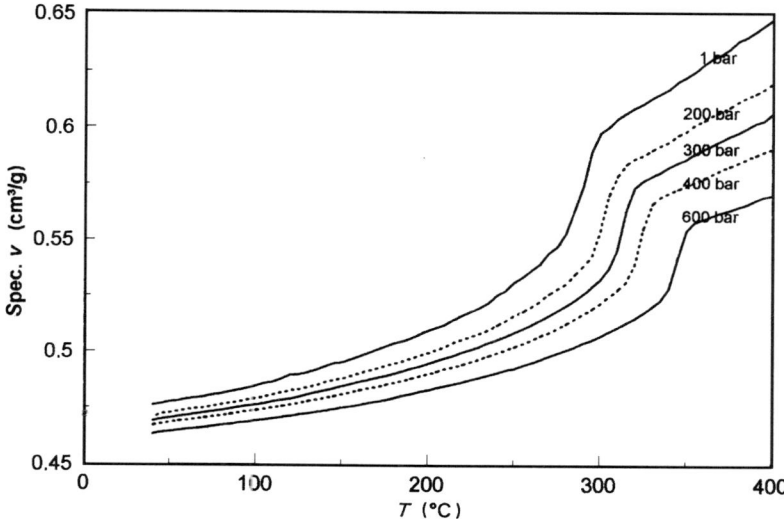

Figure 11.3. The specific volume of PFA (3.5 w% PPVE) as a function of temperature at various pressures obtained by isobaric cooling

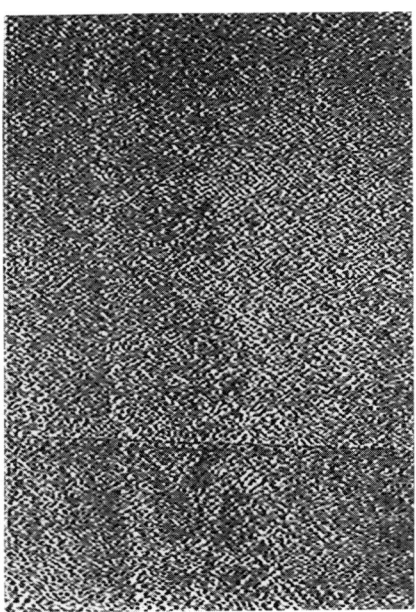

Figure 11.4. Spherulites of PFA from transfer molded articles with comparable cooling history. Left: without nucleation; right: nucleated by 5 wt % PTFE micropowder. Magnification 100×

of crystallization. Melting increases the specific volume by about 8% as with FEP [15].

In contrast to the lamellar morphology of FEP, PFA tends to crystallize in spherulites which can be easily detected by polarization microscopy (Figure 11.4). The spherulite size is sensitively influenced by nucleating agents. Their presence negatively affects the release properties in transfer moulding and the stress cracking in certain environments. PFA is compatible with PTFE micropowder which is an efficient nucleating agent as demonstrated in Figure 11.4. Addition of micropowder improves the release properties, but at the expense of mechanical properties.

4.2 END-USE PROPERTIES

Mechanical, electrical, chemical and other properties are extensively covered in excellent review articles [15,33,34] and are not addressed in this article.

In general, the properties of PFA are very similar to those of PTFE over the broad temperature range from $-200°$ to $+250°$ [33]. This may be exemplarily demonstrated by the temperature dependence of the flexural modulus as a low strain property. Table 11.1 shows a comparison of PFA with PTFE for some selected properties.

Table 11.1. Some selected properties of commercial PFA (3.5 wt% PPVE) in comparison with commercial granular non-free flowing PTFE resin. (Note: SVI denotes stretch void index.)

Property	Unit	Test method	PFA		PTFE	
Permeability			25 °C	100 °C	25 °C	100 °C
SO_2	$\dfrac{cm^3 \cdot mm}{m^2 \cdot d \cdot bar}$		190	1400	310	1600
Cl_2			110	900	160	1000
O_2		DIN 53380	240	1400	240	1400
N_2			75	700	100	700
CO_2			600	2100	800	2100
HCl			180	1800	300	1800
Water vapour	$\dfrac{g \cdot mm}{m^2 \cdot d}$	DIN 53122	0.02	4.5	0.03	3.8
Flexural modulus	MPa	DIN 53457				
		25 °C	670		600	
		100 °C	200		210	
		150 °C	120		80	
		250 °C	30		40	
Deformation under 15 MPa load	%	ASTM D 621				
		24 h	6.7		16	
		100 h	7.0		17.5	
		permanent	1.6		11	
Dielectric strength	kV/mm	DIN 53 481				
		100 μm film	125		105	
SVI value		ISO 12086	30		300	

In contrast to the microporous PTFE with a measurable void content [24] the melt processed PFA is intrinsically void free. Lower permeation coefficients should result because permeation occurs via molecular diffusion. This is indeed the case, but the effect levels off at higher temperatures. The differing polymer texture clearly shows up with the stretch void index (SVI), a measure for the generating of voids under strain, and also, but to a minor extent, with the dielectric strength.

The most remarkable difference is the very much reduced deformation under load, the so-called cold flow. This property is already significantly improved by incorporating minute amounts of PPVE into PTFE as shown with *modified* PTFE [24].

The chemical inertness of the ether bond of the perfluoro propoxy group towards strong acids even at temperatures over 300 °C is remarkable, as is the molecular weight increase on long exposures at high temperatures [15].

The overall pattern of end-use properties rivalling if not exceeding those of PTFE makes PFA a most valuable raw material for high technology applications. Its continuous service temperature is 260 °C (the same as PTFE).

4.3 PROCESSING PROPERTIES

PFA is processed by standard thermoplastic processing techniques such as extrusion or injection moulding at temperatures up to 425 °C. High temperatures are required due to its relatively high melt viscosity with a low activation energy of 50 kJ/mol in comparison to most other thermoplastics. For these processing techniques PFA grades of higher MFI values (lower molecular weights) are used.

Although PFA is a very stable resin, thermal degradation is unavoidable. Its extent is a function of temperature, residence time and shear rate. Thermal degradation occurs mainly from the end groups (equation 2–4). Chain scission becomes evident at temperatures above 400 °C depending on the shear rate. In analogy to their molecular weights, the thermal stability of the PFA melt is about two orders of magnitude lower than with PTFE. Accurate control of temperature, residence time and shear rate is required. Thermal degradation can result in discoloration and bubbles.

As with many other thermoplastics, melt fracture occurs above certain critical shear rates [34–37] leading to rough extrudate surfaces. The allowable flow rates are constrained at the low end by thermal degradation because of longer residence times and at the high end by melt fracture and chain scission. Higher extrusion rates can be achieved by *broadening* the MWD as with all thermoplastics.

Transfer moulding uses temperatures of 350–380 °C and lower shear rates. At these conditions chain scission does not occur. The gaseous products evolving from the thermal degradation of the end groups (equations 2–3) are practically entirely dissolved in the melt since the components are cooled under pressure of some 100 bar. Problems may arise at the release of the components from the mould. The release properties are significantly improved by *narrowing* the MWD. Such resins crystallize to larger spherulites with a noticeable supercooling. The mould thus becomes more transparent with a bluish hue.

PFA grades with lower melt flow indices (higher molecular weights) are preferably used for transfer moulding. Even such grades are prone to rapid nucleation even with narrow MWDs. Additional incorporation of PMVE [12] or HFP [10] improves the mold release properties. The kind of end groups might also influence the nucleation. Furthermore the mould release seems to be negatively affected by the unusual reaction increasing the molecular weight at temperatures close to or above the melting point, in particular, if the melt is produced in a melt pot and not by an extruder or an injection moulding machine.

5 APPLICATIONS

Like PTFE, PFA is a technical resin with an unique and unrivaled combination of outstanding properties. It has the best performance profile among the perfluoropolymers like FEP and partially fluorinated polymers like ethylene–TFE copolymers. Unlike these copolymers, it does not exhibit thermal stress cracking. All attempts in former years to substitute PFA as a perfluoroalkoxy polymer failed because of unavoidable sacrifices of the property profile. Compromises in terms of costs and performances were unavoidable.

Since industrial needs currently only met by PFA are growing more and more demanding, the consumption of PFA estimated for 1995 to reach 3000 tons

Figure 11.5. Heat exchanger for desulphurization of smoke gas in a coal power plant. Reproduced with kind permission of GEA Wärme und Umwelttechnik GmbH, D-44625 Herne, Germany

worldwide, is steadily increasing with annual growth rates of 10% over the last two years [38]. Such high growth rates also are expected for the next decade. Consumption is fuelled by the wire coating industry for nonflammable, environmentally inert data transmission lines of medium to high frequency at elevated temperatures, by the semiconductor industry needing high purity and chemically inert engineering resins of long endurance and by the chemical industry for reliable, safe and environment-protecting equipments including heat exchangers for desulphurization in coal power plants. Increasingly stringent regulatory, environmental, and safety regulations affecting key end-use industries are creating additional demands.

All these applications together with antistick applications for dispersion-bound coatings and heavy corrosion protection applications via rotomolding and powder coating techniques are detailed in excellent review articles [12,15,35] to which the reader is referred.

REFERENCES

1. US Pat. 3, 132, 123 (May 5, 1964), J. F. Harris, D. I. Mc. Cane (to E. I. DuPont de Nemours & Co., Wilmington, Del. USA).
2. US Pat. 3, 642, 742 (Feb. 15, 1972), D. P. Carlson (to E. I. DuPont de Nemours & Co., Wilmington, Del. USA).
3. US Pat. 3, 635, 926 (Jan. 18, 1972), W. F. Gresham and A. F. Vogelpohl (to E. I. DuPont de Nemours & Co., Wilmington, Del. USA).
4. US Pat. 3, 180, 895 (April 27, 1965), J. F. Harris and D. I. Mc. Cane (to E. I. DuPont de Nemours & Co., Wilmington, Del. USA).
5. US Pat. 3, 250, 808 (Oct. 10, 1966), E. P. Moore, A. S. Milian and H. S. Eleuterio (to E. I. DuPont de Nemours & Co., Wilmington, Del. USA).
6. US Pat. 3,358,003 (Dec. 12, 1967), H. S. Eleuterio and R. W. Meschke (to E. I. DuPont de Nemours & Co., Wilmington, Del. USA).
7. US Pat. 4, 900, 872 (May 26, 1990), G. Guglielmo and G. P. Gambaretto (to Ausimont SpA, Milan, Italy).
8. EP 0 338 755 A2 (Apr. 17, 1989), A. Nakohara, Y. Iseki and K. Murata (to TOKUYAMA Soda Kabushiki Kaisha, Japan).
9. US Pat. 4, 029, 864 (June 14, 1977), D. P. Carbon (to E. I. DuPont de Nemours & Co., Wilmington, Del. USA).
10. US Pat. 4, 262, 101 (Apr. 14, 1981), R. Hartwimmer and J. Kuhls (to Hoechst AG, Frankfurt, Germany).
11. US Pat. 4, 864, 006 (Sept. 5, 1989), E. Giannetti and M. Visca (to Ausimont SpA, Milan, Italy).
12. M. Pozzoli, G. Vita and V. Arcella, Melt processable perfluoropolymers in *Modern Fluoropolymers*, (ed. J. Scheirs) Wiley & Sons (1997).
13. EP 0 075 312 B1 (Dec. 16, 1987), T. Nakagawa, T. Amano, S. Yamaguchi and K. Asano (to Daikin Kogyo Co. Ltd, Osaka, Japan).
14. EP 0 362 868 B1 (Feb. 22, 1995), K. Ishiwari and T. Noguchi (to Daikin Industries, Limited, Osaka, Japan).
15. S. V. Gangal, *Encyclopaedia of Polymer Science and Engineering*, Wiley and Sons, Vol. 16, p. 614 (1985).

16. US Pat. 4, 588, 796 (May 13, 1986), R. C. Wheland (to E. I. DuPont de Nemours & Co., Wilmington, Del. USA).
17. US Pat. 5, 276, 261 (Jan. 4, 1994), L. Mayer and G. Löhr (to Hoechst AG, Frankfurt, Germany).
18. EP 0 030 663 B1 (Nov. 28, 1980), J. Kuhls, F. Mayer and H. Fitz (to Hoechst AG, Frankfurt, Germany).
19. US Pat. 4, 743, 658 (May 10, 1988), J. F. Imbalzano and D. L. Kerbow (to E. I. DuPont de Nemours & Co., Wilmington, Del. USA).
20. US Pat. 4, 439, 385 (Mar. 27, 1984), J. Kuhls, E. Weiß and G. Burgstaller (to Hoechst AG, Frankfurt, Germany).
21. DE 1 957 909 A1 (to be published by June 97), G. Goldmann et al. (to Hoechst AG, Frankfurt, Germany).
22. EP 0 457 255 B1 (Nov. 23, 1994), K. Ihara et al. (to Daikin Industries Limited, Osaka, Japan).
23. DE 1 901 872 (Nov. 4, 1971), C. H. Manwiller (to E. I. DuPont de Nemours & Co, Wilmington, Del. USA).
24. K. Hintzer and G. Löhr, Modified polytetrafluoroethylene, the second generation in *Modern Fluoropolymers*. (ed. J. Scheirs) Wiley & Sons (1997).
25. EP 0 178 935 A1 (Sept. 17, 1985), D. P. Carlson, D. L. Kerbow, T. J. Leck and A. H. Olson (to E. I. DuPont de Nemours & Co., Wilmington, Del. USA)
26. US Pat. 3, 085, 083 (Apr. 9, 1965), R. C. Schreyer (to E. I. DuPont de Nemours & Co., Wilmington, Del. USA).
27. W. H. Tuminello, *Polym. Eng. Sci.*, **26**, 1339 (1986).
28. S. Wu, *Polym. Eng. Sci.*, **25**, 122 (1985).
29. W. H. Tuminello, Polymer flow engineering in *Encyclopedia of Fluid Mechanics*, Gulf Publishing Co., Vol. 9, Chapter 6, p. 234 (1990).
30. M. Fleißner, *Makromol. Chem., Macromol. Symp.*, **61**, 324 (1992).
31. J. J. Weeks, I. C. Sanchez, R. K. Eby and C. I. Poser, *Polymer*, **21**, 325 (1981).
32. V. Villani, R. Pucciariello and R. Fusco, *Colloid Polym. Sci.*, **269**, 477 (1991).
33. A. Marigo, C. Marega, R. Zannetti and G. Ajroldi, *Macromolecules*, **29**, 2197 (1996).
34. S. V. Gangal, *Kirk–Othmer: Encyclopedia of Chemical Technology*, 4th edn, Vol. 11, p. 671 (1994).
35. J. Frados, Modern Plastic Encyclopedia **46**, 974 (1969).
36. E. E. Rosenbaum, and S. G. Hatzikiriakos, *Annual Techn. Conf.-Soc. Plast Eng.*, 53rd (Vol. 1), p. 1111 (1995).
37. E. E. Rosenbaum, S. G. Halzikiniakos and C. W. Stewart, *Intern. Polymer Processing*, **10**, 204 (1995).
38. Chemical Economics Handbook, *Marketing Research Report Fluoropolymers*, SRI International (1995).

12

Modified Polytetrafluoro-ethylene – the Second Generation

KLAUS HINTZER AND GERNOT LÖHR
Dyneon GmbH, D-84504 Burgkirchen, Germany

1 INTRODUCTION

Over a period of 50 years, the unique properties of polytetrafluoroethylene (PTFE) have established it to be an indispensable material for industry. Its demand had an average annual growth rate of 3–5% over the past 20 years. In the mid-1990s, annual consumption reached about 50 000 tons worldwide [1].

Two-thirds of this total is the so-called granular PTFE, polymerized in aqueous suspension, the rest emulsion polymerized is classified as fine resin powder for paste extrusion and PTFE dispersions for coating systems. This chapter solely addresses suspension polymerized PTFE.

Among its most remarkable properties which are detailed in excellent review articles [2–5] are the virtually universal chemical resistance, the high thermal stability, the very wide service temperature range, the absence of aging and of embrittlement at very low temperatures down to 4 K [2], the exceptional nonadhesion and antifrictional characteristics and the extremely low dielectric loss combined with high dielectric strength. Furthermore, PTFE may claim to be the purest polymer material since its manufacturing requires only minute amounts of adjuvants, and the processing of granular PTFE does not allow for any processing aids. Therefore PTFE is an indispensable material for the semiconductor industry.

On the other hand, PTFE as an engineering material mainly suffers from three short-comings, namely, the low creep resistance (cold flow), the very difficult weldability and its insufficiently dense polymer structure caused by its poor melt processability.

Modern Fluoropolymers. Edited by John Scheirs
© 1997 John Wiley & Sons Ltd

The reason for these weaknesses is the great tendency of the entirely linear polymer chains to crystallize from the melt. The crystalline phase typically amounting to $60-70$ wt% in end articles exhibits a ductile behavior with practically no mechanical strength. Its poor mechanical performance shows up with low molecular weight PTFE, so-called micropowders with molecular weights of some 10^5 g/mol. Mechanical strength is provided by tie-molecules bridging adjacent lamellae. The entangled tie-molecules constitute the amorphous phase.

To achieve mechanical strength and toughness the molecular weight of PTFE is required to be in the range 10^7-10^8 g/mol in order to partially suppress the crystallization and to provide sufficiently high tie-chain concentrations. The high molecular weight results in an extremely high melt viscosity in the range $10^{11}-10^{12}$ Pa s, making the polymer not melt processable. Therefore special process technologies like the preform/sintering technique or ram extrusion of the melt have been developed. Nevertheless the extreme melt viscosity impairs a perfect particle coalescence. Thus processed PTFE shows a measurable content of microvoids resulting in a microporous polymer texture.

The intrinsic shortcomings are largely overcome by significantly reducing the molecular weight and hence the melt viscosity, and simultaneously, by incorporating into the linear polymer chain bulky side groups preventing enhanced crystallization.

Perfluoro(alkyl vinyl ethers), in particular, n-perfluoro propylvinylether (PPVE) prove to be proper comonomers due to their excellent thermal stability at processing temperatures [6]. Other comonomers [7], like hexafluoropropylene and related olefins, lack the required thermal stability. The side group of the copolymerized PPVE is not incorporated in the crystal lattice. Hence the degree of crystallinity is reduced. The crystallization behavior is altered resulting in much smaller lamellae sizes.

Small amounts of copolymerized PPVE in the range $0.01-0.1$ mol% are sufficient to reduce the crystallinity of PTFE. Such modified products are classified as PTFE in accordance to ISO 12086.

This chapter deals only with PPVE-modified PTFE radically polymerized in aqueous suspensions and tries to physically elucidate its significantly better performance profile whilst not sacrificing the aforementioned unique properties of unmodified PTFE. In contrast to PTFE, literature on modified PTFE apart from patents is very scarce.

2 MANUFACTURING OF MODIFIED PTFE

2.1 POLYMERIZATION

2.1.1 Reaction Conditions

The copolymerization is carried out in aqueous suspension under practically the same reaction conditions as the homopolymerization of tetrafluoroethylene (TFE), that is, at TFE pressures from 5 to 20 bar and in a temperature range of 35–90 °C. Usually temperature and pressure are kept constant over the polymerization time. Polymerization techniques are disclosed in several patents [6,8–10].

2.1.2 Initiators and Polymerization Additives

The same initiators or redox systems [11] as with the homopolymerization can be used. Preferred are ammoniumpersulfate [6,8], alkalipersulfate [10] and potassium permanganate [12] in particular for lower polymerization temperatures. The polymerization is mainly carried out in the presence of buffers like NH_3, $(NH_4)CO_3$ [6] and ammonium or alkali oxalates [10].

The initiators mentioned lead to carboxyl end groups which are beyond the detection limits of IR spectroscopy [13,14]. In the presence of sufficient free NH_3, carboxylamide groups can also be produced.

The presence of up to 200 ppm ammonium perfluorooctanoate in the aqueous phase considerably improves the end properties [6]. By such minute amounts of the perfluorinated emulsifier the amorphous content of the polymer particles is significantly reduced, apparently resulting in a better deformability of the molding powder and hence a better performance profile.

2.1.3 Mode of Comonomer Dosage

The comonomer is either entirely fed into the reaction vessel before the start of the polymerization [6] or partially fed before the start and continuously fed during the polymerization [8–11] as requested by the copolymerization kinetics. Details of the copolymerization with hexafluoropropene are described in reference 7. Thus a higher chemical homogeneity is accomplished.

2.1.4 Control of the Chemical Homogeneity

Chemical homogeneity can be controlled by first determining the copolymerization parameter r_1 according to the familiar equation:

$$m_1/m_2 = 1 + r_1[M_1]/[M_2] \tag{1}$$

Hereby, TFE and PPVE are assigned the suffices 1 and 2 respectively while $[M_1]/[M_2]$ and m_1/m_2 are the corresponding monomer ratios in the gaseous phase and in the copolymer respectively. For simplicity equation (1) assumes that PPVE does not homopolymerize ($r_2 = 0$). The apparent copolymerization parameter $r_1 = k_{11}/k_{12}$ has a value of about 15 at 70 °C [15] and is by definition independent of the TFE pressure. It slightly decreases with lower polymerization temperatures. From equation (1) the amount of comonomer which has to be charged before polymerization can be calculated.

2.1.5 Control of the Molecular Weight

The radical copolymerization in aqueous suspension is not mass transfer controlled with standard reactor designs and agitation conditions. Therefore, at any instant of the polymerization the molecular weight is controlled by the ratio v_{Br}/v_I, whereby v_{Br} is the polymerization rate and v_I the decomposition rate of the initiator.

To keep v_I constant over the reaction time, either initiators with sufficiently large half lifetimes like persulfates below $70\,°C$ are chosen, or the consumed initiator has to be continuously replenished over the reaction time as is the case with potassium permanganate [12].

The polymerization rate v_{Br} is strongly dependent on the concentration of PPVE. The comonomer profoundly slows the polymerization even at the minute concentrations considered here.

The decrease of the polymerization rate as function of the PPVE concentration can be approximately described by the following equation [15]:

$$v_{Br} = v_{Br,o}/(1 + k \cdot x_2) \tag{2}$$

whereby $v_{Br,o}$ is the homopolymerization rate at given reaction conditions, x_2 the mole fraction of copolymerized PPVE and k the ratio k_{11}/k_{21}. The value of k is about 2.5×10^3 [mol fraction]$^{-1}$. Obviously, the copolymerization rate decreases due to the slow addition reaction of TFE to the perfluoropropylalkoxy radical. This is illustrated in Figure 12.1 by polymerization rate–time curves at different comonomer concentrations. The steadily increasing rate is due to the

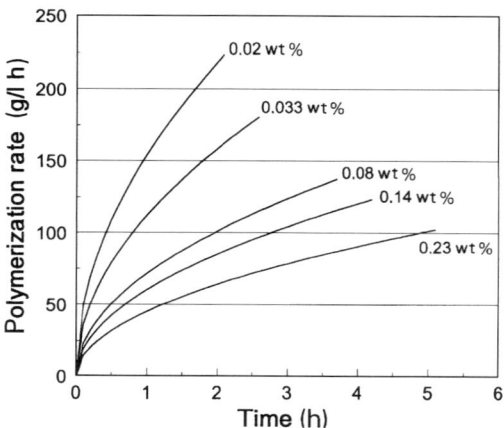

Figure 12.1.   The decrease of the polymerization rate with increasing comonomer contents at otherwise identical reaction conditions. Temperature 70 °C; TFE pressure 10 bar; initiator concentration 0.13 mmol/l (APS); final solid content 25%

simultaneously growing surface of the polymer particles which is the locus of the polymerization.

If the comonomer decreases during the reaction, the polymerization rate increases according to equation (2) and consequently the molecular weight. To keep the molecular weight under control, a constant ratio $[M_1]/[M_2]$ has to be accurately maintained over the polymerization time.

Another consequence of equation (2) is the inevitable decrease of the molecular weight by the increase of the comonomer content under otherwise equal reaction conditions (Tab. 12.1).

2.2 WORK UP

The work up of modified PTFE is the same as with unmodified PTFE and is detailed in the quoted patents [6,8–10,12]. The suspension polymerisate is carefully washed, dried and ground to a particle size of 10–50 μm for nonfree flow molding powders or granulated by help of organic solvents for free flow molding powders [8]. For ram extrusion, the reactor beads, after removing the fine and coarse particles by sieving, are used [9] in unsintered or sintered form.

Chemicals from the polymerization have to be carefully removed to avoid brownish or grayish discoloration [10] at the processing stage. Decarboxylation of the end groups may occur at the practiced drying temperatures that are above 180 °C.

3 CHARACTERIZATION OF MODIFIED PTFE

3.1 COMONOMER AND AMORPHOUS CONTENT

These quantities are conveniently measured via IR spectroscopy [6,14]. The determination of the comonomer content is sensitive down to 0.005 mol%.

3.2 ESTIMATION OF THE MOLECULAR WEIGHT

Currently, there still is no reliable method available to absolutely determine the molecular weight of high molecular weight PTFE because of its insolubility in organic solvents below its melting point. Dissolution occurs above the melting point in perfluorinated solvents [16] but is prone to chain degradation [15].

Typically, the molecular weight is indirectly estimated by the standard specific gravity (SSG) [5,17] or the heat of crystallization [18]. Both methods are founded on the increasing amorphous content with increasing molecular weight. Calibrating these methods naturally cannot avoid questionable assumptions [2]. In particular, as to the mostly used SSG method, the SSG depends on the void content which can be significantly influenced by the reaction conditions. For instance, the lower the polymerization temperature and the higher the TFE pressure, the better is the deformability of the finely ground powder, and consequently the lower the void content. Since the copolymerized PPVE alters profoundly the crystallization and deformability, these methods cannot be applied for modified PTFE.

A more realistic approach seems to be the calculation of the so-called stoichiometric molecular weight from kinetic considerations. Since the polymerization of TFE is essentially free of any side reactions like transfer to monomer or polymer, the mean number average molecular weight M_n is given by the balance equation:

$$M_n = 2 \cdot (m_p/I) \cdot 100 \qquad (3)$$

where m_p denotes the mass of polymer in base moles produced by the consumption (decomposition) of I moles of initiator. The factor 2 arises from termination by combination. The molecular weight of TFE is 100. Initiator efficiency is assumed to be 1. If the initiator decomposes into two radicals as in the case of persulfates, the factor 2 cancels. Potassium permanganate decomposes into only one initiating species.

Equation (3) is independent of the reaction conditions like temperature and pressure, but does not take into account transfer reactions.

It has been shown [19] that the copolymerization of PPVE is accompanied by a transfer (side) reaction, namely the decomposition of the perfluoropropoxyl radical:

$$\begin{array}{c} {\sim}{\sim}{\sim}{\sim}CF_2\text{--}CF_2\text{--}CF^{\boldsymbol{\cdot}} \\ | \\ CF_3\text{--}CF_2\text{--}CF_3\text{--}O \end{array} \longrightarrow \quad {\sim}{\sim}{\sim}{\sim}CF_2\text{--}COF + CF_3\text{--}CF_2\text{--}CF_2^{\boldsymbol{\cdot}} \qquad (4)$$

This reaction results in a decrease of the molecular weight which can be accounted for by modifying equation (3)

$$M_n = 100 \cdot m_p(2I + X)/(I + X)^2 \qquad (5)$$

whereby X denotes the moles of COF groups produced via reaction (4) with polymerizing m_p base moles under given reaction conditions.

Little is known about this side reaction. It is favored by higher comonomer concentration, lower TFE pressures and higher polymerization temperatures. With iso-perfluoropropyl vinylether the transfer reaction is significantly accelerated [20]. Consequently, the used PPVE has to be essentially free from this isomer.

The extent of the side reaction can be estimated by carrying out the copolymerization in an inert organic solvent like CFC-113 at much higher comonomer concentrations. Under such reaction conditions the COF groups which easily hydrolyze to carboxyl groups in the presence of water can be quantitatively determined. From polymerization investigations of PFA [15] — a TFE copolymer with 1.5 mol% PPVE — one can estimate that 1% of the incorporated PPVE units undergo the side reaction at reaction conditions as given in Table 12.1.

Table 12.1. Sample characterization of modified PTFE with varying PPVE content polymerized under equal reaction conditions: 70°C, 10 bar, $I_0 = 1.3 \times 10^{-5}$ mol/l (ammoniumpersulfate) [21]

PPVE (mol%)	SSG (g/cm^3)	Crystallinity (wt%)	Lamella[a] thickness (nm)	Melt[b] viscosity 10^{-9} Pa s	$M_n \cdot 10^{-7}$ [c] eq (3) (g/mol)	$M_n \cdot 10^{-7}$ eq (5) (g/mol)
0.0075	2.163	67	98	24	1.8	1.6
0.012	2.166	65	—	20	1.5	1.25
0.030	2.163	68	82	25	1.3	1.0
0.053	2.162	64	—	28	1.1	0.8
0.086	2.163	64	66	33	0.9	0.5

[a] Determined from electron microscopic images such as Figure 12.2 [22].
[b] Determined at 350°C according to [23].
[c] Decomposition rate constant $k_d = 2 \times 10^{-5}$ s^{-1} [24].

4 THE INFLUENCE OF THE PPVE CONTENT ON THE PHYSICAL PROPERTIES

4.1 SAMPLE CHARACTERIZATION AND GENERAL FEATURES

Table 12.1 lists some characteristics of modified PTFE samples with a 10-fold variation of the comonomer content prepared under the same reaction conditions. The polymerization rate–time curves in Figure 1 suggest a very high chemical homogeneity and justify the estimation of the molecular weight according to equation (3). Within the series a decrease of the molecular weight at least by a factor of 2 at the highest comonomer content has to be expected. Tentatively the stoichiometric molecular weight is also given according to equation (5) to demonstrate the significant influence of the side reaction (4) on the lowering of the molecular weight. The M_w/M_n can be estimated from the polymerization rate–time curves of Figure 12.1 to be about 2.5 for all samples.

Specific standard gravity (SSG) and degree of crystallization appear to be basically unchanged. This might be fortuitous since the crystallinity decreasing with higher comonomer content may be compensated by the simultaneous decrease of the molecular weight within this series. Indeed, one can establish SSG stoichiometric molecular weight calibration curves [15] for given comonomer contents as shown in Figure 12.2. These curves clearly demonstrate the influence of the molecular weight on the degree of crystallization.

The lamella thickness determined from electron microscopic images such as Figure 12.3 of surfaces of specimens fractured in liquid nitrogen decreases with increasing comonomer content at roughly the same degree of crystallization. The decrease of the lamella thickness is also reflected by an increased transparency and by a depression of the melting point (second melting) by 3 K for the highest comonomer content according to the Thomson–Gibbs equation [25,26].

The remarkable influence of even minute comonomer contents on the crystallization behavior is best demonstrated when comparing with unmodified

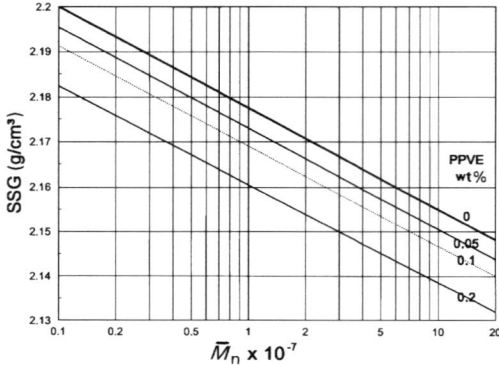

Figure 12.2. Calibration curves for standard specific gravity (SSG) with stoichiometric molecular weight according to equation (3) for different PPVE contents

Figure 12.3. Electron microscopic images of casts from surfaces of fracture (50 000 × magnification). The lamella size of modified PTFE (right) is three times smaller than with unmodified PTFE (left)

PTFE (Figure 12.3) having a lamella thickness of 200 nm even for much higher molecular weights.

At the same degree of crystallinity, the smaller crystallite size results in more tie-molecules bridging the adjacent lamellae [27] and furthermore, lower molecular weights can contribute to the tie-chain concentration [28]. This phenomenon is reflected in low strain properties such as modulus and cold flow.

Another typical feature of modified PTFE is the *intrinsic* increase of the melt viscosity. As Table 12.1 shows, the melt viscosity is slightly increasing despite a decrease of the molecular weight by a factor of 2. With linear polymers the melt viscosity is proportional to $M_w^{3,4}$. Therefore the melt viscosity should decrease by a factor of 10 within this series.

The observed increase is due to the bulky side group of the polymer chains. The intrinsic viscosity increase at a given chain length is proportional to the volume fraction of the comonomer multiplied by $(V_2/V_1)^{7.5}$ whereby v_1 and v_2 are the molar volumes of the backbone unit and the side group respectively [29]. The disentangling of the polymer chains thus appears to be drastically hindered. This phenomenon shows up with large strain properties such as ultimate tensile strength and modulus at break [6].

4.2 LOW STRAIN PROPERTIES

4.2.1 Torsion Pendulum Measurements

Figure 12.4 exemplarily illustrates the dependence of the torsion modulus and the internal friction (logarithmic decrement) on temperature for unmodified and modified PTFE. The familiar four relaxation regions [21,30] remain practically unchanged by the modification with respect to temperature.

4.2.2 Creep Resistance

Figure 12.5 shows creep resistance and recovery measurements according to ASTM D 621 for the samples of Table 12.1 together with a (unmodified) reference sample of comparable melt viscosity. The deformation under load is reduced by 50% and the recovery enhanced by a factor of 2–3 with very low comonomer contents. The effect levels off at comonomer contents above 0.05 wt%.

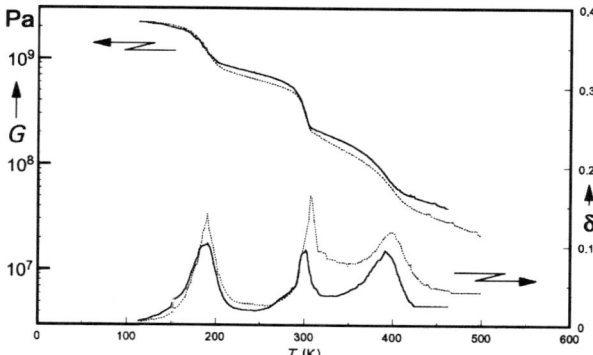

Figure 12.4.   The torsion modulus G and the internal friction δ of PTFE (···) and modified PTFE (———), 0.14 wt% PPVE

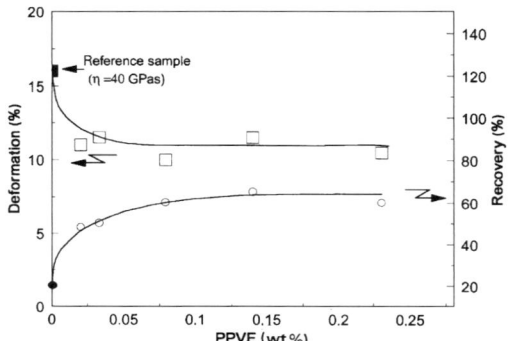

Figure 12.5. Creep resistance and recovery with varying comonomer content. Deformation after 50 h under 15 MPa load; recovered deformation after 24 h unloaded; $T = 25\,^{\circ}\text{C}$

   Clearly the phenomenon can be attributed to the much smaller lamellae size (Table 12.1 and Figure 12.2) originating from the modification and providing many more tie-molecules bridging adjacent lamellae. Such a reduction of the cold flow by 50% can only be achieved with unmodified PTFE via 25 wt% filled glass or coal PTFE compounds. But such compounds suffer from a significant deterioration of the overall performance profile such as chemical, mechanical and electrical properties.

4.2.3 Flexural Modulus

Figure 12.6 shows the temperature dependence of the flexural (elastic) modulus measured according to DIN 53 457 for a commercial modified grade in comparison with an unmodified PTFE grade. The modulus of the latter considerably decreases at the $125\,^{\circ}\text{C}$ transition region of the amorphous phase, resulting in a significantly higher modulus for modified PTFE at higher temperatures.

   The phenomenon may be rationalized by the hindered disentanglement of the tie molecules due to both a higher physical crosslink density by the smaller lamella size and to the bulky side groups.

4.3 *LARGE STRAIN PROPERTIES*

Figures 12.7 and 12.8 show the tensile strength and elongation at break of the samples of Table 12.1 in comparison with the unmodified reference sample for "skived" films from a 2 kg billet and for a ram extruded 23 mm rod. The very beneficial influence of the modification clearly shows up.

   For both specimen preparations the ultimate elongation and the tensile strength continuously increase with increasing comonomer content. The tensile strength

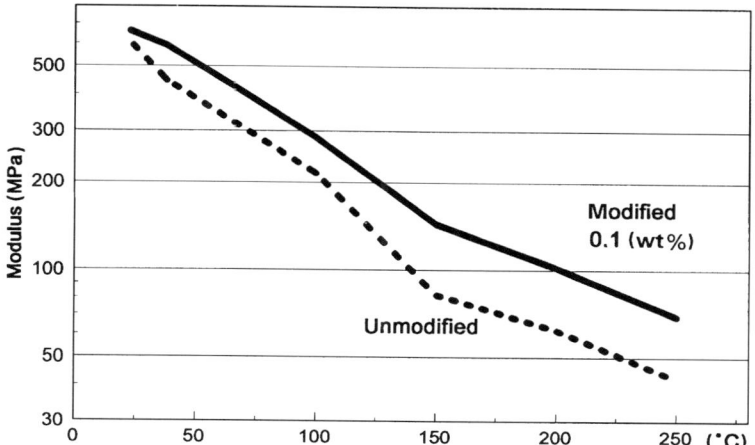

Figure 12.6. The temperature dependence of the flexural modulus of modified and unmodified commercial PTFE grades

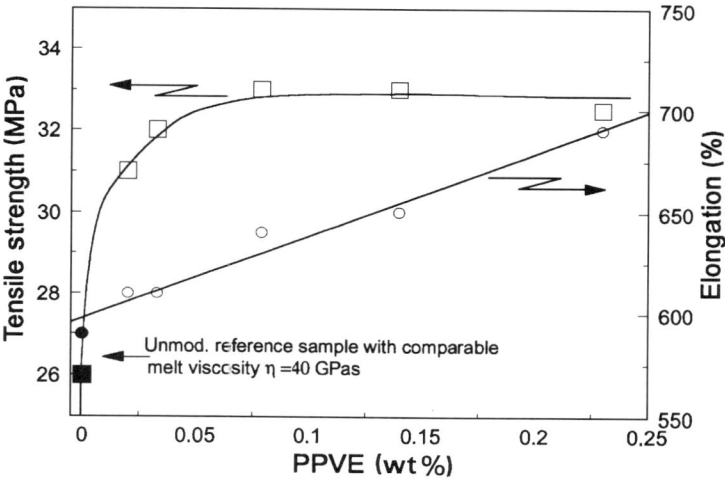

Figure 12.7. Ultimate tensile strength and elongation at varying comonomer contents. Specimen are 100 μm thick skived films from a sintered 2 kg billet. Elongation speed: 50 mm/min, $T = 23\,°C$

for the films appears to level off with higher comonomer contents, possibly due to a different cooling history, revealing the simultaneous decrease of the molecular weight with increasing modification (Table 12.1). The combination of high tensile strength with high elongation can be explained by the altered crystallization behavior of modified PTFE as demonstrated in Table 12.2.

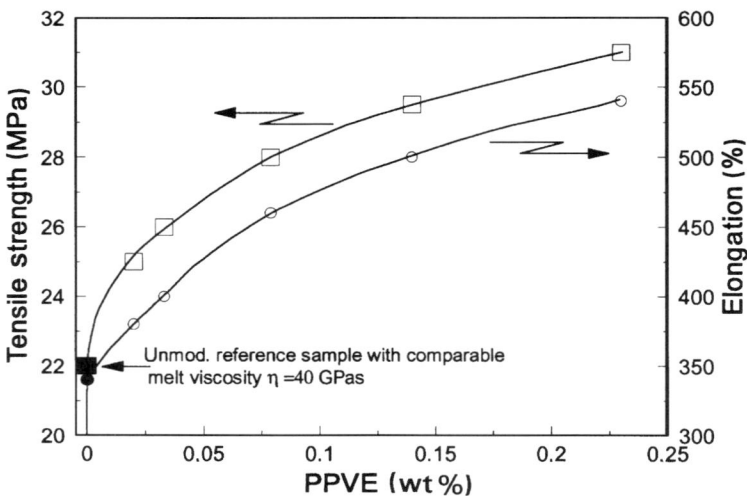

Figure 12.8. Ultimate tensile strength and elongation at varying comonomer contents. Specimen from a ram extruded rod (∅ = 23 mm). Elongation speed: 50 mm/min, $T = 23\,°C$

Table 12.2. Change of ultimate properties by quenching 100 μm thick films from the melt (TS = tensile strength at break, EL = elongation at break, η = melt viscosity at 350 °C)

	Unmodified PTFE $\eta = 40$ GPa.s			Modified PTFE 0.23 wt% PPVE, $\eta = 33$ GPa s		
	Crystal wt%	TS (N/mm²)	EL (%)	Crystal (wt%)	TS (N/mm²)	EL (%)
Unquenched	70	23	400	64	36	780
Quenched	54	40	700	51	40	550

Quenching the unmodified specimen improves the ultimate properties dramatically to the level of the modified (unquenched) specimen which is almost unaffected by quenching [6]. This leads to a higher crack resistance in large billets and facilitates the processing.

5 THE INFLUENCE OF THE REDUCED MELT VISCOSITY ON THE PHYSICAL PROPERTIES

5.1 THE DENSER POLYMER TEXTURE

Compared to commercial unmodified PTFE grades the melt viscosity of modified PTFE is lower by an order of magnitude. Thus a more efficient particle

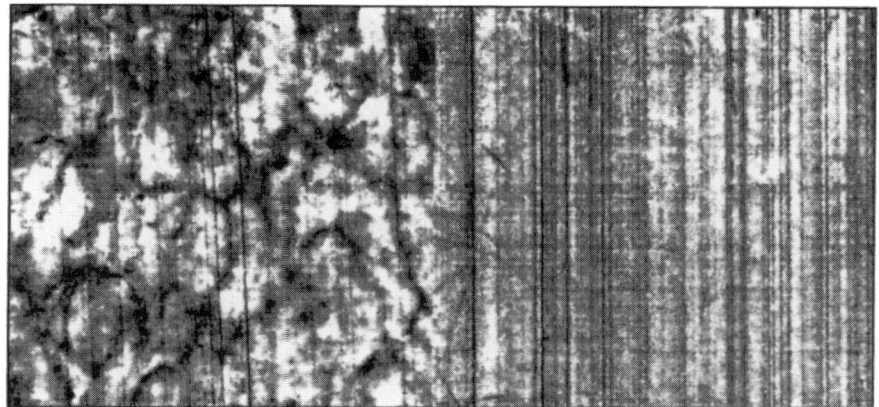

Figure 12.9. Micrographs of 100 μm skived films (25 × magnification) from free flowing (agglomerated) grades ($d_{50} \sim 400$ μm). Left: unmodified PTFE; right: modified PTFE. The polymer particle boundaries of modified PTFE are virtually indiscernible because of better particle coalescence

coalescence in the sinter process is accomplished as demonstrated in Figure 12.9 for (agglomerated) free flowing grades. The denser polymer texture results in much lower void contents, lower permeability coefficients, a higher dielectric strength in particular for free-flowing grades, and a much reduced stretch void index (Table 12.3).

WELDABILITY

Figure 12.10 illustrates the excellent weldability of modified PTFE on the basis of the stress–strain behavior of original and welded specimens in comparison with conventional PTFE. For the latter, the interdiffusion of the polymer chains within the weld surfaces does not result in sufficient chain entanglements. Failure occurs at the yield point without strain hardening.

For the welded modified PTFE specimen the stress–strain curve follows that of the unwelded specimen Thus the successful establishment of effective chain entanglements is revealed. The interdiffusion as a self-diffusion process is facilitated not only by the lower melt viscosity but also, and possibly much more, by the lower molecular weight. The self-diffusion coefficient is expected to be inversely proportional to ($\eta \cdot M_n^2$) [31]. The molecular weights of the compared specimens differ by a factor of 10. Consequently the interdiffusion of the modified polymer may be enhanced by orders of magnitude.

Furthermore the weld strength seems to be beneficially influenced by the altered crystallization behavior, leading to more physical crosslinkings which prevent the polymer chains being pulled out of the interphase during cooling.

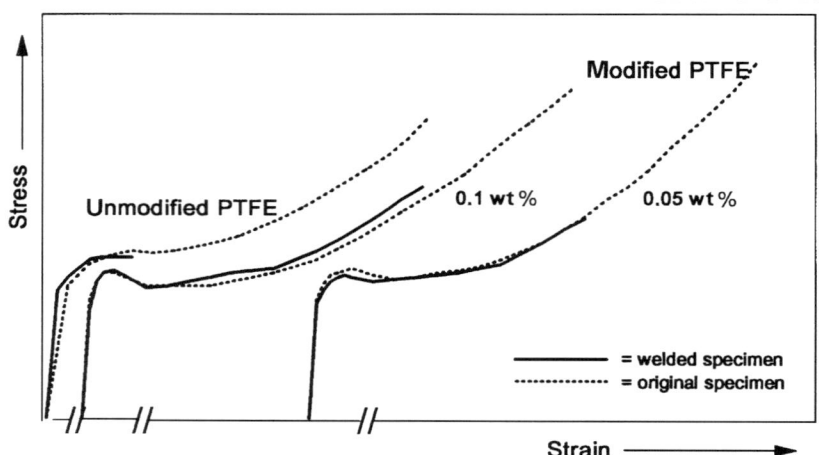

Figure 12.10. Comparison of stress–strain curves of original and welded specimens of modified and unmodified commercial PTFE grades

6 COMPARISON WITH COMMERCIAL PTFE RESINS

6.1 PHYSICAL PROPERTIES

Table 12.3 demonstrates the improved property profile of commercial modified PTFE in comparison with conventional PTFE resins for both free and nonfree-flowing grades with the same base polymers. By the modification, virtually all technically important properties are significantly improved without any evident compromise on other properties [32]. The dielectric loss is almost unaffected by the polarizable oxygen atoms of the modifier. The thermal stability of the incorporated perfluoropropyl vinylether is surprisingly high. Even on exposure to a temperature stress of 380 °C for 500 hs, no significant changes in product properties can be detected. Modified PTFE may thus justifiably be called *second-generation* PTFE.

In particular, the modification partially overcomes the disadvantages introduced by the agglomeration for achieving flowability. The agglomeration of the finely ground polymer powder results in a less dense polymer texture in molded and sintered PTFE. As a consequence, permeabilities, mechanical and electric properties are considerably affected, in particular with conventional PTFE [33]. Due to the better particle coalescence of modified PTFE the negative influence of the agglomeration is significantly lessened as already evidenced by the greatly reduced void content and stretch void index SVI — a measure of particle coalescence — for both grades.

Table 12.3. Typical physical properties of molded and sintered commercial modified granular PTFE in comparison with commercial unmodified PTFE resins

Property	Unit	Test method	Modified 0.1 wt% PPVE		Unmodified	
			Nonfree flowing	Free flowing	Nonfree flowing	Free flowing
SSG	g/cm^3		2.165	2.165	2.160	2.155
Void content	v‰	ISO 12086[a]	2.6	4.0	7.5	8.0
SVI value		ISO 12086[b]	85	85	300	300
Dielectric Strength	kV/mm	DIN 53481 100 μm film	105	100	105	80
Permeability						
SO$_2$		DIN	200	—	310	—
HCl	$\dfrac{cm^3 \cdot mm}{m^2 \cdot d \cdot bar}$	53380	210	—	300	—
Cl$_2$		25 °C	100	—	160	—
O$_2$			210	—	240	—
N$_2$			80	—	100	—
CO$_2$			540	—	800	—
Water vapor	$\dfrac{g \cdot mm}{m^2 \cdot d}$	DIN 53122 100 °C	2.6	—	3.8	—
Tensile strength	MPa	ISO 12086	41	36	41	33
Ult. elongation	%	100 μm film	630	600	450	400
Deformation under 15 MPa load	%	ASTM D 621 24 h permanent	8 4	10 5	16 11	16 11
Flexural modulus	MPa	DIN 53457 23 °C 100 °C 250 °C	650 290 60	630 255 60	600 210 40	555 210 40

[a] Void content is determined by measuring gravimetrically the (apparent) density of a specimen and the intrinsic density via its IR spectroscopically determined amorphous content [6]. The required densities of the amorphous and crystalline phase are assumed to be 1.966 g/cm^3 and 2.340 g/cm^3 respectively.
[b] The stretch void index (SVI) is calculated from the difference of gravimetrically determined densities of the unstretched and stretched specimen until break multiplied by 1000.

6.2 POTENTIAL OF SECOND-GENERATION PTFE

Modified PTFE does not only overcome the main shortcomings of the conventional *first-generation* PTFE, like pronounced cold flow, difficult weldability and microporous polymer texture, but also has a greater potential to be tailored to suit the needs of diverse applications and innovations. The greater potential is provided by the modification, as a new parameter which apparently does not imply a negative influence on the end properties. By properly controlling the degree of modification, the chemical inhomogeneity and the molecular weight distribution, for instance, the rheology of melt can be adapted to new processing techniques like blow molding or thermoforming in hollow molds. Second-generation PTFE thus may bridge the gap between first-generation PTFE and melt-processable perfluoro copolymers like the thermoplastic PFA.

7 PROCESSING AND APPLICATIONS

7.1 PROCESSING

Second-generation PTFE is nonmelt-processable and is processed with the same techniques as conventional PTFE [3,5], e.g. the preform/sintering technique like compression, automatic and isostactic molding, and ram extrusion. Since the powder properties of modified PTFE do not greatly differ from conventional PTFE molding powders no major adjustments of processing conditions are required.

In general, second-generation PTFE molding powders are more deformable and allowance for a proper release of air at the molding is advisable.

Due to its lower melt viscosity, modified PTFE is better suited for coined molding, blow molding and other thermoforming techniques for the manufacture of laminates and hollow molds [34].

7.2 APPLICATIONS

Second-generation PTFE cover all the electrical, mechanical and chemical applications [5] of first-generation PTFE. It is used to advantage where new engineering designs, higher safety and reliability standards are required, e.g. in the seals and gaskets sector, plant and equipment manufacture, mechanical engineering and the semi-conductor industry.

Due to the reduced cold flow it is specially suited for replacing asbestos in gaskets and also as a matrix material for PTFE compounds.

Its excellent weldability allows the manufacture of large, welded components with complicated geometry, flangeless pipe linings, crack-free encapsulation of metal parts, bags and containers for handling high-purity chemicals.

Its denser polymer texture results in ultra-smooth surfaces of machined components with exceptionally low ion extractables and ion retentions, making these components especially useful in semiconductor manufacturing applications.

REFERENCES

1.   Chemical Economics Handbook, *SRI International* (1995).
2.   C. A. Sperati and H. W. Starkweather, Jr., *Fortschr. Hochpolymer-Forsch.*, **2**, 465 (1961)
3.   M. Reiher, *Kunststoff-Handbuch*, Vol. XI, p. 271, Carl Hanser, München (1971)

4a. J. E. Fearn, *High Polymers, Fluoropolymers*, Vol. XXV, p. 1 Wiley-Interscience, (1971)
4b. T. W. Bates, *High Polymers, Fluoropolymers*, Vol. XXV, p. 451 Wiley-Interscience, (1971)
5. S. V. Gangal, *Encyclopedia of Polymer Science and Engineering*, Vol. 16, p. 577 Wiley & Sons, (1985).
6. US Pat. 3 855 191 (Dec. 17, 1974), T. R. Doughty, Jr. and C. A. Sperati (to E. I. du Pont de Nemours and Co., Wilmington, Del. US).
7. US Pat. 3 655 611 (Apr. 11, 1972), M. B. Mueller *et al.* (to Allied Chemical Corporation, New York, NY).
8. US Pat. 4 439 385 (March 27, 1984); US 4 774 304 (Sep. 27, 1988), J. Kuhls, E. Weiβ and G. Burgstaller (to Hoechst AG, Frankfurt, Germany)
9. US Pat. 4 379 900 (Apr. 12, 1983), R. A. Sulzbach (to Hoechst AG, Frankfurt, Germany).
10. EP Pat. 0 645 404 A 1 (Sep. 25, 1993), B. Felix, K. Hintzer, G. Löhr, Th. Schöttle (to Hoechst AG, Frankfurt, Germany).
11. US 4 078 134 (Mar. 7, 1978), J. Kuhls, A. Steininger, H. Fitz (to Hoechst AG, Frankfurt, Germany).
12. EP Pat. 0 649 863 B 1 (Oct. 10, 1994), B. Felix, K. Hintzer, G. Löhr (to Hoechst AG, Frankfurt, Germany).
13. EPA 0 222 945 A 1 (Nov. 8, 1985), M. D. Buckmaster, R. A. Morgan (to E.I. du Pont de Nemours and Co., Wilmington, Del., US)
14. T. Ogawa, Y. Iitsuguri and S. Yenemori, *Reports Res. Lab. Asahi Glass Co, Ltd*, **40** [1], 75 (1990).
15. G. Löhr and K. Hintzer, unpublished results.
16. B. Chu, J. Polymer Sci., *Part B, Polymer Physics*, **31**, 2019 (1993).
17. R. E. Moynihan, *J. Am. Chem. Soc.* **81**, 1045 (1959).
18. T. Suwa, M. Takehisa and S. Machi, *J. Appl. Polym. Sci.* **17**, 3253 (1973).
19. US Pat. 3 635 926 (Jan. 18, 1972), W. F. Gresham and A. F. Vogelpohl (to E.I. DuPont de Nemours and Co., Wilmington, Del., US).
20. DOS 2 128 256 (June 7, 1971), P. R. Resnick (to E. I. DuPont de Nemours and Co., Wilmington, Del., US).
21. J. Brandrup, G. Löhr and W. Michel, *Third Chemical Congress of North America*, Toronto, Canada, June 1988.
22. B. Heise and H. G. Kilian, University of Ulm, Germany, unpublished results.
23. G. Ajroldi, C. Garluglio and M. Ragazzini, *J. Appl. Polym. Sci.* **14**, 79 (1970).
24. J. Brandrup and E. H. Immergut, *Polymer Handbook*, 3rd edn, Wiley & Sons (1989).
25. J. G. Fatou and L. Mandelkern, *J. Phys. Chem.* **69**, 417 (1965).
26. L. Ferry, G. Vigier, R. Vassoille and J. L. Bessede, *Acta Polymer* **46**, 300 (1995).
27. R. M. Patel, K. Sehanobish, P. Jain, S. P. Chum and G. W. Knight, *J. Appl. Polym. Sci.* **60**, 749 (1996).
28. Y. Huang and N. Brown, *J. Polym. Sci., Polym. Phys.* **29**, 129 (1991).
29. M. Hoffmann, *Angew. Makrom. Chem.* **89**, 773 (1977).
30. N. G. Crum, *J. Polym. Sci.* **34**, 355 (1959).
31. J. -F. Joanny and S. J. Candan, *Hydrodynamic Properties* in *Comprehensive Polymer Science*, Pergamon Press, Vol. **2**, p. 199 (1989).
32. R. A. Sulzbach and M. Tschacher, Angew. Makrom. Chem. **109/110**, 113 (1982).
33. US Pat. 3 265 679 (Aug. 9, 1966), M. B. Black, E. E. Faust, W. S. Barnhart and R. Netsch (to Pennsalt Chemicals Corporation, Philadelphia, PA, US).
34. W. Michel, *Kunststoffe* **79** (10), 984 (1989).

13

THV Fluoroplastic

D. E. HULL, B. V. JOHNSON, I. P. RODRICKS and
J. B. STALEY
Fluorothermoplastic Business Unit, Dyneon, St Paul, MN, USA

1 BACKGROUND

In the early 1980s Hoechst AG developed a commercial production process for a unique melt processable fluoroplastic consisting of tetrafluoroethylene, hexafluoropropylene, and vinylidene fluoride (Figure 13.1). The impetus for this product was a contract to develop an outdoor fabric coating that could provide the protection of typical fluoroplastics like PTFE and ETFE, but additionally could be used with PVC-coated polyester fabric without significantly compromising overall flexibility. PTFE and ETFE could not be used since their high melt fusion temperatures destroyed PVC-coated polyester fabric. Hoechst developed the product and conducted a restricted test market (primarily fabric coating within Germany) under the brandname Hostaflon™ TFB X. In January 1993 Hoechst granted worldwide marketing rights to 3M and the product reached full commercial status under the name 3M™ THV Fluoroplastic. With the start-up of the Dyneon 3M/Hoechst joint venture in August 1996, the product is now known as Dyneon™ THV fluorothermoplastic.

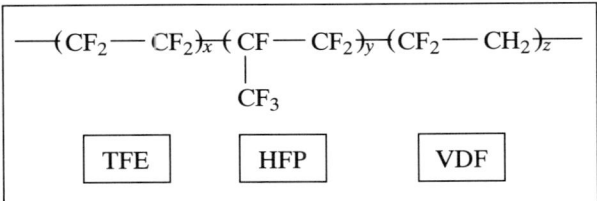

Figure 13.1. Chemical structure of THV terpolymer. (Reproduced with permission)

Modern Fluoropolymers. Edited by John Scheirs
© 1997 John Wiley & Sons Ltd

2 MANUFACTURING

THV Fluoroplastic is produced in Gendorf, Germany. THV Fluoroplastic is poly-merized under aqueous emulsion conditions and the resulting dispersions can be sold and used directly or following concentration with an emulsifier. Dry forms result from coagulation, washing, filtering and drying of the dispersion followed by extrusion (pellets) or grinding (powder) operations. Additives are generally not added to THV Fluoroplastic since the product is inherently very stable and easy to process. An electrostatic dissipative compound has been developed and was commercialized in 1995.

Products are available in 25 kilogram bags or larger containers (pellets), 50 kilogram drums (powders) or 35 kilogram (wet weight dispersions) barrels.

No chlorofluorocarbons are employed at any step during the production process and the relatively recent construction of the production facilities allowed imple-mentation of state of the art technology.

3 PROPERTIES

In general, THV Fluoroplastic is set apart from other melt-processable fluoroplastics by a combination of properties that include relatively low processing temperatures; bondability (to itself and other substrates); high flexibility; excellent clarity and low refractive index; and efficient E-beam crosslinking. One grade, THV-200, is soluble in common organic solvents. THV Fluoroplastic also retains properties common to other commercial fluoroplastics such as chemical resistance, weatherability, and low flammability.

There are four commercial THV Fluoroplastic grades (three dry and one aqueous dispersion) differing in the monomer ratios that subsequently influence melting points, chemical resistance and flexibility. Table 13.1 shows the basic properties of the three dry grades. THV-200 has the lowest melting point, the least chemical resistance of the THV Fluoroplastic grades (soluble in common solvents such as ethyl acetate, ketones, etc.), is the easiest to E-beam crosslink and is the most flexible. THV-500 has the best chemical and permeation resis-tance of the THV Fluoroplastic grades. THV-400 has a slightly lower melting point than THV-500 to meet specific processing requirements.

THV Fluoroplastic is available in pellet (THV-200G, 400G and 500G); powder (THV-200P) and aqueous dispersions (THV-330R 30% solids and 350C 50% solids).

THV Fluoroplastic is processed within the fluoroplastics fabricator base, but because of its relatively low processing temperature it is also processed by olefinic processors. Virtually all of the melt-processable plastic processing methods are used with THV Fluoroplastic including extrusion, co-extrusion and

Table 13.1. Typical THV properties (nominal values, not for specification purposes). (Reproduced with permission)

Property	ASTM method	THV grade		
		THV 200	THV 400	THV 500
Specific gravity	D792	1.95	1.97	1.98
Melting range (°C)	D3418	115–125	150–160	165–180
Thermal decomposition in air (°C)	TGA	420	430	440
Limiting oxygen index (LOI)	D2863	65	NA	75
Tensile strength at break (psi)	D638[a]	4200	4100	4100
Tensile strength at break (MPa)	D638[a]	29.0	28.3	28.3
Elongation at break	D638[a]	600%	500%	500%
Flexural modulus (psi)	D790	12000	NA	30000
Flexural modulus (MPa)	D790	82.7	NA	206.7
Hardness, Shore D	D2240	44	53	54
Dielectric constant at 23 °C				
100 kHz	D149	6.6	5.9	5.6
10 mHz	D149	4.6	4.1	3.9
Melt flow index (gm/10 min @ 260 °C 5 kg)	D1238	20	10	10
E-beam cured high temperature Resistance (°C)	NA	>150	NA	NA

[a]Property measured on extruded film

tandem extrusion; blow and coblow molding (including blown film); injection molding, skived film; vacuum forming; and solvent casting (only for THV-200). These processes produce film, tubing, containers, profiles and molded shapes.

Many of these products involve multilayer constructions where THV Fluoroplastic provides chemical, barrier or other properties in a relatively thin layer bonded to thicker layers of structural plastics or elastomers of various types.

3.1   THV FLUOROPLASTIC DISTINGUISHING FEATURES

As mentioned earlier, THV Fluoroplastic has a unique combination of properties that set it apart from other melt-processable fluoroplastics. These features are beneficial not only to end-users but to manufacturers faced with part production and design challenges.

3.1.1   THV Fluoroplastic Processing Temperatures

During most extrusion processing, THV Fluoroplastic melt temperature at the die is in the 230–250 °C range. This relatively low processing temperature presents many new options for co-processing (e.g., coextrusion, cross-head extrusion, co-blow molding) with plastics as well as with various elastomers.

The low processing temperature of THV Fluoroplastic is well below its degradation temperature and corrosion protected extruders may not be necessary. Consequently, many processors who do not have corrosion-protected equipment may be able to process THV Fluoroplastic. As with any fluoropolymer, care must still be taken to prevent long residence times in equipment and to purge equipment after process operations; attention should also be given to appropriate ventilation.

3.1.2   THV Fluoroplastic Bondability

THV Fluoroplastic can be readily bonded to itself and its unique structure presents the opportunity to develop strong, durable chemical bonding to other plastics and elastomers. Unlike many other fluoroplastics, it is not necessary to surface treat THV Fluoroplastic (i.e. chemical etch or corona) to obtain a good bond to other materials. Tie layers are generally used to bond THV Fluoroplastic to another plastic. Adhesion promoters are generally compounded into the elastomer substrate to bond THV Fluoroplastic to elastomers. The ability to form strong, durable bonds is of course an essential element of multilayer structures which utilize THV Fluoroplastic.

3.1.3   THV Fluoroplastic Flexibility

With flexural moduli values between 83 MPa and 207 MPa (THV-200 and THV-500), THV Fluoroplastic is the most flexible of the melt-processable fluoroplastics (Figure 13.2). This is beneficial when THV Fluoroplastic is used by itself such as in various film, tubing and wire and cable applications in areas where flexibility and high elongation are important. However, this high flexibility is also appreciated when THV Fluoroplastic is used to make multilayer structures with

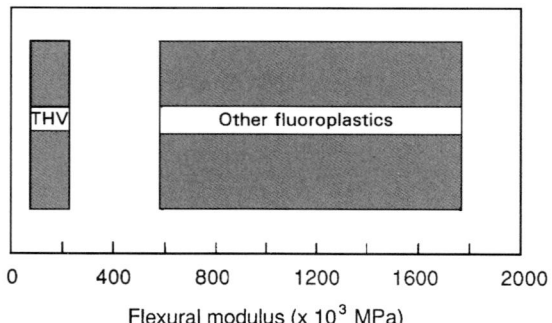

Figure 13.2. Flexural modulus values of THV relative to other commercial melt-processable fluoropolymers. It is apparent that THV polymers have much lower flexibilities relative to other fluoropolymers. THV's flexibility is a distinct advantage in multilayer constructions and tubing applications. (Reproduced with permission)

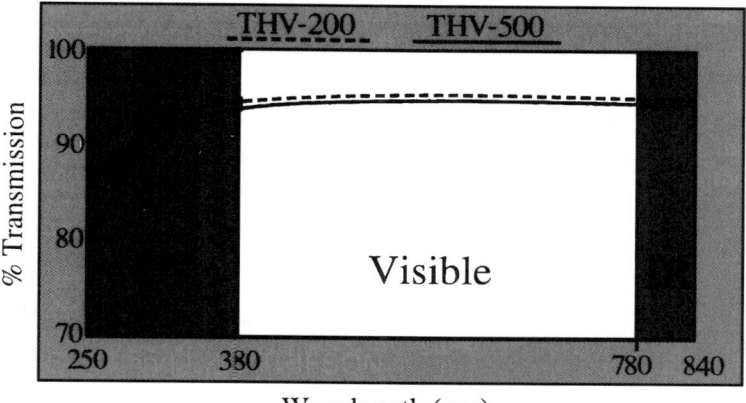

Figure 13.3. Transmission spectrum of THV-200 and THV-500 fluoropolymers. (Reproduced with permission)

other plastics, elastomers or fabrics and can either retain or improve (by allowing thinner layers of a stiffer material) the flexibility of the substrate.

3.1.4 THV Fluoroplastic Clarity and Refractive Index

THV Fluoroplastic is transparent to a broad band of light energy (UV to IR) with extremely low haze (Figure 13.3). It also possesses a very low index of refraction (i.e. 1.355 depending on grade). This low index of refraction, combined with high flexibility makes THV Fluoroplastic desirable for light management tubes as well as fiber optic data transmission constructions.

3.1.5 THV Fluoroplastic Solubility

One grade of THV Fluoroplastic, THV-200, is soluble in common solvents such as ethyl acetate and ketones. This provides the ability to cast very thin films or apply very thin coatings over appropriate substrates.

3.1.6 THV Fluoroplastic E-Beam Crosslinking

THV Fluoroplastic can be efficiently E-beam crosslinked if desired to enhance its high temperature performance. Most of the flexibility inherent in THV Fluoroplastic is retained as is its clarity. THV-200 is the easiest to E-beam crosslink of the THV Fluoroplastic grades.

In addition to its unique combination of properties, THV Fluoroplastic also provides most of the features commonly associated with fluoroplastics including chemical resistance, weatherability (including ability to shed contaminants such as dirt), low friction and low flammability.

4   PROCESSING

4.1   EXTRUSION PROCESSING

THV Fluoroplastic extrudes at lower temperatures than most fluoroplastics. The typical feed zone temperature for THV Fluoroplastic is 160–180 °C with downstream melt temperatures adjusted to obtain a die inlet temperature of 230–250 °C. This relatively low processing temperature provides two primary benefits: (1) in many cases corrosion-protected equipment is not required to extrude THV Fluoroplastic; and (2) it is easier to co-process THV Fluoroplastic with hydrocarbon plastics and elastomers (i.e. co-extrusion or crosshead inline extrusion).

As with any linear polymer like THV Fluoroplastic, melt fracture can occur under some processing conditions [1]. The traditional adjustments used to avoid melt fracture also work with THV Fluoroplastic including adjustments to die gaps and corresponding drawdown ratios, processing temperature and line speed. The onset of melt fracture in THV Fluoroplastic is generally seen as the extrudate changes from a crystal-clear appearance to an increasingly hazy appearance [1].

4.1.1   THV Fluoroplastic Coextrusion

The relatively low processing temperature of THV Fluoroplastic combined with its ability to develop strong chemical bonds to many other materials make it an ideal candidate for multilayer constructions including tubing [1]. For instance, coextruded tubes of THV Fluoroplastic on the inside (for chemical and permeation resistance) with a less expensive plastic on the outside (for strength and protection) can be easily made. Or the reverse can be done — THV Fluoroplastic on the outside for flammability or chemical resistance to protect an inner tube or wire insulation of less expensive plastic.

THV Fluoroplastic can also be sequentially inline extruded with an elastomer to produce a tube which provides the flexibility and strength of an elastomer combined with the chemical and permeation resistance of THV Fluoroplastic.

4.2   BLOW MOLDING

THV Fluoroplastic's flexibility and high elongation are ideal for blow molding difficult shapes. THV Fluoroplastic can be blow molded by itself, but it is generally co-blow molded with olefinic plastics which provide a container's structural integrity while THV Fluoroplastic provides enhanced chemical and permeation resistance.

4.3   INJECTION MOLDING

Although developed specifically for extrusion processing, current pellet grades of THV Fluoroplastic can also be injection molded [2]. Similar to extrusion processing, THV Fluoroplastic is injection molded at lower temperatures than most other fluoropolymers (200–300 °C). Generally, standard plastics injection molding equipment is suitable for fabricating this material. Research is ongoing to develop optimum processing parameters for molding THV Fluoroplastic.

Currently, typical guidelines may include:

Temperature

Melting point	115–180 °C
Process melt	200–300 °C
Mold	60–100 °C

Screw design

Geometry	Single-flighted square pitch
Barrel sizing	L/D of 15 : 1 to 28 : 1
Diameter	28 mm
Compression ratio	2 : 1 to 3 : 1 (low compression screw)
Length/diameter	23 : 1
Feed/transition/metering/profile	50% / 25% / 25%
Feed depth	4.5 mm
Metering depth	2.3 mm

Pressures

Injection	25 000 psi
Clamp force	Adequate to contain injection pressure
Back pressure	Low

Injection speed	Low 3–10 mm/sec
Orientation/shrinkage	Nominal mold shrinkage is 1.5–3.0%, depending on mold design, processing parameters, and material characteristics

Material handlings

Moisture Content	<1%
Drying	Recommended, but not always required
Coloring	THV Fluoroplastic can be pigmented
Reprocessing	THV Fluoroplastic can be reprocessed at various ratios of regrind material to virgin material

Note: Guidelines are general. Supplier should be contacted before injection molding THV Fluoroplastic.

4.4 APPLICATIONS

Table 13.2 shows a representative sampling of the various applications for THV Fluoroplastic and the properties of THV Fluoroplastic that led to its selection. Because of its unique processing and performance properties, THV Fluoroplastic

Table 13.2. Key THV features and applications. (Reproduced with permission)

Applications \ Features	Low temp. processing	Chemical and permeation resistance	Flexibility	Optical clarity	Bondability and weldability	E-beam curable	Weather-ability	Self-extinguishing
Automotive fuel systems	✓	✓	✓		✓		✓	✓
Wire cable and heat Shrink tubing	✓	✓	✓	✓	✓	✓	✓	✓
Outdoor fabric	✓	✓	✓	✓	✓		✓	✓
Safety glass	✓		✓	✓	✓		✓	✓
Flexible windows		✓	✓	✓	✓		✓	
Chemical protection garments	✓	✓	✓		✓			✓
Tank and Pipe liners	✓	✓	✓		✓		✓	✓
Solar panels	✓	✓	✓	✓	✓		✓	✓

does not generally compete with other fluoroplastics for existing applications. As this chapter is being written, THV Fluoroplastic is the newest commercial fluoroplastic and it is generally being used to develop new types of products and applications that were previously not possible (given the properties of existing fluoroplastics).

4.4.1 Chemical/Permeation Resistant Hose and Tubing

THV Fluoroplastic is used as a permeation barrier in various types of automotive fuel, liquid and vapor, hose and tubing [3] (Figure 13.4). THV Fluoroplastic is specified because of its flexibility (beneficial for assembly and chassis routing); ability to conform to various couplings (improve sealing); ease of processing for multilayer hose and tubing combined with its strong chemical bonding to various elastomeric and plastic substrates, and chemical resistance to a wide variety of fuel compositions. THV Fluoroplastic is also available in an electrostatic dissipative compound (required for some types of fuel hose and tubing) which is also chemically resistant to a wide variety of fuels including high peroxide number (i.e. PN 180) or 'sour' gasoline.

There are also a wide variety of other applications for multilayer tube or hose in which the THV Fluoroplastic fulfills the same function as in the automotive hose and tubing, but the hose or tube itself is used for various other applications such as chemical transfer hose.

4.4.2 Wire and Cable Jacketing/Insulation

THV Fluoroplastic is specified as wire and cable jacketing primarily because of its flexibility (important in aircraft, appliance, automotive and other wiring) and

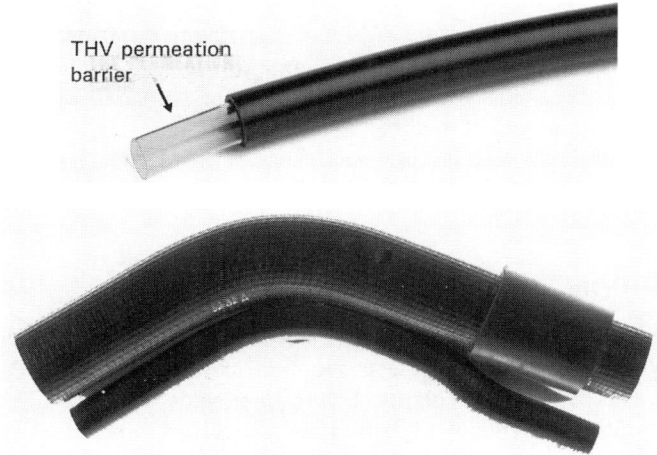

Figure 13.4. Typical examples of automotive hose with ultra-low permeation THV fuel and vapour tubing. (Reproduced with permission)

resistance to burning. E-beam crosslinking is commonly employed in this industry and such crosslinking significantly improves the high-temperature performance of THV Fluoroplastic. As noted previously, THV Fluoroplastic's relatively low processing temperature also makes it ideal for coextrusion with other less expensive plastics, and in this case THV Fluoroplastic can be coextruded as a jacket over less expensive plastic insulation to provide chemical and flammability resistance.

Although THV Fluoroplastic's electrical properties are not suitable for use as primary insulation in high-speed data communication networks (i.e. CAT 5 plenum cable); it can be used as primary insulation in less demanding applications where its flexibility again provides benefits.

4.4.3 Light Management Tubes and Fibers

THV Fluoroplastic's low refractive index (i.e. 1.355) makes it suitable for light tubes or data communication optical fiber applications particularly where high flexibility is desired.

4.4.4 Blow Molded Containers

THV Fluoroplastic's flexibility, high elongation, relatively low processing temperature and bondability make it an ideal choice for blow molded chemical, pharmaceutical or cosmetic containers where it forms the chemical/permeation barrier, while less expensive plastics (i.e. HDPE) provide the structural integrity.

4.4.5 Injection Molding

Commercial and proposed uses for THV Fluoroplastic injection molded configurations include gaskets, diaphragms, clamps, and nozzle seals in the automotive industry; gaskets, valves and seals in the chemical processing industry; and instrumentation in the testing/optical markets.

End users require the temperature and chemical resistance, the flexibility, and the excellent optical properties of THV Fluoroplastic for many of these applications.

4.4.6 Safety Glass

Safety glass is typically used in automotive and architectural window and door applications where the glass laminate is required to withstand some degree of impact as well as provide optical clarity. Examples include windows and doors in psychiatric institutions, correctional institutions as well as windshields in transportation vehicles. Each application has different impact strength requirements and in some instances multilayer laminates of glass or polycarbonate are used. In most cases a film of polyvinyl butyral (PVB) is used as an interlayer sandwiched between the glass sheets. PVB has very good optical clarity, provides the overall laminate with the necessary impact resistance and serves to bond the two sheets of glass together. However, in situations where the laminate cracks due to thermal stress caused by the heat of a fire, the PVB interlayer melts and burns relatively easily when exposed to a flame, causing the fire to spread. As a consequence, safety glass laminates using PVB as an interlayer are not a good fit in situations where there is a need for both impact resistance and a need to minimize the risk of propagating a fire.

THV Fluoroplastic film, when used as an interlayer in glass laminates, has all the features necessary for a safety glass laminate such as excellent optical clarity and impact resistance. In addition, THV Fluoroplastic does not burn spontaneously nor does it support combustion, thereby making it appropriate to use in safety glass applications where fire and its propagation are a major concern. In cases where the safety glass is exposed to light, THV Fluoroplastic film also has excellent stability to visible light and UV radiation. The ability to process THV Fluoroplastic–glass laminates using conventional laminating equipment and autoclaves is an additional manufacturing advantage.

4.4.7 Flexible Liners

Flexible liners fabricated from plastics, also referred to as drop-in liners or bag liners, are widely used in the chemical process industries as well as other industries. The typical application requires the liner to act as a flexible holding tank. The primary function of this flexible tank is to contain corrosive fluids without liner degradation from chemical attack or loss of fluid due to permeation through the liner. The structural support for the liner is provided by the tank whose walls and surfaces are protected from the corrosive fluid that is being stored.

In tank corrosion protection applications, several alternative technologies can be used to protect a tank from a corrosive fluid. Each approach uses certain types of materials that are best suited to a specific coating or lining technology; for example, spray coating a tank wall under ambient conditions can only be performed successfully using certain types of coatings. The specifics of the application ultimately determine the optimum solution — choice of material and how it is coated or lined — for protecting a tank from corrosive fluids. Key factors to be met before a material may be considered for a flexible liner application are a high degree of flexibility and resistance to flex cracking, chemical and permeation resistance, and heat sealability and weldability of the material to itself. Flexible liners, in addition to providing a chemical and permeation barrier, obviate the need to contend with the common problems of delamination and cracking of coatings. Sensing devices can also be placed between the flexible liner and the tank in order to provide an early warning system in the event of a liner leak. These factors coupled with the performance advantages of an appropriate liner material — flexibility and resistance to flex cracking, chemical and permeation resistance, ease of fabrication and installation — make flexible liners an attractive solution to tank corrosion protection problems.

THV Fluoroplastic liners embody all the key features that make it an optimum choice for 'high-end' flexible liner applications. It has the best combination of flexibility, resistance to flex cracking, and overall chemical resistance of any flexible liner material available today. Other applications that share many of the same requirements as flexible tank liners, such as bulk container liners, drum liners, collapsible tanks and floating tank covers, are also very good uses for THV Fluoroplastic film.

4.4.8 Chemical Protective Clothing

Chemical protective clothing such as garments and gloves are used in diverse segments of the chemical process industries. The primary function of these garments is to protect workers from harmful chemicals. Protective clothing such as this is also used in applications such as municipal and fire department emergency response situations where human exposure to harmful chemicals is also of concern.

A material or composite used as protective clothing needs to have the desired level of flexibility, chemical and permeation resistance. To the extent that a material must be coated or laminated to a temperature-sensitive substrate, it is critical then that this be done at the appropriate low temperatures. In some situations the risk of a fire requires that the clothing also be fire resistant and self-extinguishing.

A large array of materials are used in chemical protective clothing including various plastics and rubbers. Typically, one or more materials are coated or laminated together in order to form a composite that has the desired protective properties. THV Fluoroplastic, by virtue of its key features of flexibility, chemical and permeation resistance, low coating and laminating temperatures, and self-extinguishing nature, can bring substantial performance benefits to a chemical protective clothing barrier material. As a result of THV Fluoroplastic's combination of properties, THV Fluoroplastic can often be used as a substitute for several layers of a composite material, thereby saving on coating, laminating, and other manufacturing costs.

4.4.9 Outdoor Fabric Applications

Coated and laminated fabrics are used in many outdoor fabric applications where varying degrees of weatherability and flexibility are required. Applications include all forms of protective covers and tarpaulins, awnings and signage.

Manufacturers and users of outdoor fabric have a large spectrum of protective coatings and laminates to choose from including certain types of fluoropolymers. Typical properties that are considered to be necessary for a coating or laminate are overall weatherability — stability to light and other radiation, chemically and physically resistant to degradation by chemicals and abrasive particles, permeation resistance, especially to water and water vapor — and flexibility over a wide range of temperature conditions. In many instances the base substrate, such as PVC, requires coating or lamination to be performed at relatively low temperatures; such cases require that the coatings and laminate materials also be processable at low temperatures.

THV Fluoroplastic provides all the key properties expected of a coating or film laminate such as overall weatherability, barrier properties, flexibility and low-temperature processability for purposes of coating and laminating temperature-sensitive substrates such as PVC and polyester. In certain situations where the barrier film needs to be optically clear, THV Fluoroplastic's inherent optical clarity is an additional feature, particularly in signage applications.

Solar Applications

Certain applications for solar modules require that the module be highly flexible. An extreme case of this is a situation where the module is stored as a roll and then deployed as a flat solar panel when needed. Optically clear films are used in order to protect the solar cell surface without compromising the cell's optical transmittance.

Typical requirements for the film include a high degree of optical clarity, overall weatherability (including stability to UV radiation) and flexibility. The ability to easily laminate or coat the protective film at low temperatures to the module surface is often an added requirement.

Several types of materials are used in solar module constructions including fluoroplastics, other plastics and glass. THV Fluoroplastic meets the typical requirements of a flexible solar module protective film especially in applications where the module is required to be rolled up and stored when not in use.

5 SUMMARY

THV Fluoroplastic is a melt-processable fluoroplastic with a unique combination of properties including high flexibility; relatively low processing temperature; excellent permeation, weatherability and chemical resistance; good transparency to a wide band of light energy (UV through IR); bondability to various substrates; E-beam crosslinking; and very low flammability. Among the expanded processing options available to fabricators are coextrusions with olefinic and other temperature-sensitive plastics; in-line extrusion with elastomers; and co-blow molding, lamination and coating (aqueous and solvent) at relatively low processing temperatures.

As the marketplace exposure of this new material increases, these processing options are expected to result in a variety of new and unusual THV Fluoroplastic applications.

REFERENCES

1. Lavallée, Claude, *The 2nd International Fluoropolymers Symposium, SPI*, 'Expanding fluoropolymer processing options' (1995).
2. Peterson, Rob W. & Horns, John H. *ANTEC'95*, 'Process characterization of a fluorinated thermoplastic terpolymer using a multi-channel high speed data acquisition system' (1995).
3. Hull, Dennis E., Balzer, James R. & Tuckner, Paul F. *SAE International Congress & Exposition*, 'Unique fluoroplastic for low permeation fuel system applications' (1995).

14

Fluoropolymer Coatings for Architectural Applications

ROBERT A. IEZZI

Elf Atochem North America, Inc., Research Center, King of Prussia, PA, USA

1 BACKGROUND

1.1 GENERAL

Fluoropolymers are a unique family in the large choice of organic materials available to the coatings industry. Fluoropolymers are most often present in applications where exceptionally high performance is needed. There have been a number of fluorine-based polymers available over the years. Several of them have found their way into specialty applications other than coatings. These include: chemical process fluid handling systems, containers, computer wire insulation, electrical wire jacketing and components, piezo- and pyro-electric applications, monofilaments, membranes, and microporous filters.

Among these uses, coatings are accepted as one of the most important since they benefit most from the versatile, advantageous properties of fluoropolymers. Consequently, the technology for coatings has been extensively developed. The advances in formulation, application, processing, substrate preparation, and new polymer synthesis have resulted in optimum performance and economy.

The three fluoropolymers most commonly used for architectural coatings are polyvinyl fluoride (PVF), polyvinylidene fluoride (PVDF), and polytetrafluoroethylene (PTFE). While PVF, PVDF, and PTFE all possess similar properties, their individual performance characteristics can be better understood by considering the molecular structure of each material. The carbon/fluorine bond of all three fluoropolymers is the key to their thermal, chemical, and ultraviolet resistance properties. The number of fluorine atoms present has a direct bearing on the performance properties of each type of fluoropolymer. The unique combination of

Modern Fluoropolymers. Edited by John Scheirs

© 1997 John Wiley & Sons Ltd

properties of fluoropolymers is attributed to two intrinsic characteristics of fluorine atoms — their extremely high electro-negativity, and small atomic radius. The atomic structure of fluorine gives rise to some of the strongest chemical bonds known (Figure 14.1).

PVF contains only one fluorine atom (Figure 14.2a). Because its fusion and decomposition temperatures are so close, PVF can decompose during the baking process when used as a coating. Consequently, the baking cycle operating range is very small and requires close control.

PTFE with four fluorine atoms (Figure 14.2b) has no crystalline melting point *per se*, has a high sintering point, and consequently forms a relatively porous surface. The sintering point is well above the temperature that typical coating substrates can withstand before losing their mechanical properties. In addition,

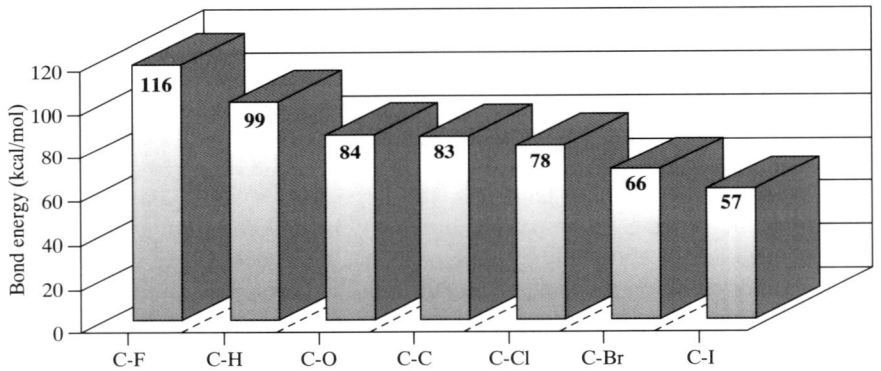

Figure 14.1.   Carbon bond energies. (Reproduced with permission)

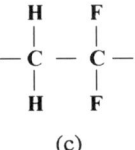

Figure 14.2.   (a) PVF; (b) PTFE; (c) PVDF. (Reproduced with permission)

PTFE has no known commonly used solvents that could be used to prepare a practical formulation.

The structure of PVDF (Figure 14.2c) contains alternating carbon/fluorine and carbon/hydrogen bonds which provide a polarity that enables the formulation of a practical coating that resists environmental degradation and dirt retention. This structure enables PVDF to resist oxidation, photochemical deterioration, fading, chalking, cracking, and airborne pollutants. Thus, PVDF has a balance of properties that makes it particularly suitable for use in coatings, especially for architectural uses.

1.2 PVDF PROPERTIES

PVDF is a high molecular weight, semi-crystalline polymer which has many unique properties such as:

- Exceptional weathering resistance
- Resistance to ultraviolet light
- High thermal and chemical resistance
- Resistance to nuclear radiation
- Good abrasion resistance
- High mechanical strength and toughness
- High purity
- Good moisture and fungus resistance
- High electrical resistivity
- Low surface energy
- Low coefficient of friction (maintenance-free, non-staining coating surface characteristics)
- Low refractive index

Crystallinity can vary from about 35 to 70%, depending on the method of preparation and thermo-mechanical history. The degree of crystallinity is important because it affects toughness and mechanical strength. The characteristics of PVDF depend on molecular weight, molecular weight distribution, extent of irregularities along the polymer chain (including main-chain defect structures and side groups), and crystalline form.

PVDF exhibits a complex crystalline polymorphism not observed in other synthetic polymers. There are four distinct crystal forms: alpha, beta, gamma, and delta. The polymorphs are present in different proportions, depending on processing conditions during polymerization. The alpha and beta forms are predominant in industrial situations.

1. The *alpha* form prevails in coatings and normal melt processing of structural parts. It is the most common form of PVDF. The alpha form is the most thermodynamically stable. Therefore, it is the most readily obtained under

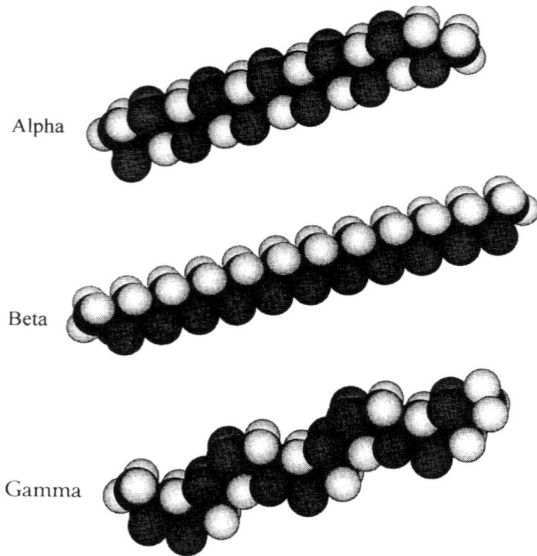

Alpha

Beta

Gamma

Figure 14.3. PVDF crystalline forms. Black = carbon; light gray = hydrogen; dark gray = fluorine. (Reproduced with permission)

a variety of conditions. The chain configuration of the alpha form is trans-gauche, placing the fluorine and hydrogen atoms alternately on each side of the chain (Figure 14.3) — the so-called 'crankcase' chain structure.

2. The *beta* form develops under mechanical deformation of melt-processed materials, usually at temperatures approaching the melting transitions. The beta form configuration consists of all the fluorine atoms on one side of the chain, and the hydrogen atoms on the other side (Figure 14.3) — the 'zig-zag' chain structure. This structure is the key to high piezo- and pyro-electric activity because the net dipole moment is very high and perpendicular to the chain direction.

3. The *gamma* form arises infrequently, and only under special circumstances (Figure 14.3).

4. The *delta* form is obtained by a distortion of one of the other phases under high electric fields.

1.3 PVDF POLYMERIZATION

Polyvinylidene fluoride is the addition polymer of 1,1-difluoroethene, $CH_2=CF_2$, commonly known as vinylidene fluoride (abbreviated VDF or VF_2). It is produced by suspension or emulsion polymerization, but most frequently by emulsion polymerization. Vinylidene fluoride is polymerized readily by free-radical initiators to form a high molecular weight, partially crystallized polymer that contains 59.4 wt% fluorine and 3 wt% hydrogen. The spatially symmetrical disposition of the hydrogen and fluorine atoms along the polymer chain gives rise to a unique polarity that affects solubility, dielectric properties, and crystal morphology. The dielectric constant is unusually high.

In addition to the PVDF homopolymer, many copolymers of vinylidene fluoride have been prepared. Among the numerous co-monomers, hexafluoropropylene ($CF_3CF=CF_2$) has assumed an important commercial role. High-performance fluoroelastomers based on vinylidene fluoride copolymers with approximately 15–40 mol% hexafluoropropylene (HFP) have been produced. Also, a tough, flexible copolymer of PVDF and tetrafluoroethylene (TFE) has been produced, as well as a terpolymer of PVDF, TFE, and HFP (hexafluoropropylene). These copolymers and terpolymers are used in the manufacture of organic coatings with higher solubilities and lower bake temperatures than the PVDF homopolymer.

2 PRODUCT TERMINOLOGY

Polyvinylidene fluoride has been commercially available since 1960. Elf Atochem was one of the first companies to commercialize PVDF for both coatings and melt-processed applications. Elf Atochem's trademark for coatings containing PVDF is KYNAR®. Elf Atochem's KYNAR 500® resin is used extensively to produce architectural coatings. The KYNAR 500 trademark is recognized worldwide, and is regarded as being synonymous with architectural coatings having outstanding long-term durability in even the most severe environments.

Subsequent to 1965, Elf Atochem developed copolymers and terpolymers based on PVDF to provide unique coating properties, particularly lower bake temperature and increased solubility (Table 14.1). Despite the success of these new materials, KYNAR 500 resin remains the predominant coating resin in terms of usage. However, the weathering properties of the PVDF copolymers and terpolymers are comparable to those of KYNAR 500 resin when properly formulated.

Table 14.1. Coating trademarks. (Reproduced with permission)

KYNAR 500®	PVDF homopolymer	Solvent dispersion
KYNAR 500® PLUS	PVDF homopolymer	Water or solvent dispersion
KYNAR 500 VLD®	PVDF homopolymer	Solvent dispersion
KYNAR 500 PC®	PVDF homopolymer	Powder coating
KYNAR SL®	PVDF–TFE copolymer	Solvent dispersion or solution
KYNAR ADS®	PVDF–TFE–HFP terpolymer	Solvent solution

2.1 DESCRIPTION OF VARIOUS COATING RESINS

2.1.1 KYNAR 500

KYNAR 500 resin is a PVDF homopolymer. It is almost always used in dispersion coatings in organic solvents. Its melting point is 320 °F (160 °C). Its significant properties, when properly formulated into a coating, include: outstanding exterior durability, chemical resistance, flexibility, and impact resistance. Abrasion resistance is also very good.

2.1.2 KYNAR 500® PLUS

KYNAR 500 PLUS resin is also a PVDF homopolymer. Its melting point is 340 °F (171 °C). The unique features of KYNAR 500 PLUS resin are:

1. It can be used in water-borne formulations, in addition to solvent dispersion formulations.
2. When properly formulated, it has higher hardness, better MEK resistance, and higher gloss (typically 10–20 gloss units: ~60° measurements) than KYNAR 500 based coatings

Coatings based on KYNAR 500 PLUS resins have the same outstanding exterior durability as those based on KYNAR 500 resins.

2.1.3 KYNAR 500 VLD®

KYNAR 500 VLD resin is a PVDF homopolymer which is used in solvent dispersion coatings. Its melting point is 340 °F (171 °C). KYNAR 500 VLD resin gives a coating with slightly higher hardness, gloss, and MEK resistance than KYNAR 500 based coatings. Its exterior weatherability is comparable to KYNAR 500 based coatings.

2.1.4 KYNAR 500 PC®

KYNAR 500 PC is a PVDF homopolymer powder coating suitable for electrostatic application (340 °F (171 °C) melting point). The properties are comparable to KYNAR 500 based coatings.

2.1.5 KYNAR SL®

KYNAR SL resin is a PVDF–TFE (tetrafluoroethylene) copolymer. TFE lowers the melting point to 255 °F (124 °C). KYNAR SL resin can be formulated as an organic solvent solution coating with a lower bake temperature (~150 °F (65 °C)) than KYNAR 500 based coatings. It can also be formulated as a solvent dispersion coating with a ~300 °F (149 °C) bake temperature. Consequently, KYNAR SL based coatings are suitable for use with heat-sensitive substrates. KYNAR SL based coatings have the same excellent properties as KYNAR 500 based coatings, but have slightly lower solvent resistance.

2.1.6 KYNAR ADS®

KYNAR ADS (air dry system) resin is a terpolymer of PVDF–TFE–HFP. Its melting point is 195 °F (90 °C). KYNAR ADS resin is used to produce air-dry touch-up coatings which can be applied in the field. Also, it is an ideal resin to produce coatings for building restoration. The weathering properties of KYNAR ADS based coatings are comparable to KYNAR 500 based coatings. However, the coating is softer than KYNAR 500 based coatings, and has lower solvent resistance.

3 KYNAR COATINGS FORMULATION

KYNAR based coatings can be formulated as solvent solution or dispersion coatings, water-borne coatings, or powder coatings. However, most KYNAR resins are used as dispersion coatings in organic solvents.

The primary components of a KYNAR based coating are as follows:

- KYNAR resin
- Acrylic modifier
- Pigments
- Organic solvents
- Other additives

3.1 KYNAR RESIN

The KYNAR resin is the primary binder component which provides the key properties of the coating. Elf Atochem does not manufacture coatings themselves; instead, they sell the various homopolymer and copolymer resins world-wide under licensing agreements. A license is granted only to quality coating companies, and only after a rigorous testing program is completed which encompasses both outdoor exposure testing and extensive laboratory testing. The license grants the licensee the right to identify their products formulated from KYNAR resins with the KYNAR trademark.

For KYNAR 500 resin, a licensee formulation must meet the following criteria to identify the product as a KYNAR 500 based coating:

1. At least 70 wt% of the total resin content must be KYNAR 500.
2. At least 40 wt% of the total solids must be KYNAR 500.

3.2 ACRYLIC MODIFIER

The acrylic modifier is usually a thermoplastic acrylic based on methylmethacrylate. The primary purpose of the acrylic is to: (a) improve pigment dispersion;

(b) increase adhesion to the substrate. The acrylic also improves the phase stability of the final coating. The inert characteristics of PVDF, while a benefit in terms of exterior durability and chemical resistance, is a detriment when producing a coating formulation. The inertness of PVDF makes pigment dispersion difficult, and inhibits reaction with the substrate to achieve good adhesion. Consequently, acrylic modifiers are used to improve pigment dispersion and coating adhesion.

The acrylic used can also be a thermoset, but thermosets are much less common than thermoplastics. Several world-wide KYNAR resin users produce their own proprietary acrylics for use with KYNAR based coatings. These proprietary acrylics provide unique properties to the coating, such as higher hardness or gloss.

3.3 PIGMENTS

Pigments are added to KYNAR based coatings for three main reasons: (1) coating aesthetics; (2) color stability; (3) ultraviolet (UV) light opacity. The pigment effects on coating aesthetics (e.g. metallic appearance) and color are obvious. However, the primary functional role of pigments to provide UV opacity. As stated previously, KYNAR based coatings are completely resistant to degradation by UV light because the PVDF resin does not absorb UV radiation. However, KYNAR based coatings are transparent to UV light. Thus, UV light can pass through a KYNAR based coating and attack underlying layers (e.g. primer) if the UV energy is not absorbed or reflected. This transmission of UV light can result in coating delamination because of destruction of the underlying layer(s).

A critical consideration in the selection of pigments for KYNAR based coatings is that the pigments must have the same long-term (20–30 years) atmospheric durability as KYNAR resins. The following pigments are usually used to achieve this long-term durability:

- Calcined metal oxide and mixed metal oxides
- Rutile titanium dioxide — exterior grade
- Mica pearlescent — exterior grade

The calcined inorganic pigments are manufactured at very high processing temperatures (up to 2400 °F) (1315 °C) that stabilize the metal oxide. The calcining process provides excellent chemical and thermal stability to the pigment. This imparts excellent exterior durability, bleed resistance, and color retention to the coating in even the most severe environments.

Exterior grade rutile titanium dioxide is the most commonly used white pigment because of its non-chalking characteristics and long-term exterior durability.

Exterior grade mica pearlescent and light interference pigments are used to produce special effects such as a metallic appearance or the appearance of a different color when viewed from a different angle. These pigments function by allowing multiple light reflection from different depths throughout the coating.

The following types of pigments are not recommended for use with KYNAR based coatings because they do not match the long-term exterior durability of the KYNAR resin:

- Organic pigments
- Fluorescent pigments
- Phosphorescent pigments
- Anatase titanium dioxide
- Extender pigments (clays, talcs)
- Cadmium pigments

3.4 ORGANIC SOLVENTS

The primary functions of the organic solvents used with KYNAR based coatings are to:

1. Provide the carrier medium for solid components (i.e. disperse KYNAR resin, pigments, and other solid additives).
2. Modify the coating viscosity to match the desired application method.
3. Dissolve KYNAR resin and promote alloying with the acrylic modifier during the baking cycle (i.e. coalescence aid).

There are three general classes of solvents associated with KYNAR resins.

1. *Active* solvents which dissolve KYNAR resin at room temperature: (a) polar solvents; (b) amides, phosphates, lower ketones.
2. *Latent* solvents which do not dissolve KYNAR resin at room temperature, but do at elevated temperature: (a) higher ketones; (b) esters; (c) glycol ethers; (d) glycol ether esters.
3. *Non-solvents* which do not dissolve KYNAR resin at any temperature: hydrocarbons, alcohols, chlorinated solvents.

A complete listing of the solvents in each category is given in Table 14.2.

Latent solvents are the most common solvents used for KYNAR resin. They produce dispersion coatings which allow the solids content of the coating to be in the range 40–50 wt%. In these dispersion coatings, the KYNAR resin is suspended as a fine powder. The resin is carried as a stable fluid dispersion which is unaffected at room temperature. When heat is applied during the baking cycle, the KYNAR resin solubilizes in the solvent, and coalesces to form a uniform film as the solvent evaporates.

Active solvents can be used to produce a solution coating. However, the solids content of solution coatings is generally limited to about half that of a

Table 14.2. Solvent classification for KYNAR resins. (Reproduced with permission)

Active solvent	Latent solvents (approx. dissolution temperature in °C)	Nonsolvents
Acetone	Butyrolacetone (65)	Hexane
Tetrahydrofuran	Isophorone (75)	Pentane
Methyl ethyl ketone	Methyl isoamyl ketone (102)	Benzene
Dimethyl formamide	Cyclohexanone (70)	Toluene
Dimethyl acetamide	Dimethyl phthalate (110)	Methanol
Tetramethyl urea	Propylene glycol methyl ether (115)	Ethanol
Dimethyl sulfoxide	Propylene carbonate (80)	Carbon tetrachloride
Trimethyl phosphate	Diacetone alcohol (100)	o-Dichlorobenzene
N-methyl pyrrolidone	Glycerol triacetate (100)	Trichloroethylene

dispersion coating because of the high viscosity which results from dissolution of the KYNAR resin.

Non-solvents are used in KYNAR based formulations to act as diluents.

3.5 OTHER ADDITIVES
Several other additives are often added to KYNAR based formulations in small quantities to impart various properties without affecting long-term weathering resistance. Examples of these additives include:

- Anti-settling agents
- Defoamers and antifoams
- Dispersion and emulsifying agents
- Preservatives and fungicides
- Surfactants
- Flatting agents
- Drying agents
- Anti-skinning agents
- Rheology modifiers
- UV absorbers

3.6 TYPICAL KYNAR 500 COMPONENT QUANTITIES
The following gives an example of the typical components of a KYNAR 500 based formulation (given in wt %).

KYNAR 500 resin (20–25%)
  minimum 70% of resin fraction
  minimum 40% of total solids
Acrylic resin (8–11%)
Pigments (12–16%)
Solvents (50–60%)

4 APPLICATION OF KYNAR BASED COATINGS

4.1 APPLICATION TECHNIQUES

KYNAR based coatings can be applied by coil coating, spray coating, or electrostatic powder coating. Coil coating is the predominant method used, but spray coating is also very common. Powder coating is used to only a limited degree. Typical application conditions are given in Table 14.3.

4.2 PRIMERS

KYNAR based coatings are usually used with a thin organic primer. The primers are used to improve adhesion to the substrate, and increase resistance to underfilm corrosion and delamination. Typical primers include: solvent based epoxy, solvent based acrylic, and water-borne acrylic. These primers are sometimes enhanced with a small amount of KYNAR resin to increase primer–topcoat compatibility. The primers are almost always doped with passivating pigments (e.g. strontium chromate) to protect exposed metal substrates at cut edges and voids in the coating.

4.3 SUBSTRATES

KYNAR based coatings are suitable for application to all common architectural substrates. These include aluminum, galvanized steel, Galvalume, Galfan, and stainless steel. The required surface preparation is similar to that used for application of any other organic coating. That is, the metal surface must be cleaned to remove dirt, oils, oxides, and other surface contaminants. The cleaned surface must then be pretreated with a thin inorganic conversion coating to maximize coating performance. For example, the preferred pretreatment for aluminum is chromium chromate. For zinc surfaces, zinc phosphate or complex cobalt oxide pretreatments are preferred. Proper cleaning and pretreatment is a critical step to ensure excellent long-term adhesion and corrosion resistance, as with any paint system.

Table 14.3. Application conditions of typical KYNAR based coatings. (Reproduced with permission)

Conditions	Coil coating	Spray coating	Powder coating
Metal temp. (°F)(°C)	450–480 (232–249 °C)	430–480 (221–249 °C)	430–480 (221–249 °C)
Baking time	30–60 sec	10–20 min	10–20 min
Primer thickness (mil)	0.2–0.3	0.2–0.4	0.2–0.4 - liquid 1.0–1.2 - powder
Topcoat thickness (mil)	0.8–1.0	1.0–1.2	1.5–2.0

4.4 FILM FORMATION MODEL OF KYNAR 500 BASED COATINGS

KYNAR 500 resin is most commonly used in dispersion coatings. The KYNAR 500 powder resin is present as dispersed particles in the latent organic solvent (Figure 14.4a). No dissolution or swelling of the KYNAR 500 particles occurs at room temperature because of the use of latent solvents. During the bake cycle, partial swelling and dissolution of the KYNAR 500 particles occurs at ~175 °F (79 °C) (Figure 14.4b). As the temperature is increased further, the partially dissolved KYNAR 500 particles begin to fuse during solvent evaporation (Figure 14.4c). Further increases in temperature (up to ~430–480 °F (221–249 °C)) cause complete evaporation of solvent, and fusing of the KYNAR 500 resin into a smooth, continuous film.

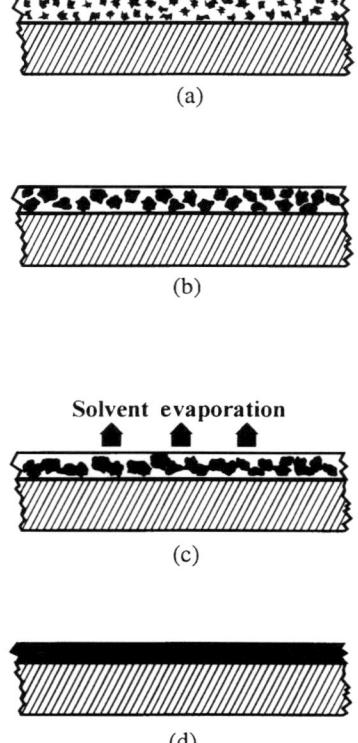

Figure 14.4. KYNAR 500 dispersion coating – film formation mechanism. (a) Room temperature - KYNAR 500 particles dispersed in latent organic solvent; (b) ~175 °F (79 °C) – KYNAR 500 particles begin to dissolve; (c) >170 °F (79 °C) partially dissolved KYNAR 500 particles begin to fuse; (d) ~430–480 °F (221–249 °C) – completely fused coating. (Reproduced with permission)

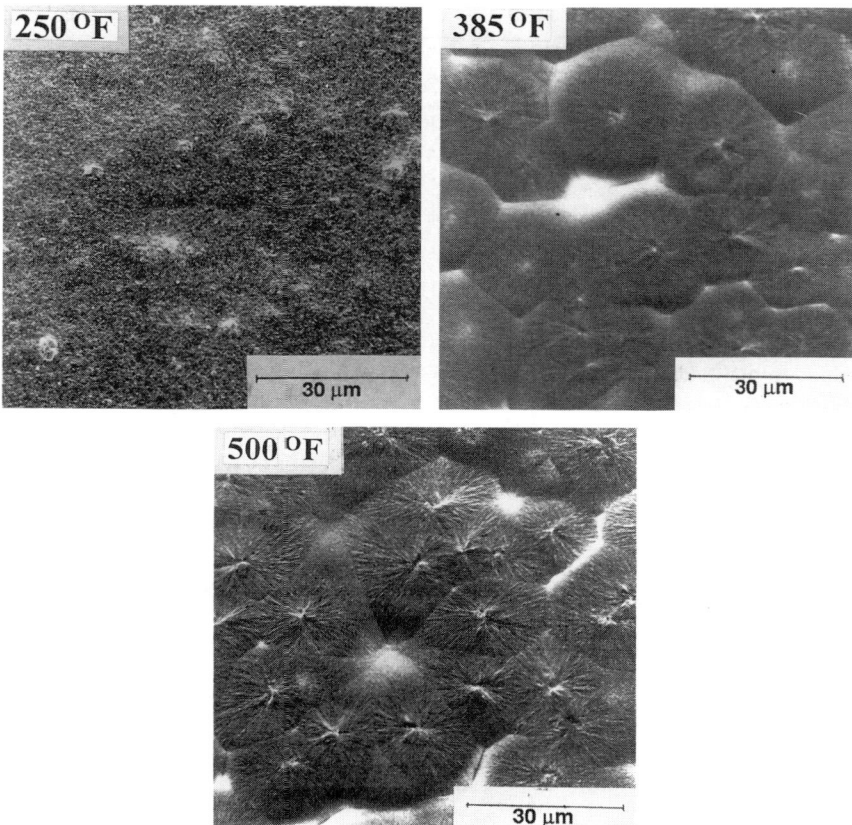

Figure 14.5. KYNAR 500 crystallization and spherulitic growth. (Reproduced with permission)

Also, during the higher temperature regions of the bake cycle ($\sim$300–500 °F), ($\sim$149–260 °C) crystallization and spherulitic growth of the KYNAR 500 resin occurs (Figure 14.5).

5 COATING PROPERTIES

The most significant property of KYNAR based coatings is their outstanding exterior durability. The exceptional weatherability is a result of the strength of the carbon–fluorine bond (116 kcal/mol), which is one of the strongest chemical bonds known. The bond strength provides a chemically inert coating, with complete resistance to UV light degradation. UV radiation is one of the major causes of deterioration of a coating exposed in the atmosphere.

Table 14.4. Properties of KYNAR based coatings. (Reproduced with permission)

Desirable properties of a coating	Intrinsic properties of KYNAR based coatings
Exterior durability	Resistant to UV degradation Long-term color and gloss retention High chalk resistance
Resistant to atmospheric pollutants, gaseous and liquid corrosives	Excellent chemical resistance — acids and liquid alkalies Not attacked by ozone
Low maintenance — low dirt pickup — nonstaining surface	Hydrophobic surface Low surface energy Low coefficient of friction
Low mildew and bacterial staining	Good moisture resistance Nonnutrient for fungal growth
Resistant to mechanical damage and wear	Good abrasion resistance Good impact resistance (tension or compression mode)
Good corrosion resistance	Excellent chemical resistance Low permeability to oxygen, moisture, and corrosive ions High electrical resistivity Good adhesion
Good formability after coating	Good mechanical properties, flexibility, adhesion under stress

In addition to exterior durability, KYNAR resins also possess several intrinsic properties which are ideally suited to produce coatings with desirable properties. These are summarized in Table 14.4.

These properties are discussed in detail in the following sections.

5.1 EXTERIOR DURABILITY

There is only one method that can accurately evaluate the exterior durability of coatings. That method is actual outdoor exposure. Accelerated weathering methods such as QUV and Xenon or Carbon Arc weatherometers are useful for screening materials under controlled conditions, but are not a substitute for actual outdoor weathering. We regularly conduct QUV and Carbon Arc weathering tests to screen variables, and detect extreme anomalies in a coating formulation. However, we rely on our outdoor exposure data to determine the true properties of a coating.

We currently have about 5000 samples of KYNAR based coatings and other architectural coatings on exposure in Miami, Florida, and Phoenix, Arizona. These samples are comprised of materials coated on commercial coil coating lines and spray lines, along with samples prepared in the laboratory. Miami and

Table 14.5. Miami, Florida outdoor exposure results. (Reproduced with permission)

Panel no.	KYNAR resin	Type coating	Color	Exp time (years)	% Gloss retention	ΔE^a
617	K-500	Solvent	White	24.0	24	2.5
717	K-500	Solvent	Dk blue	23.0	42	5.3
547	K-500	Solvent	Brown	24.0	76	6.1
475	K-500	Solvent	Green	25.0	73	7.1
2289	K-500+	Solvent	Green	12.5	100	3.9
2389	K-500+	Solvent	Brown	11.5	78	8.8
2380	K-500+	Solvent	Blue	11.5	92	4.4
2529	K-500+	Solvent	White	10.0	31	3.2
2295	K-500+	Water	Green	12.5	37	6.4
2301	K-500+	Water	White	12.5	64	2.3
2081	K-500+	Water	Yellow	13.5	90	6.6
2063	K-500+	Water	Black	13.5	74	7.9
1057	K-SL	Solvent	Green	20.0	28	4.3
931	K-SL	Solvent	Black	22.0	46	12.2
1127	K-ADS	Solvent	Green	19.0	10	2.6
2839	K-ADS	Solvent	Red	8.5	70	6.1
3492	K-500PC	Powder	White	5.0	84	1.4
3484	K-500PC	Powder	Green	5.0	83	2.8
2279	K-500VLD	Solvent	Green	10.5	52	3.5
2275	K-500VLD	Solvent	White	11.5	32	2.9

ΔE^a - CIE L*a*b* color space. (a uniform colour space measures small colour differences: higher ΔE^a values = larger colour differences)
Note: Panels exposed at 45° facing south (ASTM G7) — panels lightly washed to remove surface dirt.

Table 14.6. Arizona outdoor exposure results. (Reproduced with permission)

Panel no.	KYNAR Resin	Type coating	Color	Exp time (years)	% Gloss retention	ΔE^a
617	KYNAR 500	Solvent	White	22	24.2	5.45
848	KYNAR 500	Solvent	Amber gold	19.3	30.4	9.11
949	KYNAR 500	Solvent	Blue	19.3	22.9	4.55
1009	KYNAR 500	Solvent	Dark blue	19.3	36.2	6.19
2289	KYNAR 500 Plus	Solvent	Green	11.6	106.1	3.67
2389	KYNAR 500 Plus	Solvent	Brown	10.6	83.0	7.67
2380	KYNAR 500 Plus	Solvent	Blue	10.6	101.4	3.23
2529	KYNAR 500 Plus	Solvent	White	8.0	97.0	1.23
2295	KYNAR 500 Plus	Water	Green	11.6	69.6	2.83
2301	KYNAR 500 Plus	Water	White	11.6	51.1	2.36
2081	KYNAR 500 Plus	Water	Yellow	12.0	100.0	6.20
2063	KYNAR 500 Plus	Water	Black	12.0	61.9	6.53
933	KYNAR SL	Solvent	Green	21.0	68.4	1.46
931	KYNAR SL	Solvent	Black	21.0	41.6	11.92
2684	KYNAR ADS	Solvent	Green	7.4	63.4	1.11
2699	KYNAR ADS	Solvent	White	7.4	68.7	0.52
2279	KYNAR 500 VLD	Solvent	Green	11.0	78.2	1.58
2275	KYNAR 500 VLD	Solvent	White	11.0	58.0	3.12

ΔE^a - CIE L*a*b* color space.

Phoenix represent extreme conditions of UV radiation, heat, and humidity. Some of the samples have been exposed for up to 25 years! These samples consistently demonstrate that KYNAR based coatings have excellent color and gloss retention, low chalking, and maintain overall coating integrity. Tables 14.5 and 14.6 provide data of representative samples of various KYNAR based coatings after long-term exposure in both Florida and Arizona.

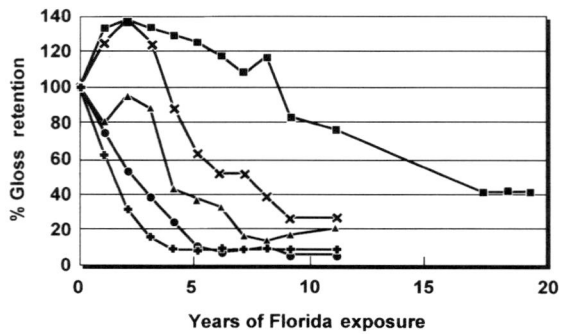

Figure 14.6. Florida exposure – gloss retention. ■ KYNAR 500; • silicone polyester; × PVF film; △ acrylic; + vinyl plastisol. (Reproduced with permission)

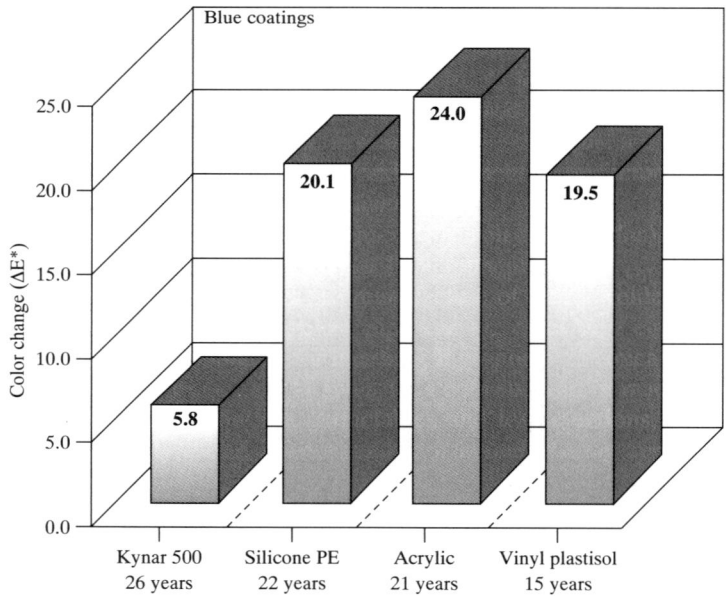

Figure 14.7. Florida exposure data (ΔE^a - CIE L*a*b* color space). (Reproduced with permission)

Also, some outdoor exposure series were initiated several years ago to provide a direct comparison of KYNAR 500 based coatings to other architectural coatings. Figures 14.6 and 14.7 give gloss retention and color change data of KYNAR 500 based coatings versus these competitive coatings. In Figure 14.6, the competitive coatings were removed from exposure after 11 years because they were badly deteriorated.

Photographs of KYNAR 500 based coatings and other architectural coatings are given in Figures 14.8 and 14.9. The top sections of the panels were

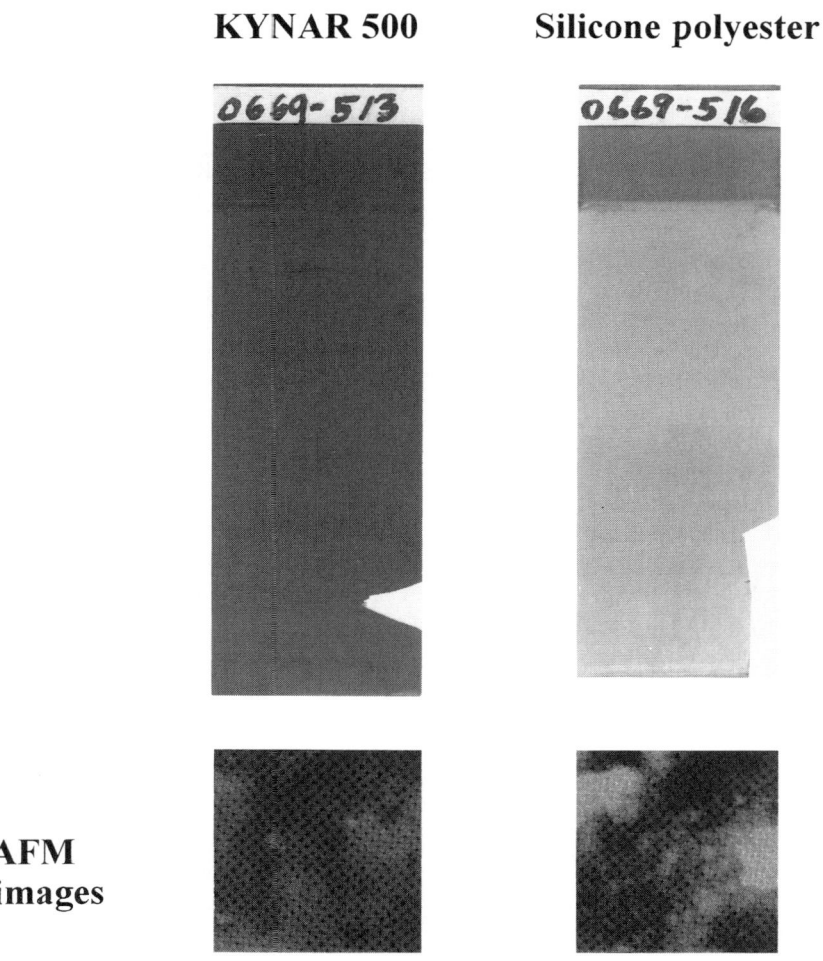

Figure 14.8. 27 years' exposure – Florida. KYNAR 500 and silicone polyester. (Reproduced with permission)

KYNAR 500   Silicone   Vinyl
                   polyester   plastisol

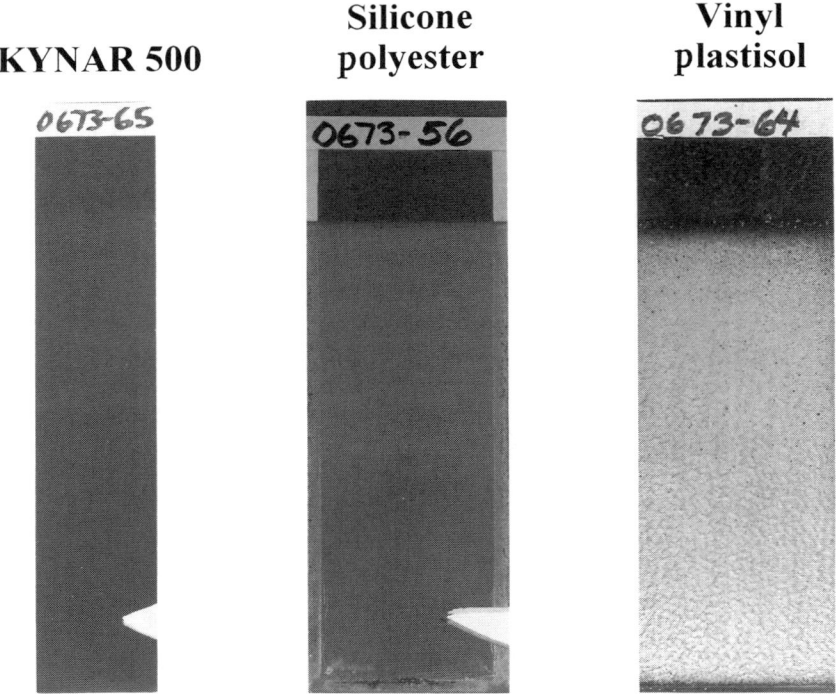

Figure 14.9. 23 years' exposure – Florida. KYNAR 500 and other coatings. (Reproduced with permission)

not exposed to retain the original appearance of the coating. Also included in Figure 14.8 are images of the exposed surfaces obtained with atomic force microscopy (AFM) operating in an oscillatory mode. AFM is a powerful, relatively new technique to examine minute details of a surface without destructive sample preparation. AFM is preferred over scanning electron microscopy (SEM) because AFM provides more precise surface characterization.

The AFM images show the KYNAR 500 based coating is relatively unaffected by natural weathering, whereas the silicone polyester based coating is severely degraded. In Figure 14.8, the KYNAR 500 based coating surface is relatively smooth, and the surface features appear to be less sharp. This is because the surface has changed only a slight amount during exposure, and the pigment particles are still encapsulated by resin. Conversely, the silicone polyester coating shows distinct regions where the coating has been completely destroyed by weathering (dark voids). Also, the pigment particles are easily visible, an indication that the resin has been depleted and unprotected pigment particles remain on the surface.

AFM also provides surface roughness data without disturbing the surface. The RMS (root mean square) roughness for the KYNAR 500 based coating was 162 nm. The silicone polyester coating had a surface roughness of 422 nm. These values further demonstrate the difference in surface degradation between the two coatings.

5.2 FLUORINATED ETHYLENE VINYL ETHER

A relatively new coating for architectural uses is based on a copolymer of fluoro-ethylene alkyl vinyl ether (FEVE). FEVE was not developed by nor used by Elf Atochem in its coating resins. The fluoro-ethylene reportedly confers weathering resistance and durability to the polymer. Alkyl vinyl ether units provide solubility in various organic solvents, transparency, gloss, hardness, and flexibility [1]. Coatings based on FEVE are generally two-component, thermosetting systems which require a high degree of cross-linking to achieve final coating properties.

We have evaluated FEVE based coatings on outdoor exposure in Florida. Our results show that FEVE based coatings have excellent gloss retention for about five years, but then the gloss decreases precipitously. It is our belief that the rapid decrease in weathering after five years is due to degradation of the alkyl vinyl ether groups. Figure 14.10 shows a comparison of KYNAR 500 based coatings to FEVE coatings after 13 years' exposure in Florida. The curves represent composite results of several samples of each resin.

5.3 RESISTANCE TO ATMOSPHERIC CORROSIVES

KYNAR based coatings resist attack by most acids and liquid alkalies. Resistance to both strong and weak acids is particularly good. Resistance to weak alkalies is

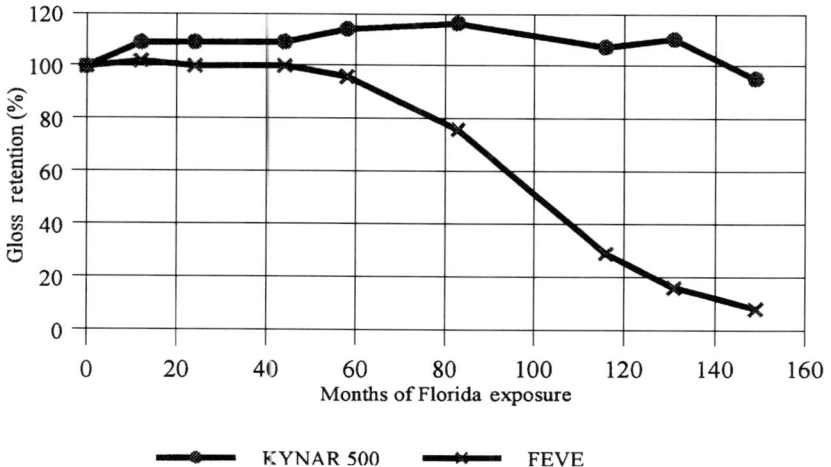

Figure 14.10.   KYNAR 500 vs FEVE. (Reproduced with permission)

Table 14.7. KYNAR resin chemical resistance. (Reproduced with permission)

Chemical	Concentration	MAX USE TEMP (°F)
Acetic Acid	10% in water	225 (107 °C)
Ammonium hydroxide	Concentrated	225 (107 °C)
Bleaching agents	—	275 (135 °C)
Brine	—	285 (140 °C)
Carbon dioxide	—	285 (140 °C)
Carbonic acid	—	275 (135 °C)
Citric acid	—	275 (135 °C)
Cresol	—	150 (65 °C)
Diesel fuels	—	285 (140 °C)
Fatty acids	—	285 (140 °C)
Ferric chloride	—	285 (140 °C)
Ferric hydroxide	—	250 (121 °C)
Ferric sulfate	—	285 (140 °C)
Formic Acid	—	250 (121 °C)
Natural gas	—	285 (140 °C)
Unleaded gasoline	—	285 (140 °C)
Hydrochloric acid	Concentrated	285 (140 °C)
Hydrofluoric acid	40% in water	250 (121 °C)
Hydrogen sulfide	—	275 (135 °C)
Nitric acid	10% in water	175 (79 °C)
Nitrogen dioxide	—	170 (77 °C)
Ozone	—	225 (107 °C)
Sodium chlorite	—	250 (121 °C)
Sodium hydroxide	10% in water	100 (38 °C)
Sulfur dioxide	—	175 (79 °C)
Sulfuric acid	60% in water	250 (121 °C)
Tar	—	250 (121 °C)
Urea	—	250 (121 °C)
Vinyl chloride	—	200 (93 °C)
Salt water	—	285 (140 °C)
Sewage water	—	250 (121 °C)

very good, but certain strong alkalies can attack KYNAR surfaces. This feature is the key to the excellent resistance KYNAR based coatings have to atmospheric pollution (e.g. acid rain), and other gaseous, liquid, and solid corrosives which can come in contact with a structure. KYNAR resins have been extensively tested by Elf Atochem for resistance to hundreds of chemicals [2]. Table 14.7 highlights some of the chemicals which are relevant to architectural uses.

5.4 LOW MAINTENANCE

KYNAR based coatings require little maintenance because of their resistance to: (a) dirt pick-up; (b) chemical and mildew/bacterial staining; and (c) mechanical damage and wear. The KYNAR resin resists dirt pick-up because of its hydrophobic, low surface energy (~23 dyne/cm), and low coefficient of friction (sliding friction to steel −0.15–0.17).

Table 14.8. Sand abrasion resistance. (Reproduced with permission)

	KYNAR 500 based coating	Silicone polyester	Baked enamel	Urethane	Plastisol
Abrasion coefficient[a]	59	23	32	44	32

[a]Liters sand/mil of coating to wear 5/32″ diameter hole in coating — ASTM D968.

Resistance to chemical staining is due to its excellent chemical inertness. The ability to resist mildew and bacterial staining arises from the fact that the KYNAR resin is a nonnutrient for fungal growth. It will not support fungal growth when tested according to Method 508 of the United States Military Standard 810B. Also, KYNAR resin has good moisture resistance with a water absorption value of 0.05% maximum per ASTM D570 method. Most organic coatings used for outdoor environments have water absorption values of about 0.1–3 wt% [3].

Resistance to mechanical damage and wear is attributable to the good abrasion resistance (Table 14.8) and impact resistance of KYNAR based coatings. The impact resistance of KYNAR based coatings is so good that usually the metal substrate can be ruptured upon impact with no cracking or loss of adhesion of the coating.

5.5 CORROSION RESISTANCE

KYNAR based coatings are recognized as having excellent corrosion resistance when exposed to even the most severe environments. They are frequently chosen for use in severe environments over other coatings and construction materials which cannot withstand such conditions over the long term.

KYNAR based coatings are chemically inert. In addition, they possess several other intrinsic properties which contribute to their excellent corrosion resistance. These are:

1. Low permeation rate of oxygen, moisture, and corrosive ions.
2. High electrical resistivity (2×10^{14} ohm-cm).
3. Good mechanical properties, flexibility, and adhesion to the substrate.

The importance of these characteristics can be seen by considering the mechanisms of paint blistering and corrosion of painted metal substrates.

5.6 ELECTROCHEMICAL CONSIDERATIONS

Corrosion of a metallic substrate in the presence of an aqueous electrolyte is an electrochemical process. In such environments, local anode and cathode cells are established on a microscopic scale. In these cells, corrosion is accelerated at the anode and slowed at the cathode. These processes are exacerbated by the presence of oxygen, moisture, and corrosive ions.

5.7   PAINT BLISTERING

All paint films are semi-permeable membranes. Water-soluble impurities commonly exist at the coating–substrate interface (e.g. salt residues from pretreating processes; hard water salts, etc.). (Figure 14.11a.) These impurities can lead to osmotic blistering. That is, water permeates through the paint film to achieve equilibrium with the impurities (i.e. osmosis). (Figure 14.11b) The water pressure that builds at the interface exceeds the necessary force to deform the film, and breaks the interfacial bonds leading to blistering (Figure 14.11c). Osmotic pressure in blisters can be as high as 25–30 atm [3].

The area under a paint blister is a likely location for an electrochemical corrosion cell to exist (Figure 14.11d). Permeation of moisture and oxygen through the paint can accelerate the corrosion process and lead to the generation of voluminous corrosion products under the paint (Figure 14.11e). These corrosion products can eventually rupture the coating and exude unsightly residues onto the painted surface (Figure 14.11f).

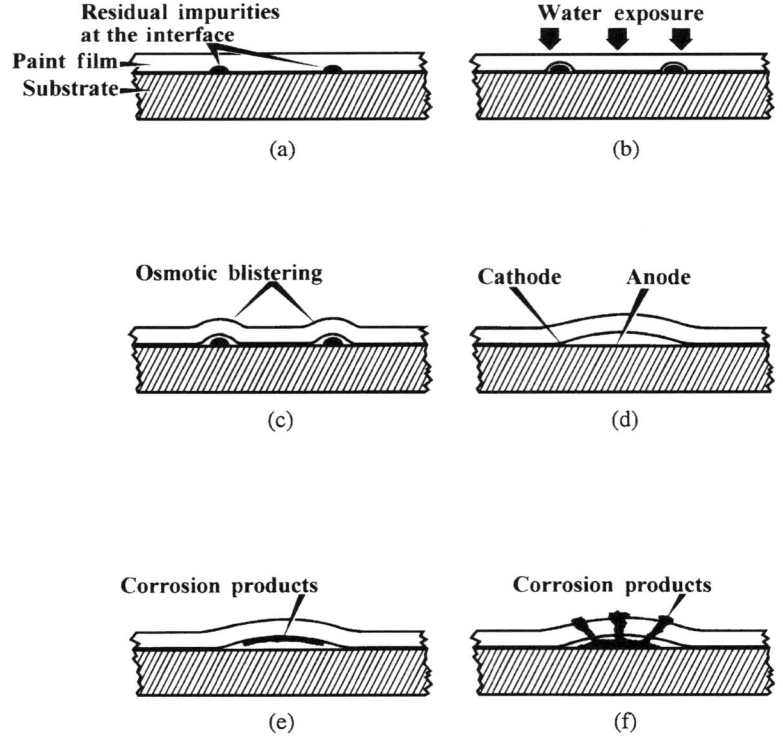

Figure 14.11.   Paint-blistering mechanism. (Reproduced with permission)

Table 14.9. Permeability of KYNAR resin to common gases. (Reproduced with permission)

Gas	Permeability (mg/mil/100 sq. in./24 h)
Water vapor	100
Oxygen	3.1
Carbon dioxide	1.7
Nitrogen	0.7
Hydrogen	55
Chlorine	0.3

The oxygen and moisture permeation rate of the coating is a key determinant in the formation of osmotic blistering. The low permeability of water and oxygen for the KYNAR resin is given in Table 14.9.

The low electrical conductivity of the KYNAR resin is also important because of the electrochemical nature of corrosion. An electrically insulating paint film helps reduce current flow in the coating system.

5.8 CATHODIC DELAMINATION

Consider the case where there is a void in the paint film over a steel or zinc-coated substrate. This can occur either as a defect from the coating operation (e.g. pore), or mechanical damage such as microcrazing after forming (Figure 14.12a). In these situations, an electrochemical cell is again created in the presence of an electrolyte (Figure 14.12b). The substrate area at the defect becomes the anode of an electrochemical corrosion cell and metal dissolution occurs ($Me \Rightarrow Me^{+x} + xe$). The area away from the pore becomes the cathode of the electrochemical cell. At the cathode, oxygen and water are reduced to form hydroxyl ions ($O_2 + 4e + 2H_2O \Rightarrow 4OH^{\ominus}$). The formation of hydroxyl ions raises the pH at the interface, destroying the paint/substrate interfacial bonds. The corrosion process continues and more substrate is exposed, depositing corrosion products on the surface of the paint (Figure 14.12c).

In the cathodic delamination mechanism, oxygen and moisture permeation through the paint film are key determinants of performance. Lower permeation of these species inhibits cathodic reduction reactions, mitigating the overall corrosion kinetics. For any electrochemical corrosion cell, the oxidation (anodic reaction) must occur simultaneously and at the same rate as the reduction reaction (cathodic reaction) so as not to violate the law of conservation of electric charge.

The mechanical properties, flexibility, and adhesion of the paint film are also important to preserve coating integrity when parts are formed. The data in Table 14.10 demonstrate the excellent flexibility and coating adhesion of KYNAR based coatings.

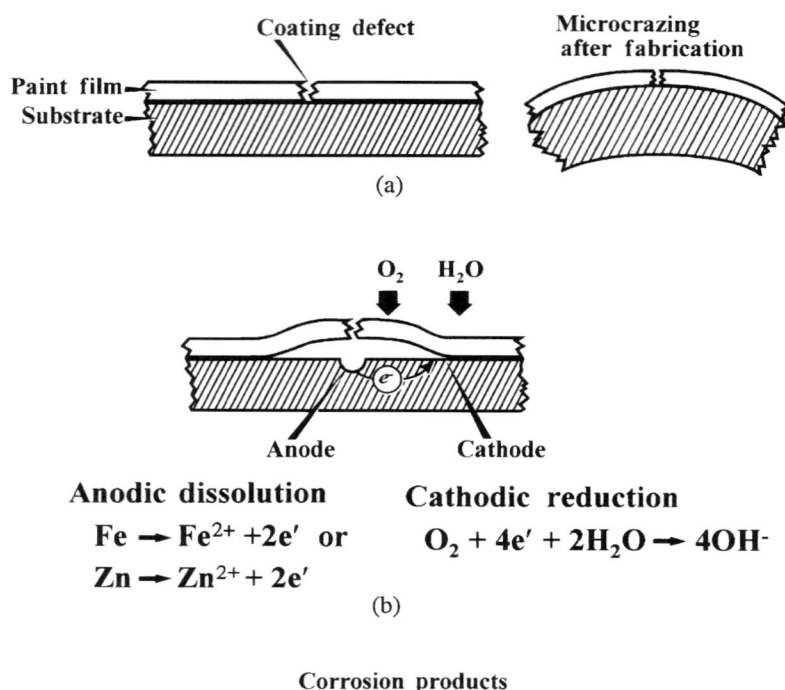

(a)

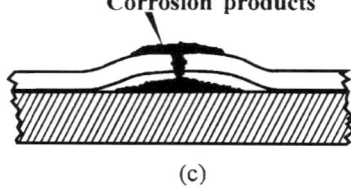

(b)

(c)

Figure 14.12.   Cathodic delamination mechanism. (Reproduced with permission)

Table 14.10.   Flexibility and adhesion of KYNAR based coatings. (Reproduced with permission)

Test	Value
Elongation 75 °F (24 °C)	50–300%
210 °F (99 °C)	200–500%
T-bend — no cracking or adhesion loss[a]	1T
Cross-hatch adhesion — reverse impact[b]	No adhesion loss
Boiling water cross-hatch — reverse impact[c]	No adhesion loss

[a]NCCA II-19 Method.
[b]NCCA II-16 Method — 0.025 in aluminum panel, 50 in.-lb.
[c]AAMA 605.2 Method — 0.025 in aluminum panel, 50 in.-lb.

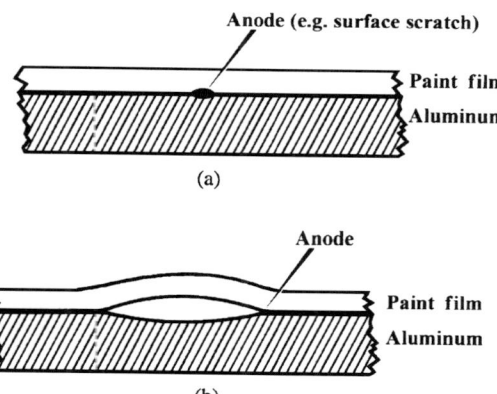

Figure 14.13. Anodic delamination mechanism. (a) Initiation; (b) propagation. (Reproduced with permission)

5.9 ANODIC DELAMINATION

Anodic delamination of a paint film is a common mechanism of paint failure on aluminum substrates. The process is again electrochemical in nature. There is an initiation step in which the corrosion originates in a microscopic corrosion cell under the paint film (Figure 14.13a). A typical initiation point is an area of high energy, such as a scratch or surface defect in the metal. The corrosion is slow initially, then accelerates to delaminate the coating (Figure 14.13b). The advancing delamination interface is the anode, not the cathode as with cathodic delamination. The advancing anode corrodes a monolayer of oxides on the aluminum surface, destroying the interfacial bonds. The cathode area is not well defined.

5.10 ACCELERATED TESTS

KYNAR based coatings (70% minimum weight KYNAR resin) routinely exhibit excellent performance in even the most severe accelerated tests designated by various technical associations. Typical performance in some of these tests is given in Table 14.11.

6 APPLICATIONS

KYNAR based coatings are usually used in applications where excellent long-term exterior durability is required, with little maintenance. Therefore, these coatings are ideal for structures such as high-rise office buildings, apartments, condominiums and sports stadia. Typical components include metal siding and

Table 14.11. Accelerated test performance of KYNAR based coatings. (Reproduced with permission)

Test	Conditions	Method	Results
Humidity	3000 h, 100 °F (38 °C), 100% rel. hum.	AAMA 605.2 ASTM D2247	No blisters
Salt spray	3000 h 5% NaCl	AAMA 605.2 ASTM B117	No scribe delamination
Muriatic acid	15 min spot test	AAMA 605.2	No blistering or change in appearance
Mortar	Wet mortar, Humidity — 100% rel. hum., 24 h, 100 °F (38 °C)	AAMA 605.2	No adhesion loss or change in appearance
Nitric acid	Contact with 70% nitric acid — 30 min	AAMA 605.2	No color change
Detergent	3% detergent solution, 100 °F (38 °C), 72 h	AAMA 605.2 ASTM D2244	No adhesion loss, blistering, change in appearance

roofing, storefronts, curtain walls, louvers, skylights, and other miscellaneous trim and extrusions. The uses are as varied as the architectural designs themselves. Components can be either post-formed from pre-coated coil stock, or spray coated after fabrication.

KYNAR SL and KYNAR 500 resins are also solution cast to produce films which are laminated onto various substrates. KYNAR SL resin is typically used to produce a solution cast printing ink, either as a solid color or as a pattern. This layer is overcoated with a KYNAR 500 based clear coat. The film is laminated to metal, plastic, or glass substrates to produce architectural materials such as wood grain finishes for aluminum or PVC residential siding and trim, stadium seating, or solar films for glass.

The three structures described in the following are examples of the versatility and typical applications of KYNAR based coatings.

6.1 CONCOURSE BUILDING, SINGAPORE (FIGURE 14.14)

The Concourse Building in Singapore was completed in 1994. It contains approximately 2 300 000 square feet of metal panels coated with KYNAR 500 based finish. It was spray coated with two shades of gray.

6.2 GEORGIA DOME (FIGURE 14.15)

The Georgia Dome was erected in 1990. It contains 300 000 square feet of metal wall and curtain wall coated with KYNAR 500 based finish. Most of the material

Figure 14.14. Concourse Building, Singapore. (Reproduced with permission)

is coil coated; the remainder is spray coated. The Georgia Dome was the venue of the 1996 Summer Olympics. It is the world's largest air-conditioned stadium with a self-supporting fabric roof.

6.3 HARRAH'S CASINO BOAT – COLORADO BELLE (FIGURE 14.16)

The white coating on this riverboat is shown to demonstrate the versatility of KYNAR 500 based coatings. This demanding application requires that the coating maintains its color and general integrity for the long term.

Figure 14.15. Georgia Dome. (Reproduced with permission)

ACKNOWLEDGMENTS

I wish to thank the following individuals for making this publication possible: the Management of Elf Atochem North America, Inc. for permission to publish this manuscript; Drs Steve Humphrey and Scott Gaboury for their technical and editorial assistance; Mr Claude Tournut for technical information; Dr Scott Gaboury

Figure 14.16. Harrah's casino boat – *Colorado Belle*. (Reproduced with permission)

and Ms Cindy Janney for providing the weathering data presented; Dr Marina Despotopoulou for providing information on the film formation model of KYNAR 500 based coatings, and AFM images; Messrs. Jack Mohnacs and Jim Hobensack for providing the photographs of buildings utilizing KYNAR 500 based coatings.

REFERENCES

1. Yamauchi, M., *et al.* 'Fluoropolymer emulsions', *European Coatings Journal*, **124**, 3/96 (1996).
2. Elf Atochem North America, Inc. 'KYNAR Chemical Resistance Chart', *CHEM CHART TR-15M*, p. 8/94.
3. Funke, W. 'Blistering of paint films', in *Corrosion Control by Organic Coatings*, H. Leidheiser (ed), p. 97 (1981).

FURTHER READING

Perillon, J. L. and E. J. Bartoszek, 'Long-life coatings with PVDF', *European Coatings Journal*, **277**, 4/95 (1995).
Dohany, J. E. and J. S. Humphrey, 'Vinylidene fluoride polymers', *Encyclopedia of Polymer Science and Engineering*, Vol. 17, p. 532, John Wiley, New York (1989).
Bartoszek, E. J., 'PVDF', *The Construction Specifier* (April 1993).

15

Ethylene/Tetrafluoroethylene Copolymer Resins

DEWEY L. KERBOW
E. I. DuPont de Nemours, Inc., Parkersburg, WV, USA

1 BACKGROUND

Copolymers of ethylene (CAS 74-85-1) and tetrafluoroethylene (TFE, CAS 116-14-3) have been known since 1946 [1]. The copolymer (CAS 11939-51-6) is comprised primarily of alternating ethylene and TFE units. It has an excellent balance of physical, chemical and electrical properties. The alternating copolymer is easily melt fabricable, but it has found little commercial utility, because it has poor resistance to cracking at elevated temperatures.

Subsequently, incorporation of certain termonomers was found to provide good crack resistance while maintaining the desirable properties of the copolymer [2]. These modified ETFE resins were commercialized in 1970 by E. I. DuPont de Nemours & Co. as Tefzel® ETFE fluoropolymer (CAS 25038-71-5).

ETFE resins are manufactured by DuPont, Asahi Glass, Daikin, Hoechst and Ausimont, under the tradenames Tefzel, Aflon, Neoflon ET, Hostaflon ET and Halon ET. Worldwide production in 1994 was about 2900 metric tons Selling price range is about USD 24–44 per kilogram.

2 MANUFACTURE

ETFE is produced under conditions similar to those used to homopolymerize TFE. Ethylene will not homopolymerize at these conditions. The energy released in thermal decomposition of TFE–ethylene mixtures can be even greater than for decomposition of TFE alone. Polymerizations must be carried out, therefore, within barricades or at suitably low pressures.

Modern Fluoropolymers. Edited by John Scheirs
© 1997 John Wiley & Sons Ltd

Suspension polymerizations are performed in inert chlorofluorocarbon or hydrofluorocarbon solvents such as 1,1,2-trichloro-1,2,2-trifluoroethane [2]. Organic peroxides, either fluorinated or hydrogenated, are used as initiators. Ionizing radiation and azo compounds have also been used. Typically, polymerizations include low levels of chain transfer agents such as methanol to control molecular weight.

Mixtures of water and inert solvents may also be used [3]. In this case, the polymerization occurs in the solvent phase, while the water serves to lower the viscosity of the mixture and to remove the heat of polymerization. Other active components of the reaction may be the same as for suspension polymerization.

Aqueous ETFE dispersions can be prepared using fluorinated surfactants such as perfluorooctanoic acid or it salts, along with inorganic initiators such as potassium permanganate or redox systems [4]. Dispersions have been commercially produced, but never developed a large market like PTFE dispersion.

For all three types, polymerization temperatures range from 30 to 80 °C, and pressures below about 1.7 MPa are used. To achieve a 50 : 50 molar ratio of monomers in the polymer, a mixture of about 75 : 25 TFE : ethylene must be initially charged to the reactor, depending on reactor pressure and temperature. Reactivity ratios have been studied as functions of these parameters [5,6]. The degree of alternation varies with the ratio of monomers fed to the reaction. At a 50 : 50 ratio, about 90% of the monomer units are alternating [7]. The effects of temperature, addition of termonomer, and ethylene/TFE ratio on degree of alternation and on molecular structure have been studied [6,8].

Copolymers of ETFE have poor stress crack resistance, even at high molecular weight. To improve cracking resistance, commercial ETFE polymers incorporate from 1 to 10 mol% of a modifier. These modifiers are derived from monomers that are chosen to have little or no chain transfer activity in the polymerization and to have bulky pendant side chains to break up crystallinity in the resulting polymer. Examples include perfluorobutyl(ethylene), hexafluoropropylene, perfluoro(propylvinyl ether), and hexafluoroisobutylene.

ETFE polymers are isolated from suspension polymerizations by evaporation of the solvent. The result is a powder with relatively low bulk density, It is typically compacted, melt extruded or otherwise agglomerated into pelletized form.

Aqueous dispersions are coagulated chemically or mechanically, filtered, washed and dried. Modified ETFE copolymers are supplied in powder, bead and melt-extruded pellet forms.

3 PROPERTIES

3.1 GENERAL POLYMER STRUCTURE

The crystal structure of alternating ETFE has been extensively studied by X-ray techniques [9–11]. The carbon chain is in a planar zig-zag orientation. Adjacent

$-CF_2-$ groups are on opposite sides of the chain, followed by a similar unit for adjacent $-CH_2-$ units. Interpenetration of adjacent chains occurs when electronic interactions cause bulky $-CF_2-$ groups of one chain to nestle into the space above smaller $-CH_2-$ groups of an adjacent chain. Each molecule therefore has four nearest neighbors, creating an orthorhombic lattice. As a result of this structure, ETFE has exceptionally low creep, high tensile strength and high modulus among fluoropolymers. Interchain forces hold this matrix until the alpha transition occurs at about 110 °C, where the physical properties of ETFE begin to decline and more closely resemble perfluoropolymer properties at the same temperature. Other transitions occur at -120 °C (gamma) and about -25 °C (beta).

As the degree of monomer alternation is reduced, this ability of chains to interlock is decreased. Thus, as ethylene content is either increased or decreased from the 50 : 50 ratio, crystalline content is reduced. Figure 15.1(a) provides a graphical representation of this phenomenon. Most copolymers of TFE with two or three-carbon alkenes show a linear relationship or a negative deviation

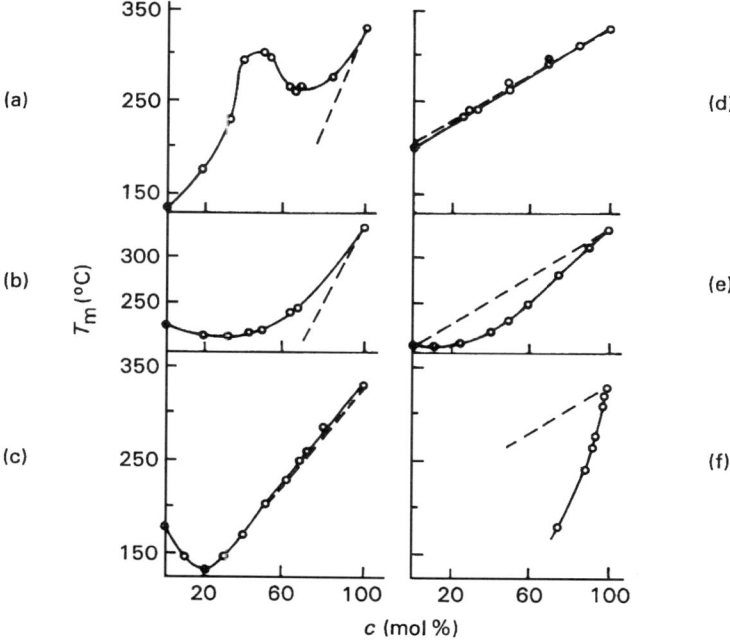

Figure 15.1. Tetrafluoroethylene copolymers. Dependence of melting endotherm maximum (degrees Centigrade) on copolymer composition. (a) $CH_2=CH_2$; (b) $CH_2=CHF$ (c) $CH_2=CF_2$ (d) $CF_2=CHF$; (e) $CF_2=CF_2Cl$; (f) $CF_2=CFCF_3$. On all graphs X-axis is mol% tetrafluoroethylene, Y-axis is °C. Reproduced by permission of *Vysokomol. Soedin.*, **19**, no. 1, 33–7 (1977)

when melting point is plotted against polymer composition. Ethylene is the only common monomer which shows a positive deviation. In highly alternating ETFE compositions, melting points of as high as 300 °C have been produced [1]. As normally produced, ETFE has about 88% alternating sequences, and the melting point is reduced to about 270 °C.

A melting point minimum is reached at about 65–70 mol% TFE. At higher TFE levels, melt processiblity is lowered and the copolymer behaves increasingly like polytetrafluoroethylene as TFE–TFE sequences dominate the properties. Below about 45 mol% TFE, ethylene–ethylene sequences increase exponentially. Properties such as chemical and oxidation resistance are degraded. Most commercial ETFE grades fall between 50 and 65 mol% TFE.

Sequence distributions of monomers can be calculated from traditional copolymer equations using reactivity ratios of

$$r_1 = k_{11}/k_{12} = 0.045 \pm 0.010$$

$$r_2 = k_{22}/k_{21} = 0.14 \pm 0.03$$

at 65 °C where k_{11} represents the reaction rate constant of comonomer 1 with radical 1, k_{12} ... comonomer 1 and radical 2, etc. [12]. Effects of temperature, pressure and solvent composition on sequence distributions have been studied [6].

Molecular weights of commercial grades have been calculated from light scattering measurements done upon solutions in diisobutyl adipate at 230 °C [13–15]. High molecular weight grades, used where high temperatures or high physical stresses are to be encountered, have molecular weights slightly over 1 million (or 1×10^6). Lowest viscosity grades have molecular weights about 500 000 (or 5×10^5).

3.2 MECHANICAL PROPERTIES

Typical physical properties are included in Table 15.1. Notable properties include exceptional toughness and abrasion resistance over a wide temperature range and a good combination of tensile strength, flex and creep resistance. Most perfluorinated polymers have modest physical properties, compared to the engineering polymers such as nylon, polycarbonates, acetals, etc. ETFE provides a good combination of the engineering properties of these hydrocarbon polymers with the chemical and thermal resistance of perfluoropolymers. It has excellent impact resistance down to −100 °C. Friction and wear properties are good; they can be improved by incorporation of fillers such as fiberglass or bronze powders. Fillers also improve creep resistance and increase softening temperature.

Physical properties such as modulus, tensile strength and high-temperature elongation are strongly influenced by degree of alternation of the comonomers. Room temperature modulus and tensile strength decrease as the ratio of TFE : ethylene is increased. High-temperature elongation is improved, however,

Table 15.1. Typical Physical Properties of Ethylene/Tetrafluoro ethylene copolymer resins

Property	Test method	Units	ETFE	PVDF	ECTFE
Melt point	DSC	°C	225–270	154–184	236–246
Specific gravity					
Melt			1.3		
Solid			1.72	1.75–1.78	1.7
Tensile strength (23 °C)	D 638	MPa	38–48	36–56	50–65
Elongation (23 °C)	D 638	%	100–350	25–500	150–250
Tensile modulus (23 °C)	D 638	MPa	830	1340–2000	
Izod impact strength (23 °C)	D 256	J/m	No break	160–530	No break
Coefficient of thermal expansion	D 696		9×10^{-5}	$\sim 10^{-4}$	5×10^{-5}
Critical shear rate		1/sec	200–10 000		
Processing temperature range		°C	300–345	200–300	260–300
Dielectric constant (1 kHz)	D 150	ε	2.6	7.5–13.2	2.6
Dielectric strength	D 149	kV/mm	59	260–950	80–90
Dissipation factor (1 kHz)	D 150		0.0008	0.013–0.019	0.0024
Limiting oxygen index	D 2863		30	44	64

and the drop off in tensile strength with increasing temperature is not as great in samples with higher TFE fraction.

The recommended upper continuous duty temperature for commercial ETFE is 150 °C. Physical strength can be maintained at even higher temperatures when crosslinking agents are incorporated and cured via peroxide or ionizing radiation [16]. Especially, cut-through resistance of thin-wall wire insulation to a hot soldering iron or resistance to physical abuse during installation or use is increased at temperatures up to 200 °C. Short-term excursions to 240 °C are possible for highly crosslinked resins.

3.3 ELECTRICAL PROPERTIES

ETFE resins are excellent insulating materials. Dielectric constant is low and essentially independent of frequency. Dissipation factor is also low, but it increases with frequency. Dielectric strength and resistivity are high and are unaffected by water. Irradiation and crosslinking increase dielectric losses.

3.4 CHEMICAL AND OXIDATIVE RESISTANCE

Modified ETFE has excellent resistance to most common solvents and chemicals [17]. It is not hydrolyzed in boiling water and weight gain is less than 0.03% in water at room temperature. It is an excellent barrier to hydrocarbon and oxygenated components of automotive fuels. Strong oxidizing acids such as nitric and some organic bases cause depolymerization under conditions of high concentration and temperature.

Thermal and oxidative degradation mechanisms of ETFE have been studied [18,19]. Weight loss in air at 180 °C is only about 0.6% per year. Main carbon chain sequences of two or more ethylene links are thought to be subject to thermal and oxidative degradation.

Although they have good thermal stability, ETFE resins may have thermal stabilizers added for high-temperature applications [20]. A wide variety of compounds, mostly metal salts such as copper oxides and halides, aluminum oxide and calcium salts act as sacrificial sites for oxidation. Addition of certain salts can alter the decomposition mechanism from oligomer formation to dehydrofluorination. Iron and other transition metal salts elements accelerate the dehydrofluorination. Hydrofluoric acid itself destabilizes ETFE at elevated temperatures; extrusion temperatures in excess of 380 °C should be avoided because degradation becomes self-accelerating at higher temperatures.

Ionizing radiation causes degradation in perfluorinated polymers. The incident energy causes loss of an atom, leaving a radical on the backbone. The primary fate of these radicals is elimination, with subsequent main chain cleavage. In ETFE resins, crosslinking is more likely than cleavage, so the rate of crosslinking approximately matches cleavage, and tensile properties are relatively unaffected at low radiation levels. The nuclear energy industry uses ETFE in wire coatings and moldings for this reason.

The copolymer will not support combustion in air. Limiting oxygen index (LOI) is about 30–31, depending on monomer ratio. LOI increases gradually as fluorocarbon content is increased up to the alternating composition (Figure 15.2). It then increases more rapidly to the index of PTFE.

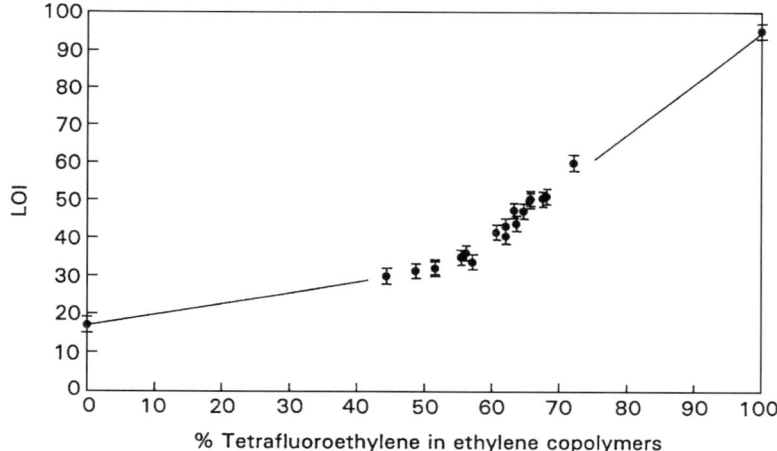

Figure 15.2. Limiting oxygen index. Effect of comonomer composition

4 PROCESSING

ETFE is readily processed by a number of melt fabrication techniques [21]. Corrosion-resistant alloys should be used for prolonged contact with the melt. It has a wide processing window (typically 280–340 °C). It can be extruded into shapes (films, rods, tubing, etc.) or as thin coatings on wire and cables. Injection molding into thin sections are possible, since critical shear rate is at least two orders of magnitude greater than for perfluorinated copolymers. Molding of thicker sections (greater than 5 mm) requires consideration of melt shrinkage which occurs during freezing. Melt density is about 1.3–1.4 g/cm^3 compared to 1.7 or higher in the solid, creating shrinkage of about 6%. Coatings can be applied by hot flocking or by dipping the heated part into a fluidized bed of ETFE powder and removing it to cool. The process can be repeated to build up thicker coatings. Powders can be sprayed electrostatically, followed by baking to coalesce the particles into a coating. Similarly, liquid suspensions in water or solvents can be sprayed, dried and baked to form coatings.

Parts produced from ETFE are easily welded by melt bonding. Spin welding, ultrasonic welding, and conventional butt welding using a flame and an ETFE rod are examples. The resin also bonds well to untreated metals. Chemical etch, corona and flame treatment have been used to increase adhesion further [22].

A large fraction of ETFE is compounded with other ingredients or otherwise modified during processing. Of these modifications, probably most significant has been crosslinking after fabrication. Di- or trifunctional monomers such as triallyl cyanurate can be melt incorporated during or after fabrication [23]. Crosslinking can then be initiated by peroxides or by ionizing irradiation, creating structures which provide improved mechanical properties, higher upper use temperatures and better cut-through resistance without significant sacrifice of electrical properties or chemical resistance. The inherent radiation resistance of the chain backbone affords the ability to undergo irradiation crosslinking.

Glass fiber is incorporated by melt blending to increase modulus, friction and wear characteristics. Typical levels are 25 and 35 wt% filler. At 25%, dynamic coefficient of friction is reduced from about 0.5 to about 0.3.

4.1 APPLICATIONS

A major use of modified ETFE copolymers is in wire and cable insulation. It can be extruded into thin coatings at high rates as primary insulation. Its lubricity aids in pulling wire through intricate pathways, and its abrasion resistance maintains insulation integrity. Crosslinking the applied insulation increases abrasion resistance, hot solder resistance and mechanical cut-through resistance. Some constructions may be rated for application at 200 °C. Its excellent chemical resistance is important for oil-well, down-hole cables. Its lower density provides advantages relative to perfluoropolymers in applications such as aerospace wiring. Radiation resistance is important in nuclear industry wiring.

Modified ETFE is extruded into film and sheet using normal flat-film techniques. Thin films are important in greenhouse applications, which utilize its radiation resistance, toughness and good light transmission characteristics. Pigmented films are applied on white boards and in outdoor advertising laminates. Biaxially oriented films have excellent tensile properties and toughness equivalent to polyester films [24]. Thicker sheets are laminated to substrates such as polyester-reinforced fiberglass to make chemical-resistant structures. The chemical process industry (CPI) uses fiber-reinforced ETFE in vessels, tanks, piping and in laminated structures for structural applications.

Injection moldings are easily fabricated, taking advantage of the low shear sensitivity and wide processing window. Lab equipment, electrical connectors and sockets, as well as distillation column plates and packings are examples. Valve body, pipe and fitting linings are also injection molded.

Tubing, sheets, piping and rod stock can be continuously extruded. Chemical resistance and permeation resistance to hydrocarbons provides important advantages in automotive fuel tubing. A high weld factor (>90%) is important in butt joints of piping and sheet lining of large vessels.

Powder or bead products are rotationally molded into structures and linings. Pump bodies, tanks and fittings for CPI are examples. Inserts can be incorporated to provide attachment points or structural members. Adhesion to steel, copper and aluminum can be up to 3 kN/m peel force [25].

Carbon-filled ETFE resins are used in antistatic or semiconductive applications, such as self-limiting heater cables. Carbon levels of about 20% are required to produce antistatic properties.

4.2 HEALTH AND SAFETY CONSIDERATIONS

ETFE resins should be processed with adequate ventilation. Acute inhalation tests have determined that the approximate lethal temperature for a four-hour exposure of rats to off-gases is 335–350°C. Volatile products below these temperatures were nontoxic. No confirmed cases of 'fume fever' have been reported for ETFE.

4.3 COMPARISON OF ETFE AND POLY(VINYLIDENE FLUORIDE) (PVDF)

ETFE copolymers and PVDF [9002-58-1] are isomeric; that is, both are comprised of only $-CH_2-$ and $-CF_2-$ groups. The alternating groups provide some similarities in properties, so in many applications the two can be used to similar effect. However, important nuances derive from the difference in repeating unit.

The major difference is in melting behavior. PVDF has a melting point in the range of 154–184 °C, compared to about 260–270 °C for modified ETFE. The discussion of chain structure in section 3.1 above describes the causes of this difference.

In both, the alternating units can crystallize with larger CF_2 groups adjacent to smaller CH_2 units on an adjacent chain. This interpenetration gives rise to high moduli for both ETFE and PVDF. Perfluoropolymer chains have no such interpenetration, so they slide past one another, leading to higher creep and lower modulus values. In general, PVDF has a higher modulus and yield strength than ETFE, while ETFE has higher impact strength and elongation although individual grades can show exceptions.

The shielding effect of the fluorines adjacent to all CH_2 groups provides good chemical and thermal stability to both PVDF and ETFE. Both are resistant to a wide range of chemicals and therefore find utility in chemical processes.

ETFE decomposes above about 340 °C to oligomers. The major decomposition route appears to be cleavage of the main carbon chain, probably within diads or higher runs of ethylene. PVDF decomposes above about 300 °C to produce primarily hydrofluoric acid (HF) and carbon char. The latter mechanism is thought to contribute to superior resistance to burning in air; PVDF has a LOI of 44, compared to 30 for ETFE.

The propensity to liberate HF causes PVDF to be subject to attack by nucleophiles, particularly organic bases. ETFE also undergoes base attack, but at much more stringent conditions of concentration and temperature. Both are attacked by strong oxidizing acids. PVDF is soluble at room temperature in polar solvents such as tetrahydrofuran, acetone and many esters. This allows casting of PVDF films from solution, but it precludes use in these chemical environments. ETFE is soluble only in certain solvents at elevated temperatures.

PVDF can be crystallized with most hydrogens on one side of the main chain and fluorines on the other. This creates a polarity which is manifest in electrical properties such as piezoelectricity. Electrets are a major application of this form of PVDF. Since normal random crystallization, which occurs during most fabrication processes, creates some fraction of this polar structure, the dielectric properties of PVDF are inferior to those of ETFE, which offers no opportunity for such polarity.

4.4 COMPARISON WITH POLY(ETHYLENE/CHLOROTRIFLUORO-ETHYLENE) ALTERNATING COPOLYMERS

Ethylene and chlorotrifluoroethylene also tend to form alternating copolymers (ECTFE) [25101-45-5]. In many respects, ECTFE is more similar to ETFE than is PVDF. Melting points are more alike. Physical properties are very similar. Both tend to decompose by chain cleavage. Limiting oxygen index of ECTFE is high, typical of chlorinated materials. However, as in the case of PVDF, this is at the expense of overall thermal stability. Maximum processing temperature of ECTFE is lower than for ETFE since the decomposition mechanism is altered by the incorporation of the chlorine into the repeating unit. Moisture barrier of ECTFE is slightly better, but halogenated solvent resistance is slightly worse. Shrinkage of ECTFE is slightly lower than for ETFE.

Stress crack resistance of the copolymer is adequate for many applications. Newer terpolymers, analogous to modified ETFE, have been introduced for more demanding applications.

REFERENCES

1. US Pat. 2 468 664 (Aug. 16, 1949), W. E. Hanford and J. R. Roland.
2. US Pat. 3 624 250 (Nov. 30, 1971), D. P. Carlson.
3. US Pat. 4 513 129 (Apr. 23, 1985), S. Nakagawa and K. Ihara.
4. US Pat. 4 338 237 (July 1982), R. A. Sulzbach et al.
5. Iuliano M, De Rosa C, Guerra G, and Petraccone V, *Makromol. Chem.* **190**, 827–835 (1989).
6. Kostov F and Nikolov A, *J. Appl. Poly. Sci.* **57**, 1545–1555 (1995).
7. Modena N, Garbuglio C and Ragazzini M, *Poly. Lett.* **10**, 153 (1972).
8. Naberezhnykh R, Sorokin A, Galperin E, Volkova E and Simakina A, *Vysokomol. Soedin.* **19**, no. 1: 33–37 (1977).
9. Wilson F and Starkweather H, *J. Poly, Sci.* Part A-2 **11**, 919 (1973).
10. Tanigami T, Yamamura K, Matsuzuwa S, Ishikawa M, Mizoguchi K and Miyasaka K, *Polymer*, **27**, 999–1006 (1986).
11. Pieper T, Heise B, and Welke W, *Polymer*, **30**, 1768–1775 (1989).
12. Modena N, Garbuglio C and Ragazzini M, *J. Polym. Sci. Part B* **10**, 153 (1972).
13. Chu B, and Wu C, *Macromolecules*, **1986**, 19, 1285.
14. Chu B, and Wu C, *Macromolecules*, **1987**, 20, 93.
15. Chu B and Wu C *Macromolecules*, **1987**, 20, 98.
16. US Pat. 4 155 823, May 22, 1979, A. J. Gotcher and P. B. Germeraad.
17. TEFZEL *Chemical Use Temperature Guide*, Bulletin E-18663-1, E. I. DuPont de Nemours & Co, Wilmington, DE (1990).
18. Morelli J, Fry C, Grayson M, Lind A and Wolf C, *J. Appl. Poly. Sci.*, **43**, 601–11 (1991).
19. Malkevich S, Tarutina L and Chereshkevich L, *Plasticheskie massy*, **1960**, 5–7, no. 6.
20. US Pat. 4 390 655 (June 28, 1983) J. C. Anderson.
21. *Extrusion Guide for Melt Processible Fluoropolymers*, Bulletin E-85783, E. I. DuPont de Nemours & Co, Wilmington, DE.
22. TEFZEL *Fluoropolymer Design Handbook*, E-31301-2, 7/90, E. I. DuPont de Nemours, Wilmington, DE (1973).
23. US Pat. 3 738 923 (June 12, 1973), D. P. Carlson and N. E. West.
24. US Pat. 4 510 310 (Apr. 9, 1985), S. B. Levy.
25. TEFZEL *Properties Handbook*, E-31301-4, E. I. DuPont de Nemours, Wilmington, DE (1993).

FURTHER READING

1. Sperati, C. A., in *Polymer Handbook*, vol. 1, John Wiley & Sons, New York (1988).
2. Carlson, D. P. and Schmiegel, W., in *Ullmann's Encyclopedia of Industrial Chemistry*, vol. A11, VCH Verlagsgesellschaft mbH (1988).
3. Gangal, S. V., in *Kirk-Othmer: Encyclopedia of Chemical Technology*, vol 11, John Wiley & Sons, New York (1980).

16

Fluoropolymers for Chemical Handling Applications

PRADIP R. KHALADKAR

Materials Consultant, DuPont Company, Wilmington DE, USA

Fluoropolymers are well established as materials for many applications in chemical processing equipment. Owing to their broad resistance to corrosives at high temperatures, they can help reduce downtime and maintenance and extend the life of equipment. In addition, they are used to protect product purity and prevent process materials from sticking to surfaces. Fluoropolymers are frequently considered as replacements for exotic alloys.

Because they lack strength and have high creep rates, fluoropolymers are usually employed as linings supported by substrates, typically carbon steel or FRP (fiberglass reinforced plastic). However, some components are made entirely of fluoropolymer, particularly when sizes are relatively small and there is a risk of corrosive attack from outside the component.

The application areas for fluoropolymers in chemical handling are:

- linings for vessels and columns
- linings for pipe
- linings for fittings, pumps and valves
- seals and gaskets
- internal components for columns and other equipment
- heat exchangers
- hoses and expansion joints

The fluoropolymers used in chemical handling applications include both fully and partially fluorinated types. The fully fluorinated polymers are polytetrafluoroethylene (PTFE), fluorinated ethylene propylene (FEP) and perfluoroalkoxy (PFA). The partially fluorinated polymers are ethylene–chlorotri-

---

Modern Fluoropolymers. Edited by John Scheirs

Table 16.1. Fluoropolymer physical properties. Sources: DuPont (PTFE, FEP, PFA, ETFE) and Ausimont (ECTFE, PVDF)[a]

Property	Test method	Units	PTFE	FEP	PFA	ETFE	ECTFE	PVDF
Tensile Strength	D 638	psi × 1000 (MPa)	3–5 (21–35)	3.4 (23)	3.6 (25)	5.8–6.7 (40–47)	4.2–4.3 (28.9–29.6)	6.8–8.0 (46.9–55.2)
Elongation	D 638	%	300–500	325	300	150–300	200	50–250
Flexural modulus	D 790	psi × 1000 (MPa)	72 (500)	85 (600)	85 (600)	170 (1200)	240–245 (1655–1689)	165–325 (1138–2241)
Hardness	D 2240	Shore D	50–65	56	60	63–72	75	77
Melting point	DTA, E-168	°F (°C)	621 (327)	500 (260)	582 (305)	473–512 (245–267)	464 (240)	320–338 (160–170)
Upper service temp.	UL 746B	°F (°C)	500 (260)	400 (204)	500 (260)	300 (150)	300 (150)	275 (135)

[a]DuPont sells PTFE, FEP and PFA as Teflon® fluoropolymer resins and ETFE as Tefzel® fluoropolymer resin. Ausimont sells ECTFE as Halar® and PVDF as Hylar®. Atochem sells PVDF as Kynar®.

fluoroethylene (ECTFE), ethylene–tetrafluoroethylene (ETFE) and polyvinyli-dene fluoride (PVDF).

The fully fluorinated fluoropolymers (PTFE, FEP and PFA) are similar in their high resistance to chemical attack, but differ in maximum service temperature. Partially fluorinated materials often used in chemical applications (ETFE, ECTFE and PVDF) are not as resistant, and they differ from each other in levels of resistance and maximum service temperatures. However, they have higher stiffness and tensile strength than the fully fluorinated fluoropolymers. Physical properties and maximum service temperatures for all of these materials are given in Table 16.1.

1 VESSELS AND COLUMNS

Linings can be classified by thickness. Relatively thin linings up to 0.025 in. (0.6 mm) are used where the rate of corrosion of bare steel in the medium of interest is <0.010 in. (0.25 mm) per year. Thicker linings are used where the corrosion rate of steel in the medium of interest is higher. Therefore, the corrosion rate of steel is an important starting point.

There are four proven methods for applying fluoropolymer linings on steel: (1) adhesively bonded sheets, (2) loose sheets, (3) rotolining and (4) spray and baked or electrostatic powder coatings. A fifth lining method is building dual laminate structures of FRP lined with fluoropolymer sheet. Spray and bake coating is used for thin linings, powder coating for both thin and thick linings, and the other methods for thick linings. Table 16.2 summarizes information about these lining methods so that they can be compared.

1.1 ADHESIVE-BONDED SHEET LININGS

Fabric-backed sheets are bonded to steel vessel walls with neoprene-or epoxy-based adhesive. Joints are heat-welded using rods and overlying cap strips of the same polymer as the sheets. Heads are thermoformed to the required shape. Figure 16.1 shows the fabrication sequence for adhesive-bonded linings.

Vacuum ratings may vary with vessel size, although vessels lined with this method are rated to withstand full vacuum at ambient temperature. This type of lining has been successful in handling unstable sodium hypochlorite in a chlorine destruct system.

1.2 LOOSE LININGS

Fluoropolymer sheets are welded into lining shapes, folded and slipped into the housing. The lining is flared over body and nozzle flanges to hold it in place. Weep holes can be made in housings to release permeants. Loose linings have very low tolerance for vacuum. This lining system with FEP has been used in service with fluorobenzene.

Table 16.2. Fluoropolymer lining systems

Lining system	Lining materials	Maximum size	Design limits	Fabrication[a]
Adhesive bonding	Fabric-backed PVDF, PTFE, FEP, ECTFE, PFA	No limit	Pressure allowed. Full vacuum only at ambient temperature. Smallest nozzle is 2 in. (51 mm). Maximum temperature limited by adhesive, typically 275 °F (135 °C)	Neoprene or epoxy adhesive, sheets welded. Heads thermoformed or welded
Dual laminate	Same as adhesive bonding	12 ft (3.7 m) dia.	No pressure allowed. Vacuum rating not determined	Liner fabricated on mandrel. FRP built up over liner
Sprayed dispersion	FEP, PFA, PFA w/mesh and carbon, PVDF, PVDF w/glass or carbon fabric	8 ft (2.4 m) dia., 40 ft (12.2 m) length	Pressure allowed. Vacuum rating not determined	Multicoat application. Each coat is baked
Electrostatic spray — powder	ETFE, FEP, PFA, ECTFE, PVDF	8 ft (2.4 m) dia., 40 ft. (12.2 m) length	Pressure allowed. Vacuum rating not determined	Multiple coats applied by electrostatic spraying. Each coat is baked
Rotolining	ETFE, PVDF, ECTFE	8 ft (2.4 m) dia., 22 ft (6.7 m) length	Pressure allowed. Vacuum rating not determined	Rotationally molded
Isostatic molding, paste extrusion	PTFE		Pressure allowed. Vacuum rating depends on lining thickness	PTFE preformed under pressure or paste extruded, then sintered
Loose lining	FEP, PFA	Determined by body flange	Pressure allowed. No vacuum. Gasketing required between liner and flange face	Liner with nozzles hand or machine welded

[a]Nondestructive spark testing should be used, along with visual inspection for all systems except loose linings. Adhesive bonding can be done in the shop or field; other systems are shop only

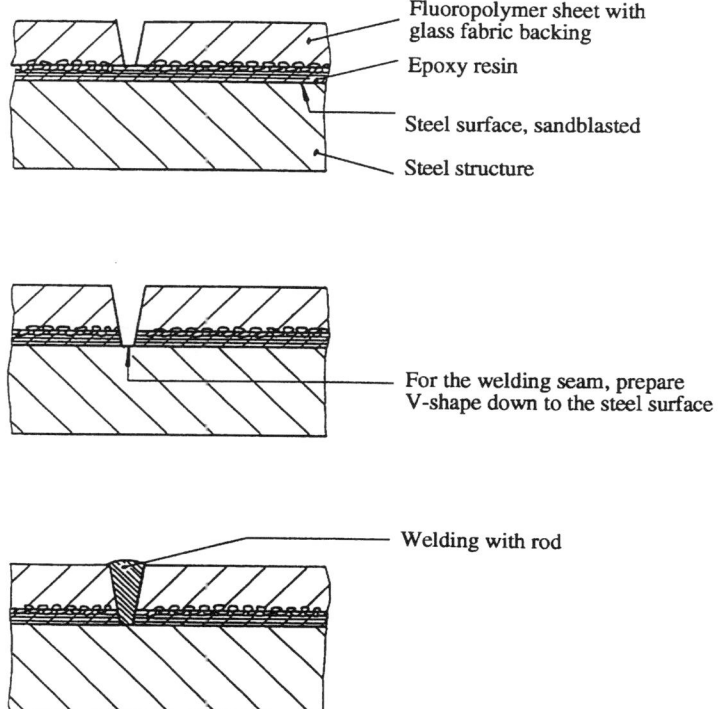

Fluoropolymer sheet with glass fabric backing

Epoxy resin

Steel surface, sandblasted

Steel structure

For the welding seam, prepare V-shape down to the steel surface

Welding with rod

Figure 16.1. Adhesive bonded linings are fabricated by thermoplastic welding of fluoropolymer sheets backed with glass fabric. Drawing courtesy of Symalit Company, Ltd, Lenzburg, Switzerland

1.3 ROTOLINING

In rotolining, a steel vessel is loaded with sufficient powdered fluoropolymer to cover the lined area at the specified thickness. With openings covered, the piece is rotated in three dimensions in an oven. Polymer melts and covers the interior to form a seamless lining. Thicknesses of 0.1–0.2 in. (2.5–5.1 mm) are most common. The size of the coated piece is limited by oven size; the largest oven today is 8 ft. (2.4 m) in diameter and 22 ft (6.7 m) long.

Rotolining is highly effective for column sections. It has been used to line scrubbers and neutralization equipment used to manufacture chlorofluorocarbons. Figure 16.2 shows a column with a rotomolded ETFE lining.

1.4 SPRAY-AND-BAKE AND ELECTROSTATIC POWDER COATING

There are two techniques for applying FEP, PFA and ETFE fluoropolymers as coatings. The first method, spray-and-bake coating, uses sprayed waterborne

Figure 16.2. This column is rotolined with ETFE to withstand corrosives. Manufacturer: RMB Products, Inc., Fountain, Colo. USA

fluoropolymer suspensions. For the second method, electrostatically charged powders are blown on a hot surface. For either method, surfaces are primed, and each coat is baked before the next is applied. These technologies are both energy and labor intensive.

Electrostatic coating with PFA has been used to reline glass-lined equipment. Fluoropolymer coatings have performed well in protecting equipment that removes organic contaminants from ground water. ETFE coatings are used to protect centrifuge baskets in the pharmaceutical and other industries.

In considering the use of coatings, note that fluoropolymer coatings can be susceptible to delamination in applications where temperatures cycle frequently from ambient to steam.

There are two proprietary fluoropolymer lining systems with reinforcement for greater strength and durability. The first, PVDF with glass- or carbon-fiber fabric, has shown good performance in service with sulfuric acid and with bromine. The second, a newer system using wire mesh and activated carbon with PFA dispersion, has worked well in service with HCl at 212 °F (100 °C). Lining thicknesses for both these systems are 0.040–0.080 in. (1–2 mm).

1.5 DUAL LAMINATES

Structures are made of FRP built up on ETFE, PVDF, PFA or FEP sheeting that is formed as a lining. This lining is fabricated by machine and hand welding

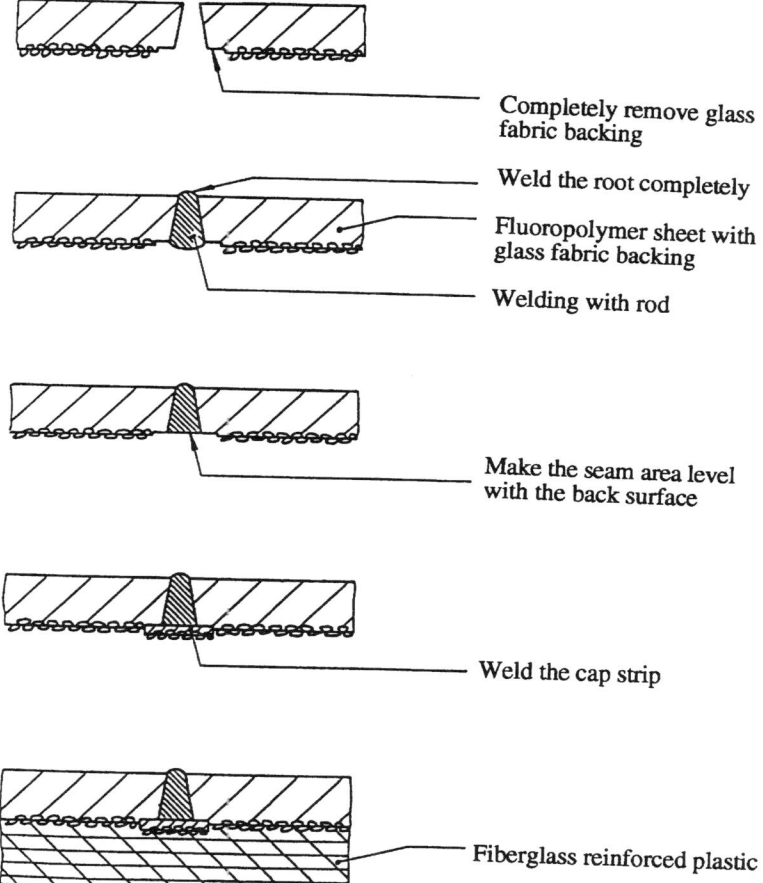

Figure 16.3. Dual laminates are fabricated by thermoplastic welding of glass-fabric-backed fluoropolymer sheets followed by building up layers of glass-fiber-reinforced plastic. Drawing courtesy of Symalit Company, Ltd, Lenzburg, Switzerland

of glass-fabric-backed fluoropolymer sheets like those used for adhesive-bonded systems. The fabric aids in bonding the sheeting to the FRP. Figure 16.3 shows the fabrication sequence for dual laminate structures.

Unlike steel, dual-laminate structures resist external corrosion without external painting.

Dual-laminate tanks lined with ETFE have been successfully used in chlorine neutralization service. One of the tanks is shown in Figure 16.4.

The labels accompanying the figure read:

Completely remove glass fabric backing

Weld the root completely

Fluoropolymer sheet with glass fabric backing

Welding with rod

Make the seam area level with the back surface

Weld the cap strip

Fiberglass reinforced plastic

Figure 16.4. A dual-laminate tank lined with ETFE is used to contain sodium hydroxide solution used to neutralize chlorine. The worker is spark-testing the lining. Manufacturer: C. P. F. Dualam Inc., Montreal, Canada

2 INSPECTION AND QUALITY ASSURANCE

Visual inspection is still the predominant method for inspecting linings.

For welded linings and dual laminates, it is extremely important to train workers in the required procedures. Workers should be qualified with tests to assure that they understand the necessary techniques.

Spark testing can reveal the quality and condition of a lining, but if not done properly, it can damage linings. Special care should be taken in spark testing older linings, and the test should not be repeated.

3 PIPE

Steel or FRP pipe is lined with extruded or molded fluoropolymer tubing. With PFA, FEP, ETFE and PVDF, tubing is extruded using conventional thermoplastic technology and inserted in pipe sections. Heat and pressure are used to flare linings over flange faces so pipe sections are self-gasketing. PTFE, however, cannot be processed with conventional melt processing technology.

There are four methods for manufacturing PTFE pipe linings: paste extrusion, ram extrusion, tape wrapping and isostatic molding. In paste extrusion, the resin is mixed with a hydrocarbon extrusion aid and extruded as tubing. The extruded shape is sintered at a high temperature to drive off the hydrocarbon and coalesce the resin. In ram extrusion, PTFE resin is forced through a heated die. Tape wrapping involves building up layers of tape on a mandrel to the required thickness followed by sintering. In isostatic molding, the resin is pressed within a cylindrical mold, usually by isostatic pressure applied by inflating rubber tooling. The resulting 'green form' is then removed from the mold and sintered.

These linings on steel are not bonded with adhesive. They are well supported by an exterior shell, and they are rated by manufacturers for various amounts of vacuum, depending on the lining material and its thickness. Some pipes are swaged so the shell fits tightly against the liner.

Connections for lined pipe are usually made with flanges. Complex piping configurations are used to reduce the number of flanged connections and the potential for emissions. Such piping can be rotolined, or bent into the desired shape after lining using proprietary technology. Figure 16.5 shows a PTFE-lined pipe that has been bent to eliminate the need for flanges.

Straight and complex piping can also be fabricated with dual laminates as described above for vessels. Some straight dual laminate pipe is built on seamless extruded PFA, PVDF or FEP tubing with glass fabric embedded in the exterior wall.

Smaller-diameter pipe and tubing is available for use without support. PFA, for example, is widely used in tubing for aggressive fluids used in processing

Figure 16.5. Pipe sections lined with PTFE are bent to eliminate the need for additional flanged connections that would need to be monitored for leaks. Manufacturer: Dow Plastic-Lined Piping Products, Bay City, Mich. USA

semiconductors. PFA is extremely resistant to attack by these chemicals, and it helps prevent contamination of processes. Unsupported piping of PVDF is sometimes used in chemical processing plants.

4 FITTINGS, PUMPS AND VALVES

Steel fittings, pumps and valves can be lined with PFA, FEP, ECTFE, ETFE and PVDF by thermoplastic transfer molding. A removable core supported within the component during molding defines the interior shape and thickness of the lining. Sizes are limited by the capacity of the transfer molding equipment, polymer rheology and other factors.

Fittings and pumps are also made with proprietary dual laminate technology as described above in section 1. Because the lining is securely attached to the component's FRP shell, manufacturers rate these components for relatively high levels of vacuum. One pump manufacturer's design uses a FRP shell over a rotationally molded PFA lining.

Lined valves for chemical service, typically butterfly and plug designs, often have internal components encapsulated in fluoropolymer. To help keep them in position, linings and encapsulations are keyed into recesses in the metal substrates. A valve lined with PFA is shown in Figure 16.6.

Figure 16.6.   Using PFA to line this valve and encapsulate its plug allows service in hot 56% nitric acid at up to 248 °F (120 °C). Manufacturer: Xomox Corp., Cincinnati, Ohio. USA

Isostatic molding is used to form PTFE linings for components. The process is described above in section 3. Isostatic molding is also used to make PTFE shapes for pumps and other components. These shapes usually must be machined to meet required tolerances.

Valves made principally of corrosion-resistant metal often have seats or seals of PTFE. For these uses, PTFE provides chemical resistance and conforms well to metal components for tight sealing. Metal pumps also use PTFE seals, and pump diaphragms are available with PTFE facing the fluid media. An elastomer structure supports the PTFE element.

Some relatively small components are made of solid fluoropolymer. PFA, FEP, ECTFE, ETFE and PVDF are injection molded to form fittings, valves, regulators and other devices. Like the PFA tubing described above, these are used in manufacturing processes where corrosion and contamination cannot be tolerated. Small components are also machined from blanks of PTFE or other fluoropolymers.

5 SEALS AND GASKETS

PTFE is a standard material for gaskets and seals because it provides an extremely broad range of chemical resistance and service temperatures from cryogenic to 500 °F (260 °C). Because they are compatible with virtually any media, PTFE gaskets can reduce inventory requirements for plants with multiple processes.

For gaskets, PTFE offers conformability, and for seals, inherent lubricity. For these applications, PTFE is often compounded with other materials to help reduce creep and for seals, to increase stiffness.

There are many types of PTFE gaskets. The simplest design is cut from sheet skived from a billet of PTFE that has been molded and sintered. It is suitable for relatively low temperatures and pressures. A variation of this type is an envelope design with thin PTFE sheet formed over a resilient core so that only the formed PTFE sheet is in contact with the flanges and exposed to the media.

For more demanding temperature and pressure conditions, PTFE is filled with materials such as barium sulfate to reduce creep under thermal and mechanical stress. Another approach to reducing creep uses special manufacturing technology that produces gasketing with a strong fibrillated and expanded microstructure. For the most extreme conditions, PTFE is contained by an alloy metal strip in a spiral-wound or envelope gasket design.

PTFE gasketing is supplied as preformed gaskets, and as sheeting and tape that can be cut to size. Cutting gaskets from sheeting as needed helps reduce stocking requirements.

Like gaskets, seals of PTFE come in a range of types. Most seals used in the chemical processing industry protect shaft bearings in pumps, motors and other rotating equipment. These seals have non-contacting labyrinth designs and can both retain lubricants in bearings and block intrusion of contaminants. PTFE lip seals are sometimes used, but labyrinth types are usually selected because they are more robust.

Should contact accidentially occur between seal elements, the lubricity of PTFE reduces friction and possible overheating. The nonsticking preformance of PTFE helps prevent buildup of contaminants. And its broad chemical resistance allows use of such seals in a wide range of harsh chemical environments.

6 INTERNAL COMPONENTS FOR COLUMNS AND OTHER EQUIPMENT

Fluoropolymers are used for a wide variety of packings, trays, dip tubes, spargers and distributors. They provide resistance to chemical attack during contact distillation, absorption, stripping and other processes. Designs take into account the relatively low mechanical strength of fluoropolymers so that packings, for example, will not be crushed by overloading trays.

These components can be molded from PFA, ETFE or FEP, or machined from PTFE shapes. They also can be made by coating metal parts with fluoropolymer or using FRP dual laminate technology (see above in section 1).

Compared with ceramic components such as packings, fluoropolymers offer greater resistance to damage by thermal shock during temperature cycling.

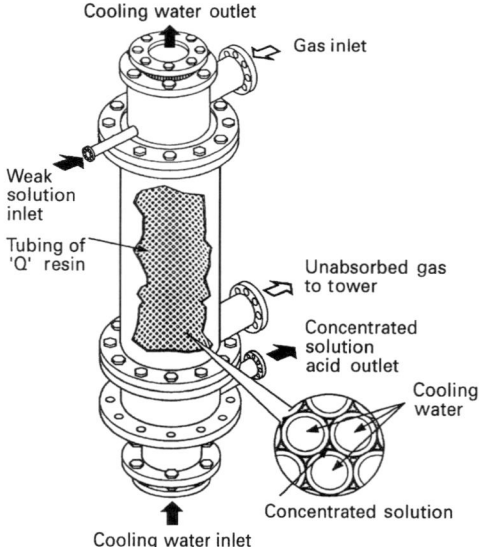

Figure 16.7.   This cooler/absorber functions like a column with the tubing acting as temperature-controlled packing that can add or remove heat. The tubing is PFA and the tube sheet is PTFE. Manufacturer: Ametek, Inc., Haveg Division, Wilmington, Del. USA

7 HEAT EXCHANGERS

Heat exchangers with all wetted surfaces made of fluoropolymer offer extensive resistance to chemicals. One design has internal tubes of PFA or FEP. The outer shell is a FRP or steel column lined with PTFE, and the honeycomb-like PTFE tube sheet is joined to the tubes by a bonding process.

These heat exchangers are often used in removing heat from strong acids, and the tubing bundles without the shells are used to heat acid baths and other aggressive solutions.

The units can also be configured as cooler/absorbers for stripping HCl, SO_2 and other aggressive gasses from process streams. For this application, the internal PFA tubing is compounded with graphite for increased thermal conductivity. Figure 16.7 is a drawing of a cooler/absorber.

8 HOSE AND EXPANSION JOINTS

Hose lined with fluoropolymer is used to transfer aggressive and ultrapure fluids in plants and during distribution. It is made by building up a flexible supporting shell of wire and elastomer over tubing of PTFE, FEP or PFA. The tubing is convoluted to increase flexibility.

Fluoropolymer expansion joints are installed to limit transmission of vibration caused by pumps, agitators and other equipment. They also allow piping systems to expand and contract during thermal cycling without stressing gaskets and other components.

These joints can be convoluted designs machined from PTFE stock. Such units can serve for long periods of time since PTFE has a long flex life. Others are molded from PFA or FEP and have flexible elastomeric jackets for support.

9 TESTS FOR MATERIALS SELECTION

Permeation, vacuum and pressure/temperature cycling endurance are major issues for all linings. A reliable predictor of success is good experience with similar applications coupled with laboratory and field testing.

Laboratory screening can be performed at pressures above atmospheric with a Roberts cell, or with an Atlas cell at atmospheric pressure (ASTM C868). Figure 16.8 shows a schematic drawing of an Atlas cell.

Field testing should follow laboratory screening. Such tests can be done with lined blind flanges on process piping or equipment. If seaming is involved in the lining system being studied, specimens should have seams. Tests of lined coupons that have been suspended in process streams provide useful information about the chemical compatibility of the lining. Exposure for lined flanges and coupons is typically for three months, but may be longer.

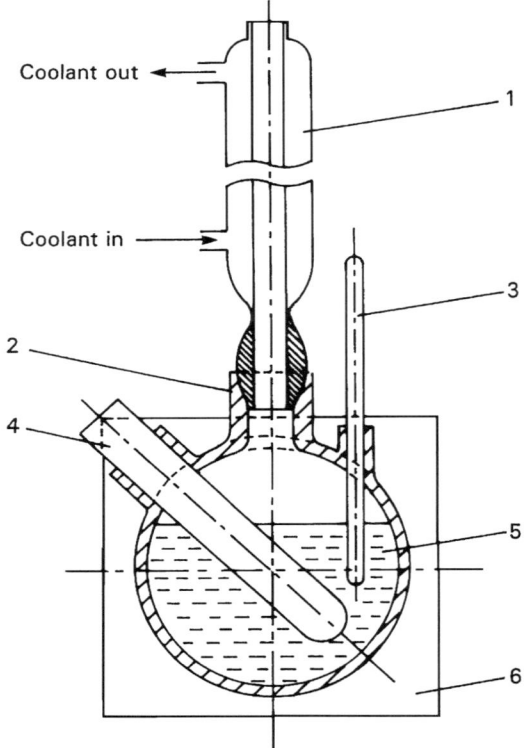

Figure 16.8. An Atlas cell, a schematic representation of which is shown here, is used for laboratory screening with a blind flange lined with the candidate material. (1) Condensation column; (2) Atlas cell; (3) thermometer, (4) heater sheath; (5) text medium; (6) fluoro shield coated test plate, ASTM C868

Degradation and discoloration should be noted and loss of adhesion for adhered linings should be determined with a peel pull test. Spark testing will reveal small failures such as pinholes or cracks.

As for any polymer, permeation is an issue for fluoropolymer linings. It is advisable to test in advance of installing equipment. Efforts are under way to develop a standard test method for permeation rate determinations.

10 FAILURE ANALYSIS AND REPAIR

If a lining or component fails, careful analysis can reveal the cause and suggest ways to prevent future problems. The first step in analysis is identification of the polymer backbone, if it is not already known, through laboratory analysis.

Failure of fluoropolymers usually involves pinholes, cracking, blistering or delamination, depending on the construction. Failure is often not detected until the substrate is severely corroded. Failures are typically caused by exceeding design conditions or poor fabrication.

For failures where risks are great, it may be necessary to simulate the failure in the laboratory. Test specimens are usually blind flanges or pipe spools.

Visual analysis is still the most useful method for studying failures. Failed linings should be examined for loss of adhesion and discoloration or other evidence of permeation and absorption. Surface etching or cracking indicates that the polymer has been chemically attacked. Swelling is evidence of absorption of organics. How well the substrate surface was prepared for lining can be determined by microscopic examination of the back of the lining specimen.

The difficulty of repair depends on the age and quality of the material and the severity of the damage. For linings, a test patch should be spark- and adhesion-tested. If thermoplastic welding or localized spray and bake coating is required, testing for weldability or adhesion is necessary.

11 ISSUES IN USING FLUOROPOLYMERS

Permeation, increasing regulatory requirements and emerging technologies should be considered in applications for fluoropolymers. To make informed decisions about technology, engineers must stay abreast of these issues.

11.1 PERMEATION

Permeation is a complex phenomenon. Currently, there are no usable test methods to support engineering decisions. Data reported in the literature is based on tests with thin films and represents steady-state values. These values may be useful in determining steady-state losses such as those involved in fugitive emissions.

For chemical handling applications, the time required for the permeant to reach the substrate and its effect on the lining-to-substrate interface may be even more important than steady-state loss values. Efforts are now under way to create, validate and standardize a test protocol for permeation information. At present, in the absence of useful data, engineers must use increased lining thicknesses to minimize the probable effects of permeation.

11.2 INCREASING REGULATORY REQUIREMENTS

Nondestructive and nonintrusive testing will need to be developed to meet emerging requirements for the prevention of unexpected failures. In the USA, for example, the Occupational Safety and Health Regulations are becoming more stringent and demand improved management of process safety. Currently, visual inspections or destructive examination are the only available methods for assessing the condition of lined and other equipment.

11.3 NEW TECHNOLOGIES

Lined steel and dual laminate fittings with complex shapes (see section 3 above) are used to reduce the number of flanged connections in piping systems. Technology is now being developed to eliminate the need for flanges in joining fluoropolymer-lined steel pipe. Thermoplastic fluoropolymer linings that extend beyond the ends of steel pipe shells are connected by melt bonding. These connections are protected by slipping a short piece of larger steel pipe over the joined lined sections and welding it in place.

In planning for the use of fluoropolymers, it is advisable to keep informed about new technologies. Methods for applying fluoropolymers continue to evolve. For example, work with laser and plasma systems in the application of fluoropolymer coatings is showing promise.

BIBLIOGRAPHY

Buxton, L. W. and D. R. Goldsberry, 'Fluoropolymer lined chemical systems and permeation', *Proceedings of Managing Corrosion with Plastics Symposium*, NACE International, PO Box 21830, Houston, TX 77218–8340, USA, November 1993.

Glein, G. A., 'Dual laminate for difficult corrosion problems — selection criteria and techniques', *Proceedings of Managing Corrosion with Plastics Symposium*, NACE International, PO Box 21830, Houston, TX 77218–8340, USA, October 1991.

Glein, G. A., 'Dual laminate selection criteria and techniques', *Materials Performance Magazine*, NACE International, PO Box 21830, Houston, TX 77218–8340, USA, March 1992.

Hall, N. L., 'Fluoropolymer-lined reinforced thermosetting plastic chemical process equipment', *Proceedings of the Society of the Plastics Industry, Western Section, Composites Institute*, April 1988.

Hall, N. L., 'Manage corrosion with fluoropolymer dual laminates', *Chemical Engineering Progress Magazine*, June 1994.

Heffner, D. K., 'Fluoropolymer linings in the transportation industry', *Materials Performance Magazine*, NACE International, PO Box 21830, Houston, TX 77218–8340, USA, July 1992.

Imbalzano, J. F., D. N. Washburn and P. H. Mehta, 'Permeation and stress cracking of fluoropolymers, *Chemical Engineering Magazine*, New York, USA, January 1991.

Khaladkar, P. R., 'A comparison of fluoropolymer linings', *Materials Performance Magazine*, NACE International, PO Box 21830, Houston, TX 77218–8340, USA, February 1994.

Rapra Technology Ltd, MTI Publication No. T-4: *Prediction of Service Performance of Equipment Made of or Lined with Polymer Materials*. Published for the Materials Technology Institute of the Chemical Process Industries, Inc, by NACE International, PO Box 21830, Houston, TX 77218–8340, USA, 1994.

Tatnall, R. E., MTI Project 84: *Review of Spark Testing Practices*. Materials Technology Institute of the Chemical Process Industries, 1994.

Tatnall, R. E., MTI Project 99: *Manual on Inspection of Linings*. Materials Technology Institute of the Chemical Process Industries, 1996.

17

Stereoregular Fluoropolymers via Ring Opening Metathesis Polymerisation

E. KHOSRAVI
University of Durham, Durham, UK

1 INTRODUCTION

In chain growth polymerisations a measure of control over the molecular weight and molecular weight distribution can be obtained using 'living' polymerisation chemistries in which there are no spontaneous termination reactions [1–4]. Some chain growth polymerisations are stereoselective and some give stereoregular polymers [5–7]. The control of molecular weight, molecular weight distribution and microstructure in functionalised polymers would present many potential opportunities for technological exploitation. Living and stereoregular polymerisations present themselves as an obvious approach for the preparation of such well-defined materials.

It has been shown that fluorinated bicyclic olefins of the type shown in Figure 17.1 undergo ring opening metathesis polymerisation (ROMP) using classical initiators based upon a transition metal chloride and a Lewis acid cocatalyst, e.g. WCl_6/Me_4Sn. The process is poorly characterised and gives largely atactic products. All of the polymers produced displayed rather broad molecular weight distributions, generally showing values of M_w/M_n much greater than most probable value for a well-behaved chain growth polymerisation, i.e. $M_w/M_n = 2$ [8]; an observation which is consistent with the presence of several kinds of initiator and/or propagating chain end [9]. The advent of well-defined single component transition metal alkylidenes has made it possible to polymerise fluorinated bicyclic olefins in a well-controlled living manner allowing the synthesis of polymers with a molecular weight distribution M_w/M_n of 1.05 [10–12] and, for the first time, it was possible to prepare stereoregular fluoropolymers and stereoblock copolymers.

Modern Fluoropolymers. Edited by John Scheirs
© 1997 John Wiley & Sons Ltd

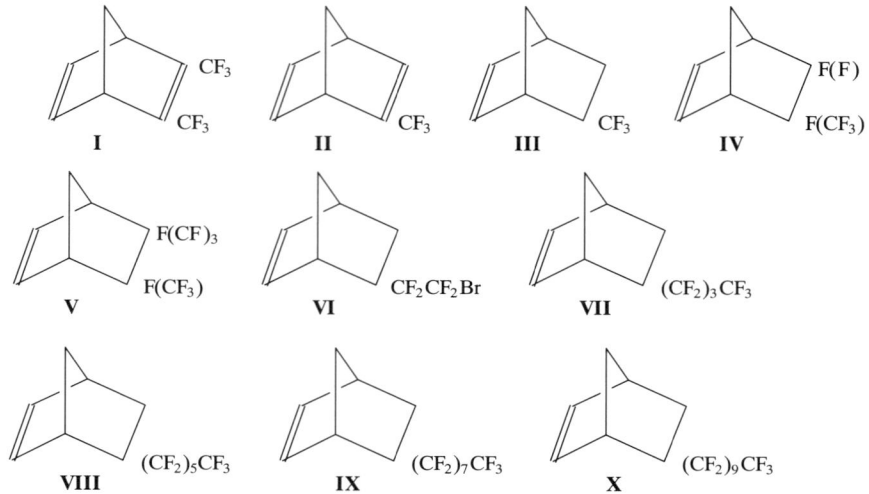

Figure 17.1

2 STEREOREGULAR FLUORO HOMOPOLYMERS

The key to controlled polymerisation of fluorinated bicylic olefins using well-defined Schrock initiator is that while it is inactive towards ordinary internal double bonds, it reacts rapidly with the strained double bonds in the monomer to give a linear polymer (Figure 17.2). The living nature of these polymerisations can be conveniently monitored by ^{1}HNMR [13]. The reactions are carried out in solvents such as benzene, toluene or THF and the living polymerisation reactions are terminated by the addition of aldehydes (e.g. benzaldehyde). The living polymer reacts with aldehydes readily to give metal oxide in a Wittig-like capping reaction.

Polymerisation of bis(trifluoromethyl) norbornadiene, BTFMND, (**I**, Figure 17.3) initiated by Schrock alkylidenes $Mo(CH-t-Bu)(NAr)(OR)_2$ gives all *trans* poly(BTFMND) when R = *t*-butyl [5,6] (i.e. **IIa** initiation, see Figure 17.3) and all *cis* poly(BTFMND) when R = hexafluoro-*t*-butyl [14] (i.e. **IIb** initiation).

All *trans* polymer shows behaviour that one would expect for a semicrystalline thermoplastic. DSC studies reveal a well-defined T_g at 97 °C and a broad melting endotherm at 200 °C, the shape and area of which is dependent on sample history. In samples precipitated from solution the T_g transition is not particularly marked and the melting peak is consistent with the presence of multiple melting points. In melt-quenched samples the T_g transition is marked and the melting peak area is small, the melting peak area increases on prolonged annealing at 180 °C (72 h), indicating a slow ordering process that is consistent with low chain mobility in the solid state. All earlier atactic samples were amorphous and only exhibit a well-defined T_g at 125 °C.

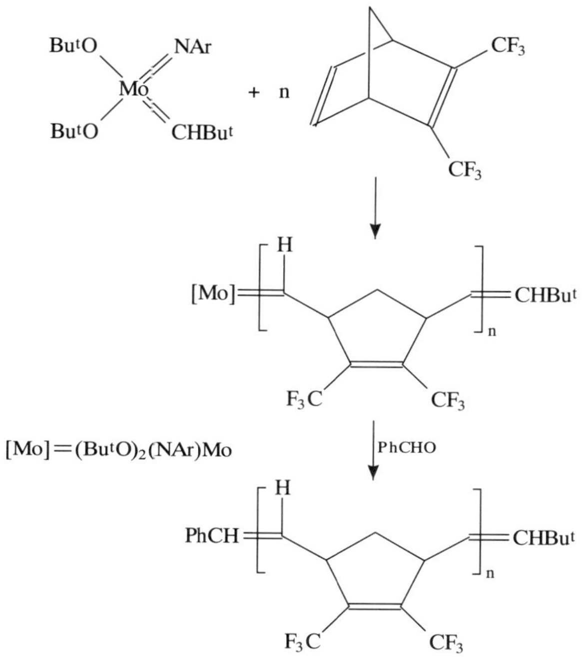

Figure 17.2

Dynamic mechanical thermal analysis of solution-cast films of both tactic and atactic samples show no energy dissipation peaks below T_g, indicating that in solid state both are relatively stiff polymers with little or no motion below T_g [6]. All *cis* polymer exhibits a well-defined T_g at 145 °C and no melting endotherm is observed.

Fibres can be drawn from the melt exhibiting 500% elongation on stretching, whereas fibres drawn from atactic polymer prepared by WCl_6/Me_4Sn are weak and cannot be stretched.

Detailed ^{13}C analysis showed that the all *trans* polymer is 92% tactic and the all *cis* polymer is 75% tactic. It has recently been shown that an initiator with a binaphthol replacing the two alkoxides and the arylimido isopropyl groups replaced by methyls gives poly(BTFMND) with 100% *cis* vinylenes and which is also 100% tactic [7]. Schrock and coworkers [15] have reported that all *trans* norbornadiene polymers are syndiotactic and that all *cis* norbornadiene polymers are isotactic regardless of the nature of the substituents in the monomers. ^{13}C NMR analysis of all *trans* poly(BTFMND) did not allow an assignment of its microstructure and, despite its high tacticity, its degree of crystallinity was too low to obtain X-ray diffraction data capable of reliable analysis. The all *trans* polymer displayed a remarkably high relaxed dielectric constant ε_R (greater than

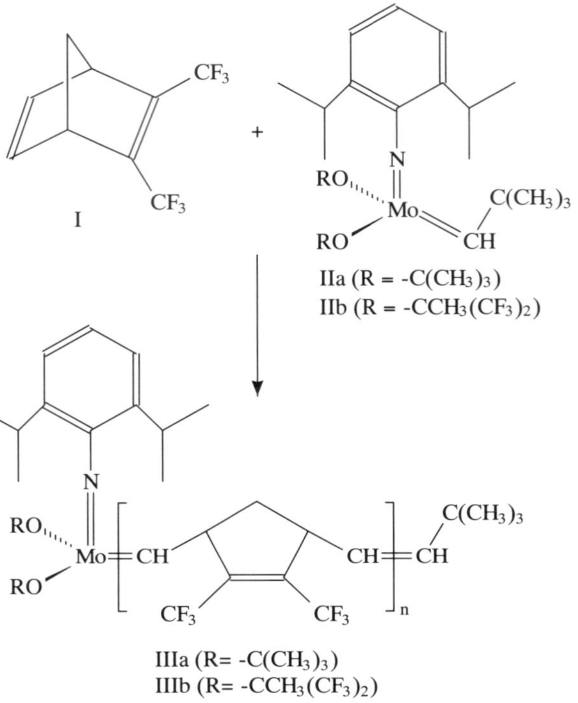

Figure 17.3

40, cf. PVDF ≈ 15), the all *cis* polymer displayed a relatively low relaxed dielectric constant ($\varepsilon_R \approx 6$), whereas the atactic polymer made via WCl_6/Me_4Sn initiation (54% *trans* vinylene content) showed an ε_R value of ≈16 [16]. The explanation of this large difference requires that in the *trans* polymer the polar bis(trifluoromethyl)cyclopentenyl rings in the chain (Figure 17.4) can act in colla- borative reinforcing sense in response to an electric field, whereas in the *cis* polymer their individual effects tend to cancel each other out [17]. This allowed an assignment of syndiotacticity unambiguously to all *trans* and, with a degree of uncertainty, to all *cis* polymers. The assignment of the tacticity of the all *cis* polymer has proved to be more problematic since, although there is only one way to obtain a very high ε_R value (cooperative reinforcement of individual ring effects) there may be several ways of effecting cancellation of dipoles and consequently low ε_R value [16].

The potentially high polarity of fluorinated polymers led to the investigation of the pyroelectric properties of these materials for possible application as active components in various types of electrical devices, particularly heat sensors, electromagnetic radiation detectors and thermal imaging systems. Much of the

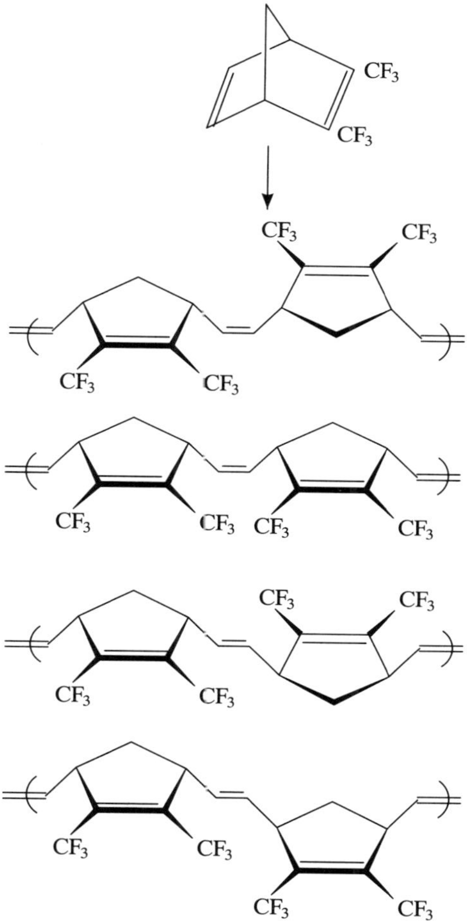

Figure 17.4

work on piezoelectric and pyroelectric behaviour of polymers has been focused on poly(vinylidene fluoride)(PVDF). *Trans*-syndiotactic poly(BTFMND) has a high permittivity above T_g greater than 40 and a saturation polarisation approaching 20 mC m^{-2} with a pyroelectric coefficient approaching 6 µC m^{-2} K^{-1}. While these values are less than for PVDF (50 mC m^{-2} and 30 µC m^{-2} K^{-1} respectively), the low tan δ (<0.001) and permittivity at ambient temperature ($\varepsilon_U = 2.6$) allows favourable comparison with PVDF. One figure of merit, F_D, ($F_D = \gamma/(\varepsilon \tan \delta)^{1/2}$) used to compare the pyroelectric response of detectors [18] suggests that in this respect poly(BTFMND) with a value of 118 µC m^{-2} K^{-1} at 20 °C is better than PVDF ($F_D \approx 64$ µC m^{-2} K^{-1}; data taken at 1.69 Hz from

Davies [19]). For the *cis* poly(BTFMND), the low ε_R of 5.7 does not allow one to obtain high γ or P at moderate fields.

PVDF works as a composite in which the crystalline phase provides the locked-in polarisation and the soft deformable amorphous phase responds to the applied stress [19,20]. It might be possible to mimic this mechanism by synthesising two-phase systems comprising a high-T_g highly polar phase dispersed in a low-T_g high permittivity amorphous phase, which allow changes of phase composition and morphology to optimise the piezoelectric and pyroelectric response [21].

3 STEREOBLOCK COPOLYMERS

Living chain growth polymerisation allows the possibility of making block copolymers which in turn can allow control of supramolecular organisation via the phase separation of incompatible blocks [22]. Blocks derived from the same monomer but having different microstructures may be incompatible, leading to the possibility of morphology control and hence bulk property control in a material derived from one monomer; such stereoblock copolymers have been prepared via anionic [23] and metallocene [24] methods. It has also been demonstrated that this is possible using ring opening metathesis polymerisation [25]. Poly(BTFMND) has been synthesised as a stereoblock copolymer containing *cis* and *trans* vinylene blocks via ligand exchange in living stereoselective ROMP initiated by a well-defined Schrock type initiator. The living all *trans* polymer **IIIa** (see Figure 17.3), formed via initiation of the polymerization **I** with **IIa**, undergoes ligand exchange with hexafluoro-*t*-butanol to give a living chain end which initiates the polymerisation of **I** to give a stereoblock copolymer with *cis* and *trans* vinylene blocks. The entire course of the reaction has been monitored by ^{1}HNMR. In NMR tube scale experiments initiation of **1** (10 equivalents) with **IIa** gave all *trans* polymer with a living chain end characterised by a doublet at δ 11.34 arising from propagating alkylidene **IIIa**. Upon the completion of the exchange reaction a new doublet appears at δ 12.42 due to the chain end alkylidene **IIIb** carrying two hexafluoro-*tert*-butoxy ligands. The process has also been repeated on a larger scale (2×100 equivalents of **I**). The relatively narrow polydispersity, $M_w/M_n = 1.16$, observed is as expected for the product of a well-defined living polymerisation process. Differential scanning calorimetry revealed two transitions at $\approx 95\,^\circ$C and 145 $^\circ$C as expected for the *trans* and *cis* blocks respectively.

4 CONCLUSIONS

This brief review demonstrates that stereoregular fluoropolymers have recently become available via well-defined living ROMP and that the product materials are of potential technological interest.

REFERENCES

1. Odian, G., *Principles of Polymerisation*, 3rd edn. Wiley, 1991.
2. Georges, M. K., Veregin, R. P. N., Kazmaier, P. K., and Hamer, G. K., *Macromolecules*, **26**, 2987 (1993).
3. Szwarc, M., *Carbanions, Living Polymers and Electron-Transfer Processes*, Wiley-Interscience, New York, 1968.
4. Kennedy, J. P., *Makromol. Chem. Macromol. Symp.*, **32**, 119 (1990).
5. Bazan, G. C., Schrock, R. R., Khosravi, E., Feast, W. J., and Gibson, V. C., *Polymer Commun.*, **30**, 258 (1989).
6. Bazan, G. C., Khosravi, E., Schrock, R. R., Feast, W. J., Gibson, V. C., O'Regan, M. B., Thomas, J. K., and Davis, W. M., *J. Am. Chem. Soc.*, **112**, 8378 (1990).
7. McConville, D. H., Wolf, J. R., and Schrock, R. R., *J. Am. Chem. Soc.*, **115**, 4413 (1993).
8. Koeing, J. L., Chapter 2 in *Chemical Microstructures of Polymer Chains*, Wiley, New York, 1980.
9. Bin Alimunar, A., Blackmore, P. M., Edwards, J. H., Feast W. J., and Wilson, B., *Polymer*, **27**, 1281 (1986).
10. Gilliom, L. R., and Grubbs, R. H., *J. Amer. Chem. Soc.*, **108**, 733 (1986).
11. Schrock, R. R., Feldman, J., Grubbs, R. H., and Cannizzo, L., *Macromolecules*, **20**, 1169 (1987).
12. Murdzek, J. S., and Schrock, R. R., *Macromolecules*, **20**, 2640 (1987).
13. Feast, W. J., and Khosravi, E., 'Recent development in ROMP', Chapter 3 in *New Methods of Polymer Synthesis*, Vol 2, eds Ebdon, J. R. and Eastmond, G. C., Blackie Academic and Professional, London, 1995.
14. Feast, W. J., Gibson, V. C., and Marshall, E. L., *J. Chem. Soc., Chem. Commun.*, 1157 (1992).
15. O'Dell, R., McConville, D. H., Hofmeister, G. E., and Schrock, R. R., *J. Am. Chem. Soc.*, **116**, 3414 (1994).
16. Davies, G. R., Feast, W. J., Gibson, V. C., Hubbard, H. V. St A., Khosravi, E., Marshall, E. L., Ward, I. M., *Polymer*, **36**, 235 (1995).
17. Davies, G. R., Feast, W. J., Gibson, V. C., Hubbard, H. V. St A., Khosravi, E., Petty, M., Petty, M. C., Tsibouklis, J., Ward, I. M., and Wellings, S. C., *Proceedings of the Symposium Functionele Polymeren-de Polymeren van Morgen*, TU Delft, November 1992.
18. Whatmore, R. W., *Rep. Prog. Phys.*, **49**, 1335 (1986).
19. Davies, G. R., in *Physics of Dielectric Solids* (Isnt. Phys. Conf. Ser. No. 58), Institute of Physics, Bristol, 1981, p. 50.
20. Broadhurst, M. G. and Davies, G. T., *Ferroelectrics*, **60**, 3 (1984).
21. Feast, W. J., Gibson, V. C., Khosravi, E., Marshall, E. L., and Wilson, B., unpublished results.
22. Woodward, A. E., *Atlas of Polymer Morphology*, Hanser, Munich, 1988.
23. Poshyachinda, S., Edwards, H. G. M., and Johnson, A. F., *Polymer*, **32**, 334 (1991).
24. Coates, G. W. and Waymouth, R. M., *Science*, **267**, 217 (1995).
25. Broeders, Y., Feast, W. J., Gibson, V. C., and Khosravi, E., *J. Chem. Soc., Chem. Commun.* 343 (1996).

18

The Radiation Crosslinking of Fluoropolymers

BERNARD J. LYONS

22 Hallmark Circle, Menlo Park, CA USA

1 BACKGROUND

A recent review by this author [1] contains a bibliography of references, up to mid-1988, that deal with various aspects of the radiation crosslinking of fluoropolymers.

Fluoroelastomer gums, like other gum rubbers, are crosslinked to provide them with the physical strength necessary for their intended use. In the case of the partly crystalline thermoplastic fluoropolymers, the reasons why crosslinking would be desired are not so obvious. Thermoplastic fluoropolymers are, generally speaking, selected for their mechanical strength, chemical inertness, excellent electrical properties, mechanical strength at higher temperatures and relative insensitivity to oxidation (both thermal and UV initiated). This combination of properties makes them suitable as electrical insulation for wire and cable and other electrical applications. The properties desired of an electrical insulation include the following:

1. Excellent electrical properties, for example, high electrical resistivity and dielectric strength (AC and DC), low power factor and dielectric constant. Most, but not all, fluoropolymers have excellent electrical properties, but even the exceptions, under suitable conditions, have found many uses in electrical applications.

2. Excellent mechanical properties, maintained over the operating temperature range and service life of the insulation. It is well known that many thermoplastics have long-term mechanical properties that are nothing like as good as their short-term properties. This is manifested as relatively poor stress relaxation and creep behavior. Moreover, many exhibit thermal stress cracking or reduced elongation at temperatures between a second-order transition temperature and

Modern Fluoropolymers. Edited by John Scheirs
© 1997 John Wiley & Sons Ltd

their crystalline melting points. This reduced elongation occurs, for example, in some poly(vinylidene fluoride–PVDF) resins, between about 400 K, and in some ethylene–tetrafluoroethylene copolymer (ETFE) and ethylene–chlorotrifluoroethylene copolymer (ECTFE) resins, between about 430 K and their crystalline melting points. Crosslinking reduces or even largely eliminates non-oxidative stress relaxation and creep and, even at relatively low levels, essentially eliminates thermal stress cracking and improves the elongation properties at temperatures near to but below the melting point. The cut-through resistance, especially at higher temperatures, of polymeric insulations can also be significantly improved by crosslinking.

3. Maintenance of structural integrity under sustained thermal or electrical overload situations. In many applications, an insulation that melts off the conductor under conditions of electrical or thermal overload is not considered to be functional. Crosslinking maintains the integrity of insulations under sustained electrical overload or during exposure to higher temperatures (for example during soldering operations).

4. Non-flammable or at least highly flame retarded. Many fluoropolymers are or can be made to be non-flammable or highly flame retarded in air.

5. Form a non-conductive char under pyrolysis conditions. A requirement in many applications (for example, in signal or control cables) is that, if an electrical insulation is subjected to pyrolysis conditions, as in a fire situation, enough of the insulation should remain around the conductor as a char that some electrical integrity is preserved. Such applications include electrical utilities (especially nuclear), chemical processing and aircraft, especially military aircraft. Significant degrees of crosslinking can modify the pyrolysis behavior of an insulation and improve char formation, especially if the formulation also includes, for example, antimony oxide.

The cost combined with the high density of fluoropolymers (raw materials are priced by the pound or kilogram, but it is the properties per cubic inch or centimetre that are the selection criteria used in product applications) means they are only likely to be considered as candidates for a commercial application if the service conditions or some other factor rule out more common polymers. Thus, crosslinked fluoropolymers tend to find applications generally in chemically hostile and/or demanding environments (usually at higher service temperatures). Typical of such applications are wire and cable harnesses in commercial or military aircraft, electrode leads in impressed current corrosion protection, control cables in power plants, especially nuclear, and coatings in chemical process plant.

2 CROSSLINKING METHODS

Although peroxides can be used to crosslink fluoro-rubber gums, processing temperature requirements rule them out for many thermoplastic fluoropolymers.

The preferred crosslinking method for extruded fluoropolymer products is by exposure to ionizing radiation. Ionizing radiations comprise X-rays, gamma rays from radioactive materials and high-energy electrons from radioactive materials or from accelerators. Only two types of radiation are used commercially: gamma radiation from radioactive cobalt-60 (which is produced in specialized atomic reactors by neutron irradiation of cobalt) and accelerated electrons. Cobalt-60 gamma rays are very penetrating, enabling relatively large assemblies to be irradiated, but their use has the disadvantage that the dose rates are low. Because of this the workpiece should be protected from contact with oxygen during the irradiation (to prevent radio-oxidation). High-energy electrons have a relatively low penetrating power (a 1 MeV electron, that is, one accelerated through a voltage gradient of 1 MV, has a range in water of only about 0.6 cm). On the other hand, the dose rate attainable is very high, many orders of magnitude higher than for gamma irradiation, so that there is normally no need to exclude oxygen during irradiation except when very thin coatings are processed (much less than 100 microns). Details of the various types of equipment used have been summarized, for example, by Spinks and Woods [2].

Some dosage and radiolytic chemical yield units used in radiation crosslinking should be mentioned here. The internationally approved unit of absorbed dose is based on the gray (Gy); in polymer processing the kilogray (kGy) is a convenient unit of measurement. However, many sources still continue to use the now obsolete unit, the rad, where 1 megarad (Mrad), which is equal to 10 kGy, is an energy deposition of 100 calories per gram of irradiated substrate. Radiolytic chemical yields (the G value, which is the amount of a particular chemical change or group of changes per unit of energy absorbed), are now expressed in micromoles per joule (μ mol J^{-1}). The older and obsolete unit for the G value, that is, the yield of molecular events per 100 electron volts (eV) of energy absorbed, is still used by some authors. A G value of 1 μ mol J^{-1} is equivalent to a G value of 10.4 events per 100 eV.

3 RADIATION CHEMISTRY

A detailed discussion of the radiation chemistry of these polymers is given in reference [1]. A brief summary of pertinent features follows. However, in commercial crosslinking applications, the use of unsaturated monomers as crosslinking promoters is ubiquitous so these will be discussed in more detail.

4 ENHANCEMENT OF CROSSLINKING

4.1 GENERAL

Crosslinking promoters (sometimes called 'prorads') were first used to enhance radiation crosslinking in certain rubbers, polyvinyl chloride and polyethylene in the 1950s. Bernstein, Odian and Binder [3] list some of the requirements for the successful use of polyfunctional unsaturated monomers in polymers:

1. Their boiling point should be high enough not to cause appreciable volatilization at mixing and processing temperatures.
2. They should be compatible with the polymer at normal and at elevated temperatures. It follows that, as the additive does not penetrate into crystalline regions of polymers, such crosslinking only occurs in the amorphous regions.
3. Compatibility of a promoter is more important than reaction efficiency;
4. No dose rate effect was observed during the irradiation of polyethylene-polyfunctional monomer mixtures.

Other factors, cited by Lyons [1], include the unsaturation content (usually, but not always, the higher the better); the importance of unsaturation type and stability (allyl groups although somewhat less reactive than vinyl or acrylic groups are much less prone to homopolymerize at higher temperatures or during irradiation); the structure of the molecule (for example, whether it contains other types of reactive groups and whether the structure is rigid or flexible), and its polarity (symmetrical and/or low polarity structures are more soluble in non-crystalline low polarity polymers).

Lyons and Crook [4] studied the decay of carbon–carbon unsaturation either present initially in polyethylene or in the form of unsaturated additives and found the same decay kinetics were involved in all instances. Lyons and Cross [5] showed that low concentrations of triallyl cyanurate (TAC) in polyethylene, poly(vinyl acetate) and poly(ethyl acrylate) lowered the gelling dose (increased crosslinking) but higher concentrations increased the gelling dose (reduced crosslinking) even though much higher gel fractions were observed with the latter at higher doses. This can be explained in terms of a sequential decay of allyl groups in each additive molecule, in which, at high additive concentrations, competition between TAC molecules for the available free radicals reduces the chances of further links involving already (partially) reacted TAC molecules. At very high doses the 'crosslinked unit' equivalent of each TAC molecule in polyethylene was calculated to be about 2.5. For 'bridged crosslinking' through each TAC molecule the equivalent would be 3 and the similarity of the experimentally determined equivalent suggests that bridge crosslink formation rather than homo polymerization is highly favored with allyl compounds. However, the maximum effect of added TAC is only seen at dose levels far higher than those needed to eliminate all the allyl groups, suggesting that appreciable 'clustering' of the reacted TAC molecules occurs as a result of the radiolytic reaction [1].

5 FLUORINATED ELASTOMERS

The original commercially available fluorinated elastomers are copolymers of vinylidene fluoride with hexafluoropropylene (VDF–HFP) or tetrafluoroethylene (VDF–TFE). Harrington [6] and Dixon *et al.* [7] independently disclosed that they could be crosslinked by radiation. Although VDF–HFP and VDF–TFE have been reported to be crosslinked by radiation, most references in the patent literature disclose compositions that include unsaturated monomers as crosslinking promoters, to reduce the dose needed and presumably to minimize radiation damage to the elastomer.

Since about the early 1970s, copolymers of tetrafluoroethylene and propylene (TFE–P), first disclosed to be crosslinked by radiation by Yamamoto, Uchijima and Ito [8] and, more recently, copolymers having ethylene–tetrafluoroethylene (ETFE) and TFE–P copolymer blocks have become available. These copolymers are frequently mentioned in the patent literature as being crosslinked by radiation (in the presence of prorads). Also mentioned in the patent literature and in this context are graft copolymers. Ito [9] found the G (values) for crosslinking and chain scission in TFE–P derived from radiolytic stress relaxation under nitrogen to be 0.152 and 0.023 μ mol J^{-1} respectively, in good agreement with the values (0.106 and 0.021 μ mol J^{-1}) previously obtained from gel analysis [10].

Allyl compounds such as diallyl maleate are effective crosslinking promoters for TFE–P copolymers [11]. The effectiveness of triallyl isocyanurate (TAIC) and trimethylolpropane trimeth-acrylate (TMPTM) in Viton GLT™ (VDF–HFP–TFE) has been compared by Kaiser *et al.* [12]. Incorporation of TAC, TAIC or N, N'-(m-phenylene)-bismaleimide (MPBM) reduced the dose required for development of acceptable properties in PVDF and VDF–HFP to 100 kGy [13]. The patent literature reviewed in preparing this chapter has numerous references to the use of monomers with fluorinated elastomer gums. Major applications for these elastomers include cable jackets and heat shrinkable tubing, mainly in demanding environments where high-temperature resistance is required.

6 THERMOPLASTIC FLUORINATED POLYMERS

6.1 POLY(TETRAFLUOROETHYLENE-CO-HEXAFLUORO-PROPYLENE)

This copolymer (TFE–HFP) was disclosed by Bowers [14] and Bowers and Lovejoy [15], and later confirmed by Lovejoy, Bro and Bowers [16] to undergo crosslinking by ionizing radiation if irradiated in nitrogen at temperatures above the 'glass transition temperature'. Zhong, Sun and Zhang [17] studied gel formation in an FEP sample containing 14 mol% hexafluoropropylene on irradiation at various temperatures using the Charlesby–Pinner equation, which relates the sol fraction s to the radiation dose r:

$$s + \sqrt{s} = \frac{p_0}{q_0} + \frac{1}{q_0 u_1 r} \tag{1}$$

where p_0 is the chain scission probability per monomer unit for unit radiation dose, q_0 the crosslinking probability per monomer unit for unit radiation dose and u_1 the number average degree of polymerization. This equation assumes that the initial molecular weight distribution in the polymer is random, but if p_0 is significant compared with q_0 the experimental results for polymers with non-random distributions will approach this relation at higher doses. As they point out, their results show a linear dependence of $s + \sqrt{s}$ on $1/r$. Table 18.1 includes the values they obtained for the chain scission to crosslink ratio (p_0/q_0) and the (extrapolated) gelation dose (the dose at which a gel just starts to form). I also present in this table estimates of crosslinking and chain scission yields relative to that at 423 K, measured from Figure 2 of their paper. Although such estimates are only approximate, they yield the interesting but tentative conclusions that from 423 to 493 K the crosslinking yield does not change very much but increases sharply at 513 K, whilst the chain scission yield drops sharply between 423 and 483 K but remains almost constant at the higher temperatures. Ueno has disclosed the use of TAC [18] and of TAIC [19] to increase radiolytic crosslinking in FEP.

6.2 POLY(TETRAFLUOROETHYLENE)

For many years this polymer (PTFE) has been considered to undergo only chain scission on exposure to ionizing radiation. In fact, a significant recycling of scrap PTFE (mainly residues from machining operations) uses radiation processing, employing doses up to about 400 kGy, to reduce the molecular weight of this polymer to a range in which it can be comminuted to a fine powder for use as a dry lubricant. Zhong, Sun and Zhang [20] concluded from a study of the depression of the melting and crystallization temperatures of PTFE, irradiated at various temperatures and followed by heat treatment at 653 K, that radiation branched

Table 18.1. Values of selected radiation parameters of FEP irradiated at various temperatures (taken from Zhong, Sun and Zhang [17])

Irradiation temperature (K)	p_0/q_0	q_0 normalized to that at 423 K[a]	p_0 normalized to that at 423 K[a]	Gelation dose (kGy)
423	1.35	1	1	123
483	0.85	0.81	0.7	100
493	0.8	1.0	0.6	80
513	0.42	2.2	0.7	2.5

[a]Estimates by the author from Figure 2 of their paper.

structures (in addition to chain scissions) resulted from such treatment, but only chain scissions resulted from irradiations not followed by such heat treatment. Some recent papers [21-23] have shown that irradiation of PTFE in vacuum at temperatures above its melting point (for example, 603-613 K), results in significant improvements in tensile strength and elongation at 473 K and in the tensile modulus at room temperature. These improved physical properties strongly contrast with the profoundly deteriorated properties that result from irradiation at lower temperatures, and clearly indicate that crosslinking or endlinking occurs in the molten polymer. As the temperature is increased above about 623 K, thermal depolymerization of PTFE is increasingly accelerated by irradiation and predominates over the crosslinking at yet higher temperatures. The predominant macromolecular radicals observed in PTFE are: $-CF_2-C \cdot (F)-CF_2-$ **(I)**, resulting from the removal of a fluorine atom from a $-CF_2$-group, and $-CF_2-C \cdot (F_2)$ **(II)**, formed by scission of backbone chains [22]. The fate of the fluorine atom which accompanies the formation of **I** has not yet been disclosed by any of these workers. Charlesby [24], in his listing of covalent bond energies shows that the C−F bond strength is 4.42 eV, greater than the C−H bond strength (4.28 eV) and, especially, the C−C bond strength (3.44 eV for a paraffin). Hence it seems likely that any fluorine atom formed during the radiolysis would cleave a backbone chain to yield, as one of the scission fragments, radical **II**. However, Siegel and Hedgpeth [25] indicate that the ratio of the radical **I** to radical **II** is 10 : 1, which would seem to rule out this possibility, although they do state that this ratio is much smaller (actual value not given) for irradiations in air. Analysis of gases evolved from PTFE during radiolysis might answer this question although there are significant experimental obstacles to such measurements.

6.3 POLY(VINYL FLUORIDE), POLY(VINYLIDENE FLUORIDE) AND POLY(TRIFLUOROETHYLENE)

Although Timmerman and Greyson [26] disclosed the crosslinking of poly(vinyl fluoride) (PVF) and poly(vinylidene fluoride) (PVDF) by ionizing radiation in 1962, the evidence they presented is ambiguous, The first comprehensive study of PVDF and poly(trifluoroethylene) (PTrFE) radiolysis was published by Yoshida, Florin and Wall [27] in 1965.

In 1966, Wall, Strauss and Florin [28] reported on the pyrolysis of unirradiated and irradiated PVF and PVDF, concluding that irradiation increases both the evolution rate of volatiles and the residue after pyrolysis. For PVDF the thermal stability of the polymer varies inversely with the radiation dose. Lyons [29] also found that the temperature of exothermic dehydrofluorination of PVDF in nitrogen as measured by temperature programmed DSC is significantly decreased by prior irradiation. The decrease in stability is even more marked if the samples are heated up in air and/or contain certain metal oxides (reconfirmed by Pukshanskii *et al.* [30]). Thus, PVF and PVDF, like polyvinyl chloride, are significantly thermally destabilized by radiation doses, for example, in excess of 100 kGy.

Figure 18.1. A technician at a well-known commercial aircraft company prepares a wiring harness from wiring insulated with a radiation crosslinked ethylene–tetrafluoroethylene copolymer in a two-layer construction. Published by permission of Raychem Corporation

For this and other reasons, it is very advantageous to incorporate crosslinking promoters such as polyunsaturated monomers into such polymers, as they can provide high levels of crosslinking at a dose that does not compromise the thermal stability of the polymer to an excessive extent. Recently, Zhao *et al.* [31] have shown that gel formation in PVDF that has been quenched from the melt with icewater is greater than that in the polymer cooled slowly from the melt. Thus its behavior resembles that of slow and shock cooled high-density polyethylene (see, for example, Lyons [32]).

A patent to Lanza [33] first disclosed that TAC and other polyfunctional unsaturated monomers were compatible with and enhanced the radiation crosslinking of crystalline PVDF. The effectiveness of polyfunctional monomers in PVDF increases with the number of unsaturated groups in the molecule (Makuuchi, Yoshi and Abe [34]). This reference also indicates that, for monomers with the same functionality, those with better compatibility with PVDF are more effective. Lyons and Kim [35] found that the addition of unsaturated monomers to PVDF containing finely divided carbon black improved the performance of the irradiated material as a self-limiting heater under high voltage stresses. Some years ago the writer carried out a survey of the effect of monomers in PVDF, reported

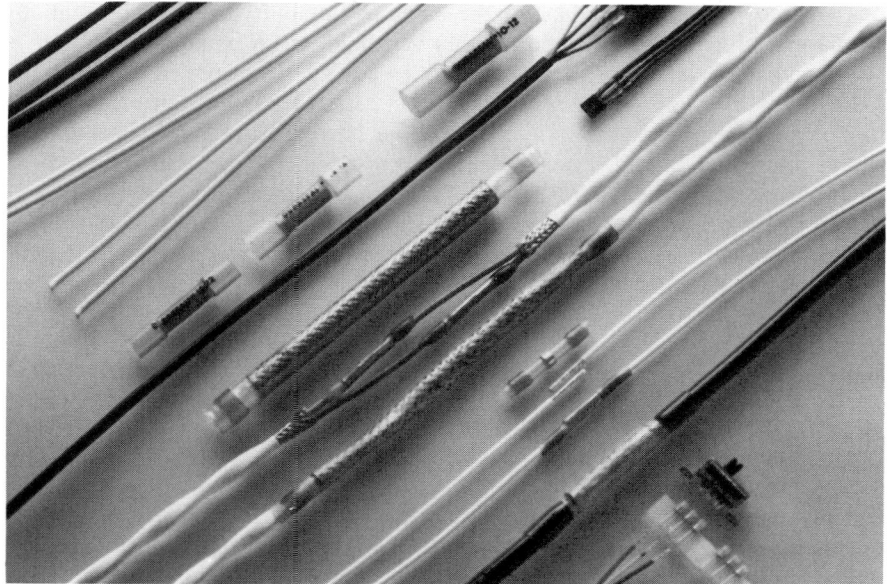

Figure 18.2. Radiation crosslinked poly(vinylidene fluoride) heat-shrinkable tubing containing meltable solder inserts, showing the wires or braid to be soldered, together with the tubings, before and after shrink-down. Published by permission of Raychem Corporation

in part in ref. 32. Monomers found to be particularly effective in PVDF (in addition to TAIC and TAC) include diallyl itaconate (three unsaturated groups), triallyl citrate, bis(maleimido-methyl)ether and ethylene bis-maleimide (the latter two being very effective). In decreasing order of effectiveness are *m*-phenylene bis-maleimide (but this was indicated by Klier [36] to perform better than TAIC in PVDF copolymers), 3,9-divinyl spiro-bis-*m*-dioxane, tetra-allyl pyromellitate. dipentaerythritol hexamethacrylate, trimethylol-propane trimethacrylate, diallyl malate methacrylate, diallyl adipate and diallyl suberate. Triallyl aconitate (four unsaturated groups), furfuryl methacrylate (three unsaturated groups) and divinyl suberate were all found to be almost ineffective. Interestingly diallyl adipate is significantly more effective than divinyl suberate, indicating that the poor performance of the vinyl compound is probably not due to incompatibility but rather to homopolymerization. Evidently, the reactivity of the unsaturated group to free radical attack and to polymerization and the presence of other groups in the monomer sensitive to free radical attack are among the factors that influence its response to free radical attack. It should be noted here that small differences in relative rankings of monomers between various workers are not unexpected especially with the more reactive monomers. They can be ascribed to differences in the preparative methods used and to the inherent variability of what is essentially a radiation initiated chain reaction with a very high chain transfer ratio.

A dual layer insulated wire and cable with a polyethylene core and a poly(vinylidene fluoride) primary jacket (both radiation crosslinked) was introduced to the military and commercial aircraft market some 30 years ago and is now finding use in modern automobiles, where temperatures in the engine compartment can reach quite high levels. Another interesting use for crosslinked poly(vinylidene fluoride) is in the production of heat shrinkable tubing with solder inserts which melt and flow around a substrate (such as two wires, or a wire and a termination, to be electrically connected) when the tube is heated to enable it to recover towards its original shape (Fig. 18.2).

6.4 ETHYLENE–TETRAFLUOROETHYLENE (ETFE) AND ETHYLENE–CHLOROTRIFLUOROETHYLENE (ECTFE) COPOLYMERS

Carlson and West [37] disclosed that irradiation improves that tensile properties at elevated temperatures of ETFE and ECTFE alternating copolymers, especially the ultimate elongation. Irradiation could be carried out at room temperature or at and above 423 K. Irradiation at room temperature followed by heat treatment at 435 K in nitrogen for 20 minutes was indicated to be the most effective. As irradiation was also shown to increase the solder iron resistance of the polymers, they presumed that crosslinking was occurring. Room temperature irradiation in air results in very little crosslinking in ECTFE [1]. Similar studies by the author and E. Scalco, reported in ref. 1 have shown that this is also the case for ETFE even when electron irradiation is followed within a few minutes by annealing at 433 K. Somewhat better results were obtained if air was excluded during the irradiation. However, irradiation in nitrogen of both ECTFE and ETFE at a temperature of at least 423 K did result in elastic modulus values greater than 330 kPa above their melting points. This indicates that these two copolymers have a much smaller tendency to crosslink at room temperature when compared with PVDF. The solubility behavior of an ECTFE copolymer has been studied by Luo, Pang and Sun [38]. Sun *et al.* [39] derive a p_0/q_0 ratio of 0.3 for an ECTFE copolymer (F-30) and a G (crosslink) value of 0.13(5) μ mol J^1.

As was found with PVDF, monomers much improve the crosslinkability of both these copolymers. TAC was early suggested for use in crosslinking ECTFE and ETFE copolymers (Carlson *et al.* [37], Robertson and Schaffhauser [40]), as was decamethyleneglycol dimethacrylate (Aronoff, Dhami and Shieh [41]). Other suggested monomers for ETFE and ECTFE copolymers include triallyl trimellitate and diallyl tridecanedioate (Aronoff and Dhami [42]) and unsaturated bis-imides or ester-imides (Gotcher, Germeraad and Jansons [43]). Substantial amounts of TAIC may be absorbed into shaped articles of ETFE copolymers by imbibing (to saturation), for example, at 473 K and the resultant composition radiation crosslinked (Gotcher and Germeraad [44]). Several other monomers have been suggested as crosslinking additives for ECTFE and ETFE (Aronoff, Dhami and Shieh [45], Aronoff and Dhami [46]; Dhami [47]; Gotcher, Germeraad and White [48] and Kojima [49]).

Despite the suggestion that TAC and TAIC are so volatile at the processing temperature of ETFE as to substantially preclude incorporation into this polymer [44], many more recent patents disclose the use of TAIC in amounts up to 10% in ETFE and several address the important issue of monomer stability at processing temperatures, suggesting the use of metal oxides (for example, Seki, Ando and Yagyu [50]) and Irganox® 1010, an antioxidant (Ishii and Moshiki [51]).

Radiation-crosslinked dual layer ethylene-tetrafluoroethylene copolymer insulated wire and cable is used extensively in the military and commercial aircraft market (Fig. 18.1).

7 CLOSING COMMENTS

We have already noted that the C$-$F bond strength is 4.42 eV, greater than the C$-$H bond strength (4.28 eV) and, especially, the C$-$C bond strength (3.44 eV for a paraffin), which makes it very unlikely that molecular fluorine is produced in the radiolysis of PTFE and FEP (see also the comments of Dole [52]). With regard to the mechanism of crosslinking of other fluoropolymers, it should be noted that the H$-$F bond strength is 5.82, indicating that hydrogen atoms (H) would be effective to abstract fluorine atoms from a fluoropolymer chain. However, hydrogen fluoride has not been found to be an important radiolytic product in the irradiation of ETFE polymers (some is formed during irradiation of ETFE polymer in air, but it is believed that this results from the hydrolysis of carbonyl fluoride, which is known to be produced when this polymer is irradiated in the presence of oxygen). Hydrogen fluoride is produced during the irradiation of PVF, PVDF and copolymers containing significant amounts of VDF and, by inference, those which also contain significant amounts of VF, but this very likely arises from a molecular dehydrofluorination reaction similar to the dehydrochlorination reaction which occurs in their chlorinated analogues (PVC and PVDCl) rather than by an abstraction reaction. Hydrogen gas, which is formed in polyethylene during radiolysis, has been shown to readily participate in exchange reactions with secondary alkyl main chain radicals (the H$-$H bond strength of 4.52 eV and the C$-$H bond strength of 4.28 are similar enough for this to occur). Hydrogen fluoride, having a much greater bond strength, would very probably not participate in such reactions. One is therefore led to the conclusion that if radical migration occurs in hydrogen containing fluoropolymers, it involves only exchange reactions of hydrogen atoms or main chain radicals, resulting from the scission of C$-$H bonds, with further C$-$H bonds or a random walk hydrogen or fluorine atom exchange along polymer chains. This would lead us to expect that free radical migration would be more difficult in the more highly fluorinated polymers so that crosslinking would occur more readily in the most highly hydrogenated fluoropolymers, yet one of the most readily crosslinked fluoropolymers is polytrifluoroethylene! Clearly, either the above logic is faulty or there are other more important factors which determine the response of these polymers. One such factor, described in ref. 1, is whether these polymers, during irradiation, are above or below certain second-order transitions, such as I have alluded to above in the case of PVDF and ETFE, that are not necessarily glass transition temperatures and may be similar to the gamma transition seen in high-density polyethylene (HDPE). This transition in HDPE occurs at about 343–353 K and is associated with changes in chain mobility at the edges of the lamellae. Radiolysis of HDPE at or above this temperature range, or prompt annealing of the polymer irradiated at lower temperatures results in significantly higher yields of crosslinks than are seen at lower irradiation temperatures without prompt annealing (Lyons and Weir [32]). Enhanced radiation yields for polymers irradiated above certain transition temperatures are, indeed, often noted.

REFERENCES

1. Lyons, B. J., *Radiat. Phys. Chem.*, **45**, 158 (1994).
2. Spinks, J. W. T. and Woods, R. J., *An Introduction to Radiation Chemistry*, Chapter 2, p. 14, 3rd edn, Wiley, New York (1990).
3. Bernstein, B. S., Odian, G. and Binder, S., NY0-9104 (1960); *J. Polymer Sci.*, **A2**, 2835 (1964).
4. Lyons, B. J. and Crook, M. A., *Trans. Faraday Soc. (Lond.)*, **59**, 2334 (1963).
5. Lyons, B. J. and Cross, P. E., *Trans. Faraday Soc. (Lond.)*, **59**, 2350 (1963).
6. Harrington, R., *Rubber Age*, **77**, 865 (1957); **83**, 472 (1958).
7. Dixon S., Rexford, D. R. and Rugg, J. S., *Ind. Eng. Chem.*, **49**, 1687 (1957).
8. Yamamoto, T., Uchijima, K. and Ito, Y., *Jpn. Kokai* JP 73 37 982 (1973).
9. Ito, M., *Radiat. Phys. Chem.*, **31**, 615 (1988).
10. Ito, M. *1st SPSJ, IPC*, Kyoto 184 (1985).
11. Tabata, Y. and Kojima, G., *Jpn. Kokai* JP 73 38 465 (1973).
12. Kaiser, R. J., Miller, G. A., Thomas, D. A., Sperling, L. H., *J. Appl. Polym. Sci.*, **27**(3), 957 (1982).
13. Vokal, A., Pallanova, M., Cernoch, P., Klier, I. and Kopecky, B., *Radioisotopy*, **29** (5–6), 426–34, (1988); CA112(14):119976u.
14. Bowers, G. H., UK 898 410, (1962); US Pat. 3 116 226 (1963).
15. Bowers, G. H. and Lovejoy, E. R., *Ind. Eng. Chem. Prod. Res. Devel.*, **1**, 89 (1962).
16. Lovejoy, E. R., Bro, M. I. and Bowers, G. H., *J. Appl. Polymer Sci.*, **9**, 401 (1965).
17. Zhong, X., Sun, J., and Zhang, Y,. *Polymer*, **33** 5341 (1992).
18. Ueno, K., *Eur. Pat. Appl.*, EP 164 201 A1, Sumitomo Electric Industries, Ltd. (11 Dec 1985).
19. Ueno, K., *Jpn. Kokai Tokkyo Koho*, JP 61025822 A2 Showa, Sumitomo Electric Industries, Ltd. (4 Feb 1986); *Jpn. Kokai Tokkyo Koho*, JP 61 155 410 A2 Showa, (15 Jul 1986).
20. Zhong, X., Yu, L., Zhao, W., Zhang, Y. and Sun, J., *Polym. Degrad. Stab.*, **41** 233 (1993).
21. Tabata, Y. *Solid State Reactions in Radiation Chemistry, Taniguchi Conf.*, p. 118, Sapporo, Japan (1992).

22. Sun, J., Zhang, Y. and Zhong, X., *Polymer*, **35** 2881 (1994).
23. Oshima, A., Tabata, Y., Kudoh, H. and Seguchi, T., *Radiat. Phys. Chem.*, **45** 269 (1995).
24. Charlesby, A., *Atomic Radiation and Polymers*, p. 19, Pergamon Press, London, England (1960).
25. Siegel, S. and Hedgpeth, H., *J. Chem Phys.*, **46** 3940 (1967).
26. Timmerman, R. and Greyson W., *J. Appl. Polymer Sci.*, **22**, 456 (1962).
27. Yoshida, T., Florin, R. E. and Wall, L. A., *J. Polymer Sci.*, **A3**, 1685 (1965).
28. Wall, L. A., Strauss, S. and Florin, R. E., *J. Polymer Sci.*, **A4**, 349 (1966).
29. Lyons, B. J., US Pat. 3 582 518 (1971).
30. Pukshanskii, M. D., Shirinyan, V. T., Bobrovskii, A. P. and Sirota, A. G., *Zh. Prikl. Khim. (Leningrad)*, **54**(1), 2608–10 (1981).
31. Zhao, Z., Luo, Y., Jiang, B., Yang, H., Teng, F. And Chen, X., *Radiat. Phys. Chem.*, **41**, 467 (1993).
32. Lyons, B. J. and Weir, F. E., 'The effect of radiation on the mechanical properties of polymers', Chap. 14 of *The Radiation Chemistry of Macromolecules*, (Ed. M. Dole), Vol. II, p. 294, Academic Press, New York (1974).
33. Lanza, V. L., UK Pat. 1 108 990 (1968); US Pat. 3 580 829 (1971).
34. Makuuchi, K., Yoshii, F. and Abe, T., *Nippon Kagaku Kaishi* (10) 1828 (1975).
35. Lyons, B. J. and Kim, J., US Pat. 4 188 276 (1980).
36. Klier, Ivo *et al. Plasty Kauc.* **28** 172 (1991).
37. Carlson, D. P. and West, N. E., Ger. Offen No.: 1957993 (1970); UK Pat. 1 280 653 (1972); US Pat. 3 738 923 (1973); US Re 28 628 (1975).
38. Luo, Y. X., Pang, F. C. and Sun, J. X., *Radiat. Phys. Chem.*, **18**(3–4), 445–57 (1981).
39. Sun J. Z., Zhang Y. F., Zhong X. G. and Zhang W. X., *Radiat. Phys. Chem.*, **42**, 139 (1993).
40. Robertson, A. B. and Schaffhauser; R. J., US Pat. 3 947 525 (1976).
41. Aronoff, E. J., Dhami, K. S. and Shieh, T. C., US Pat. 3 894 118 (1975).
42. Aronoff, E. J. and Dhami, K. S., US Pat. 3 763 222 (1973); US Pat. 3 840 619 (1974).
43. Gotcher, A. J., Germeraad, P. B. and Jansons, V., US Pat. 4 121 001 (1978); US Re 31 103 (1982).
44. Gotcher, A. J. and Germeraad, P. B., US Pat. 4 155 823; 4 353 961 (1979).
45. Aronoff, E. J., Dhami, K. S. and Shieh, T. C., US Pat. 3 894 118 (1975).
46. Aronoff, E. J. and Dhami, K. S., US Pat. 3 911 192 (1975).
47. Dhami, K. S., US Pat. 3 970 770 (1976); US Pat. 3 985 716 (1976); US Pat. 3 995 091 (1976).
48. Gotcher, A. J., Germeraad, P. B., White, L. J., US. Pat. 4 176 017, (1979).
49. Kojima, M., US Pat. 4 258 121 (1981).
50. Seki, I., Ando, Y. and Yagyu, H. *Jpn. Kokai Tokkyo Koho*, JP 60260634 A2, 23 Dec 1985 JP 60260635 Showa, Hitachi Cable, Ltd. (23 Dec 1985).
51. Ishii, T., Moshiki, K., *Jpn. Kokai Tokkyo Koho* JP 06 231 616 (1994).
52. Dole, M. 'Fluoropolymers', ••• Chap. 9 of *The Radiation Chemistry of Macromolecules*, Vol II, p. 169, Academic Press, New York (1973).

19

Kalrez®-Type Perfluoro-elastomers – Synthesis, Properties and Applications

JOHN B. MARSHALL

DuPont Dow Elastomers L.L.C. Wilmington, DE, USA

1 INTRODUCTION

Perfluoroelastomers offer the highest protection against high service temperatures and oxidative environments by combining the resilience and sealing force of an elastomer and the chemical inertness and thermal stability of PTFE. This high-performance elastomer is most commonly employed in severe thermal and/or chemical applications. Perfluoroelastomers can handle the widest variety of chemical reagents of any elastomer available today, resisting attack by amines, ketones, ethers, esters, acids, alkalis, oxidizers, fuels, and oils [1]. In addition to the broad range of applications in the chemical processing industry, perfluoroelastomers are used in a variety of industries including aircraft/aerospace, semiconductor chip manufacturing, chemical transportation, petrochemical, nuclear power, and analytical and process instrumentation. The near universal chemical resistance and high-temperature capabilities make perfluoroelastomers ideal for mechanical seals, pump housings, valves, gaskets, diaphragms, and other custom parts. DuPont invented the first perfluoroelastomer in 1968 and later commercialized it solely as Kalrez® perfluoroelastomer parts.

The expensive monomers used and the complex processing for perfluoroelastomers result in a material that costs more than any other elastomeric material. The cost of a perfluoroelastomer is often justified by the reduction of downtime due to seal failure, elimination of contaminants in a system due to seal degradation, reduction of maintenance costs associated with scheduled preventative maintenance shutdowns, and the reduction of the risks of the possibility of environmental emissions. In many applications, the cost of the

Modern Fluoropolymers. Edited by John Scheirs
© 1997 John Wiley & Sons Ltd

elastomer is often more than offset by the longer operating life and assurance against failure.

The chemistry of polymerization and crosslinking is briefly discussed in the sections that follow. Two of the referenced articles [2,3] provide a more detailed description of the polymerization and crosslinking chemistry used to synthesize perfluoroelastomers.

2 COMPOSITIONS

Perfluoroelastomers are similar to Teflon® PFA fluoropolymers in that they have a fully fluorinated molecular backbone. However, the lack of interpolymer bonds, or 'crosslinking', in PTFE causes it to behave as a plastic, thereby limiting its practicality in sealing applications. Conversely, perfluoroelastomers employ a crosslinking network, or curing system, which provides the molecule with the combination of elastomeric 'memory' and chemical resistance.

The chemical and thermal resistance of perfluoroelastomers is determined by the stability of the crosslinking network and the tetrafluoroethylene (TFE) to perfluoro(alkyl vinyl ether) ratio. The most thermally stable perfluoroelastomer commercially available allows a continuous upper service temperature in excess of 300 °C [1]. The outstanding performance of fluoroelastomers relative to other elastomers is largely due to the relative stability of the carbon–fluorine bond, which has a bond dissociation energy of 536 kJ/mol. Other elastomers have less stable bonds, such as carbon–hydrogen, with a bond energy of 339 kJ/mol. or carbon–chlorine, with a bond energy of 389 kJ/mol. Fluoroelastomers, which contain 1–2% carbon–hydrogen bonds have a continuous upper service temperature limit of close to 200 °C.

In perfluoroelastomers, all of the hydrogen atoms have been replaced by fluorine. The absence of the weaker carbon–hydrogen and carbon–chlorine bonds provide perfluoroelastomers with unparalleled chemical resistance for almost any fluid.

3 POLYMERIZATION CHEMISTRY

The commercially available perfluoroelastomers are based on a copolymer of tetrafluoroethylene and perfluoro(alkyl vinyl ether). The most commonly used ether is perfluoro(methyl vinyl ether) (PMVE), due to its favorable polymerization properties. Higher molecular weight perfluoro(alkyl vinyl ethers) may also produce elastomeric copolymers with tetrafluoroethylene, however, at considerably slower polymerization rates. Thus it is more difficult to synthesize high molecular weight polymer chains with higher molecular weight perfluoro(alkyl vinyl ethers) needed to ensure good elastomeric properties.

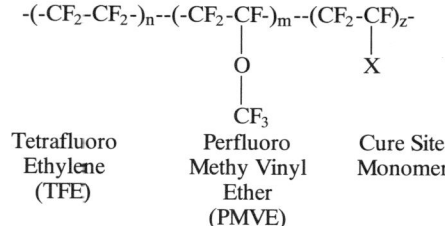

| Tetrafluoro Ethylene (TFE) | Perfluoro Methy Vinyl Ether (PMVE) | Cure Site Monomer |

Figure 19.1. Perfluoroelastomer backbone

A small amount (about 0.5 mol%) of a third monomer (cure site monomer) is added in the polymerization to create a crosslinking site on the base polymer. This monomer must readily copolymerize with perfluoro(alkyl vinyl ether) and tetrafluoroethylene in a random fashion. In addition, this cure site monomer must facilitate the forming of high molecular weight chains, and should facilitate chain transfer.

The polymer chain illustrated (Figure 19.1) is made by using either redox or thermally generated free radicals. In both cases the reaction is carried out in a stirred reactor under pressure. The redox initiator is more effective in providing high molecular weight polymers. However, the redox system creates sulfonic acid end groups which have strong ionic domains, making polymer processing difficult. The alternative method of thermally initiated free radical polymerization yields carboxylic acid end groups which form weaker ionic domains, thereby making the polymer easier to process. The higher temperature of polymerization causes more chain transfer, therefore higher molecular weight polymer chains are more difficult to form. In either polymerization method, the reaction may be carried out in batch or continuous-feed systems. The perfluoroelastomers thus produced are amorphous with no detectable crystallinity.

4 CURING CHEMISTRY

Perfluoroelastomers may be largely differentiated by the curing chemistry employed [2]. The most commercially employed curing systems are highlighted as follows.

4.1 NITRILE CURE SYSTEMS

The most thermally stable perfluoroelastomer available employs a nitrile-containing cure site monomer. Compounds employing this crosslinking mechanism (i.e. Kalrez 4079), have a continuous upper service temperature limit exceeding 300 °C. The crosslinking is initiated by a catalytic interaction of alkyl tin compounds with the nitrile group, thereby creating a triazine crosslinked structure. Perfluoroelastomers based on this type of curing system are not recommended for use in hot aliphatic amines, ethylene oxide, propylene oxide, and hot water/steam applications [4].

4.2   PEROXIDE CURE SYSTEMS

Perfluoroelastomers containing bromine or iodine cure sites can be cured with aromatic or aliphatic peroxides in the presence of co-agents, such as triallyl isocyanurate. This type of cure system is typically used in applications where a more universal chemical resistance is of primary concern and thermal stability is less important. The continuous service temperature limit of perfluoroelastomers employing this crosslinking system is limited to 218 °C. Perfluoroelastomers of this type are typically used in service applications with organic acids, inorganic acids, esters, ketones, aldehydes, ethylene oxide, and propylene oxide [5]. Perfluoroelastomers with this curing system offer the most universal overall chemical resistance.

4.3   DINUCLEOPHILIC CURING SYSTEMS

Perfluoroelastomers containing perfluorophenoxy or hydrogen cure sites may be cured using stable bisphenols. Dinucleophilic cured perfluoroelastomers offer a more universal chemical resistance at the expense of upper service temperature capabilities. The continuous upper service temperatures of perfluoroelastomers employing a dinucleophilic crosslink is 288 °C. Typical applications employing this type of curing system range from low sealing force/high elongation applications to high-pressure applications requiring resistance to extrusion [6].

4.4   IRRADIATION CURING SYSTEMS

Perfluoroelastomers may be cured without a curing agent through the use of high-energy radiation. Through the use of electron beam or γ-rays, a clear perfluoroelastomer can be crosslinked that is free from impurities since no other fillers or chemical reagents are necessary. Perfluoroelastomers made with other curing systems contain fillers and chemicals which increase the chance of introducing contaminants into a sensitive sealing application. The combination of chemical resistance and the lack of extractables make this material best suited for applications where contamination is of major concern. The lack of reinforcing fillers and the low upper use temperature of 150 °C may limit the use of this type to special sealing applications only.

   The curing process for the chemical crosslinked materials consists of molding for 10–60 minutes at 150–200 °C, followed by a step postcure in air or under a nitrogen blanket for as low as 4 hours or up to 120 hours at temperatures ranging from 200 to 300 °C. In the case of the irradiation cured materials, the molding is done at 100–120 °C and the molded parts are then subjected to electron beam or γ-rays.

Table 19.1. Typical physical properties of perfluoroelastomers [1]

Specific gravity (g/cm³)	2.00
Tensile strength at break (MPa)	16.9
Elongation at break (%)	150
100% modulus, (MPa)	7.2
Durometer hardness (Shore A)	75
Compression set resistance, 70 hours at 204 °C (%)	25
Clash–Berg stiffness temperature (°C) (66.6 MPa torsional modulus)	−2
Brittle point (°C)	−50
Temperature of retraction (°C) TR-10	−1

5 RELATIVE PERFORMANCE OF ELASTOMERS

Perfluoroelastomers overall provide the best combination of chemical and high-temperature resistance among all commercial elastomers available today (Table 19.1). Figure 19.2 shows that perfluorocarbons (i.e. perfluoroelastomers) outperform the fluoroelastomers in both chemical and thermal stability.

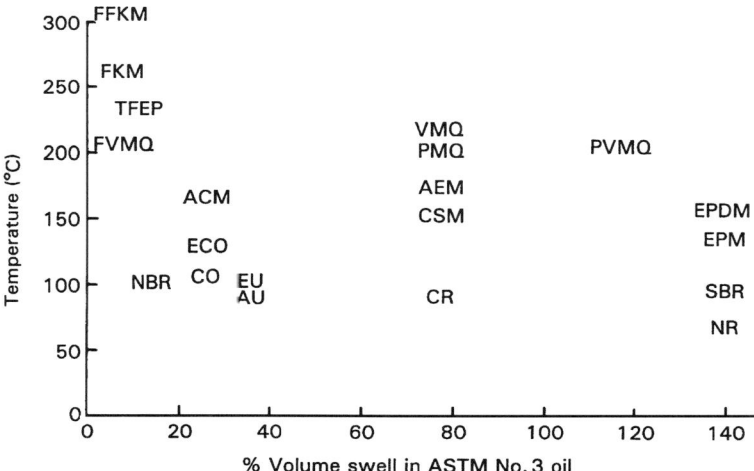

Figure 19.2. Retention of elastomeric properties (ASTM D2000) [7]. Fluoro-elastomers: vinylidene fluoride fluorocarbon (FKM); perfluorocarbon (FFKM). Non-fluoroelastomers: natural rubber (NR); acrylonitrile butadiene (NBR); chlorobutadiene (CR); acrylate (ACM); ethylene acrylate (AEM); ethylene propylene (EPM); ethylene propylene diene (EPDM); chloro-sulfonyl-polyethylene (CSM); tetrafluoroethylene propylene (TFEP); fluorovinylmethyl polysiloxane (FVMQ); epichlorohydrin (CO); epichlorohydrin ethylene oxide (ECO); polyester urethane (AU); poylether urethane (AU); phenyl methyl siloxane (PMQ); vinyl methyl siloxane (VMQ); phenyl vinyl methyl siloxane (PVMQ); styrene butadiene (SBR). (Reproduced with permission of NACE International)

6 PERFLUOROELASTOMER APPLICATIONS

6.1 CHEMICAL PROCESSING INDUSTRY

The chemical processing industry exposes elastomers to a large variety of chemical reagents and operating conditions. The upper service temperature applications, temperature cycling, chemicals used together or in succession, and operating pressures require extreme capabilities of elastomeric materials needed in the chemical processing industry. Table 19.2 shows the outstanding

Table 19.2. Comparative chemical stability of fluoroelastomers [7]. (Reproduced with permission of NACE International)

Fluid type	Typical example	FFKM[a]	FKM[b]	TFEP[c]	FVMQ[d]
Aliphatic hydro-carbons	Propane	Excellent	Excellent	Poor–Fair	Excellent
Aromatic hydro-carbons	Benzene	Excellent	Excellent	Poor–Good	Good–Excellent
Organic acids	Acetic acid	Excellent	Fair–Good	Good–Excellent	Fair
Inorganic acids	Sulfuric acid	Excellent	Excellent	Poor–Excellent	Fair–Good
Alcohols	Methanol	Excellent	Fair–Excellent	Good–Excellent	Fair–Good
Inorganic base	Sodium hydroxide	Excellent	Poor–Fair	Excellent	Poor–Fair
Organic base	Aniline	Excellent	Fair	Excellent	Poor–Fair
Ethers	Methyl t-Butyl ether	Excellent	Fair–Good	Poor	Fair
Ketones	Acetone	Excellent	Poor	Poor–Fair	Poor
Nitrous oxidizers	Ammonium nitrate	Excellent	Good	Excellent	Fair
Halogen oxidizers	Chlorine	Excellent	Excellent	Fair–Good	Fair–Good
Heat transfer fluids	Ethylene glycol	Excellent	Excellent	Excellent	Excellent
Oils and greases	Mineral Oil	Excellent	Excellent	Good–Excellent	Good–Excellent
Halogenated refrigerants	HFC-134a	Fair–Good	Poor–Fair	Poor	Poor–Fair
Polar solvents	Water	Excellent	Excellent	Excellent	Excellent

[a] Perfluorocarbon.
[b] Vinylidine fluoride fluorocarbon.
[c] Tetrafluoroethylene propylene (TFEP).
[d] Fluorovinylmethyl Polysiloxane (FVMQ).

performance of perfluoroelastomers in a variety of solvents as compared to other types of fluoroelastomers.

6.2 SEMICONDUCTOR INDUSTRY

Perfluoroelastomers are the most suitable material available for use in semiconductor processing equipment [8,9]. In these applications the elastomer comes in direct contact with dry process chemicals and reactive plasmas such as O_2, C_2F_6/O_2, CF_4/O_2 and NF_3. In addition, the elastomer is typically exposed to aggressive wet chemical environments such as sulfuric acid/hydrogen peroxide mixture, ammonium hydroxide/hydrogen peroxide/ultrapure deionized water mixture, 49% (vol) hydrofluoric acid and ultrapure deionized water. In such applications, limiting the contaminants to only parts per billion is very important. The use of perfluoroelastomers that combine chemical resistance and minimum liberation of contaminants becomes the most cost-effective sealing material for the semiconductor chip fabrication process.

6.3 AIRCRAFT/AEROSPACE APPLICATIONS

Aerospace propellant systems often utilize very aggressive fuels and oxidizers which are incompatible with all currently available elastomers, including fluoroelastomers. Perfluoroelastomers are often used in these systems due to their oxidative stability and compatibility with both oxidizers and hydrazine-type fuels. Applications range from the Space Shuttle to ballistic missiles to rockets for the delivery of scientific payloads into orbit.

Advances in gas turbine engine technology have lead to a need for elastomeric seals which are able to function for increasingly long times between engine overhauls in spite of higher operating temperatures and the use of increasingly aggressive aircraft fluids. Perfluoroelastomers play a very important role in sealing applications in both commercial and military aircraft engines. The vast majority of modern military engines incorporate such seals in their lubrication systems and in their air handling systems. Aircraft which depend on these seals include the F14, F15, and F16 fighters, as well as the B1 and B2 bombers.

6.4 OIL/GAS PRODUCTION

In response to the 1990 Clean Air Act Amendment, fuel suppliers are forced to use aggressive additives in gasoline to meet new minimum oxygen weight percent standards. Oxygenates, such as methyl tertiary butyl ether (MTBE) have led materials specialists in the fuel supply industry to specify higher performance elastomeric materials to prevent equipment leakage, equipment failure, and costly downtime. In MTBE environments, perfluoroelastomers perform considerably better than fluoroelastomers [10,11], leading to a longer seal life and improved protection against seal failure.

6.5 CHEMICAL PLANT VALVE APPLICATIONS

The 1990 Clean Air Act amendment requires valves used in contact with 149 volatile organic compounds conform to fugitive emissions of less than 500 parts per million by 1997, which represents a 20-fold reduction in previous allowable emission rates [12]. In many applications, chemical plants must retrofit existing valves or replace them with new valves that conform to the more stringent standards. Many valves currently use PTFE, with a loading force that 'energizes' the seal. The lack of compression recovery inherent in PTFE combined with the loading force necessary for effective sealing causes the seal to wear much quicker than a perfluoroelastomer. The elastomeric memory inherent to perfluoroelastomers allow effective sealing with only 30–50% of the sealing load required in PTFE applications. The chemical resistance provided by a perfluoroelastomer combined with the lighter sealing load results in longer service life and reduced fugitive emissions [13]. In these applications, the valve stem packing employed uses PTFE as a backup material which supports the less rigid perfluoroelastomer.

7 SEAL DESIGN CONSIDERATIONS WITH PERFLUOROELASTOMERS [14]

In most sealing applications, an elastomeric seal is the most susceptible component of the seal assembly. The sealing performance of an elastomer is a function of the stability of the material in the seal environment (e.g. swell behavior), the mechanical properties of the elastomer, and the mechanical design of the seal installation.

Before choosing the type of elastomer, the following factors must be considered:

1. The minimum and maximum operating temperatures, thermal cycling and potential short-term temperature excursions should be identified.
2. The pressure range that the elastomer will be subject to.
3. The fluid or fluids to be sealed.
4. The seal environment (static vs dynamic).

If the choice is made to use a perfluoroelastomer, there are certain design factors that must be considered. The linear coefficient of thermal expansion is approximately 50% greater than that for fluoroelastomers. This often becomes important for high-temperature applications (i.e. temperatures over 204 °C). When perfluoroelastomers are used at the upper service temperature limit (up to 316 °C), seal design consideration becomes extremely critical to accommodate the increased volume of the elastomer. If the gland size is insufficient to accommodate the large thermal expansion, then extrusion may

Table 19.3. Perfluoroelastomer linear and thermal expansion

| Operating temperature | | Percent expansion | |
(°C)	(°F)	Linear	Volumetric
25	77	0.0	0.0
38	100	0.3	0.9
93	200	1.5	4.8
149	300	2.9	8.8
204	400	4.1	12.9
260	500	5.4	17.1
316	600	6.7	21.5

occur. The linear expansion at the higher operating temperatures (see Table 19.3) is an intrinsic property of the perfluoroelastomer and should not be mistaken for absorption or degradation, which renders other elastomer types useless at the higher temperature ranges.

Installed compression for static seals should be in the range 12–18%. Over-compression will result in increased compression set and premature failure of the O-ring, due to cracking, splitting, or extrusion. Correct compression is extremely important in high-temperature applications. At temperatures below 0 °C it may be necessary to apply additional compression to the ring. An installed compression of 15–21% is suggested for these lower temperature applications. However, this higher compression may cause problems if the O-ring is cycled between very low and very high temperatures (>300 °C); an approximate engineering balance must be reached.

The installed stretch for perfluoroelastomer O-rings should be in the range 1–3%; If stretched over 5%, the internal stresses in the O-ring could cause premature failure.

Selection of a lubricating fluid is not typically a concern with perfluoroelastomers. Their general inertness allows almost any lubricating material to be used.

Perfluoroelastomers are capable of handling the widest selection of chemicals of any elastomer. However, perfluoroelastomers will typically swell in halogenated solvents such as Suva®. Although the base polymer is not degraded, the product will swell and soften due to the compatibility of the two materials. Aggressive amines, such as hexamethylene diamine at high temperatures, may chemically attack perfluoroelastomers.

ACKNOWLEDGMENTS

The author would like to thank Dr. Richard Cella, John Legare, Dr Anestis Logothetis, and Russ Schnell for their guidance and assistance in preparing this chapter.

TRADEMARKS

Kalrez® is a registered trademark of DuPont Dow elastomers
Teflon® is a registered trademark of E. I. duPont de Nemours and Co.

REFERENCES

1. *Physical Properties and Compound Comparisons*, DuPont Dow Elastomers Product Literature # E-94428-3, DuPont Dow Elastomers, Newark, DE, USA, 19711 (1994).
2. Logothetis, A. L., 'Chemistry of fluorocarbon elastomers', *Progress in Polymer Science*, Pergamon Press, Great Britain, pp. 251-295 (1989).
3. Logothetis, A. L., 'Fluoroelastomers', *Organofluorine Chemistry: Principles and Commercial Applications*, (Ed. R. E. Banks) Plenum Press, New York, USA, pp. 373-396 (1994).
4. *Physical Properties and Compound Comparisons*, DuPont Dow Elastomers Product Literature # E-94428-3, DuPont Dow Elastomers, Newark, DE, USA, 19711 (1994)
5. *Kalrez® 2035 Ethylene Oxide and Propylene Oxide Service*, DuPont Dow Elastomers Product Literature # H-58375, DuPont Dow Elastomers, Newark, DE, USA, 19711 (1994).
6. Logothetis, A. L., 'Perfluoroelastomers', *IRC95 Proceedings*, Kobe, Japan (1995).
7. Ferber, E., 'Fluoroelastomers for harsh environments', Corrosion 96 Conference, Paper No. 408, Nace International, PO Box 218340, Houston, Tex. USA (1996).
8. Bletsos, I. and Legare, J., 'The role of additives in perfluoroelastomer sealing parts exposed to reactive plasmas', *Microcontamination'93 Conference Proceedings*, pp. 625-639, Canon Communications, Santa Monica, Ca., USA (1993).
9. Legare, J., Thomas, E., Fulford, K., and Cargo, J., 'Characterization of elemental extractables in perfluoroelastomer and fluoroelastomer sealing parts', *Microcontamination'93 Conference Proceedings*, pp. 36-46, Canon Communications, Santa Monica, Ca., USA (1993).
10. Alexander, J., Ferber, E., Stahl, W., 'Avoid leaks from reformulated fuels', *Fuel Reformulation*, **4**, No. 2, (1994).
11. *Leak Prevention of Reformulated Fuels and Oxygenates*, DuPont Dow Elastomers Product Literature # H-42581, DuPont Dow Elastomers, Newark, DE, USA, 19711 (Oct. 1993).
12. Paul, B., Wolz, D., and Winkel, L., 'Controlling fugitive emissions by retrofitting valves', *Chemical Processing* (April, 1995).
13. *Technical Guidelines and Design Information for Using KALREZ® in Fugitive Emissions and Other Related V-Ring Packing Systems*, DuPont Dow Elastomers Product Literature # H-22159-1, DuPont Dow Elastomers, Newark, DE, USA, 19711 (Sept. 1995)
14. Schnell, R., *Seal Design Considerations Using Kalrez® Parts'*, DuPont Dow Elastomers Product Literature # E33808-3, DuPont Dow Elastomers, Newark, DE, USA, 19711 (1992).

20

Fluorosilicones

M. T. MAXSON, A. W. NORRIS and M. J. OWEN
Dow Corning Corporation, Midland, MI, USA

1 BACKGROUND

Fluorine in fluoropolymers is almost always present in partial or complete replacement of hydrogen in hydrocarbon groups. Substitution of the larger, more electronegative fluorine atom for hydrogen generally decreases solubility and surface energy and increases thermal stability and resistance to chemical attack. It is the strong C−F bonds and the inaccessibility of the carbon chain to potential reactants which produce the thermal and chemical resistance of fluorocarbon polymers. However, silicone materials already have these advantages by virtue of their inorganic siloxane backbone so major enhancement of such properties on fluorine incorporation is not expected. The challenge is to devise fluorosilicone structures that preserve these inherent benefits. Accordingly, much of the commercial interest in fluorosilicones centers around other aspects of substitution by fluorine such as solvent resistance, oleophobicity and low surface tension.

Only materials with a siloxane backbone are surveyed here. Other organosilicon polymers such as carbosilane and polysilane with fluorine substitution have been investigated but are little more than research curiosities at present. The loose term 'fluorosilicone' means polymers containing C−F bonds rather than Si−F bonds. This latter functionality is too reactive and has utility only in intermediates. Fluorinated carbon groups directly attached to silicon atoms are likewise insufficiently hydrolytically and thermally stable so no commercially promising fluorosilicone material can be totally fluorinated, there must be a hydrocarbon bridging group or spacer between the two entities. Thus the repeating structure of interest here is: $[R_fX(CH_2)_n]_x(CH_3)_ySiO_z$.

The X group is a consequence of the chemistry chosen to link the R_f fluorocarbon group to the hydrocarbon spacer. It can be oxygen or sulfur, for example, but is not present in commercial materials. The R_f group could be linear or branched, aliphatic or aromatic, but in practice has been limited to the CF_3

Modern Fluoropolymers. Edited by John Scheirs
© 1997 John Wiley & Sons Ltd

group until recently when longer aliphatic groups such as $CF_3(CF_2)_3$ have been introduced. The length of the hydrocarbon spacer, n, is optimally two [1]. Any longer and the fluorocarbon benefits are diluted. If the spacer is shorter (i.e. $n = 1$) similar hydrolytic and thermal deficiencies arise as when n is zero. When z is two and x and y are unity, linear polymers that are usually fluids result; but as z increases, or cross-linking sites are introduced along the polymer chain, a range of gels, coatings, sealants and elastomers can be produced up to resinous structures, $R_f(CH_2)_n SiO_{3/2}$, with no methyl groups. Linear polymers with no methyl (i.e. $x = 2$, $y = 0$) are very difficult to prepare and are not available commercially.

The original and by far the most widely available fluorosilicone since its introduction in the 1950s is polymethyltrifluoropropylsiloxane [PMTFPS], more rigorously known as poly[methyl(3,3,3-trifluoropropyl)siloxane] or poly[methyl(1H,1H,2H,2H-trifluoropropyl)siloxane]. Unless specifically mentioned, the unfluorinated carbons are those nearest to silicon, i.e. polymethylnonafluorohexylsiloxane [PMNFHS] is poly[methyl(1H,1H,2H,2H-nonafluorohexyl)siloxane]. PMTFPS was chosen for industrial production [2] because it combines excellent solvent resistance with adequate thermal stability for the minimum degree of fluorination, single C−F bonds being unable to provide the stability and solvent resistnace offered by the CF_3 cluster. Essentially, it contains the least amount of fluorine (least added cost) consistent with marked property improvement. The cost/benefit balance can also be varied by copolymerization with polydimethylsiloxane [PDMS]. This can be done with other polymers although only the PMTFPS–PDMS and PMNFHS–PDMS copolymers have any current commercial utility.

2   MANUFACTURE

2.1   MONOMERS

Hydrosilylation is by far the most important route for obtaining monomers and other precursors to fluorinated polysiloxanes, as it is for organofunctional silanes generally. Hydrosilylation [3] is the addition of a silicon hydride moiety across an unsaturated linkage. SiH functional silanes are thus important intermediates. They are produced industrially as by-products of the direct process for manufacturing methylchlorosilanes. The hydrosilylation reaction is usually performed with transition metal complexes of platinum or rhodium such as Speier's catalyst, hexachloroplatinic acid in isopropanol. The originality of the use of this reaction with fluorinated species lies more in the nature of the olefin selected rather than in reaction conditions, which are very similar to those employed with unfluorinated reagents:

$$CF_3CH{=}CH_2 + CH_3SiHCl_2 \xrightarrow{\;H_2PtCl_6\;} CF_3(CH_2)_2Si(CH_3)Cl_2$$

The preparation of 3,3,3-trifluoropropene is described in the patent literature [4]. The resultant silane, methyl(3,3,3-trifluoropropyl)dichlorosilane, is the precursor to the industrially most important polymethyltrifluoropropylsiloxane [PMTFPS] polymers. More generally, the fluorinated olefin can be represented by $R_f-Q-CH=CH_2$. Boutevin and Pietrasanta [5] have reviewed the possible Q groups which include ether, ester, amine, amide, sulfonamide, urethane and acetal functionality. Fluorinated silanes and cyclic siloxanes can also be prepared by other methods used for the synthesis of silicone intermediates such as the use of Grignard reagents or organic derivatives of alkali metals. An example of monomer preparation by Grignard reagent is given by Zhao and Mark [6]:

$$C_6F_5MgBr + CH_3Si(OC_2H_5)_3 \longrightarrow C_6F_5Si(CH_3)(OC_2H_5)_2 + BrMgOC_2H_5$$

2.2 POLYMERIZATION

The synthesis of polymeric siloxanes can be conducted either through condensation of linear silanol-ended oligomers or through the ring-opening polymerization of cyclics [7,8]. The most common route for the preparation of PMTFPS is through the base catalyzed ring-opening polymerization of the cyclic trimer. The cyclic species is obtained through hydrolysis of the corresponding dichlorosilane. Although the cyclic tetramer is the most thermodynamically preferred state, it is too stable to ring open under typical base catalyzed conditions and, therefore, the cyclic trimer is normally used. When it is fully equilibrated at 383 °K with KOH as the catalyst, 82% cyclics and 18% polymer is formed. When diluted with cyclohexanone solvent, the mixture approaches 100% cyclics [9]. The polarity of the trifluoropropyl group and its size both contribute to the ring-chain equilibrium lying so high on the cyclic species side. Consequently, the ring-opening polymerization is not normally taken to full equilibrium for PMTFPS, but is quenched instead when maximum molecular weight [MW] is achieved. The kinetics of this process have been studied in detail with the rate of polymerization identified for a variety of temperature, catalyst type, and catalyst concentration reaction conditions [10,11]. Hydroxides or silanolates of alkali metal ions are typically used as catalysts. The strength of the metal ion [K+ > Na+ > Li+] must be considered when determining polymerization conditions. These reactions can be accelerated by use of strong promoters such as acid amines, glycol ethers, and dimethylsulfoxide, or weaker ones such as tetrahydrofuran [12].

Li is the only ion which does not rearrange the siloxane bonds. This suppression of redistribution reactions allows high conversion to high polymer with narrow polydispersity to be achieved [13]. Use of either Na or K results in a combination of ring-opening and condensation polymerization mechanisms occurring simultaneously, producing polydispersities in the range of 2–2.5. These polymerizations can be carried out to produce a wide range of MW from low viscosity liquids to high MW polymer gums. Recently, stereoregular PMTFPS has been prepared by polymerizing the pure *cis* and *trans* isomers of the cyclic trimer under conditions that suppress siloxane redistribution reactions [14]. Polymers made from 96% *cis*-isomer were solid and crystalline at room temperature, with a melting temperature of 48 °C. This stereoregularity results in the cured PMTFPS exhibiting strain-induced crystallization which dramatically improves the mechanical properties over the conventional material which does not exhibit this phenomenon [15].

Condensation polymerization reactions are a much less common route to high MW PMTFPS. The reactions of silanol-terminated methyltrifluoropropylsiloxane catalyzed by stannous octoate and alkali metal carbonates have been studied as well as the syntheses and characterization of *p*-fluorophenyl, *p*-trifluoromethylphenyl and pentafluorophenyl substituted siloxanes via condensation polymerization [16].

Polymerization routes available for copolymers are the same as for homopolymers. When ring-opening mixtures of cyclic trimers, random copolymers will only result if the relative reactivity ratios of the cyclic species are similar. The most common copolymers obtained this way are poly(methylvinylsiloxane-co-methyltrifluoropropylsiloxane), a cross-linkable polymer, and poly(dimethylsiloxane-co-methyltrifluoropropylsiloxane). Block copolymers can be obtained by sequential addition of cyclic monomers, or by having two monomers of very different reactivity ratios.

3 PROCESSING

Fluorosilicones are cross-linked in much the same manner as conventional silicones. The subject has been comprehensively reviewed by Thomas [17]. The three main reactions for cross-linking commercial fluorosilicones are: peroxide-induced free radical reactions, condensation reactions, and hydrosilylation addition. UV is also used to cure certain gels.

The peroxide-induced free radical reaction is an adaptation of conventional organic chemistry relying on the availability of C—H bonds. These are already present in the fluorosilicone (hydrocarbon spacers and methyl groups) but unsaturated vinyl groups are incorporated in practice into the basic polymer structure for maximum effectiveness. The precise mechanisms involved are not wholly clear, but the formation of a trimethylene link between vinyl and methyl is widely accepted. Conventional peroxides such as benzoyl peroxide are commonly used. Typical cure cycles are 5–10 min at 115–170 °C, depending on the choice of peroxide, with a postcure recommended for complete reaction of the unsaturated groups and removal of peroxide decomposition by-products to optimize long-term thermal aging properties.

There are a variety of condensation reactions that can be employed. One of the earliest room-temperature curing reactions that is still widely used for commercial fluorosilicone sealants is the acetoxy-functional condensation system where cross-linking is brought about by exposure to atmospheric moisture. Silanol (SiOH) groups are formed which can then further react with the starting material or each other to produce a siloxane cross-link. Thicker-section sealant applications usually rely on hydrosilylation cure. This is the same chemistry used to produce fluorosilicone monomers with the vinyl functionality present on silicon. Conventional organosiloxane polymerization techniques are used to incorporate SiH or SiCH=CH$_2$ groups either as a pendent functionality along the polymer chain, or as a terminal functionality. A big advantage of this type of cure is the absence of volatile by-products. One disadvantage is that currently it is only available as a two-part formulation.

Fluorosilicone elastomer compounds can be molded, extruded or calendered by any of the conventional methods used in the industry for PDMS elastomers. Compression molding is widely used for fabricating parts. Injection molding is important for high production operations. Transfer press molding is useful for molding complex parts in multi-cavity molds. Extrusion curing is used to make tubings and gaskets. Calendering is used to produce thin sheets and to provide coatings on metal, fabric and other substrates.

4 PROPERTIES

4.1 FLUID AND CHEMICAL RESISTANCE

Fluorosilicones are widely recognized for their fuel and fluid resistance, and used in many applications where this is required. The volume swell in solvents is related to the fluorine content. PMTFPS is 36.5% F with an estimated Hildebrand solubility parameter of 8.8 (cal/cm^3)$^{1/2}$. Cured fluorosilicone elastomers have good resistance to jet fuels, oils, and hydrocarbon solvents and fuels as shown in Table 20.1. However, higher swells are observed in ketones and esters [18]. Fluorosilicones also perform well with low volume swells in alcohol/fuel blends, once the solvents are removed the physical properties return nearly to the original unswollen state [19].

Table 20.1. Fluid and chemical resistance (ASTM D 471)

Fluid	Immersion conditions	Hardness change (points)	Volume change (%)
ASTM No. 1 oil	3 days/150 °C	−5	0
Crude oil 7 API	14 days/135 °C	−10	+5
JP-4 fuel	3 days/25 °C	−5	+10
ASTM Ref. Fuel B	3 days/65 °C	−5	+20
Benzene	7 days/25 °C	−5	+25
Carbon tetrachloride	7 days/25 °C	−5	+20
Methanol	14 days/25 °C	−10	+4
Ethanol	7 days/25 °C	0	+5
Hydrochloric acid (10%)	7 days/25 °C	−5	0
Nitric acid (70%)	7 days/25 °C	0	+5
Sodium hydroxide (50%)	7 days/25 °C	−5	0

4.2 HEAT RESISTANCE

Silicones in general are known for their excellent retention of properties at elevated temperatures. Fluorosilicone elastomers are no exception although they have slightly reduced high-temperature stability compared to PDMS [20]. The ultimate temperature stability is dependent on cure conditions and environment. A typical PMTFPS elastomer heat aged for 1350 hours at 200 °C will show a two-point reduction in durometer hardness, a 40% reduction in tensile strength and a 15% reduction in elongation. Typically, two degradation mechanisms occur, either reversion (occurs in confinement) or oxidative cross-linking. The reversion mechanism is accelerated if residual basic polymerization catalysts remain in the material and results primarily in the formation of cyclic tetramer. Oxidative cross-linking typically occurs by radical abstraction of protons which can then recombine to form additional cross-linking sites, a process that ultimately results in embrittlement of the cured elastomers.

4.3 LOW-TEMPERATURE PROPERTIES

PMTFPS has a glass transition temperature (T_g) of -75 °C. Unlike PDMS it does not exhibit a low-temperature crystallization at -40 °C, due to the inability of the polymer chains to pack into a crystalline lattice. Owing to the bulkiness of the trifluoropropyl groups, the T_g is 50 °C higher than PDMS. Because of this low T_g and absence of a low-temperature crystallization, the fluorosilicone remains quite flexible at very low temperatures. The brittleness temperature by impact (ASTM D 746B) has been measured as -59 °C and the temperature of retraction* (ASTM D 1329) as -50 °C for commercial PMTFPS elastomers [19]. These glass transition and brittlepoint temperatures are considerably lower for fluorosilicone elastomers than fluorocarbon elastomers.

4.4 SURFACE PROPERTIES

Surface energy is the fundamental surface property of polymers and can be considered either a surface tension (force per unit length, mN/m) or a surface free energy (free energy per unit area, mJ/m^2). These quantities are identical for liquids but not for solids. Moreover, the surface energies of liquid and solid polymers are determined in totally different ways so agreement between such values for the same polymer (e.g as a liquid and as a cross-linked film) is not expected. Polymer liquid surface tensions (σ_{LV}) shown in Table 20.2 are directly measured, whereas the solid surface energies are inferred from contact angle measurements using semiempirical equations, with values dependent on the choices made [21]. σ_{LV} is a function of MW and temperature. These data in Table 20.2 are extrapolations to infinite MW except for polyheptadecafluorodecylsiloxane where data are too scanty for extrapolation and the data for M_n 19 600 is used.

---

*Temperature of retraction is the temperature at which an elongated frozen specimen retracts a fixed amount, usually 10%.

Table 20.2. Surface properties of polymers

Polymer	σ_{LV} (mN/m)	σ^d (mJ/m^2)	σ_S (mJ/m^2)
Polyheptadecafluorodecylmethylsiloxane	18.5	12.7	7.0
Polymethylnonafluorohexylsiloxane (PMNFHS)	19.2	14.6	9.5
Polymethyltrifluoropropylsiloxane (PMTFPS)	24.4	18.3	13.6
Polydimethylsiloxane (PDMS)	21.3	22.6	22.8
Polytetrafluoroethylene (PTFE)	25.6	19.8	19.1

No temperature-dependent studies have yet been carried out for fluorosilicones, all data in Table 20.2 are at room temperature (20–25 °C). The solid surface energy values listed are σ_s, from the Owens and Wendt approach using water and methylene iodide, and σ^d, the dispersion force component of surface free energy using n-hexadecane. These values reflect the familiar decrease in surface energy with longer fluorocarbon side-chain length observed with other fluoropolymer families. Note the unexplained reversal of order of PDMS, PMTFPS and PTFE between the liquid and solid states.

Contact-angle-derived data are also available for self-assembled fluorosiloxane monolayers [21]. Chaudhury and Whitesides [22] have developed the Johnson, Kendall and Roberts technique to directly measure the solid surface free energy of such monolayers including one derived from heptadecafluorodecyltrichlorosilane. Their value of 15.5 mJ/m^2 corresponds well with contact angle data using a fluorinated liquid but not so well with more common liquids such as methylene iodide and n-hexadecane.

4.5 ELECTRICAL PROPERTIES

Ku and Liepins [23] separate the response of polymers to an electric field into two main parts, dielectric behavior and bulk conductive behavior. Of the four fundamental properties, those characterizing dielectric behavior are the dielectric constant (ε) representing polarization and the tangent of dielectric loss angle or dissipation factor ($\tan \delta$) representing relaxation phenomena. Those properties characterizing bulk conductive behavior are dielectric strength (E_B) representing breakdown phenomena and conductivity or its inverse, resistivity (ρ_v), representing electrical conduction.

Ku and Liepins have summarized these key properties for various polymers. Table 20.3, extracted from their summary, gives data for LDPE (low-density polyethylene), NR (natural rubber — cross-linked cis-polyisoprene), PDMS, and Viton (copolymer of vinylidene fluoride and hexafluoropropylene). The PMTFPS data are our own. The two fluoropolymers are quite similar with lower dielectric strength, higher dielectric constant, higher tangent of dielectric loss angle and lower volume resistivity than PDMS and the organic polymers. Beevers [24] has specifically reviewed the dielectric properties of siloxanes including several

Table 20.3.   Electrical properties of polymers

Polymer	E_B (60 Hz) (V/mil)	ε (100 Hz)	$\tan \delta$ (100 Hz)	ρ_v (ohm.cm)
Low-density polyethylene	742	2.2	0.0039	2.5×10^{15}
Natural rubber	665	2.4	0.0024	1.1×10^{15}
PDMS	552	2.9	0.00025	5.3×10^{14}
PMTFPS	350	7.0	0.020	1.0×10^{14}
Viton	351	8.6	0.040	4.1×10^{11}

references to PMTFPS materials. Technical interest in the electrical properties of PMTFPS is a consequence of its relatively low dielectric loss and high electrical resistivity coupled with non-electrical advantages such as low levels of water absorption and retention of these properties in harsh environments.

4.6 SAFETY AND TOXICITY

PMTFPS is a relatively inert material under normal conditions. Skin tests on albino rabbits have shown no dermal irritation or toxicity. In over 40 years of industrial usage no problems have been reported related to human dermal contact with uncured or cured fluorosilicone products. However, when exposed to elevated temperatures in air or when burned, toxic fluorinated compounds are formed and inhalation of such vapors is harmful. One toxic species is 3,3,3-trifluoropropionaldehyde (TFPA). The no-effect temperature limit for heating in air is 150 °C. The amount of TFPA generated, while being extremely small, is dependent upon temperature, percentage of fluorosilicone in the sample, sample surface area, and presence of oxygen.

In addition to TFPA generation, fluorosilicones generate formaldehyde vapor when heated over 150 °C in air in the same way as PDMS and many organic compounds. Formaldehyde generation is very small — micrograms per hour per gram of sample at 200 °C. Even though TFPA and formaldehyde are generated in trace amounts (below detection limits in most commercial processes and applications) care must be taken to minimize human exposure to vapors in processes such as curing and post-curing where temperatures exceed 150 °C. No concern has been expressed by fabricators of peroxide-cured fluorosilicone compounds under such conditions. Peroxide decomposition products are present in much higher amounts and fabricators are familiar with the need for proper process venting.

5 APPLICATIONS

Apart from release coatings, the commercial applications of fluorosilicones are largely those of PMTFPS. MW and type of reactive functionality play a major role in the application as do additional materials needed in most cases to achieve products with useful features. For example, the largest use of PMTFPS is in peroxide cured high-strength elastomers where the addition of 20–40 pph of high-surface-area fumed silica increases strength of the cured elastomer by a factor of 10.

5.1 FLUIDS

The traditional uses of PMTFPS fluids have been as antifoams for organic liquids such as crude oil, hydrocarbon fuels, solvents, etc. and as lubricants in bearings that are subject to exposure to such organic fuels and solvents. Newer uses include providing cosmetics and other skin formulations with long-lasting water and oil repellency.

Non-aqueous antifoams are generally insoluble in the medium in which they function and of lower surface tension than it. In these antifoam applications the reduced organic compatibility conferred by the trifluoropropyl group is more important than surface tension lowering benefits. Even though fluid PMTFPS has a higher surface tension than PDMS it is still sufficiently low to function as an antifoam in many organic systems where PDMS is quite ineffective on account of its solubility. Examples are in dry cleaning solvent recovery where PDMS fluids are foam promoters but PMTFPS fluids are effective antifoams [25] and in diesel fuel. Novel non-aqueous fluorosilicone antifoams obtained by polymerization of vinyl-functional PDMS in the presence of diacyl peroxides containing fluoroether groups have also been reported [26]. Antifoam products for aqueous systems are more complex and usually include a dispersed hydrophobic solid to promote film rupture by dewetting mechanisms. This mixture of a fluid and a solid, known in the industry as a compound, is often provided in an emulsion form to enhance dispersibility. Fluorosilicone-based emulsions and compounds are effective against a variety of silicone surfactants and some other fluorosurfactants.

Lubrication is a complex topic involving surface and bulk effects. Fluorosilicone fluids are chosen as lubricants when the combination of factors they offer meets the needs of the application. These include good lubrication even in the presence of harsh chemicals and fuels, wide serviceable temperature range, excellent fire, flash and oxidation resistance, high viscosity index and relatively flat viscosity–temperature dependence. Common uses are as a lubricating oil in vacuum pumps handling reactive gases, and in bearings subjected to washing by fuels or solvents. Typically for such bearing application the fluorosilicone base oil is formulated into a grease. In most liquid lubricant operations only a portion of the interface between parts is separated by a fluid film, the rest is in boundary contact. The low intermolecular forces in PDMS are not conducive to good boundary lubrication as highly condensed, difficult-to-displace boundary films are not developed. The considerable effectiveness of PMTFPS as a steel-on-steel lubricant results from the bulkiness of the fluorocarbon side-group [27], but at the highest use temperatures may also be due to decomposition and surface reaction enhancing boundary lubrication in a similar manner to the mode of action of halogen-containing extreme pressure additives.

5.2 SURFACTANTS

Fluorocarbon-substituted analogs of methylsiloxane aqueous surfactants [28] and new surfactants whose hydrophobes are co-oligomers of fluoroalkylated acrylates and methacrylates with dimethylsiloxanes [29] have been reported, but no commercial aqueous uses have yet developed although some of the latter type materials were found to be potent selective inhibitors of HIV-1 *in vitro*. Proprietary fluorosilicone surfactants are used to stabilize silicone foam, similar to the major use of conventional silicone surfactants to stabilize polyurethane foam. Fluorosilicone–PDMS block copolymers as copolymeric surfactant additives can confer fluoropolymer benefits to conventional silicone surfaces. This concept can be extended to virtually any polymer as fluorosilicones are amongst the lowest known surface energy materials. For example, Inoue and co-workers [30] have examined the surface characteristics of fluoroalkylsilicone–poly(methyl methacrylate) block copolymers and their poly(methyl methacrylate) blends and shown beneficial enhancement of surface hydrophobicity, oleophobicity and reduced adhesion, particularly with longer fluoroalkyl chain substitution in the silicone portion.

5.3 GELS

Soft, clear tacky gels can be obtained by lightly cross-linking low to medium MW PMTFPS [31] usually by platinum catalyzed hydrosilylation addition. Gel hardness is determined by cross-link density. Such addition-cured gels are available as either one- or two-part products. In the two-part product part A contains the platinum catalyst and vinyl functional PMTFPS and part B the SiH and vinyl functional PMTFPS. Chemical inhibitors or encapsulated catalysts are used to achieve one-part products [32]. Heat is required to cure one-part products and is often used with two-part products as well.

Fluorosilicone gels are used to protect electronic/electrical devices from thermal stress, dust, dirt, moisture, chemicals and hydrocarbon fuels such as gasoline, diesel fuel and jet engine fuels [33]. It is the resistance to these nonpolar fluids that allows PMTFPS gels to perform in environments where PDMS gels would fail. In addition to fuel resistance, PMTFPS gels offer other useful features including good electrical properties, good high- and low-temperature properties ($T_g - 75\,^\circ$C, remain soft after long exposure to $135–150\,^\circ$C), and an easy-to-apply low viscosity of $500–1000$ cP curing in minutes at $125–150\,^\circ$C.

5.4 COATINGS

Self-assembled fluorosilicone monolayers formed by adsorption and reaction of suitable chlorosilane and other monomers were first investigated by Sagiv and co-workers [34]. Such monolayers and multilayers formed *in situ* and their corresponding preformed polymers have been considered for a variety of coating applications including low-friction, wear-resistant coatings, liquid-repellent fabric finishes, and antifouling or anticontamination coatings in areas as diverse as marine coatings and solar energy devices in desert environments. Many patents claim fluorosilane and fluorosilicone coatings for lubrication and protection of magnetic recording media and optical lenses in equipment such as cameras and photocopier components.

The main coatings application is the use of fluorosilicone–PDMS copolymers as release coatings for PDMS-based pressure-sensitive adhesives (PSAs). A combination of bulk and surface properties is required including inertness to at least one of the separating surfaces at the temperature of the release process and low surface energy. This latter property helps in good wetting of the release liner substrate, paper of plastic, to achieve a smooth, uniform coating. Conversely, it causes poor wetting of the release coating by the adhesive and is also a direct reflection of the low intermolecular forces necessary for effective release. Fluorosilicones provide these needed features, particularly in the case of PDMS-based silicone PSAs. These adhesives are known for their durability, wide temperature use range, long service life, and ability to adhere to lower surface energy substrates such as human skin. This has led to several medical uses including attachment of transdermal drug-delivery patches. Release liners are required for facile use of such products and PMNFHS-based materials satisfy this need.

5.5 ADHESIVES AND SEALANTS

An adhesive's primary function is to bond two materials together, whereas a sealant must seal, protect or prevent intrusion of unwanted substances into or past a barrier. In practice, despite these different functions, sealant and adhesive products can be interchanged. They generally consist of medium MW PMTFPS vinyl or hydroxyl end-blocked, a small amount of reinforcing silica filler, and either a condensation (RTV — room temperature vulcanization) or addition cure system. One-part moisture activated RTV fluorosilicone sealants have been commercially available for many years. Because of their excellent resistance to jet engine fuels and their excellent low-temperature flexibility and high thermal stability, these materials have found both military and civilian aerospace applications [35]. For example, in the US Air Force F-111 fighter bomber as the primary integral fuel tank sealant. Other uses include sealing and repairing small engine fuel systems and sealing various chemical and liquid process storage tanks. Most fluorosilicone RTV sealants utilize the acetoxy condensation cure system. While this provides efficient cross-linking and good adhesion it also liberates acetic acid. In recent years systems based on hydrosilylation addition cure technology that release no corrosive by-products have been developed.

Two-part, heat-cured fluorosilicone sealants have some military aircraft applications and are beginning to be used in automotive fuel system sealing. Addition cured materials have the advantage of not needing moisture to initiate cure and generating no by-products. While this makes them ideal for deep section curing they lack good adhesion. Priming the surface with special siloxane primers can improve adhesion, but this is still insufficient for some applications and fabricators object to the additional priming step. Recent research has led to addition cured PMTFPS sealants containing integral adhesion promoters that eliminate the priming step. Such products are finding their first applications in the microelectronic industry.

A special class of fluorosilicone sealants is referred to as 'channel sealants' or 'groove injection sealants'. These materials are sticky, putty-like compositions that do not cure. They are used to seal fuel tanks of military aircraft and missiles. Designers meet their goal of maximizing range and weapon load by reducing airframe weight and increasing fuel capacity. One way is to use the wing structure itself as a fuel container. To render such a system leak proof, a channel or groove is machined on the inner surface where the wing skin is attached to the structural support. Fluorosilicone sealant is injected under high pressure into this groove, sealing all fasteners and small gaps.

5.6 ELASTOMERS

Most commercial PMTFPS polymers are used in the manufacture of peroxide curable high consistency fluorosilicone elastomer compounds. These elastomers consist mainly of high MW (0.8–2.0 million) PMTFPS with 0.2–1.0 mol% methylvinylsiloxane to improve effectiveness of peroxide free-radical vulcanization, a reinforcing filler (usually high surface area fumed silica), a small amount of low MW fluorosilicone diol processing fluid, and a peroxide catalyst. Other additives such as pigments, thermal stability enhancers, and extending fillers are often added to meet final product requirements [36]. It is also common for fabricators to blend the uncured fluorosilicone compound with PDMS compounds to achieve a lower cost elastomer at the expense of reduced fuel resistance and mechanical strength. Experimental addition cured PMTFPS elastomers have been made, but none are currently marketed. While offering some advantages over peroxide cure such materials have not become popular in the way that PDMS addition cured elastomers have.

The properties of a fluorosilicone elastomer depend upon the base formulation and the final compounding ingredients. Table 20.4 lists representative property data for commercial fluorosilicone elastomers. They are particularly well suited to the many applications involving exposure to aircraft fuels, lubricants, hydraulic fluids and solvents. Compared to other fuel-resistant elastomers fluorosilicones offer the widest hardness range and the widest operating service temperature range of any material [19]. These attributes, coupled with the general ease of fabrication, make fluorosilicone the material of choice in this area. The automotive and aerospace industries are the two largest users of such elastomers. In the automotive industry applications include, fuel line pulsator seals, fuel

Table 20.4. Properties of typical commercial PMTFPS elastomers

Property	Typical range
Specific gravity (g/cm^3)	1.35–1.65
Hardness (Shore A)	20–80
Tensile strength (MPa)	5.5–11.7 (22 °C)
	2.4–4.1 (204 °C)
Elongation (%)	100–600 (22 °C)
	90–300 (204 °C)
Modulus at 100% (MPa)	0.5–6.2
Compression set (%) (22 h/177 °C)	10–40
Tear strength, die B^{*a} (kN/m)	10.5–46.6
Service temperature (°C)	−68–+232
Bashore resilience^{*b} (%)	10–40

^{a}* Die B refers to a particular specimen shape in ASTM D624.
^{b}* Bashore resilience is resilience measured by a falling metal plunger according to ASTM D2632.

injector O-rings, fuel line quick-connect seals, gas cap washers, vapor recovery system seals, electrical connector inserts, exhaust gas recirculating diaphragms, fuel tank access gaskets, and engine cover and oil pan gaskets. In the aerospace industry fluorosilicone O-rings, gaskets, washers, diaphragms and seals are used in fuel line connections, fuel control devices, electrical connectors, hydraulic line connections and fuel system access panels.

BIBLIOGRAPHY

1. Pierce, O. R., *Appl. Polym. Symp.*, **14**, 7 (1970).
2. Kim, Y. K., in *Kirk-Othmer Encyclopedia of Chemical Technology*, 3rd edn., Vol. 11, Wiley-Interscience, New York, 1980, p. 74.
3. Speier, J. L., *Adv. Organomet. Chem.*, **17**, 407 (1979).
4. US Patent 4 798 818 (January 17, 1989) to W. X. Bajzer, R. L. Bixler, Jr, M. D. Meddaugh and A. P. Wright (to Dow Corning Corporation).
5. Boutevin, B. and Pietrasanta, Y., in *Progress in Organic Coatings*, W. Funke, L. Valentine and G. Bierwagen (eds), Vol 13, Elsevier, New York, 1985, p. 297.
6. Zhao, Q. and Mark, J. E., *Macromol. Rep.*, **A29** (Suppl. 3), 221 (1992).
7. Saam, J. C., in *Silicon-Based Polymer Science*, J. M. Zeigler and F. W. G. Fearon (eds), Advances in Chemistry Series 224, American Chemical Society, 1990, p. 71.
8. Chojnowski, J., in *Siloxane Polymers*, S. J. Clarson and J. A. Semlyen (eds), Prentice-Hall, Englewood Cliffs, NJ, 1993, p. 1.
9. Wright, P. and Semlyen, J. A., *Polymer*, **11**, 462 (1970).
10. Yuzhelevskii, Yu., Kagan, Ye., Kogan, E. V., Klebanskii, A. L. and Nikiforova, N., *Vysokomol. Soedin. Ser.* **A11**, 1539 (1969).
11. Veith, C. A. and Cohen, R. E., *J. Polym. Sci., Part A, Polym. Chem.*, **27**, 1241 (1989).
12. Yuzhelevskii, Yu., Pchelintsev, V. and Kagan, Ye., *Vysokomol. Soedin. Ser.*, **A15**, 1795 (1973).

13. Lee, C. L., Frye, C. L. and Johannson, O. K., *ACS Polym. Prepr. Polym. Chem. Div.*, **10(2)**, 1361 (1969).
14. Kuo, C-M., Saam, J. C. and Taylor, R. B., *Polymer International*, **33**, 187 (1994).
15. Battjes, K. P., Kuo, C-M., Miller, R. L. and Saam, J. C., *Macromolecules*, **28**, 790 (1995).
16. Drechsler, L. L. Ph. D. Thesis, University of Cincinnati, 1994.
17. Thomas, D. R., in *Siloxane Polymers*, S. J. Clarson and J. A. Semlyen (eds), Prentice-Hall, Englewood Cliffs, NJ, 1993, p. 567.
18. Gomez-Anton, M. R., Masegosa, R. M. and Horta, A., *Polymer*, **28**, 2116 (1987).
19. Norris, A. M., Fiedler, L. D., Knapp, T. L. and Virant, M. S., *Automotive Polymers and Design*, **19** (April), 12 (1990).
20. Knight, G. J. and Wright, W. W., *British Polymer Journal*, **21**, 199 (1989).
21. Kobayashi, H. and Owen, M. J., *Trends in Polym. Sci.*, **3**, 330 (1995).
22. Chaudhury, M. K. and Whitesides, G. M., *Science*, **255**, 1230 (1992).
23. Ku, C. C. and Liepins, R., *Electrical Properties of Polymers*, Hanser, Munich, 1987, p. 326.
24. Beevers, M. S., in *Siloxane Polymers*, S. J. Clarson and J. A. Semlyen (eds), Prentice-Hall, Englewood Cliffs, NJ, 1993, p. 415.
25. Sawicki, G. C. and White, J. W., *Chemspec Europe* 89 BACS Symposium (British Association For Chemical Specialities, Sutton, Surrey, England).
26. Matsuda, K., Abe, M., Ogino, K., Yoshino, N. and Sawada, H., *Shikizai Kyokaishi*, **67(2)**, 88 (1994); *Chem. Abs.*, **122**, 164097.
27. Tabor, D. and Willis, R. F., *Wear*, **11(2)**, 145 (1968).
28. Kobayashi, H. and Owen, M. J., *J. Colloid Interface Sci.*, **156**, 415 (1993).
29. Sawada, H., Ohashi, A., Oue, M., Baba, M., Abe, M., Mitani, M. and Nakajima, H., *J. Fluorine Chem.*, **75**, 121 (1995).
30. Inoue, H., Matsumoto, A., Matsukawa, K., Ueda, A. and Nagai, S., *J. Appl. Polym. Sci.*, **40**, 1917 (1990).
31. Maxson, M. T. and Benditt, K. F., *SAE Technical Paper No.* 880023, Int. Cong. and Expo., February (1988).
32. US Patent 5 059 649 (October 21, 1991) to M. T. Maxson and B. VanWert (to Dow Corning Corporation).
33. Chiotis, A., *SAE Technical Paper* No. 950301, Int. Cong. and Expo., February (1995).
34. Maoz, R. Netzer, L., Gun, J. and Sagiv, J., *J. Chim. Phys.*, **85**, 1059 (1988).
35. Maxson, M. T., *Aerospace Engineering*, pp. 15–18, December (1990).
36. Maxson, M. T., *Gummi Fasern Kunststoffe*, **12**, 873 (1995).

21

Melt-processable Perfluoropolymers

M. POZZOLI, G. VITA

Melt processable fluoropolymers, Applications development and Technical Service – Ausimont SpA, Bcllate (MI), Italy

V. ARCELLA

Fluoropolymers R&D – Ausimont SpA, Bollate (MI), Italy

1 INTRODUCTION

The development of fluorine-containing polymers followed the synthesis of low molecular weight polychlorotrifluoroethylene (PCTFE) [1] and the accidental discovery of polytetrafluorcethylene (PTFE) [2] in the late 1930s.

The unique combinatior of properties of PTFE was immediately apparent, like the difficulties of transforming it into appropriate shapes. In fact the melt viscosity of PTFE is in excess of 10 billion poise, and transformation techniques similar to metal sintering and ceramics have to be used, such as sintering of granular resin from suspension polymerization grades and past extrusion of fine powders from emulsion polymerization grades.

The need for easier processable resins led to the development of different highly fluorinated plastics. Although many fluoropolymers have been prepared, since the discovery of PTFE, major commercial products are homopolymers and copolymers deriving from free radical polymerization of a limited number of fluoromonomers, such as tetrafluoroethylene (TFE), vinyl fluoride (VF), vinylidene fluoride (VF$_2$), chlorotrifluoroethylene (CTFE), hexafluoropropylene (HFP), perfluoropropylvinylether (PVE) and more recently perfluoromethylvinylether (MVE).

Melt-processable perfluoropolymers, detailed in the present chapter, derive from the need for having a resin with PTFE properties but able to be transformed by conventional melt-processing techniques.

The copolymer of TFE and HFP [3], known as fluorinated ethylene propylene (FEP) copolymer, was the first melt-processable perfluoropolymer, having a melt

Modern Fluoropolymers. Edited by John Scheirs
© 1997 John Wiley & Sons Ltd

viscosity low enough for conventional melt processing. However, FEP does not have the same heat stability and high-temperature properties as does the homopolymer of TFE, so copolymers of TFE and PVE, known as perfluoroalkoxy polymers (PFA) [4] have been developed. These polymers possess a thermal stability closer to that of PTFE, and can be used at the same continuous service temperature (260 °C).

The cost of PFA is higher than FEP due to the higher cost of the PVE monomer compared to HFP. Economically attractive perfluorovinylethers–TFE copolymers have been recently introduced on the market, due to the use of different technologies for both monomer preparation and polymerization [5,6]. These new perfluoropolymers are obtained by copolymerizing TFE with the MVE monomer, and named MFA to distinguish them from the conventional PFA's.

2 CHEMICAL STRUCTURE AND RELATED PROPERTIES

Highly fluorinated plastics have gained a great commercial importance thanks to a unique combination of high thermal stability, outstanding resistance to most chemicals, low surface energy, low dielectric constant and dissipation factor, excellent weatherability, low moisture adsorption and low flammability.

The above properties are maximized in the case of perfluoropolymers, although partially fluorinated plastics possess some their own unique properties.

Since the late 1930s, following the discovery of polytetrafluoroethylene (PTFE), it was recognized that a fluorinated macromolecular chain possesses unique properties of thermal stability and chemical inertness [7].

In PTFE these two characteristics appear to be maximized. In fact it possesses a very high melt viscosity at high temperature (10^{12} poises at 360 °C), a high crystalline melting point (327 °C) and a high thermal stability. It is notable that this polymer, if we consider the other extreme of the temperature spectrum, also possesses unusual toughness at a temperature as low as 4 K.

Furthermore perfluorocarbon polymers are practically insoluble in all common solvents and highly resistant to any chemical attack. These unusual properties can be attributed to:

- high bond energy of C−F and C−C link
- low secondary bonding forces
- relatively small size of fluorine

Bond energy of C−F and C−C in fluorocarbons is higher than that of C−H and C−C in hydrocarbons. The C−F bond energy goes from 447 kJ/mol for the CH_3F to 485 kJ/mol for the CF_4 and remains much higher than that of the C−H link [8,9]. The C−C bond energy rises from an average value of 348 kJ/mol for hydrocarbon systems to 360 kJ/mol in fluorocarbons [9,10]. The high bond energies of C−F and to a lesser extent of C−C links mostly account for the high thermal stability of fluorocarbons and their high resistance to radical assisted abstraction of F from a C$-$F bond. Van der Waals' forces for simple hydrocarbons and fluorocarbons are reported to be about 40 and 4 kJ/mol respectively, that is, fluorine-based materials show second bonding forces about 10 times lower than corresponding hydrocarbons [11].

An interesting consequence of this property is the similarity between boiling points of fluorocarbons and homologues hydrocarbons [11]. As the molecular weight of fluorocarbons is considerably higher than hydrocarbons at constant carbon number, a higher boiling point should be expected. In fluorocarbons the decreased van der Waals' attractive forces offset the increased molecular weight. Intermolecular forces in PTFE are only 3.18 kJ/mol, at least in part attributed to the crowding of fluorine atoms which protect and isolate the carbon backbone reducing the interaction between different macromolecular chains. Covalent radii of fluorine and hydrogen are respectively 0.72 Å and 0.37 Å. The increase of the number of fluorine atoms attached to the same carbon atom reduces the C$-$F bond length, which is accompanied by an increase in bond energy. The large size of fluorine accompanied by the C$-$F bond shortening can have important consequences in highly fluorinated fluorocarbon polymers. Probably the most important is the increased resistance to chemical attack which arises from the shielding action of fluorine on the carbon backbone and C$-$F bond strength. A further important consequence of the high bond energy of the C$-$F link, one of the strongest in organic compounds, is related to the polymerization chemistry of fluorinated monomers and subsequent derivable structures.

The high C$-$F bond energy prevents chain transfer reaction to polymer through the C$-$F group during the polymerization reaction, since this reaction is very unlikely on the basis of kinetic and thermodynamic factors under the usual polymerization conditions [12]. This means that generally perfluorocarbon polymers should be less branched than homologous hydrogenated polymers and the increase of the fluorine content in the fluoropolymer should reduce the degree of branching.

The most important evidence of the linear structure of perfluorocarbon polymers is demonstrated by PTFE, which has a degree of crystallinity as high as 93$-$98%. Such a high level of crystallinity can be obtained only with substantially unbranched structures [12].

The introduction of bulky groups, such as HFP, PVE or MVE, in the PTFE structure, to obtain FEP, PFA and MFA polymers, reduces the crystalline order and reduces the crystallization of the macromolecular chains. The size of the bulky group determines the amount of comonomer needed to obtain the desired degree of structural disorder. The larger the size of the bulky group, the lower is the amount of comonomer needed, and the correspondingly higher melting point that is obtained as shown in Table 21.1.

The higher stability of PFA and MFA polymers compared to HFP can be attributed to the less steric bond strain of the vinylether monomers used. Because, in this case, the bulky group is attacked to backbone by an ether link [13].

Table 21.1. Main comonomers used in TFE-based perfluoropolymers

Perfluoropolymer	Comonomer	Mol %	$T(^{\circ}C)$
FEP	$\begin{array}{c}CF_2 = CF \\ | \\ CF_3\end{array}$	5–6	265
MFA	$\begin{array}{c}CF_2 = CF \\ | \\ O \\ | \\ CF_3\end{array}$	3–4	285
PFA	$\begin{array}{c}CF_2 = CF \\ | \\ O \\ | \\ CF_2 \\ | \\ CF_2 \\ | \\ CF_3\end{array}$	1–2	305

3 MONOMER PREPARATION

3.1 TETRAFLUOROETHYLENE (TFE)–HEXAFLUORO-PROPYLENE (HFP)

The preparation of the basic fluoromonomers, TFE and HFP, can be achieved by the hydrofluorination of chloroform to give chlorodifluoromethane (CFC22) as follows: [14,15]

$$CHCl_3 + 2HF \longrightarrow CHClF_2 + 2HCl$$

The reaction proceeds at temperatures between 80 and 100 °C and a catalyst should be present such as antimony chlorofluoride. The CFC22 is pyrolyzed at high temperature and gives TFE plus by-products. In a silver tubular reactor heated by an electric furnace a selectivity of 93% for TFE and 1% for HFP with 38.4% CFC22 conversion was obtained at a temperature of 687 °C and a contact time of 1.8 seconds [16]. The reaction mechanism of the CFC22 pyrolysis is supposed to pass from an unstable intermediate, i.e. difluorocarbene, which subsequently forms TFE. The reaction scheme can be as follows:

$$CHClF_2 \xrightarrow{\text{heat}} (CF_2) + HCl$$

$$2(CF_2) \xrightarrow{\text{heat}} CF_2{=}CF_2 + SP(\text{byproducts})$$

According to the reaction conditions, i.e. thermal profiles of the pyrolyzer, or in a different apparatus, TFE can add to difluorocarbene or decomposes itself to difluorocarbene and subsequently add to it to give HFP.

This last monomer forms predominantly at about 750 °C, while higher temperature favour the formation of higher homologues, such as perfluoroisobutene (PFIB) [17]. The reaction mechanism can be outlined as follows:

$$CF_2=CF_2 \longrightarrow 2(CF_2)$$

$$CF_2=CF_2 + CF_2 \longrightarrow CF_2=CF-CF_3 + SP$$

$$CF_2=CF-CF_3 + CF_2 \longrightarrow CF_2=C(CF_3)_2$$

Great care must be taken in the design and control of the pyrolyzer, especially temperature profiles, in order to avoid troublesome problems during the process and to obtain the right products at high efficiency. For example perfluorocyclobutane and polytetrafluoroethylene are formed reversibly at about 600 °C [17]. Another problem is the controlling of PFIB formation during the process since it is highly toxic. The lethal concentration (LC50) of PFIB has been reported to be only few parts per billion [18].

3.2 FLUOROVINYLETHERS (FVE)

The classic process to produce FVE involves the chemistry of fluorocarbon epoxides discovered in 1959 [19,20], the most useful of which is hexafluoropropylene oxide (HFPO).

A more recent process prepares the same FVE and other different ones employing a different route. This involves the chemistry of perfluoroalkylhypofluorites instead of fluorocarbon epoxides.

3.3 PVE BY THE EPOXIDE PROCESS

Fluorocarbon epoxides have been prepared by oxidation of HFP with alkaline hydrogen peroxide or with oxygen in the vapour phase or in a chlorofluorocarbon solution. The vapour phase process requires the use of a free radical catalyst.

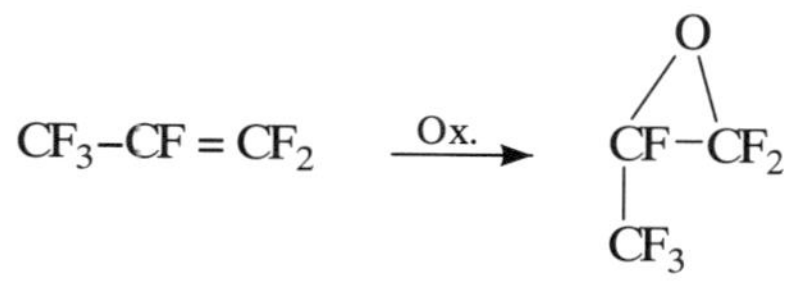

$$CF_3-CF=CF_2 \xrightarrow{Ox.} \underset{\underset{CF_3}{|}}{\overset{\overset{O}{/\backslash}}{CF-CF_2}}$$

Fluoroepoxides can be attacked by nucleophiles giving an acid fluoride after elimination of a fluoride ion [21]. The reaction with sodium carbonate and the subsequent pyrolysis produces the fluorovinyl ether [22]. The following reaction in the presence of metal fluoride can be considered:

$$\underset{\underset{CF_3}{|}}{CF-CF_2} \overset{O}{\overbrace{}} + \quad MF \quad \longrightarrow \quad C_3F_7OM$$

$$C_3F_7OM + \underset{\underset{CF_3}{|}}{CF-CF_2} \overset{O}{\overbrace{}} \quad \longrightarrow \quad C_3F_7-O-\underset{\underset{CF_3}{|}}{CF-CF_2}-OM$$

$$C_3F_7-O-\underset{\underset{CF_3}{|}}{CF-CF_2}-OM \quad \longrightarrow \quad C_3F_7-O-\underset{\underset{CF_3}{|}}{CF-C}\,OF + MF$$

The resultant acid fluoride is pyrolyzed over sodium carbonate at moderate temperature $(250\,^\circ C)$ to give the propyl vinyl ether by elimination of fluoride and loss of CO_2.

$$C_3F_7-O-\underset{\underset{CF_3}{|}}{CF-C}\,OF + Na_2CO_3 \quad \longrightarrow \quad C_3F_7-O-\underset{\underset{CF_3}{|}}{CF-COONa} + NaF + CO_2$$

$$C_3F_7-O-\underset{\underset{CF_3}{|}}{CF-COONa} \quad \overset{heat}{\longrightarrow} \quad C_3F_7-O-CF = CF_2 + NaF + CO_2$$

3.4   MVE BY THE HYPOFLUORITE PROCESS

The chemistry of perfluoroalkylhypofluorites has been known since the 1970s, but the main routes to obtain fluorovinylethers were only conducted on a laboratory scale because of difficulties in handling fluorine. Recently an industrial process has been developed on an industrial scale to obtain fluoroalkylvinylethers.

The starting material to obtain MVE is carbonyl fluoride which is fluorinated in the presence of a catalyst to give the methyl hypofluorite.

$$FCOF + F_2 \overset{CsF}{\longrightarrow} CF_3OF$$

The subsequent addition to 1,2-dichlorodifluoroethylene followed by dehalogenation gives MVE:

$$CF_3OF + CFCl{=}CFCl \longrightarrow CF_3OCFCl-CF_2Cl \overset{Zn}{\longrightarrow} CF_3OCF{=}CF_2$$

4   POLYMERIZATION TECHNOLOGY

Perfluorinated copolymers of TFE with HFP, MVE, PVE can be prepared by high-pressure emulsion polymerization techniques [23]. Since the monomers can be in the gaseous form under the polymerization conditions used, gas–liquid mass transfer can be a limiting step for the polymerization kinetics. In this case a proper reactor design has to be considered to perform useful reactions and obtain acceptable reactor productivity. Propagation rate for emulsion polymerization is given by the following expression:

$$R_p = K_p \, N n \, M_p$$

where K_p is the propagation rate constant, N the total number of polymer particles per unity volume, n the average number of free growing polymer radicals per particle and M_p the concentration of monomers in the growing polymer particle, which depends on the pressure of monomers in the reactor.

Thus the reaction rate can be increased by increasing the polymerization pressure. However, high-pressure reactors increase considerably the plant investment, maintenance and operating costs.

Fast polymerization kinetics can be obtained at relatively lower pressure by a proprietary emulsion polymerization technique [6]. In this case R_p is increased by increasing the number of polymer particles per unity volume (N). In this polymerization N can be two orders of magnitude higher than in conventional emulsion polymerization (10^{18} vs 10^{16}).

Organic or inorganic peroxy compounds, such as ammonium persulfate can be used as polymerization initiators. Inorganic initiators generally lead to ionic polymer chain ends, such as $- CF_2COOH$. These carboxylic end groups must be removed from the raw polymer to avoid polymer discoloration and bubble formation during high-temperature processing. Carboxylic end groups can be formed also from chain rearrangement process involving the PVE and, in a much less extent, MVE. These unstable end groups can be removed by a stabilization process which transforms them into more stable ones such as ester or amide [24,25]. Suitable emulsifying agents, such as ammonium perfluorooctanoate, are generally used.

When inorganic peroxy compounds are used, ionic chain ends can be controlled by minimizing the amount of decomposed initiator and using a chain-transfer agent to obtain the desired molecular weight.

An appropriate molecular weight distribution is required to obtain good processing behaviour, such as high flowability in injection molding and a fast extrusion rate for economical cycle times.

Molecular weight and composition can be regulated in a semi-batch process. In this case monomers and other chemicals are continuously added to the reactor, and after a determined time the reaction is stopped and the produced polymer discharged.

5 PROPERTIES

5.1 GENERAL PROPERTIES

Due to the high bonding strength of the carbon, fluorine and oxygen atoms, PFA, MFA and FEP demonstrate nearly the same outstanding capabilities as PTFE at temperatures ranging from extremely low to the extremely high. Therefore these fluorocarbon resins combine the processing ease of conventional thermoplastics with the excellent properties of PTFE. Furthermore, in the anti-corrosion protection and anti-stick coating fields, they show some advantages over PTFE, having a more morphologically homogeneous structure without micro-porosity.

Basic data for FEP and PFA are extensively available in the literature [26,27] while little data has been published for the new generation of MFA resins [32,33].

A comparison on the basic physical properties of these semi-crystalline thermoprocessable fluoropolymers is reported in Table 21.2. Since some of these properties depend upon crystallinity, these values vary according to fabrication conditions, particularly the cooling rate. Crystallinity and specific gravity increase as the cooling rate is reduced. The most relevant difference, among the three polymers, is in melting point. It is noteworthy to see the same difference (about 20 °C) between PFA, MFA and FEP.

The friction coefficients of PFA and MFA are slightly inferior to that of FEP. It has been verified [26] for FEP that the static friction decreases with an increase in load, and the static coefficient of friction is lower than the dynamic coefficient. MFA shows intermediate values between PFA and FEP.

The perfluoropolymers have an hydrophobic behaviour and are completely resistant to hydrolysis. They are excellent barriers to water permeation and their basic properties and dimensional stability remain unchanged even after year-long immersion in water.

The coefficient of thermal expansion is an important property in many high-temperature applications where the polymer is linked and coupled, through mechanical or chemical bonding systems, to other materials, notably metals. The coefficient of thermal expansion as a function of temperature is shown in

Table 21.2.   General properties of perfluorinated polymers

General properties	Method (ASTM)	Unit	PFA	MFA	FEP
Specific gravity	D 792	g/cm^3	2.12–2.17	2.12–2.17	2.12–2.17
Melting temperature	D 2116	°C	300–310	280–290	260–270
Coefficient of linear thermal expansion	E 831	1/K 10^{-5}	12–20	12–20	12–20
Specific heat	—	kJ/kg K	1.0	1.1	1.2
Thermal conductivity	D 696	W/K.m	0.19	0.19	0.19
Flammability	(UL 94)		V-O	V-O	V-O
Oxygen index	D 2863	%	>95	>95	>95
Hardness Shore D	D 2240		55–60	55–60	55–60
Friction coefficient (on steel)	—		0.2	0.2	0.3
Water absorption	D 570	%	<0.03	<0.03	<0.01

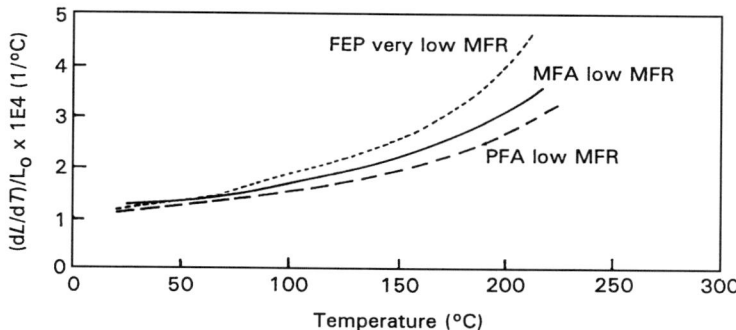

Figure 21.1. Thermal expansion coefficients as a function of temperature

Figure 21.1. Throughout this chapter the following descriptions will be used as references to typical commercial polymer grades:

	Melt flow rate (MFR)		Melt flow rate (MFR)
PFA low MFR	2–3	PFA high MFR	12–14
MFA low MFR	2–4	MFA high MFR	12–15
FEP very low MFR	1.5–2.0	FEP medium MFR	6–8
		FEP high MFR	18–22

5.2 MECHANICAL PROPERTIES

The flexibility at low temperatures, where most polymers become brittle, and the maintenance of some strength at high temperatures, where most of polymers are no longer solid, represents one of the most important characteristics of PFA, MFA and FEP. Extensive lists of mechanical properties of FEP and PFA are available in the literature [28–30].

Tensile strength data for the low MFR grades, as a function of temperature, are shown in Figure 21.2.

The mechanical characteristics of PFA, MFA and FEP are very similar at room temperature, though, at increasing temperature, differences become readily apparent. At 250 °C, while FEP does not retain any tensile resistance, both MFA and PFA still retain tensile strength in excess of 8–12 MPa. Mechanical properties of PFA at elevated temperatures, but below its melting point, generally are superior to those of PTFE at the same temperature. As with all thermoplastics, creep must be taken into consideration when a PFA, MFA or FEP component is being designed for services involving the application of continuous stresses like in lined pipes, hoses, seals, gasket, etc.

Among the partially and totally fluorinated thermoplastics, PFA, MFA and FEP exhibit (up to 100 °C) the lowest creep resistance. They show much better

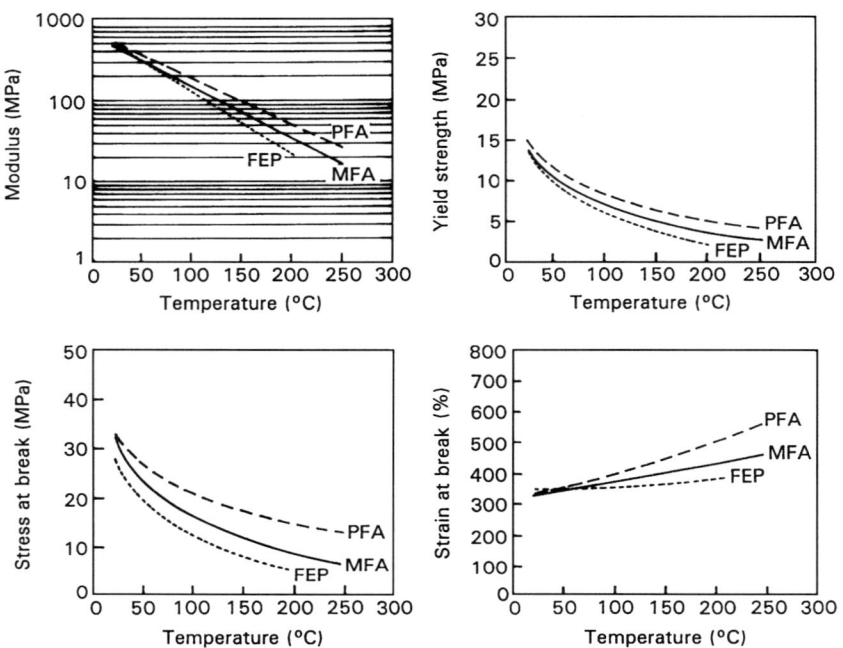

Figure 21.2. Tensile properties as a function of temperatures for high-viscosity grade perfluoropolymers (ASTM D 1708 method)

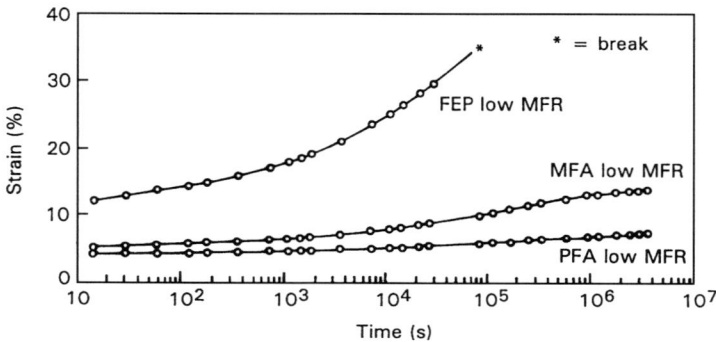

Figure 21.3. Tensile creep curves at 200 °C (applied load = 2 MPa)

creep properties above 150 °C in comparison to PVDF, ECTFE and ETFE. Creep measurement at high temperature (200 °C) have been carried out, according to ASTM D 2990 method, on PFA, MFA low MFR and FEP very low MFR (see Figure 21.3). Isochronous stress–strain curves for 2000 hours creep time are reported in Figure 21.4. They show that, according to the applied load and to the

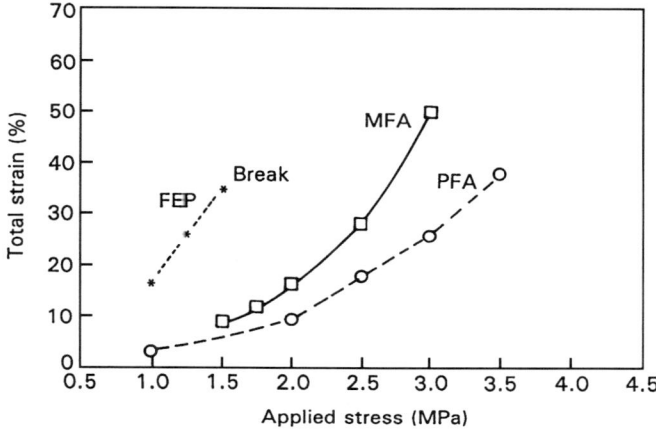

Figure 21.4. Isochronous stress–strain curves for 2000 hours creep time

test time, FEP specimens will undergo a partial and progressive embrittlement that eventually leads to cracking. This behaviour is generally known as thermal stress cracking (TSC). It represents the most important difference in behaviour between FEP on the one hand and PFA and MFA on the other.

MIT flex life can be considered another measurement to determine the polymer toughness. Furthermore, it represents a key property for applications where the items are continuously subject to folding cycles. Fluoropolymer insulated cables for dynamic services employed in high temperature or chemical aggressive environment is a typical situation. Data for PFA, MFA and FEP have been obtained at room temperature (Figure 21.5), according the to MIT test procedure (ASTM

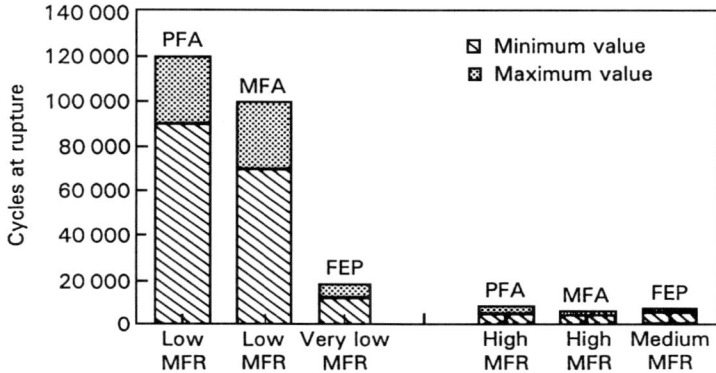

Figure 21.5. Flex-life cycles on films obtained at 23 °C (thickness = 0.3 mm). MIT method (ASTM D 2176)

Table 21.3. Abrasion Taber test data obtained at 23 °C (ASTM D 1044)

Fluoropolymers	Typical MFR (g/10 min.) ASTM D 2116	Wear index (mg/1000 cycles)
PFA low MFR	2–3	9–10.5
MFA low MFR	2–3	10–11.5
FEP very low MFR	1.5–2	14–15.5
PFA high MFR	12–15	15.5–17
MFA high MFR	12–15	15.5–17
FEP medium MFR	6–8	18.5–20

D 2176). The folding endurance is strongly dependent on the polymer molecular weight and its distribution. The toughness behaviour of these polymers is confirmed also by tensile impact test. Izod test (ASTM D 256) carried out at room temperature show that the polymers do not break.

Fluorocarbon resins perform well also at very low temperatures [27]. Tests made on PFA and MFA at liquid nitrogen temperature indicate that the polymers remain relatively flexible and tough down to −200 °C.

Wear performances of unfilled PFA, MFA and FEP have been determined (see Table 21.3) by means of the typical Taber test procedure (ASTM D 1044). Abrasion resistance is typically linked, for semicrystalline polymers, to the crystallinity. The MFA structure is, in such respect, closer to PFA therefore it shows an abrasion resistance, at room temperature, comparable to that of PFA and significantly higher than FEP.

5.3 ELECTRICAL PROPERTIES

The outstanding electrical properties distinguish the fluorocarbon resins from the traditional thermoplastic polymers. Moreover PFA, MFA and FEP in comparison to the partially fluorinated polymers are only slightly affected by temperature up to their maximum service temperature and for these reasons they are widely used as electrical insulators. In Table 21.4 the electrical properties of these fluoropolymers are reported.

The dielectric strength, volume and surface resistivity of these polymers are excellent and unaffected by temperature at least up to their glass transition temperature. Dielectric strength of these fluoroplastomers is generally superior to the values shown by PTFE. This last property is a function of the thickness (see Figure 21.6).

These resins do not form a carbonized conducting path [31] when tested by means of arc tracking resistance behaviour (ASTM D 495).

Table 21.4. Typical electrical properties of perfluoropolymers

Property	ASTM method	Unit	PFA	MFA	FEP
Volume resistivity					
(from 23 to 150 °C)	D 257	Ohm-cm	$>10^{17}$	$>10^{17}$	$>10^{17}$
Surface resistivity	D 257	Ohm	$>10^{17}$	$>10^{17}$	$>10^{17}$
Arc resistance	D 495	Second	>200	>200	>200
Dielectric strength					
(thickness = 1 mm)	D 149	kV/mm	30–32	30–32	26–30
Dielectric constant	D 150				
(50 Hz; 23 °C)			2.0–2.1	2.0–2.1	2.1
(100 kHz; 23 °C)			2.0	2.0	2.0
Dissipation factor	D 150				
(50 Hz; 23 °C)			3×10^{-4}	3×10^{-4}	3×10^{-4}
(100 kHz; 23 °C)			5×10^{-4}	5×10^{-4}	5×10^{-4}

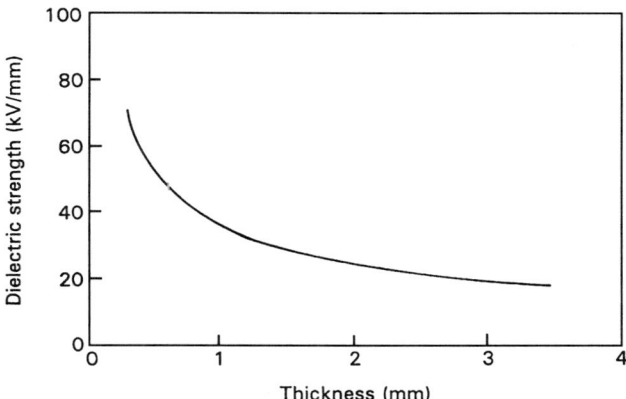

Figure 21.6. Effect of thickness on dielectric strength at 23 °C of MFA resin

The dielectric constant of PFA, MFA and FEP is about 2.04 over a wide range of temperature and frequencies (from 100 Hz to 1 GHz). The dissipation factor at low frequency (from 100 Hz to 10 kHz) decreases with increasing frequency and decreasing temperature. In the range 10 kHz–1 MHz, temperature and frequency have little influence while above 1 MHz the dissipation factor increases with the frequency. The difference between PFA/MFA and FEP at high frequencies is notable. At 1 MHz MFA has a dissipation factor of 2.67×10^{-4} while FEP shows a value of 7.25×10^{-4}. This can obviously lead to better signal attenuation behaviour of PFA/MFA in comparison to FEP in coaxial cable for high-frequency applications. Attenuation data obtained on insulated cable [32], as a function of frequencies, are shown in Figure 21.7.

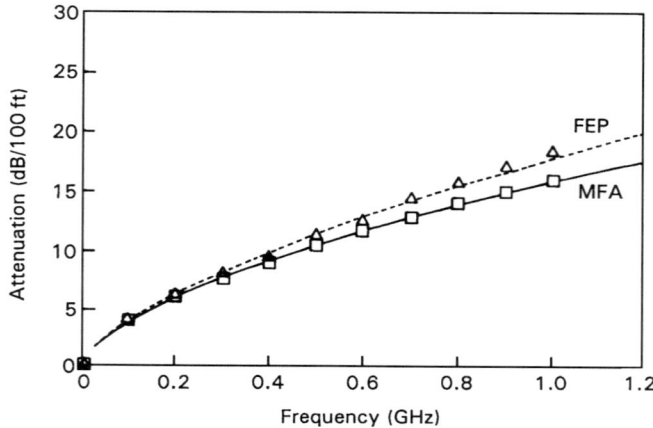

Figure 21.7. Attenuation data on RG-58 cables (thickness = 0.95 mm). (Courtesy of Comm/Scope Inc., Network Cable Div., Claremont, NC.)

5.4 OPTICAL PROPERTIES AND RADIATION EFFECTS

Fluorocarbon films have high transmittance in the ultraviolet, visible, and infrared regions of the spectrum. For instance the visible light transmission (400–700 nm wavelength) for a 0.025 mm thick PFA film is more than 90%. In the UV region (200–400 nm wavelength) transmittance data have been obtained on MFA and FEP films having a thickness of 0.2 mm (Figure 21.8). This property is strongly dependent on the crystallinity level and the crystal morphology in the polymer. The outstanding behaviour shown by MFA, is remarkable — better than values shown by both PFA and FEP. The refractive index values among these polymers are similar and close to 1.3.

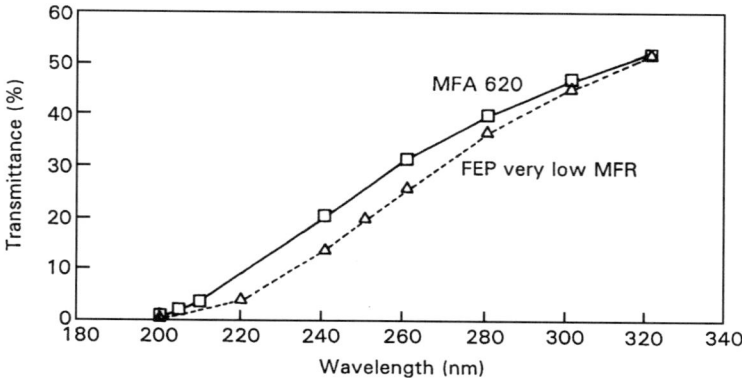

Figure 21.8. Transmittance values in the UV region obtained on 200 μm films

Like PTFE, perfluoropolymers are not highly resistant to high-energy ionizing radiation [27]. The main effect of radiation is the reduction of tensile strength, especially on elongation at break. Above 5 Mrad there is a retention of elongation lower than 5%, while at 50 Mrad PFA, MFA and FEP are degraded.

5.5 CHEMICAL RESISTANCE, PERMEABILITY AND HIGH PURITY

Perfluoropolymers have excellent chemical resistance even at elevated temperatures and pressures. They are not attacked by inorganic bases, strong mineral acids and inorganic oxidizing agents commonly used in the chemical industry. They are also inert to most of organic compounds and mixtures of them while they react with fluorine and molten alkali.

Elemental sodium (or other alkali metals) in contact with fluorocarbon polymers removes fluorine from the polymer. This reaction is widely used in anhydrous solutions to etch the surfaces of these fluoropolymers so that they can be adhesive bonded to other materials. In Table 21.5 are shown the effects of certain selected solvents and chemicals on the mechanical properties of PFA and MFA resins [33].

The water and solvent absorption of fluoropolymers is in general very low [26]. Permeation is closely related to absorption and it is a function of temperature, pressure and crystallinity. Since these resins are melt processed they are completely void free and the overall permeation is substantially lower than the permeation shown by PTFE.

Some special-purpose perfluoropolymers with additional benefits are commercially available. These resins combine many of the properties of the parent resins with enhanced purity and improved thermal stability while processing. The 'high purity grades' are especially designed for applications where improved colour, lower extractable fluorides and freedom from other foreign materials are needed (purity in the parts per billion). They are widely employed in the semiconductor industry and for the transportation of ultra-pure chemicals.

5.6 FLAMMABILITY AND SMOKE

Fluorocarbon resins, compared to traditional thermoplastics, have an outstanding fire resistance. When PFA, MFA and FEP are exposed to a naked flame, they burn, but cease burning as soon as the flame is removed. This is explained by the oxygen index (i.e proportion of oxygen necessary to sustain combustion) which is >95% (ASTM D 2863). In flammability tests according to UL (Underwriters' Laboratories) they meet the requirements for 94 V-O. Furthermore, when ignited by sustained flame from other sources, the smoke generated by PFA, MFA and FEP is very low. The fire resistance behaviour of PFA, MFA and FEP is the best in the class of fluoropolymers (highest oxygen indices) and they produce less smoke when compared to the partially fluorinated polymers like ETFE, ECTFE and PVDF.

Table 21.5. Chemical resistance data: change as percentage of initial value

Chemical	T (°C)	Stress at break		Elongation at break		Weight gain	
		PFA	MFA	PFA	MFA	PFA	MFA
Cyclohexanone	156	−5.4	0.3	−3.1	0.7	0.4	0.4
Toluene	110	−12	−3	−3	4.2	0.7	0.7
Carbon tetrachloride	77	−7.3	−3.2	3.9	5.7	2.4	2.4
Dimethyl formamide (DMF)	154	−4.7	3.2	−1.9	2.4	0.2	0.2
Methylene chloride	40	−11	−7.4	−1.2	−0.4	0.9	0.8
Nitric acid, 65%	120	−8.6	−7.1	3	4.8	1.9	1.9
Fuming nitric acid	23	0	0.6	0.3	1.1	0.1	0.1
Fuming sulfuric acid	23	−2.2	6.1	−3.7	3	0	0
Phosphoric acid, 85%	100	−0.3	3.9	−1	2	0	0
Methyl ethyl ketone	80	−0.6	−3.5	3	1	0.4	0.4
Hydrogen peroxide, 30%	23	−1.3	5.5	−1.5	2.8	0	0
Zinc chloride	100	−1.6	−0.6	−2.8	−1.9	0	0
Ethylene diamine	117	−4.1	1.9	−3.7	2.3	0.1	0.2
Sodium hydroxide	120	−3.5	−0.3	−2.6	−1.1	0	0
n-Butylamine	78	2.3	−10	0.5	−5	0.45	0.44
Ethylene Tetrachloride	121	−3	−10	2.3	0.1	2.1	2.2
Tetrahydrofuran	65	9.2	−8.7	7.7	−2.7	0.64	0.66
Meforex 123	23	0.3	−8.4	9.9	3.5	2.8	2.4
Ammonium hydroxide, 30%	66	−1.7	−2.4	−1	−0.9	0.23	0.25

5.7 PROCESSING

Standard thermoplastic processing techniques can be used to process the thermo-plastic perfluorinated resins. In their melt state, PFA, MFA and FEP are corrosive to most metals. Therefore processing equipment are constructed using corrosion-resistant nickel-based alloys (like Hastelloy®, Xaloy®, Reiloy®, Inconel®) and must be able to operate at temperatures up to 330–420 °C.

Rheological properties and thermal stability are the fundamental characteristics which govern the processing behaviour of the thermoplastic resins. While the partially fluorinated polymers show normal pseudoplastic behaviour (i.e. viscosity decreases with increasing flow rate), PFA, MFA and FEP show a sudden transition from the Newtonian behaviour to an overflow regime when a critical value of the shear rate is reached (see Figure 21.9). Processing behaviour and working temperatures of MFA are very similar to those of PFA. When a thermoplastic flows through a die orifice, both in extrusion and molding, an unstable condition called 'melt fracture' can occur giving essentially a roughness of the resin surface. Melt fracture occurs with flow rates above the polymer's 'critical shear rate'. This

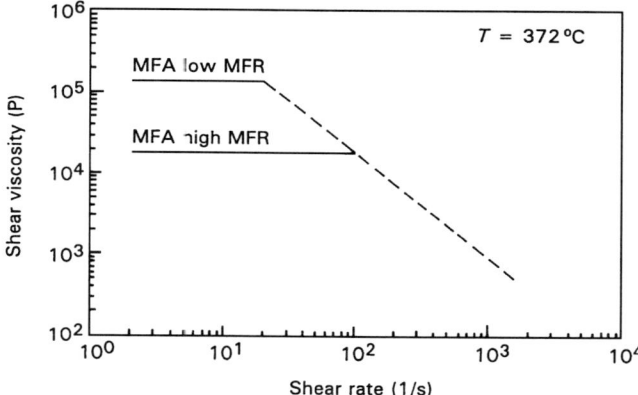

Figure 21.9. Rheological curves of MFA resins

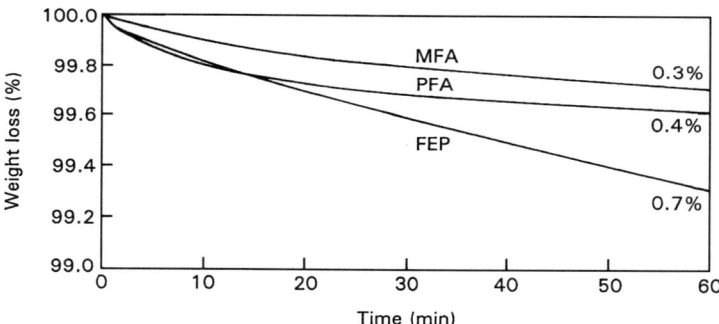

Figure 21.10. Thermogravimetric analysis in air at 380 °C for 1 hour

polymer property is a key parameter of great industrial importance since it can affect drastically the extrusion line speed. An interesting work on the melt fracture behaviour of some perfluorinated polymers is reported in ref. [34].

Perfluoropolymers also have exceptional thermal stability in the melt state at typical processing temperatures. As shown in Figure 21.10, by means of thermogravimetric analysis PFA and MFA have the highest thermal stability (different slope and lower final weight loss) in comparison to FEP.

6 APPLICATIONS

6.1 INTRODUCTION

As outlined while discussing properties, perfluoropolymers are a class of polymers whose excellence is particularly remarkable in the following key areas:

- electrical properties (both insulation properties and dielectric properties)
- thermal properties (mainly the maximum continuous service temperature)
- chemical resistance properties
- surface properties (both as friction factor and anti-adhesion)
- fire resistance and inertness properties
- purity properties

Since their early commercial development, perfluoropolymers have been challenged from time to time by other classes of polymers, where one of the above-mentioned properties was improved in comparison with the standard behaviour of FEP, MFA and PFA. However, no polymer is presently known, with a combination of all these properties, or at least some of them, which is a threat to the markets of perfluoropolymers. This is one of the reasons why the consumption of perfluoropolymers has been steadily growing over the last 30 years, even during times of recession.

Present applications of melt-processable perfluoropolymers can be divided according to the main premium property that makes these polymers the preferred ones in a given application. Therefore the same scheme as above will be followed. In particular, it is worth considering the following classes:

- applications in the wire and cable (W & C) industry
- applications in the chemical process industry (CPI)
- applications in the semiconductor industry (SI) and wherever high purity materials are required
- applications where the key property is high release or anti-stick behaviour.

6.2   APPLICATIONS IN THE WIRE COATING INDUSTRY

The W & C industry is largely using perfluoropolymers in the following two main applications:

1. Wires and cables for medium-high frequency data transmission in combination with either high temperature or fire resistance.
2. Wires and cables for connections (i.e. hook-up wires) rated at high temperature ($>200\,^{\circ}C$) or wherever a hostile environment is present and combined with high temperature.

1. In this application the premium properties are the low dielectric constant and the exceptionally low dissipation factor that make perfluoropolymers particularly suitable whenever low attenuation must be sought in data transmission cables.

However, by comparing these two properties between FEP, MFA and PFA on the one hand and a standard polyethylene, electrical grade, on the other, these two classes of polymers look quite comparable and similar. Therefore, perfluoropolymers are used only when some other minimum performance criteria must be fulfilled.

If low attenuation is to be combined with strict fire resistance requirements, fluoropolymers are the natural choice in data transmission cables. This is the case, for example, of the so-called plenum constructions, that is, cables that can run in air-conditioning ducts and open spaces between ceiling and floors, without any metal protection or conduit. In North America and countries where US standards have been fully adopted, this cable lay-down is allowed, provided that the cable can pass the UL (Underwriters Laboratories) 910 specification, that is the so-called 'Steiner tunnel test'. In this test there is a simulation of a flame propagation in an air duct and the cable must give clear evidence that neither the flame spread nor the smoke generation exceed a very strict threshold. Plenum applications are the major market for fluoropolymers, particularly for FEP. It is estimated that in 1994 about 60–65% of overall worldwide consumption of FEP was taken by the plenum market. The wide use of local area network (LAN systems), with huge amounts of cables laid horizontally for each floor of office buildings or university campuses, have boosted the need for high-performance data transmission cable that must fulfil the plenum requirements. The most common construction to date (1996), that is the 'workhorse' of these applications, is the four unshielded twisted pair construction (4 UTP), that can be used in most common cabling frames and architecture. Its specification is contained in the Telecommunications Industry Association (TIA) 568A, where both electrical characteristics up to 100 MHz and fire performance, according to UL 910, are specified.

The need to have cable structures able to carry higher and higher frequencies is leading these applications towards even higher performing applications as far as the high-frequency performance is required. While to date (1996) no firm projection can still be made, there are signs of a trend to raise the specifications from 100 MHz (the present Category 5 level, according to TIA 468A) to 350 MHz or even up to 600 MHz, as already proposed in some countries for the equivalent constructions in non-plenum design (cables in metallic conduits).

In such a case, FEP could be displaced in these new high-end applications by higher performance polymers like MFA and PFA, whose dissipation factors are considerably lower than those of FEP above 100 MHz. The other possibility could be the more wide-spread use of foamed constructions, even at very thin wall thicknesses, where the improvement in electrical properties due to the foaming would be associated with a considerably more economical design. To date, foaming of perfluoropolymers is done reasonably well in relatively thick constructions (notably coaxial cables) and active work is being done by the major resin producers and cable manufacturers to achieve high expansion levels (more than 25–30%) even at very thin wall thicknesses, without losing too much of the original mechanical strength.

In plenum applications, as mentioned, the most common construction is the 4 UTP. Beside that, coaxial cables, with the dielectric core made with perfluoropolymers, are quite common.

Coaxial cables with perfluoropolymer dielectric cores and also jackets are also widely used whenever high-temperature resistance must be met, together with the flame resistance requirements. The most important example is probably the US military specification (Mil C-17), that prescribes several coaxial cable constructions with perfluoropolymers.

Worth mentioning as further evidence of the outstanding overall properties of perfluoropolymers and of their reliability is the use of perfluoropolymers (mainly PFA and MFA) in miniature coaxial cables for data transmission in medical equipments used for endoscopic investigations.

2. Connection cables (mainly hook-up wires) for environments at 200 °C or even higher than that, or with very hostile chemicals, are the second group of cables where perfluoropolymers are widely used. Examples vary from the production of electrical appliances to the car industry. The outstanding values of dielectric strength and volume resistivity make the perfluoropolymers the ideal choice when miniaturized cables have to be developed for such difficult applications.

Up to 200 °C FEP is used, while MFA and PFA are used for higher temperatures, up to 250 °C. MFA, thanks to its balance between cost and performance, has achieved, in spite of being the last semicrystalline perfluoropolymer launched on to the market to date, 1994 a considerable in road in this kind of application.

Compliance of some grades of MFA to the UL 758 specification up to 250 °C for appliance wires with a minimum 10 mils of wall thickness has been recently reported by UL [35].

Without pretending to be exhaustive, it is worth mentioning the following applications as typical examples where perfluoropolymers are used: wires for appliances and household equipment, where heat sources are present; heating cables (self-regulating), for chemical equipment electrical tracing; connection wires in lamp-holders for halogen lamps; wires for thermocouples; and cables for special oil-drilling equipments (data-logging cables).

In the car industry there is an increasing number of applications, where perfluoropolymers are now used, sometimes in combination with PTFE or with other polymers: cables for under-hood applications; cables for automatic gears, where very aggressive oils are used; cables for gasoline level sensors; cables for brake-wearing sensors, jacket for push-pull mechanical cables.

6.3   APPLICATIONS IN THE CHEMICAL PROCESS INDUSTRY

The use of perfluoropolymers in the CPI is obviously linked to the almost universal chemical resistance shown by the three main polymers, that is, FEP, MFA and PFA. Basically, four groups of main applications can be found:

1. Lining, via extruded sheets, of vessels, tanks and equipment;
2. Thick electrostatic powder coating (min. 800 microns) or rotolining of vessels, tanks and equipment;

3. Lining, via transfer molding or injection lining, of valve bodies (and related components) and fittings, in carbon steel lined piping systems, whereas pipes are normally lined with PTFE;
4. Self-supported tubes and pipes, especially used in heat-exchanger applications;

Coating, rotolining and sheet lining of vessels are common practices in the chemical protection of exposed metal surfaces (usually made with carbon steel), with beneficial cost reduction in comparison with the use of exotic metal alloys. The use of perfluoropolymers is normally limited to applications where other, lower performance fluoropolymers, notably PVDF and ECTFE, cannot cope with the environment or temperature.

Lined piping systems made with PTFE (straight lined pipes) and melt-processable perfluoropolymers (fittings and valves) also represent, like sheet lining and coating, a typical application where the limited mechanical performance of perfluoropolymers is compensated by their use as a liner, where most of the mechanical stresses are borne by the metal backing.

In all these applications, beside their generic chemical resistance, two other factors play a major role: their permeation resistance and their thermal/environmental stress cracking resistance.

While the resistance to permeation of fluoropolymers is generally good, it must be said however, that some chemicals (HF, HCl and steam as well) are quite critical in the permeation risk they can present even for 3–4 mm perfluoropolymer lining. Proper vents are therefore provided in the piping systems with different designs according to the various manufacturers.

Crack resistance is generally an intrinsic property of the polymer. However, FEP has shown to be quite different from PFA and MFA, due to its tendency to transform, under a given combination of stress, temperature and environment, its original ductile behaviour into a quasi-brittle one (with formation of crazes, the subsequent increase of the effective local stress in the residual ligament and eventually complete plastic collapse). For this reason, lined piping systems with FEP elements are not normally recommended above 150 °C, while the other two polymers, notably PFA, are generally rated from 180 to 220 °C, according to the various manufacturers' designs.

Self-supported tubes and pipes made with perfluoropolymers are especially used in heat exchangers of desulphurization plants, especially in coal power-plants. Environmental issues, and hence the rapid growth of use of desulphurization plants in countries where coal is the most important energy source make this application one of the most important for melt-processable perfluoropolymers. In this application, however, FEP is seldom used due to its inferior thermal stress crack behaviour.

Other applications where perfluoropolymers are used in self-supported systems in the CPI are mainly linked to control and instrumentation equipment, as well as laboratory lines.

6.4 HIGH PURITY APPLICATIONS (SEMICONDUCTOR INDUSTRY AND RELATED)

In the last decade use of perfluoropolymers in applications linked to the semiconductor industry has been steadily increasing at very high growth rates. Notably, some special grades of PFA are the most popular material for self-supported piping systems where high-purity fluids must be carried from tanks to the point of use, or for wafer baskets in which silicon wafers are transported during the various phases of semiconductor production.

Two major requirements are typical in these applications: (1) chemical resistance must provide outstanding inertness to the most aggressive environment and to very exotic chemicals; (2) ions leaching (both cations and anions), or particles, from the material involved in the fluid transportation, or even more, in direct contact with wafers, must be minimal.

While the first requirement makes this application equivalent to the most traditional ones in the CPI, low leaching or release is peculiar to the so called high-purity applications. Perfluoropolymers obtained with perfluoroalkylvinylethers as comonomers have been chosen because of their universal chemical resistance and also because their thermal stability is the necessary basis to achieve a very low release from their backbone.

Special ways of preparation and handling of the polymer must be undertaken in order to obtain grades which can comply with the needs of the semiconductor industry. Particular attention must be given to the release of the fluoride ions, that can come out of the end groups of the backbone during thermal degradation and that are particularly dangerous in the various steps of the semiconductor processes. This is the reason why new grades of perfluoroalkoxy materials have been designed in order to minimize the fluoride ion release.

6.5 RELEASE AND ANTI-STICK APPLICATIONS

Within these applications, one can put all applications where surface properties of perfluoropolymers are the premium factors that make perfluoropolymers the ideal materials. Their parent material, PTFE, has been known for decades to be the best material for anti-stick and low friction factor coefficients. Now, melt-processable perfluoropolymers combine the unique properties of PTFE with considerable advantages in processability and also some improvement as far as some other properties are concerned (abrasion resistance, optical clarity, higher permeation resistance and higher dielectric strength). In some cases, as explained later, the combination of PTFE and perfluoropolymers offers the best compromise in terms of cost and performance.

Case histories of moulded items where perfluoropolymers are chosen because of their low friction factors are lengthening and involves primarily industrial parts to be used where rotary or alternate motion is present.

A typical example of extruded items where surface properties play a major role are thermoshrinkable tubes: FEP, MFA and PFA tubes can be expanded after extrusion at temperatures slightly below their melting point and then thermoshrunk over cylinders and rolls, because they keep the elastic memory to revert to their original dimension after heating. FEP and MFA offer the best expansion rates and also the easiest processing conditions during thermofitting, although expansion of PFA is more limited, and thus shrinking more difficult. Industrial rolls where low friction must be provided are generally covered with thermofit tubes made with perfluoropolymers. Photocopy machines as well can contain rolls whose surface has been lined with a perfluoropolymer liner. In such cases, FEP is generally not considered whenever rolls temperature exceeds 150 °C.

The most important group of applications where surface properties are the key factors can be found in the coating industry. Coatings made with melt-processable perfluoropolymers provide excellent non-stick and release properties, especially in the food-industry and they are becoming more and more popular where premium performance in comparison with standard PTFE-made coatings is necessary. Formulations made with both PTFE dispersions and melt-processable perfluoropolymers dispersions are available on the market and they are used especially for cookware. The improvement in abrasion resistance in comparison with standard PTFE formulations is now well known in the patent literature. Industrial applications in bakeries, due to more stressing cycles, require nevertheless, a further improvement in abrasion resistant formulations where the base polymer is a melt-processable perfluoropolymer.

As most of these applications are in the food industry, compliance with the most typical regulations for continuous contact with food is necessary. Perfluoropolymers are, generally speaking, recognized by both FDA and last EEC directives as compliant for these typical uses in the food industry.

REFERENCES

1. Farbenindustrien I. G., British Patent 465520 (1937).
2. Plunkett, R. J., US. Patent 2230654 (1941).
3. Bro, M. I., Sandt, B. W., US Patent 2946763 (1960).
4. Harris, J. F., McCane, D I., US Patent 3132123 (1964).
5. US Patent 4900872 (1990).
6. US Patent 4864006 (1989).
7. Anderson, R. F., Punderson, J. O., *Organofluorine Chemicals and Industrial Applications,* ed. Banks, R. E., Horwood, Chichester, p. 235 (1979).
8. Tatlow, J. C., *Organofluorine Chemicals and Industrial Applications*, ed. Banks, R. E., Horwood, Chichester, p. 22 (1979).
9. Moore, W. J., *Physical Chemistry*, Longman, London, 5th edn. (1972).
10. Banks, R. E., *Fluorocarbons and Their Derivatives*, Macdonald, London, 2nd edn. (1970).
11. McBee, E. T., *Ind. Eng. Chem.*, **19**, 237 (1947).
12. Sperati, C. A., Starkweather, H. W., *Fortschr. Hochpolym.*, **2**, 465 (1961).
13. England, D. C., *et al.*, *Proceedings of the R. Welch Conference on Chemical Research*, **26**, Synthetic polymers, Houston (1982).
14. Hamilton, J. M., *Advances in Fluorine Chemistry,* Butterworth (1963).

15. Banks, R. E., *Fluorocarbons and their Derivatives,* Macdonald, London (1970).
16. US Patent 3306940 (1967).
17. Kometani, Y., *High Polymer,* **27,** 503 (1978).
18. Ulm, K., *10th International Symposium on Fluorine Chemistry,* Vancouver, BC (1982).
19. Trabant, P., *et al., Fluorine Chemistry Reviews,* Vol 5, Marcel Dekker, NY (1959).
20. Resnik, P., *Kirk-Othmer: Encycl. of Chem. Technol.,* 3rd edn., Vol 10, (1980).
21. Moore, E. P., US Patent 3322826 (1967).
22. Fritz, C. G., Moore, E. P., Selman, S., US Patent 3114778 (1963).
23. Khan, A. A., Morgan, R. A., US Patent 4380618.
24. Schreyer, R. C., US Patent 3085083 (1963).
25. Carlson, D. P., *et al.,* US Patent 4599386, (1986).
26. Gangal, S. V. (with 84 references), *Kirk-Othmer: Encycl. of Chem. Technol.,* 4th edn., Vol. 11, p. 644 (1994).
27. Gangal, S. V. (with 37 references), *Kirk-Othmer: Encycl. of Chem. Technol.,* 4th edn, Vol 11, p. 671 (1994).
28. Diamond, R. J., *Plastics,* **27**, 109 (1962)
29. Frados, J., *Modern Plastics Encyclopedia,* **46**, 974 (1969).
30. Brydson, J. A., *Plastics Materials,* **10**, 203 (1966).
31. Fasing, E. W., McCane, D. I., 22nd *Proc International Wire and Cable Symposium,* Atlantic City (1973).
32. Vita, G., Pozzoli, M., 44th *Proc. International Wire and Cable Symposium,* Philadelphia (1995).
33. Vita, G., Pozzoli, M., presentation at SPI fall Conference (1995).
34. Rosenbaum, E. E., Hatzikiriakos, S. G., Stewart, C. W., *Intern. Polymer Processing,* **10**, 204 (1995)
35. UL report on Component–Appliance Wiring Material, File NC2118, Project 95ME50540 (1995).

22

Teflon® AF Amorphous Fluoropolymers

PAUL R. RESNICK
DuPont Fluoroproducts, Fayetteville NC, USA
WARREN H. BUCK
DuPont Fluoroproducts, Wilmington, DE, USA

Teflon® AF is a family of amorphous fluoropolymers based on copolymers of 2,2-bistrifluoromethyl-4,5-difluoro-1,3-dioxole (PDD) which retain the outstanding chemical, thermal and surface properties associated with perfluorinated polymers while also having unique electrical, optical and solubility characteristics. This combination of properties makes them useful for a large variety of applications.

All previously commercialized perfluorinated polymers such as Teflon® PTFE, FEP or PFA are semicrystalline materials. The first commercial perfluoropolymer, Teflon® PTFE, is highly crystalline and was discovered by Roy Plunkett in 1938 [1]. Its extremely high molecular weight estimated to be between 5 and 100×10^6 Daltons and melt viscosity of the order of 10^{13} Pa . s at 380 °C make it difficult to process. The introduction of hexafluoropropylene and perfluoroalkylvinyl ethers into the polymer chain to give Teflon® FEP and Teflon® PFA respectively, lower both the crystallinity and molecular weight allowing these polymers to be melt processed. These semicrystalline polymers do not possess good optical transparency and in general have poor solubility. Amorphous polymers usually have excellent optical properties and solubility characteristics. Teflon® AF is an amorphous fluoropolymer family which have both the properties expected of amorphous plastics and the properties expected of perfluorinated polymers. The PDD dioxole monomer in Teflon® AF yields polymers which have even further unexpected properties. Teflon® AF is a copolymer of PDD and tetrafluoroethylene, TFE. Its structure is shown in Figure 22.1 [2–5].

The dioxole monomer, PDD, is synthesized in four steps from hexafluoroacetone and ethylene oxide as shown in Figure 22.2. Hexafluoroacetone

Modern Fluoropolymers. Edited by John Scheirs
© 1997 John Wiley & Sons Ltd

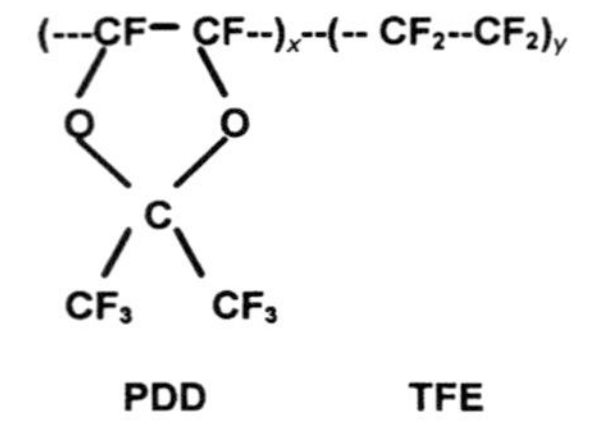

Figure 22.1. Structure of Teflon® AF

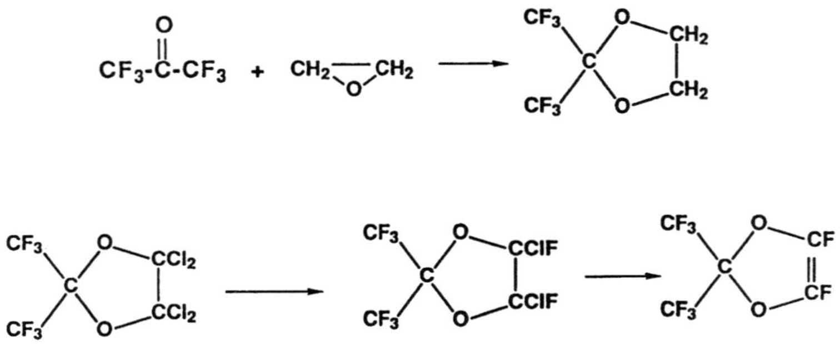

Figure 22.2. Synthesis of PDD

condenses with ethylene oxide to form a highly chemically stable dioxolane ring in quantitative yield. Exhaustive chlorination followed by chlorine-fluorine exchange yields 2,2-bistrifluoromethyl-4,5-dichloro-4,5-difluoro-1,3-dioxolane in greater than 90% yield. Dechlorination of this dioxolane with magnesium, zinc or a mixture of titanium tetrachloride and lithium aluminum hydride gives PDD monomer. PDD is a clear colorless liquid boiling at 33°. It is highly reactive and is stored at low temperature with a trace amount of radical inhibitor. Free radical initiated polymerization and copolymerizations of PDD or other fluorinated dioxoles may be carried out in either aqueous or non-aqueous media.

PDD readily polymerizes with tetrafluoroethylene and other fluorine containing monomers such as vinylidine fluoride, chlorotrifluoroethylene, vinyl fluoride and perfluoroalkylvinyl ethers. It can also homopolymerize to give an amorphous polymer with a glass transition temperature, T_g, of 335 °C. Thus the number of potential polymers of PDD is very large. This number becomes even larger when one considers the many other dioxoles with different substituents in the 2-, 4- and 5-position of the dioxole ring that can and have been prepared and polymerized. At present the commercial Teflon® AF products are copolymers of PDD and tetrafluoroethylene and have glass transition temperatures of 160 and 240 °C for Teflon® AF-1600 and Teflon® AF-2400 respectively.

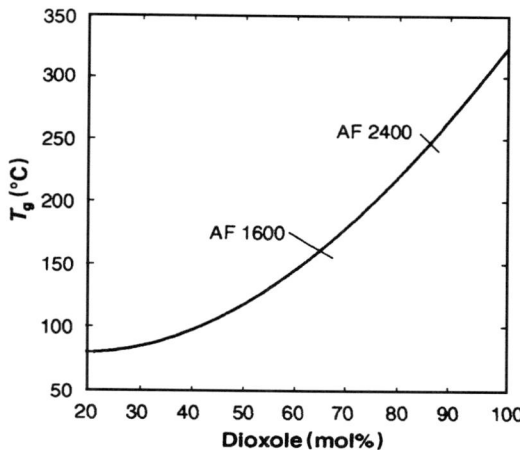

Figure 22.3. Composition of Teflon® AF vs T_g

Copolymers may be prepared which have any proportion of tetrafluoroethylene and PDD. The T_g of these polymers is a function of the PDD content and is shown in Figure 22.3. The lower limit for T_g is approximately 80° corresponding to 20 mol% PDD. Copolymers with less PDD are no longer amorphous due to the presence of tetrafluoroethylene runs of sufficient length to result in crystallization of the polymer. As noted the T_g of PDD homopolymer is 335°, but it is difficult to fabricate since it possesses limited melt flow below its decomposition temperature and is soluble to only a few tenths of 1%. Teflon® AF-1600 and 2400 contain 65 and 87 mol% PDD respectively. In general, the glass transition temperature of a PDD copolymer decreases when the PDD concentration in the polymer is lowered by addition of any comonomer.

Aqueous polymerization of PDD and tetrafluoroethylene is usually carried out in the presence of a fluorinated surfactant and ammonium persulfate or other metal persulfate initiators. Under these conditions small amounts of acid fluoride or carboxylic acid chain ends may be produced by ring opening during the polymerization. For some applications these unstable ends are removed and converted to trifluoromethyl groups by first treating the polymer with ammonia or alkyl amines followed by high-temperature fluorination with elemental fluorine. (Figure 22.4) [6].

The T_g of dioxole copolymers with tetrafluoroethylene is highly sensitive to the structure of the dioxole monomer. Replacing the trifluoromethyl groups at the 2-position of the dioxole ring with either fluorine or large fluoroalkyl groups dramatically changes the T_g of the polymer and the reactivity of the monomer (Figure 22.5). One rationale for the high T_g of Teflon® AF (Fig. 22.5, X = F; $R_1 = R_2 = CF_3$) is that steric interactions involving the two trifluoromethyl

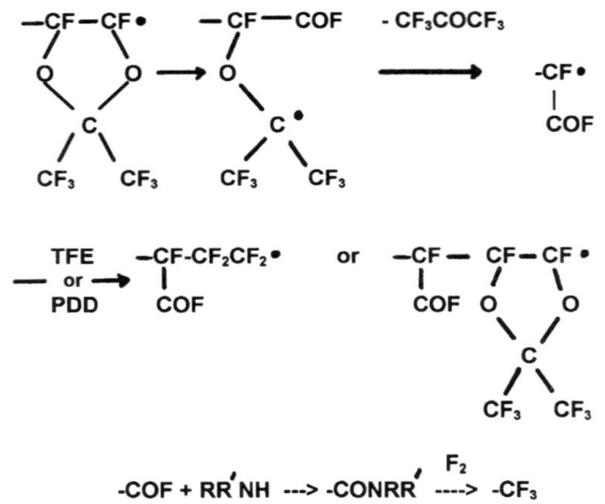

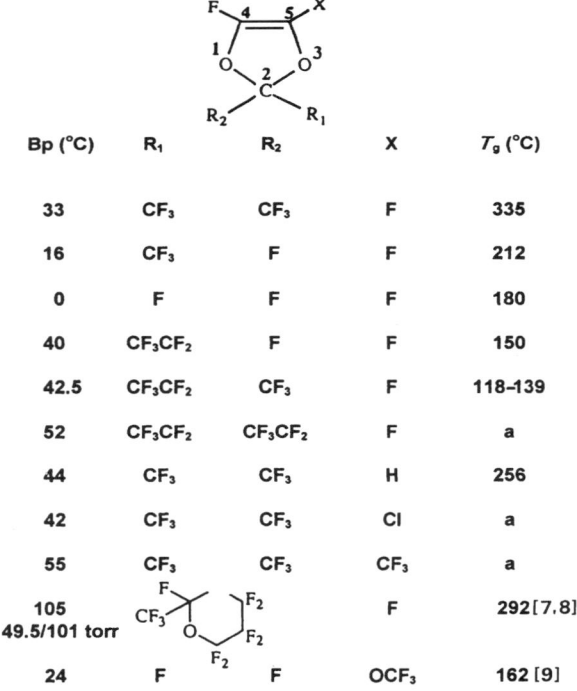

Figure 22.4. Ring opening during polymerization and removal of carboxylate side chains

Bp (°C)	R₁	R₂	X	Tg (°C)
33	CF₃	CF₃	F	335
16	CF₃	F	F	212
0	F	F	F	180
40	CF₃CF₂	F	F	150
42.5	CF₃CF₂	CF₃	F	118–139
52	CF₃CF₂	CF₃CF₂	F	a
44	CF₃	CF₃	H	256
42	CF₃	CF₃	Cl	a
55	CF₃	CF₃	CF₃	a
105 49.5/101 torr			F	292[7,8]
24	F	F	OCF₃	162 [9]

Figure 22.5. Glass transition temperature of dioxole homopolymers (a = does not form homopolymer) [7,8,9]

groups in these dioxole polymers lead to a highly congested chain structure with limited mobility. These interactions and hence the T_g of the these polymers is highly sensitive to differences in substituents at the 2-position of the dioxole ring of the monomer. Note especially the difference in T_g of the homopolymer of PDD, 335°, and that of its isomer, 2-pentafluoroethyl-2,4,5-trifluoro-1,3-dioxole, 150° [10].

Teflon® AF copolymers share many characteristics of other Teflon® types, but also have significant differences. Figure 22.6 lists some of these similarities and differences. All of the polymers in the Teflon® family have high-temperature stability due to their perfluorinated structure. This also accounts for their excellent chemical resistance, low surface energy, and low water absorption. The limiting oxygen index, LOI, of all of the Teflon® polymers is greater than 95, meaning that they require an atmosphere of greater than 95% oxygen to sustain combustion. Teflon® PTFE, FEP, and PFA are all semicrystalline. Since Teflon® AF polymers are completely amorphous, they have different properties. The semicrystalline Teflons® have high melting points and are only soluble in high boiling perfluorocarbon solvents at temperatures near their melting points. Teflon® AF polymers are soluble in several perfluorinated solvents at room temperature. The crystallites in semicrystalline Teflon® scatter light and consequently have low optical transmission. Teflon® AF is free of crystallites and absorbing functional groups and thus has high optical transmission across a broad wavelength region from the near infrared to the ultraviolet. Teflon® AF polymers also have unusually low refractive indices and dielectric constants. The stiffness imparted by the in-chain ring structure of the dioxole unit also causes Teflon® AF to be stiffer and exhibit a higher tensile modulus than other Teflons®. Finally the Teflon® AF family of polymers have extraordinarily high gas permeability compared to other polymers, both fluoropolymers and non-fluoropolymers.

Figure 22.7 shows the TFE sequence length distribution in PDD–TFE copolymers as a function of composition. These sequence lengths were calculated from

Similarities	Differences
High-temperature stability	Non-crystalline, amorphous
Excellent chemical resistance	Soluble at ambient temperature in fluorinated solvents
Low surface energy	Transparent
Low water absorption	Lower refractive index
Limiting oxygen index (LOI) > 95	Stiffer
	High gas permeability

Figure 22.6. Teflon® AF. Comparison to other Teflon® polymers

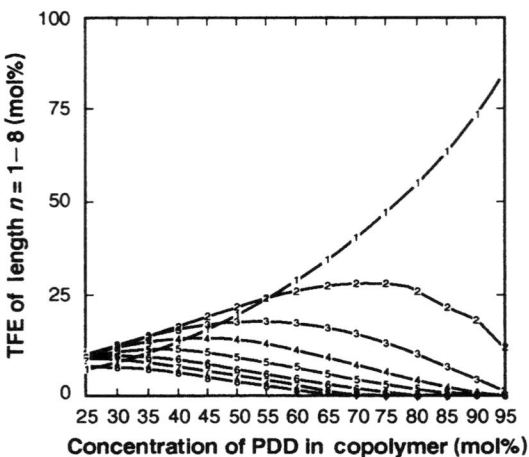

Figure 22.7. $-CF_2CF_2-$ (TFE) sequence length distribution for PDD–TFE copolymers

the experimentally measured reactivity ratios [11]. The numbers in Figure 22.7 represent the mol% of TFE present in run lengths of 1, 2, 3, etc. For example Teflon® AF-1600 has 65 mol% PDD and about 30% of the TFE present has a run length of one, i.e. as single $-CF_2CF_2-$ units flanked on both sides by dioxole units. It can also be seen from Figure 22.7 that about 25% of the TFE in AF 1600 has run length 2, 15% run length 3, 10% run length 4, 5% run length 5, and the remaining 15% has run length greater than 5. For Teflon® AF-2400, 75% of the TFE has run length 1, 20% run length 2, nearly 5% run length 3, and virtually no TFE run lengths greater than 3. Since the barrier to rotation of the carbon–carbon bond in a TFE residue is much less than that of the carbon–carbon bond in the dioxole ring, increasing TFE content should lead to the observed decrease in glass transition temperature and to increased solubility. The presence of a significant number of TFE sequences greater than five or six is necessary for TFE crystallinity to be observed. When PDD content is less than 55 mol% most of the TFE is present in runs of two or longer and when PDD content is less than 20 mol% TFE crystallinity is observed.

Amorphous Teflon® AF is a unique family of polymers. Its structure is characterized by microvoids which have been examined using positron annihilation lifetime spectroscopy [12]. These voids result in lower than expected polymer density, low dielectric constant, low refractive index, high gas permeability and low thermal conductivity. Most likely the origin of these microvoids is loose chain packing caused by the high energy for rotation and reorientation of the dioxole ring containing polymer chain. The densities of some fluoropolymer resins are shown below. The densities of Teflon® AF vary with composition or T_g and are lower than the other fluoropolymers. The densities of these other semicrystalline polymers will vary due to differences in their degree of crystallinity.

Densities of fluoropolymers

Polymer type	Density (g/cc)
AF-1600	1.8
AF-2400	1.7
PTFE	2.15–2.20
FEP	2.12–2.17
PFA	2.12–2.17

The dielectric constant for Teflon® AF is unaffected by humidity, and is believed to be the lowest dielectric constant known for a solid organic polymer and is appreciably lower than that of Teflon® PTFE. Given its low dielectric constant even very thin coatings have good insulating characteristics. Figure 22.8 shows the dielectric constant, ε, of Teflon® AF 1600 and 2400 as a function of frequency at 22 °C. The dielectric constant data were measured at the Massachusetts Institute of Technology Laboratory for Electromagnetic and Electronic Systems by W. B. Westphal using a method similar to ASTM D-150. The dielectric constant of AF 1600 has a nearly constant value of 1.93 over a frequency range of 1 MHz to 10 GHz. Similarly, AF 2400 has a dielectric constant of 1.90 over the same frequency range. For comparison, Teflon® PTFE has a dielectric constant of 2.0 and Teflon® FEP and PFA both have dielectric constants of 2.1. Figure 22.9 shows the dependence of Teflon® AF dielectric

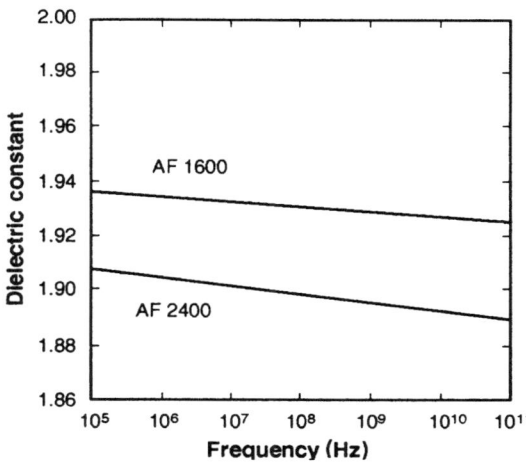

Figure 22.8. Dielectric constant of Teflon® AF at 22 °C

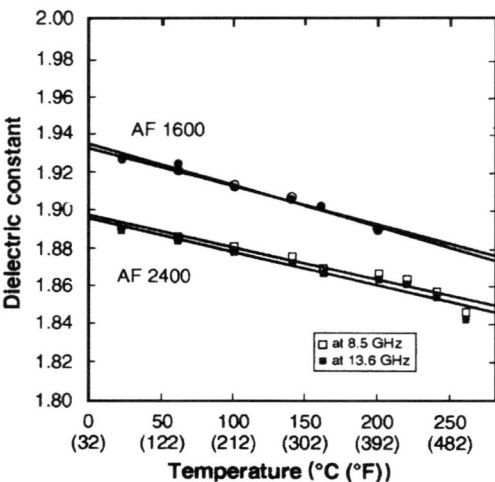

Figure 22.9. Dielectric constant of Teflon® AF vs temperature

constant on temperature at 8.5 and 13.6 GHz. The change is only approximately 0.05 units from room temperature to the glass transition temperature for both Teflon® AF 1600 and AF 2400. At temperatures above the T_g the dielectric constants decrease more rapidly. The change in ε with temperature for Teflon® AF is only about one-half the change seen for crystalline Teflon®. This is probably due to changes in the degree of crystallinity with temperature for the latter polymers. Figure 22.10 shows the dissipation factor of Teflon® AF as a function

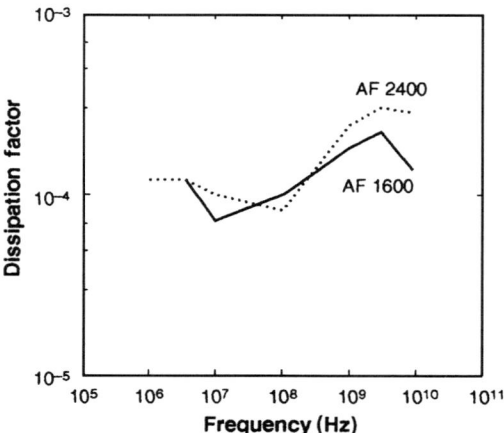

Figure 22.10. Dissipation factor of Teflon® AF at 22 °C

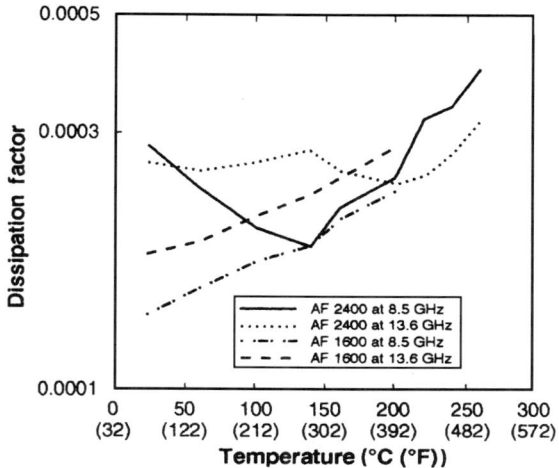

Figure 22.11. Dissipation factor of Teflon® AF vs temperature

of frequency at room temperature. Values range from 8×10^{-5} to 3×10^{-4} for AF 2400 and 8×10^{-5} to 2×10^{-4} for AF 1600. These values are essentially equivalent to the reported dissipation factor of 2×10^{-4} for PTFE [13]. Figure 22.11 shows the temperature dependence of the dissipation factor of Teflon® AF at two fixed frequencies. There is little change, 2×10^{-4} units, from room temperature to above the polymer T_g at low frequency. The T_g of 160 and 240 °C respectively for Teflon® AF 1600 and 2400, refer to values measured by differential scanning calorimetry, which has an effective measurement frequency of 10^{-2} Hz. At the GHz frequency range used for these measurements, T_g will be appreciably higher. The dissipation factor is maximum at T_g, so the increasing dissipation factor in Figure 22.11 indicates the approach to T_g at the measurement frequencies.

Starkweather *et al.* [14] investigated the low-temperature dielectric properties of Teflon® AF and other Teflon® polymers. They found that the γ relaxation found at -186 °C for PTFE is only one-third the intensity for AF 1600 and AF 2400. This relaxation is attributed to the cooperative motion of the four carbon atoms of a TFE dimer unit. Since the total TFE content and concentration of TFE units of length greater than one is small for both AF compositions, this explains the reduced intensity of the γ relaxation. Teflon® AF exhibits a very weak maximum in the low temperature dissipation factor at -200 °C which is attributed to rotation of the perfluoromethyl group on the dioxole ring.

The optical properties of Teflon® AF are outstanding and quite unique as one would expect from a perfluorinated amorphous material. Figure 22.12 shows the transmission spectrum of Teflon® AF from 200 to 2000 nm. The polymer has high transmission from the deep ultraviolet through the near infrared spectral regions. Note that the spectrum shown is for a 0.22 mm (0.009 in.) thick sample.

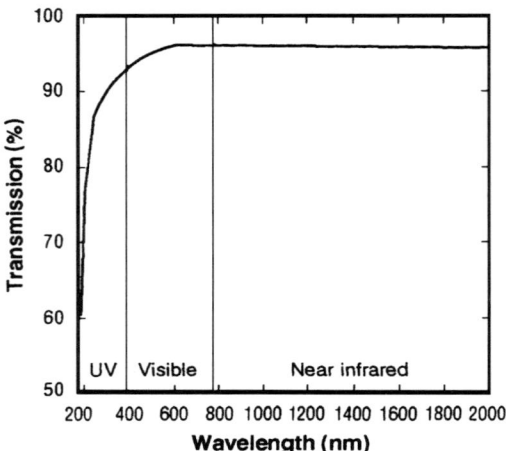

Figure 22.12.   Transmission of Teflon® AF (0.22 mm thick sample)

Rothschild and Sedlacek [15] found that Teflon® AF 1600 films of thickness between 0.2 and 2 μm showed no absorbance at 200 nm. They also showed that a Teflon® AF pellicle, which had been optimized for use at 248 nm, could withstand irradiation at 193 nm for the equivalent of 10 years of full-time exposure in a semiconductor manufacturing environment.

The refractive index of Teflon® AF shown in Figure 22.13 is the lowest known for any solid organic polymer. The refractive index of AF 1600 at 20 °C at the sodium D line is 1.31 and that of AF 2400 is 1.29. The indices were measured by an Abbe refractometer using α-bromonaphthalene as contact liquid. The refractive index of AF 2400 is near the theoretical lower limit predicted by Groh and Zimmermann [16] based on group contribution calculations. The Abbe number is a measure of dispersion or the change in refractive index with wavelength. It is defined at the bottom of Figure 22.13, where n_D is the refractive index at the sodium D line (589 nm), n_F is the refractive index at the hydrogen F line (486 nm), and n_C is at the hydrogen C line (656 nm). These Abbe numbers for Teflon® AF shown in Figure 22.13 are extremely high

AF	n_D	Abbe no.*	dn_D/dT (ppm/°C)	
			Below T_g	Above T_g
1600	1.31	92	−77	−329
2400	1.29	113	−78	−378

Figure 22.13.   Refractive index of Teflon® AF. Abbe number $= (n_D = 1)/(n_F\text{-}n_C)$

Stress optical coefficient (SOC)

Polymer	SOC (Brewster)
Teflon® AF-1600	4.5
Polystyrene	10
Poly(methyl methacrylate)	4.0

Figure 22.14. Stress optical coefficient (SOC). Brewster = 10^{-13} cm²/dyn

by comparison with other materials [17]. The Abbe number for acrylics is 57, for polycarbonate 34, for styrene 31, and for BK-7 glass 64. The variation in refractive index with temperature below Tg is -77 ppm/°C for AF 1600 and -78 ppm/°C for AF 2400. However, above their glass transition temperatures the variations in refractive index with temperature for AF 1600 and 2400 are -329 and -378 ppm/°C respectively. Other amorphous polymers have values between 300 and 500 ppm/°C above their T_g [18]. These reduced variations in refractive index with temperature are a consequence of the relatively low coefficient of thermal expansion for Teflon® AF. The values of the stress optical coefficient (SOC), of Teflon® AF 1600 [19] and several other optical polymers are shown in Figure 22.14 [20]. The SOC is a measure of the change in refractive index in different directions as a material is stressed. Low values of SOC are desirable for optical components subjected to stress during use.

As noted earlier, the density of Teflon® AF is extremely low in comparison to other perfluorinated polymers. As shown in Figures 22.15 and 22.16 the coefficients of thermal expansion are linear both below and above the glass

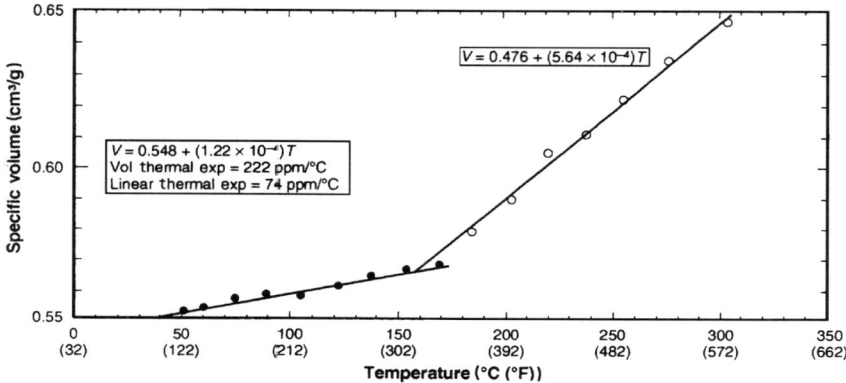

Figure 22.15. Thermal expansion of Teflon® AF-1600 (Lot P29-3032F). Extrapolated to zero pressure from high-pressure dilatometry

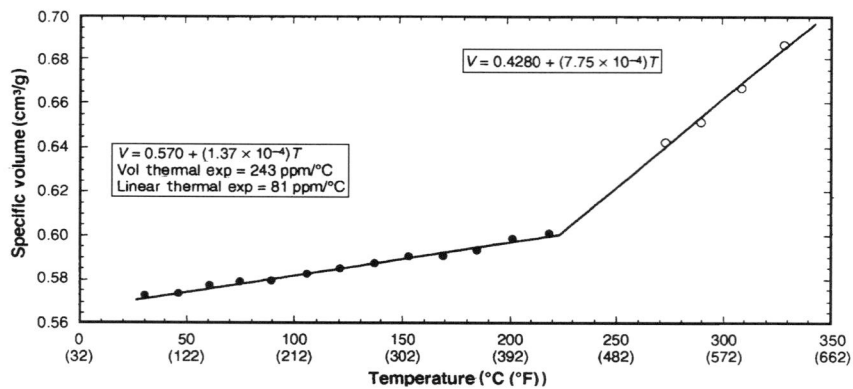

Figure 22.16. Thermal expansion of Teflon® AF-2400 (Lot P29-4012F). Extrapolated to zero pressure from high-pressure dilatometry

transition temperature but greater above T_g [21]. Thus the volume coefficient of thermal expansion for Teflon® AF 1600 below the T_g, 160°, is 222 ppm/°C and 960 ppm/°C above the T_g. The values have been extrapolated to zero pressure. For Teflon® AF 2400 the values are 243 ppm/°C below the T_g and 1280 ppm/°C above the T_g. These thermal expansion properties are in contrast to those of the semicrystalline fluorinated polymers such as PTFE, FEP and PFA whose coefficients increase, sometimes quite sharply, with increasing temperature. The rapid rate of change near the melting temperature is due to melting of smaller crystallites which are absent in Teflon® AF.

Figure 22.17 shows the extraordinarily high gas permeabilities of Teflon® AF. They are greater than two orders of magnitude higher than those of PTFE and are exceeded only by those of the most permeable polymer known, poly(1-trimethylsilyl-1-propyne). The gas permeability of Teflon® AF increases with

Gas	AF-1600	AF-2400 Melt processed	AF-2400 Solution cast	PTFE	Selectivity gas/N_2 AF-2400	
					Melt processed	solution cast
CO_2		280 000	390 000	1200	5.7	5.0
O_2	34 000	99 000	160 000	420	2.0	2.0
He		270 000	360 000		5.5	4.6
H_2		220 000	340 000	980	4.5	4.4
N_2		49 000	78 000	140		
Ethylene		35 000				
Methane		34 000	60 000		0.7	0.8
Ethane		18 000	37 000		0.4	0.5
H_2O (g)	117 000	410 000				

Figure 22.17. Gas permeability of Teflon® AF [22,23] (cB) (cB = centi-Barrer = $10^{-8} \times (cm^3 \cdot cm)/cm$ Hg s cm^2)

dioxole content and reaches a maximum with PDD homopolymer. Although PDD homopolymer is extremely difficult to work with due to very limited solubility and melt processability, several thin films have been fabricated and showed high oxygen and nitrogen permeabilities. Permeabilities of small gas molecules in Teflon® AF-2400 are higher than those of large, condensable gases, indicating that the high gas permeabilities result from very high diffusion coefficients. The selectivity shown by Teflon® AF-2400 is a function of the differences in diffusivity of the gases. It has been suggested that the low nitrogen/organic vapor selectivity and the strong dependence of selectivity on vapor activity will severely limit the use of Teflon® AF-2400 as a membrane material for the separation of air from organic vapors [22].

Teflon® AF has very low thermal conductivity [24]. The conductivities of Teflon® AF-1600 and AF-2400 and other materials at room temperature are shown in Figure 22.18. The thermal conductivity of Teflon® AF-1600 has also been shown to have weak absolute pressure and temperature dependencies in comparison with PTFE [25]. Figure 22.19 shows the thermal conductivity

Material	Conductivity (W/mK)
AF-2400	0.05
AF-1600	0.07
Polystyrene	0.12
Polyvinyl chloride	0.16
Teflon® PFA	0.19
Teflon® FEP	0.20
Teflon® PTFE	0.23
Corning glass 7740	1.14

Figure 22.18. Thermal conductivity of selected materials at room temperature

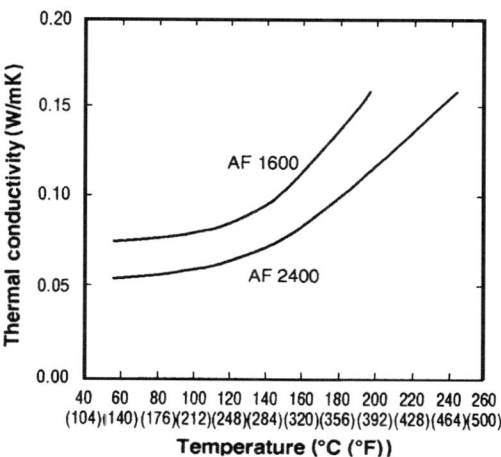

Figure 22.19. Thermal conductivity of Teflon® AF

Solvent	Boiling point (°C)	Solubility Parameter $(J/m^3)^{1/2}$	$(cal/cm^3)^{1/2}$
'Fluorinert' FC-72	60	0.0123	6.0
Perfluoromethycyclohexane	76	0.0129	6.3
Perfluorobenzene	82	0.0123	6.0
Perfluorodimethylcyclohexane	102	0.0139	6.8
Perfluorooctane	103	0.0115	5.6
'Fluorinert' FC-75	103	0.0129	6.3
Perfluorodecalin	142	0.0135	6.6
'Fluorinert' FC-40	155	0.0135	6.5
Perfluoro-1-methyldecalin	160	0.0143	7.0
Perfluorodimethyldecalin	180	0.0147	7.2

Figure 22.20. Solvents for Teflon® AF

of Teflon® AF-1600 and AF-2400 below their glass transition temperatures. The conductivity results for Teflon® AF are consistent with a low-density fluoropolymer containing microvoids.

One of the unique properties of Teflon® AF is its solubility in fluorinated solvents. Figure 22.20 lists some of these solvents and their boiling points. All of the known solvents for Teflon® AF are essentially perfluorinated. Like the semicrystalline fluoropolymers, Teflon® AF is insoluble and not swollen in hydrocarbon solvents. 'Fluorinert' FC-75 is a product of the 3M company and is nominally perfluoro(n-butyl-tetrahydrofuran). 'Fluorinert' FC-40 is a perfluoro(alkyl amine). Teflon® AF is also soluble in Ausimont's 'Galden' and Hoechst's perfluorinated ethers. The higher boiling solvents tend to be poorer solvents at room temperature, but solubility at higher temperatures is improved. Teflon® AF 1600 is more soluble than AF 2400. In general, solubility decreases as dioxole content of the polymer increases. At room temperature Teflon® AF 1600 has a solubility limit of 12–15% by weight. This is the concentration at which the solution will not flow when its container is inverted. In contrast, the solubility limit of AF 2400 is only 1.5–2 wt% at room temperature. The homopolymer of PDD is soluble at only a few tenths of 1%. Solubility can be changed by modifications to the polymer structure. Increased solubility may be obtained by substituting other fluorinated monomers such as chlorotrifluoroethylene for tetrafluoroethylene or by lowering the molecular weight of the polymer.

The chemical resistance of Teflon® AF is essentially the same as that of the other Teflon® polymers such as PTFE, FEP or PFA except for its solubility in fluorinated solvents as noted as discussed earlier. Teflon® AF is resistant to attack by most chemicals including hot acids, hot caustic, chlorine, organic esters, ketones and alcohols. Materials known to attack PTFE such as molten alkali metals and sodium naphthalide will also attack Teflon® AF. Figure 22.21 shows the chemical resistance of Teflon® AF 1600 in a variety of organic and inorganic liquids. Measurements were made by drying polymer samples to constant weight, then immersing them in the liquids for seven days. After immersion, the samples

Reagent	Temperature (°C)	% Wt change	Appearance change
Acetone	23	0	None
CCl₄	23	0	None
CFC-113	23	+6.2	Swollen
12N HCL	60	0	None
Hexanes	23	0	None
50% HF	60	0	None
98% H₂SO₄	60	0	None
MEK	23	0	None
10% NaOH	60	0	None
44% NaOH	60	0	None
CCl₂=CCl₂	23	+0.1	None
Toluene	23	0	None

Figure 22.21. Chemical resistance of Teflon® AF (seven days immersion)

Temperature (°C)	Weight loss (%)	Hours
260	None	4
360	0.30	1
380	0.50	1
400	1.90	1
420	8.80	1

Figure 22.22. Thermal stability of Teflon® AF

were rinsed with water and dried to constant weight. CFC-113, 1,1,2-trichloro-1,2,2-trifluoroethane, swells Teflon® AF appreciably and is difficult to remove completely. The other liquids have essentially no effect on Teflon® AF.

The thermal stability of Teflon® AF is high and approaches that of other perfluorinated plastics. Figure 22.22 shows the isothermal weight loss data in air for Teflon® AF as a function of temperature. These data were obtained by holding the sample at the indicated temperature for one hour in a thermogravimetric analyzer (TGA) and measuring the rate of weight loss. The rates plotted are the maximum rates observed, which were always seen at the beginning of the experiment. Initial weight loss was observed between 350 and 360 °C (Figure 22.23). At lower temperatures AF 1600 showed lower rates of weight loss than AF 2400, but at 420 °C the rates were equal. The maximum weight of weight loss for Teflon® AF is 0.15%/min at 420 °C. The rate of weight loss for Teflon® FEP at 420 °C is shown for comparison and is slightly less than that of Teflon® AF.

The commercially available grades of Teflon® AF are copolymers of PDD and tetrafluoroethylene. As previously noted many other comonomers have been copolymerized with PDD. The resulting copolymers have differing physical properties. Figure 22.24 shows the effect of changing the structure of the comonomer polymerized with PDD on refractive index. CTFE is chlorotrifluoroethylene and PMVE is perfluoromethylvinyl ether. The physical and chemical properties of

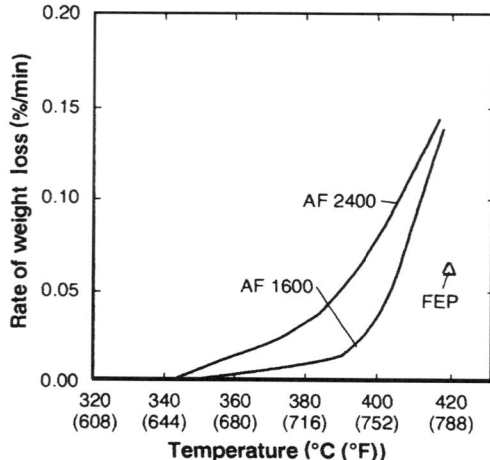

Figure 22.23. Thermogravimetric curves for Teflon® AF (FEP is shown for reference)

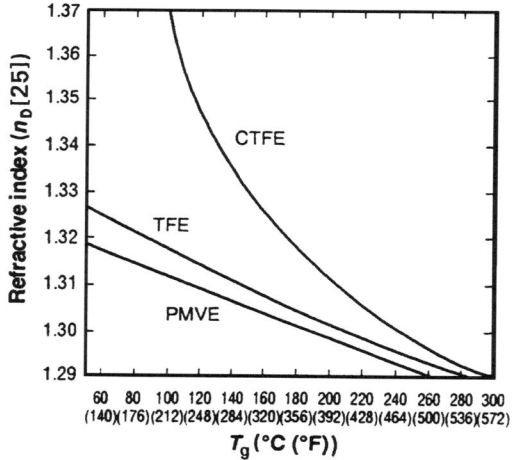

Figure 22.24. Refractive index of copolymers of PDD as a function of composition

these copolymers are generally comparable to the TFE copolymers. The refractive index has been plotted against the glass transition temperature, T_g, rather than polymer composition; T_g is related to composition in the same way as are the PDD–TFE (Teflon® AF) copolymers. A plot of polymer composition against refractive index for Teflon® AF is shown in Figure 22.25. The refractive index of PDD–CTFE copolymers changes much more rapidly with T_g or composition

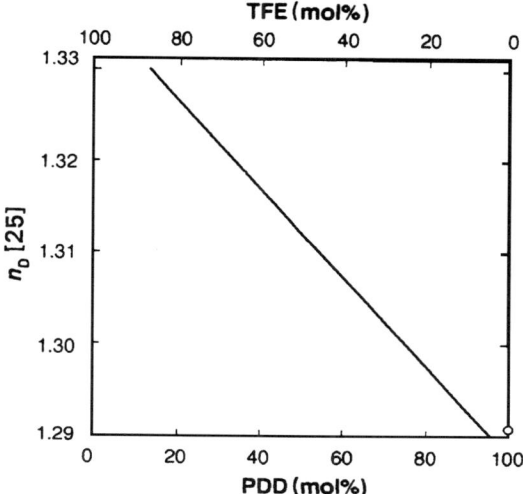

Figure 22.25.   Refractive index of Teflon® AF polymers

than does the refractive index of the PDD–TFE copolymers. The refractive index of polychlorotrifluoroethylene is 1.42–1.43 [18]. All three curves converge at a refractive index of 1.29 at a T_g of 335 °C, which is the T_g of PDD homopolymer. Since PMVE does not readily homopolymerize, it is not possible to obtain the refractive index of polyPMVE. Copolymers of PDD and PMVE have a lower refractive index at a given T_g than either PDD–TFE or PDD–CTFE copolymers. The change in refractive index with T_g is also less for PDD–PMVE copolymers than for the other copolymers. Differences in refractive index at a T_g greater than about 220 °C for these copolymers are small since the amount of comonomer is quite low.

Figure 22.26 shows the effect of the change in molecular weight on the solution viscosity of a copolymer of PDD and TFE with a T_g of 160 °C. Teflon® AF 1601 has a lower molecular weight than Teflon® AF 1600, but with identical composition. The differences in solution viscosity at room temperature in 'Fluorinert' FC-75 at a given concentration are substantial. The Brookfield viscosity of a solution of AF 1601 at 18 wt% concentration is equal to that of a solution of AF 1600 at 6.4 wt%. The lowered molecular weight results in slightly poorer physical properties, but higher solubility as shown in Figure 22.27. Further reductions in molecular weight at this composition lead to further reductions in solution viscosity, but the polymers are likely to be too brittle.

Teflon® AF may be processed by a wide variety of methods. Processing of the polymers have been performed from solution using spin coating, casting, dip coating, spraying and painting. Melt processing by extrustion, compression

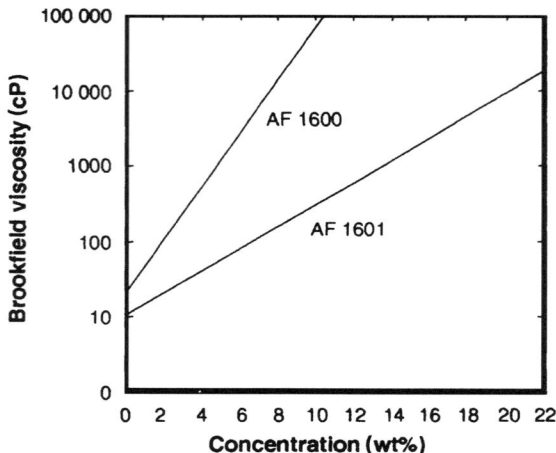

Figure 22.26. Brookfield viscosity of Teflon® AF polymers

Property	ASTM method	Units	AF-1600	AF-1601
Glass transition temp.	D3418	°C	160	160
Tensile strength	D638	MPa	27	27
Elongation at break	D638	%	>20	17
Tensile modulus	D638	GPa	1.6	1.55
Vol. coff. thermal exp.	E831	ppm/°C	260	260
Specific gravity	D792		1.78	1.78
Dielectric constant	D150		1.93	1.93
Dissipation factor	D150		0.00012	0.00012
Optical transmission	D1003	%	>95	>95
Refractive index	D542		1.31	1.31
Contact angle with H$_2$O		Degrees	104	104
Critical surface energy		dyne/cm	15.7	15.7
Water absorbtion	D570	%	<0.01	<0.01
Melt viscosity	D3835	250°/s	2657	5000
Solubility in FC-75 (ambient)		%	12-15	25-30
Brookfield viscosity in FC-75		Pa s	5500	5500
			(6.4%)	(18%)

Figure 22.27. Physical properties of Teflon® AF polymers

molding and injection molding have also been successful. In addition thin films of Teflon® AF have been prepared by either laser ablation or vacuum pyrolysis.

Solution processing methods are possible due to the room temperature solubility of Teflon® AF in perfluorocarbon solvents. Spin coating is an effective technique for making thin to ultra-thin uniform thickness coatings on flat substrates. Figure 22.28 gives typical data for spin speed and film thickness for Teflon® AF 1600 solutions in 'Fluorinert' FC-75 spun onto glass. A 6 wt% solution gives coating thickness between 1 and 3 μm while a 1% solution gives

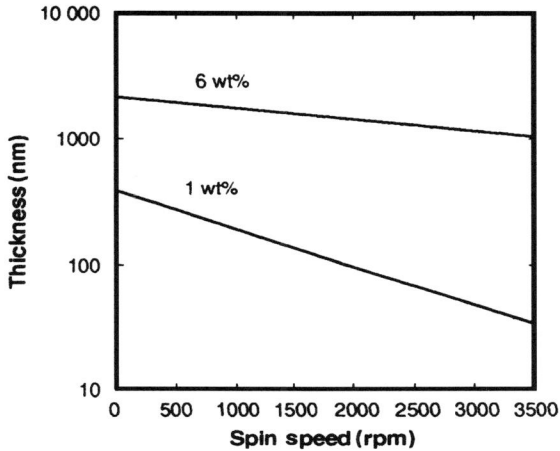

Figure 22.28. Spin coating conditions for solutions of Teflon® AF on glass

coating thicknesses between 50 and 200 nm. Microscopic examination of these films show them to have uniform thickness and no pinholes. The film thickness on other substrates will vary with the nature of the substrate, the spin speed and the concentration of the solution.

While spin coating is useful for obtaining thin uniform coatings on flat substrates (Figure 22.29), nonplanar surfaces are better coated using dip coating or spraying procedures. Dip coating has been used to produce Teflon® AF 2400 antireflective coatings on lenses [26]. Coating thickness was controlled by solution concentration and the rate of withdrawal of the lens from the polymer solution. Spraying is useful for coating planar and nonplanar substrates where thickness uniformity is not critical. Thick coatings on irregular surfaces may be applied by painting. Teflon® AF is amenable to all methods of application requiring polymer solutions.

In addition to solution-based application methods, Teflon® AF can be melt processed using all of the methods available for other melt fabricable Teflon® polymers. Compression molding is generally done at a temperature 100°C higher than T_g. Extrusion and injection molding conditions depend on the part being extruded or molded. Because Teflon® AF is optically clear, special care must be taken to insure a clean, contamination-free environment during the melt fabrication process (Figure 22.30). The low thermal conductivity of Teflon® AF may also require longer heat up times than other polymers.

Laser ablation is a technique by which a material is vaporized in a high-vacuum environment by irradiation with a high-powered laser and the vapor directed to a substrate for coating. Advantages are the ability to deposit extremely thin films, of the order of tens of nanometers without the use of solvent. This technique has

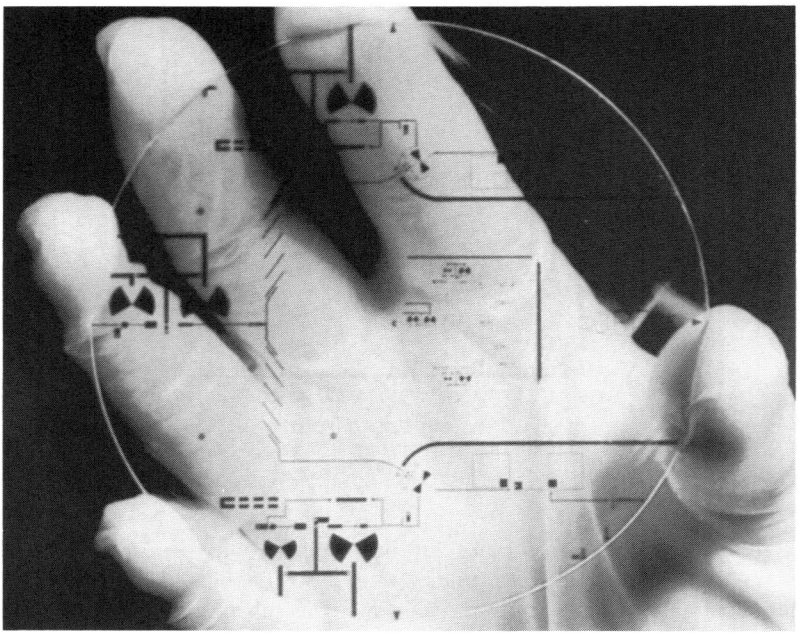

Figure 22.29. Spin coating of Teflon® AF enables the production of ultra-thin uniform thickness coatings on flat substrates

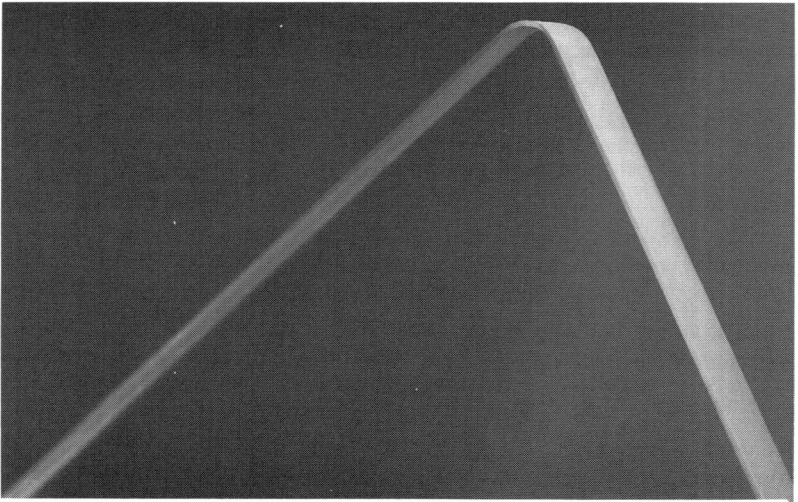

Figure 22.30. Compression Molded Sheet of 1/8″ (3.2 mm) thick TEFLON® AF. Note: exceptional clarity is a hallmark of amorphous fluoropolymers

Figure 22.31. A spin coater with a solution of Teflon® AF on a silicon wafer

been successfully applied to Teflon® AF using a Nd–YAG laser at 10^{-7} Torr [27]. The copolymers retained their composition during this process. Vacuum pyrolysis is a similar process in which the material is vaporized by thermal rather than laser irradiation. Teflon® AF has also been applied to various substrates using this technique [28,29]. The resulting coatings also retain their composition during this process. Teflon® AF coatings applied by vacuum pyrolysis are claimed to have superior adhesion to the substrate compared to Teflon® AF coatings applied from solution [28].

Teflon® AF is a unique family of amorphous copolymers possessing an unusual combination of properties. The use of these polymers exploit either one or more of these properties since at present the cost of the polymers is quite high. Teflon® AF has many uses including preparation of antireflective coatings [30], low dielectric coatings (Figure 22.31), deep ultraviolet pellicles used in electronic chip manufacturing processes [31], cladding in plastic optical fibers [32] and as a low dielectric constant insulator for high-performance interconnects [33]. These and other uses utilize the combination of chemical and thermal stability of the polymers along with the ease of fabrication especially from solution, their excellent optical and electrical properties, high permeability to gases and mechanical properties.

REFERENCES

1. R. J. Plunkett, US Patent 2 230 654 (1941).
2. P. R. Resnick, US Patent 3 865 845 (1975); US Patent 3 978 030 (1976).
3. M. -H. Hung, P. R. Resnick and E. N. Squire, *Proceedings of the First Pacific Polymer Conference*; Maui, Hawaii, Dec. 1989, p. 331.
4. P. R. Resnick, *Polym. Prepr. (Am. Chem. Soc., Div. Polym. Chem)*, 1990, **31**, 312.
5. M. -H. Hung, US Patent 4 908 461 (1990).
6. P. G. Bekiarian, M. D. Buckmaster and R. A. Morgan, US Patent 4 946 902 (1990).
7. T. Sugiyama and H. Murofushi, Jpn. Kokai JP 05 9 224; (CA 118:255546).
8. H. Murofushi and I. Kaneko, Jpn. Kokai JP 04 346 989 (CA 118:255521).
9. W. Navarrini, V. Tortelli, P. Colaianna and J. A. Abusleme, US Patent 5 498 682 (1996).
10. M. -H. Hung, *Macromolecules*, 1993, **26**, 5829.
11. J. L. Koenig, *Chemical Microstructure of Polymer Chains*, Wiley-Interscience, New York, NY, 1980, Chapter 2.
12. R. B. Gregory, *J. Appl. Phys.*, 1991, **70**, 4665; W. J. Davies and R. A. Pethrick, *Eur. Polym. J.*, 1994, **30**, 1289.
13. C. A. Sperati in *Polymer Handbook*, 2nd edn, J. Brandrup and E. H. Immergut, eds, Wiley-Interscience, New York, NY, 1975, p. V-31.
14. H. Starkweather, P. Avakian, R. Matheson, J. Fontanella and M. Wintersgill, *Macromolecules*, 1991, **24**, 3853.
15. M. Rothschild and J. H. C. Sedlacek, *SPIE*, 1992, **1674**, 618.
16. W. Groh and A. Zimmermann, *Macromolecules*, 1991, **24**, 6660.
17. J. H. Lowry, J. S. Mendelowitz and N. S. Subramanian, *SPIE*, 1990, **1330**, 142.
18. L. Bohn in *Polymer Handbook*, 2nd edn, J. Brandrup and E. H. Immergut, eds, Wiley-Interscience, New York, NY, 1975, p. III-241.
19. Measured by C. M. Paulson, Jr, DuPont Central Science and Engineering.
20. D. W. Van Krevelen, *Properties of Polymers*, 2nd edn, Elsevier, New York, NY, 1976, p. 222.
21. Measured by S. Michielsen, DuPont Central Science and Engineering.
22. I. Pinnau and L. G. Toy, *J. Membr. Sci.*, 1996, **109**, 125.
23. S. M. Nemser and I. C. Roman, US Patent 5 051 114 (1991).
24. Measured by M. Y. Keating, DuPont Central Science and Engineering.
25. S. P. Andersson, O. Andersson and G. Backstrom, in press.
26. I. M. Thomas and J. H. Campbell, *SPIE*, 1990, **1441**, 294.
27. G. B. Blanchet, *Appl. Phys. Lett.*, 1993, **62**, 479.; G. B. Blanchet-Fincher, US Patent 5 192 580 (1993); G. B. Blanchet-Fincher, US Patent 5 288 528 (1994).
28. J. Grieser, R. Swisher, J. Phipps, D. Pelleymounter and E. Hildreth, *Proceedings of the SPIE*, Optical Surfaces Resistant to Severe Environments, 1990, **1330**, 111.

29. T. C. Nason, J. A. Moore and T. M. Lu, *Appl. Phys. Lett.*, (1992), **60**, 1866.
30. H. G. Floch and P. F. Belleville, *SPIE*, 1992, **1758**, 135; H. G. Floch and P. F. Belleville, *J. Sol-Gel Sci. Technol.* 1994, **1**, 293.
31. D. E. Keys, US Patent 5 061 024 (1991).
32. E. N. Squire, US Patent 4 530 569 (1985).
33. C. -C. Cho, R. M. Wallace and L. A. Files-Sesler, *J. of Electronic Mat.*, 1994, **23**, 827.

23

Liquid Fluoroelastomers

E. WILLIAM ROSS, JR and GARY S. HOOVER
Pelmor Laboratories Inc., Newtown, PA, USA

1 INTRODUCTION

The unusual properties of fluoroelastomers were recognized from the time
E. I. Du Pont introduced their VITON in 1957. Foremost among those properties
is the ability to withstand exposure to elevated temperatures and harsh chemicals
that would seriously degrade other elastomers. The flexibility and resilient nature
of fluoroelastomers also set them apart from nonelastomeric polymers.

For most of their existence, however, fluoroelastomers have been
perceived — and seen — as compounds used almost exclusively in a solid form.
In the early years — as well as today — parts made from VITON could be found
in a number of commercial applications, notably in the automotive industry.

Although the advantages were apparent, fluoroelastomers were slow to be
adopted by industry. One drawback was cost: compared with alternative materials,
fluoroelastomers were expensive. Moreover, industry was limited in its ability
to use these products. Parts needed to be molded or extruded from VITON
compounds. Molds and dies have practical limitations in size and shape that
greatly reduce the ability of conventional fluoroelastomers to meet all possible
end-use requirements.

Today's liquid fluoroelastomer technology was born from this early need for
a product which combines all of the physical properties of solid fluoroelastomer
in a form that is easy to apply and versatile in use. Liquid fluoroelastomers are
used today as adhesives, coatings, caulks and combinations of these.

This chapter will describe variations in liquid fluoroelastomers that allow the
products to meet specific needs. We will compare fluoroelastomeric coatings and
sealants to other materials, review methods of applying liquid fluoroelastomers,
and discuss some of the more prominent industrial uses.

Modern Fluoropolymers. Edited by John Scheirs
© 1997 John Wiley & Sons Ltd

2 PRE-LIQUID PHASE

There are two primary types of liquid fluoroelastomers: solvent borne and water-borne. At this time, solvent-borne products hold a significantly larger market share than water-borne products. The two technologies share broad similarities, but differ in many details as well. For simplicity, this chapter will focus on solvent-borne systems.

Polymers should be chosen based on the requirements of the end product. Copolymers of vinylidene fluoride and hexafluoropropylene provide excellent general-purpose products. These products are flexible, resilient, chemically resistant and able to withstand exposure to elevated temperatures. Terpolymers of vinylidene fluoride, hexafluoropropylene and tetrafluoroethylene can be used where exposure to particularly harsh environments [1] is required, but some degree of flexibility, resilience and processability might be sacrificed.

Simple fillers such as carbon blacks, titanium dioxide and mineral fillers can be used as needed to meet end-use requirements. More exotic fillers can also provide interesting properties, but each should be carefully evaluated for compatibility with individual systems. Metallic fillers should be avoided. Large amounts of filler will produce harder, less flexible coatings with lower elongation and to some extent higher tensile strengths [2].

Metal oxides are generally required as acid acceptors and activators [3], but have been eliminated in some formulations to meet unique requirements. These metal oxides such as: calcium oxide, magnesium oxide, lead oxide and zinc oxide react to neutralize hydrogen fluoride that is liberated during the curing process [4]. These metal oxides also provide improved acid performance of the cured product. Cure time, curative and desired chemical performance should be considered as guidelines to determine the activity level and amount of metal oxide required.

Other additives can be used as desired. Nonblack liquids are easily pigmented. Waxes and other processing aids may be used to improve surface qualities, but could harm adhesion properties. As with all additives, end requirements and system compatibility must be carefully evaluated.

3 LIQUID PHASE

Low molecular weight ketones and esters (such as methyl ethyl ketone and ethyl acetate) tend to provide the best solvency for these fluoroelastomers. As the molecular weight of solvents rise, solvency decreases and solution viscosity tends to increase.

Higher molecular weight solvents (such as cyclohexanone and methyl isobutyl ketone) tend to have lower vapor pressures and higher boiling points. These features create slow drying coatings with smooth surfaces.

The properties of the fluid can be altered by adjusting the ratio of solid to solvent within the coating or caulk. A product that is less than 15% solid will have a very watery consistency with a viscosity typically less than 50 cps. These low solid coatings are easily sprayed. They also produce thinner, smoother films than higher solid versions. A disadvantage of a lower solids coating is its coverage — 85% of the coating weight that is applied will simply evaporate. Since fluoroelastomers have densities that are usually much higher than the solvent that dissolves them, the non-volatile volume for a 15% (by weight) solid coating will be approximately 7%. In other words, a 100 μm thick wet coating will dry to only 7 μm thick.

Products with solids of approximately 55% by weight and higher fall into the class of caulking compounds. These products will generally have viscosities in excess of 100 000 cps. Within this class, very high solids versions will be extremely stiff, and hold their shape as they dry. These products work well on vertical surfaces where other products might sag and flow. Lower solids versions within this class are self-levelling. They flow into narrow horizontal cracks better and dry with smoother, more even surfaces.

4 CURING

Fluoroelastomer solutions can be cured with amines, phenols, and (in the case of bromine modified polymers) peroxides. Heat-activated curatives provide the advantage of stability at ambient temperatures.

The amount of curative used in a formulation can be varied, based on the desired properties of the finished liquid. Excess accelerator will produce harder, less flexible products with higher tensile strengths and lower elongations. Minimal amounts of accelerator will produce easily deformed products that will be soft and weak and will not have memory after being stretched.

A relatively quick curing product is desirable, but a fast cure occurs at the expense of pot life. For example, curatives at levels designed to provide 80% cure after 48 hours, yield a pot life of 8 hours. After that time the viscosity will increase, and the product will eventually 'gel'. A gelled product might have most of its original solvent present, but the curative crosslinks the polymer creating a firm, springy sample. It is possible to break the gelled sample into soft chunks, but it will not go back into solution.

The applicator must be careful to avoid solvent loss during application of the fluoroelastomer because pot life could be dramatically reduced. The applicator should also take note of atmospheric conditions. Low temperatures will increase the viscosity and change the coating characteristics. High temperatures will decrease viscosity and increase the rate of solvent evaporation. Humidity will change the rates at which water (generated during the curing process) escapes and could change the cured characteristics.

If post cures are possible, they can improve the cure of vulcanizates by eliminating any residual water or other volatiles that might be present. Ideally all solvent should be gone before the product is exposed to any heat. Forced flashing of solvent at elevated temperatures can cause blistering. Even if all of the solvent has been eliminated, care should be taken to gradually increase the temperature no more than 38 °C per hour from a maximum 65 °C start temperature. This stepped increase will allow volatiles to escape gradually and reduce blistering.

5 CURED PRODUCT PROPERTIES

Fluoroelastomers offer the best combination of chemical and temperature resistance when compared to all other elastomers [5].

Cured fluoroelastomer coatings and caulks have properties that vary greatly depending on formulation. For example, tensile strength values can range from 600 PSI (4.13 MPa) to 2000 PSI, (13.79 MPa) and elongations can range from 150 to 600%.

Experimentation has shown very little loss in elastomeric properties after exposure to temperatures as high as 260 °C. Typically, a cured fluoroelastomer coating that has been post-cured will, after exposure to 260 °C for 70 hours, gain approximately 20% tensile strength and lose approximately 20% elongation. These values will vary greatly, depending on the formulation.

Cured fluoroelastomer coatings can withstand constant exposure to a wide variety of chemicals with little or no degradation. This broad chemical resistance is important to an engineer solving a problem involving unknown chemical exposures. For example an automotive engineer may choose a coating specifically to resist gasoline but later find that motor oil is also present.

In an experiment conducted in our laboratories, three elastomers — two types of liquid fluoroelastomer and one polysulfide — were immersed in varying concentrations of hydrochloric acid to compare the degree of volume change after exposure for seven days. The results are summarized in Table 23.1.

All fluoroelastomer samples and the polysulfide samples in 5% HCl showed good surface quality and no deterioration after the immersions. The samples were firm, solid and looked and felt nearly identical to the samples before immersion.

Table 23.1. Volume changes (%) after seven day immersion

	PELSEAL	PELSEAL PG	Polysulfide
5% hydrochloric acid	+0.9	−1.1	+3.6
15% hydrochloric acid	−1.4	−1.1	Unmeasurable semi-solid
35% hydrochloric acid	+9.7	+7.2	Completely dissolved

Note: % HCl (by volume)
PELSEAL is a registered trademark of Pelmor Laboratories, Inc. PELSEAL and PELSEAL PG are fluoroelastomer-based caulks

Table 23.2. Comparison of chemical resistance of selected elastomers

	Fluoroelastomer (VDF–HFP)	Polysulfide	Silicone	EPDM
Ammonium chloride	E	E	F	E
Ammonium nitrate	E	E	F	E
Aromatic hydrocarbons	E	ND	F	NR
ASTM fuel B	E	E	NR	G
ASTM No. 3 oil	E	G	NR	G
Automatic trans. fluid	E	ND	NR	NR
Automotive gasoline	E	ND	NR	NR
Aviation gasoline	E	ND	NR	NR
Beer	E	NR	E	E
Bromine gas	E	G	NR	NR
Butane	E	E	NR	NR
Butyl alcohol	E	G	G	G
Chlorinated solvents	E	NR	NR	NR
Chromic acid (70%)	E	NR	F	NR
Crude oil	E	G	NR	NR
Cyclohexane	E	E	NR	NR
Diesel oil	E	E	NR	NR
Fatty acids	E	NR	F	F
Fluorobenzene	E	ND	NR	NR
Fuel oil	E	E	NR	NR
Hexane	E	E	NR	NR
Hydrochloric acid	E	NR	G	G
Hydrogen gas (60 °C)	E	NR	F	E
Hydrofluoric acid (60%)	E	NR	NR	G
Jet fuel (JP1)	E	G	NR	NR
Kerosene	E	G	NR	NR
Mineral oil	E	G	G	NR
Naphtha	E	G	NR	NR
Nitric acid (90%)	E	NR	NR	NR
Phosphoric acid (85%)	E	NR	NR	E
Sulfuric acid (95%)	E	NR	NR	NR
Toluene	E	NR	NR	NR
Trichloromethane	E	NR	NR	NR
Trifluoroethane	E	NR	NR	NR

E — excellent;
G — good;
F — fair;
NR — not recommended;
ND — no data available.

The samples of polysulfide that were immersed in the 15% HCl looked blistered before removal. The samples disintegrated when they were removed from the acid. The polysulfide turned into a soft jelly, and it was impossible to measure its volume. The samples immersed in 35% HCl completely dissolved within hours of immersion and created a light gray puddle of sediment in the bottom of the flask.

This experiment indicates polysulfide should not be used in any application that could require even incidental exposure to hydrochloric acid above extremely low concentrations. For example a concrete secondary containment area around a hydrochloric acid tank should be sealed with a fluoroelastomer rather than a polysulfide.

The Table 23.2 rates fluoroelastomer, polysulfide, silicone and EPDM on their abilities to withstand exposure to a number of common chemicals.

6 APPLICATION SUBSTRATES

Liquid fluoroelastomers adhere readily to many substrates. Surfaces commonly used include metals such as: stainless steel, aluminum, copper, brass, cast iron and most alloys. Nonmetallic substrates include but are not limited to: glass, ceramic,

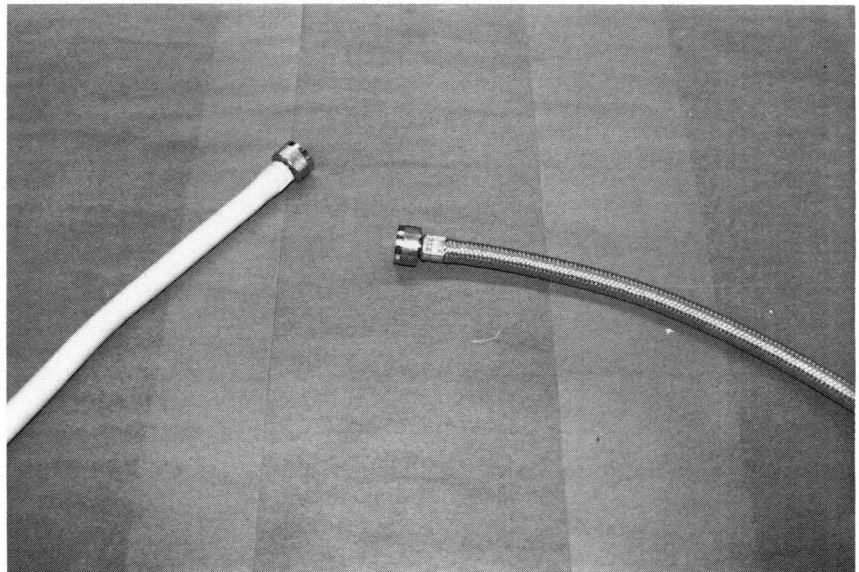

Figure 23.1. Sample on the left is a braided metal hose (as pictured on the right) that has been coated with a liquid fluoroelastomer. The hose now has resistance to hot acids and corrosive liquids and gases that would otherwise have attacked the metal hose. This is a low-cost option which combines the strength and flexibility of the metal with the chemical resistance and economy of the fluoroelastomer coating

wood, fabric, concrete, rubber, plastic and other liquid top coating materials. All of these substrates can be protected from harsh service environments with a fluoroelastomer barrier layer. (Figure 23.1). The key to obtaining superior results is in the surface preparation work done prior to the application of the fluoroelastomer.

7 SUBSTRATE PREPARATION

Like most comparable products, fluoroelastomers adhere best to rough, clean surfaces free of any particulate matter or grease [6]. Surface blasting with sand, water or other abrasive substance will help create the desired surface on most metallic substrates. Sandpaper can be used on rubber, glass, fabric and other surfaces that do not react well to blasting or are too complex for blasting. Once the surface has been abraded, all surface grease and contaminants should be removed. Solvents such as acetone, methyl ethyl ketone and toluene are preferred as long as they are compatible with the surface. These solvents will aggressively break down grease and oils without leaving residues of their own.

Other substrates such as PTFE require surface preparation with chemicals called etchants. If a chemical etching material is used to prepare a surface for adhesion, care should be taken to ensure that all of the chemical residue is removed prior to the application of the coating.

Once the substrate is prepared to receive the coating, it is time to work with the fluoroelastomer itself. Traditionally, liquid fluoroelastomers are two-part systems consisting of a base fluoroelastomer and a catalyst or accelerator. This two-part system was developed from the need to have an ambient curing liquid. Originally, a one-part fluoroelastomer was not stable enough to be stored for any reasonable period of time. As technology changed and new curatives appeared, however, practical one-part systems were developed. Currently there are aerosol versions of liquid fluoroelastomers and single-part caulking tubes of liquid fluoroelastomers. This chapter discusses the two-part systems.

8 APPLICATION OF LIQUID FLUOROELASTOMER

After the liquid fluoroelastomer is mixed with a catalyst, the material has a working pot life ranging from four hours to several days. Variables that affect pot life are viscosity of the product, the type and amount of curative and atmospheric conditions.

The viscosity of the liquid fluoroelastomer and its end-use requirements will dictate how it is applied to the substrate. Low-viscosity products can be sprayed, brushed, dip-coated or rolled onto the substrate. High-viscosity products are limited to trowelling or caulk gun applications.

The addition of a catalyst to a low-viscosity product is relatively easy. The catalyst is simply stirred into the base fluoroelastomer. High-viscosity products are more difficult to stir. This fact has made the need for one-part fluoroelastomer caulks more apparent in today's marketplace.

Some of the pitfalls associated with the application of a liquid fluoroelastomer include: blistering, orange peeling, air entrapment and shrinkage, to name a few. Blistering is a common problem with many products that have some volatile content. The problem becomes pronounced if the product is heated before all solvent has been allowed to escape. The rapidly expanding solvent forms blisters that become trapped in the coating.

Orange peeling is the term used to describe a coating that dries with dimples and ridges in its surface (thereby taking on the appearance of the skin of an orange) [7]. The term is common to the adhesives and coatings industry and applies equally to fluoroelastomer coatings. If orange peel is suspected while a coating is drying, a quick spray of solvent on the coating surface helps to 'open' the skin of the fluoroelastomer and dry the coating in a more uniform manner.

Air entrapment is a common problem with two-part products and liquid fluoroelastomers are no exception. As the surface skins over, air that was mixed into the product becomes trapped and causes voids in the cured sample. The two components should be mixed slowly to prevent air entrapment. A slow, steady, circular stirring motion is preferred. If the catalyst is stirred into the fluoroelastomer too quickly, a frothy mixture will result. The mixing time might be shorter, but the cured coating will be very rough looking and have voids that can limit performance. Air entrapment is more common with the higher solids fluoroelastomer caulks and sealants.

Shrinkage of the liquid fluoroelastomer is always a concern especially when a solvent medium is used. Care should be taken when calculating the overall volume requirements for an application. If shrinkage related to solvent loss is not considered, the applicator may have to apply the material more than once, or possibly run out of material for the job. There will always be some shrinkage as long as there is some volatile content to the product. The problem is reduced in water-based liquid fluoroelastomers which have higher solid contents at lower viscosities.

9   USES FOR LIQUID FLUOROELASTOMER

Liquid fluoroelastomers are used in a wide range of applications and are often the material of choice for harsh environments. They are most often used to coat, seal, adhere to, and encapsulate substrates. They are used as protective coatings on elastomers, metals and other surfaces that need to be protected from harsh service conditions. A flexible fluoroelastomer coating can extend the service life of the coated substrate by separating it from the harmful environment and absorbing shock and stress. In adhesive applications, the bond can flex without failure, effectively increasing the strength of the bond. Applications for liquid fluoroelastomers are limited only by the imagination of the engineers and problem solvers tasked with correcting today's design flaws. Liquid fluoroelastomers are useful in virtually any application where temperature degradation or chemical attack are present.

9.1 AEROSPACE

Performance requirements of the aerospace industry are much greater than the demands of more traditional applications. Due to the severe demands placed on parts and components, molded fluoroelastomers are a popular material among engineers in the aerospace industry. Liquid fluoroelastomers have similar properties to molded fluoroelastomers and have provided solutions to unusual problems.

Aerospace parts and components are engineered to perform in harsh environments that subject them to chemicals, temperature changes, varying pressures, high speeds and extreme forces. Aerospace engineers must also design systems with the knowledge that even a minor failure can be catastrophic. Engineers at the National Aeronautic Space Administration (NASA) in Cape Canaveral, Florida had all of these concerns in mind when they were looking for an adhesive to bond a fluoroelastomer up-stop bumper to the inside of one of the shuttle vehicles. They determined that the adhesive had to withstand constant temperature cycling and vibration stress while remaining flexible enough to prevent bond failure during takeoff and landing situations. After extensive testing of the various materials it was determined that a liquid fluoroelastomer based adhesive was best situated for this application.

Aerospace technology continues to evolve in an effort to make air transportation faster and more efficient. As performance demands increase, limitations of traditional materials will become more apparent, and engineers will be forced to look at novel methods of manufacture and more exotic designs. Liquid fluoroelastomers will play a key role in that continuing effort to improve aircraft capabilities.

9.2 AUTOMOTIVE

In the automotive industry, fluoroelastomers have been used since their development in the 1950s. Fuel lines, O-rings, seals, gaskets, hoses and pressure regulators are examples of their use. With the introduction of liquid fluoroelastomers, these same applications are made even more effective. Liquid fluoroelastomers bond very well to other solid fluoroelastomers (see Figures 23.2–23.4). When a fluoroelastomer gasket is in place on a metal surface, minute gaps need to be filled to prevent any leakage of fuel or oil. Liquid fluoroelastomers seal these gaps easily and prolong the life of many seals and gaskets. One application where fluoroelastomers are needed is in the area of fuel intake.

Almost all of today's cars are equipped with a fuel injection system. The system consists of pumps, hoses and seals. All of the components must withstand fuel contact. The typical fuel injection system uses a fluoroelastomer hose with a metal stem attached to it. The metal stem comes in direct contact with the

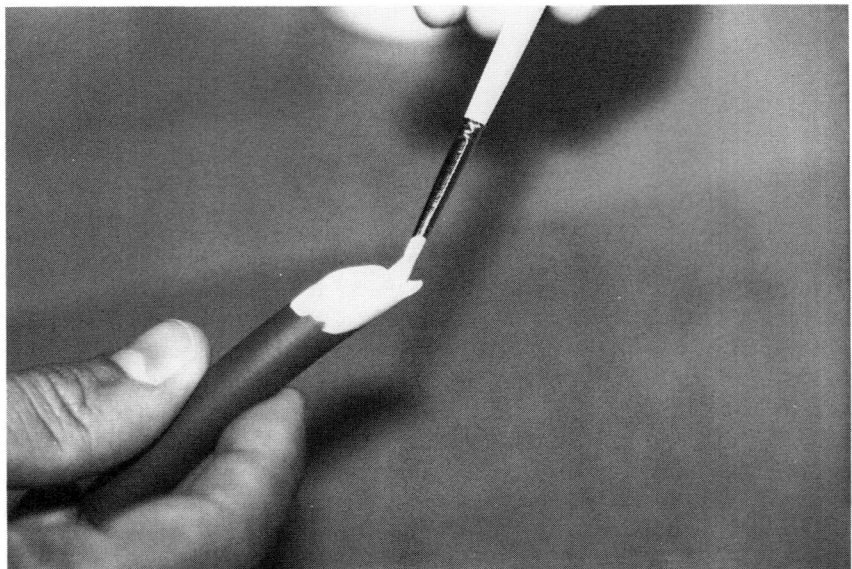

Figure 23.2. A liquid fluoroelastomer (based on VDF–HFP in a polar solvent mixture) being applied to a fluoroelastomer extrusion

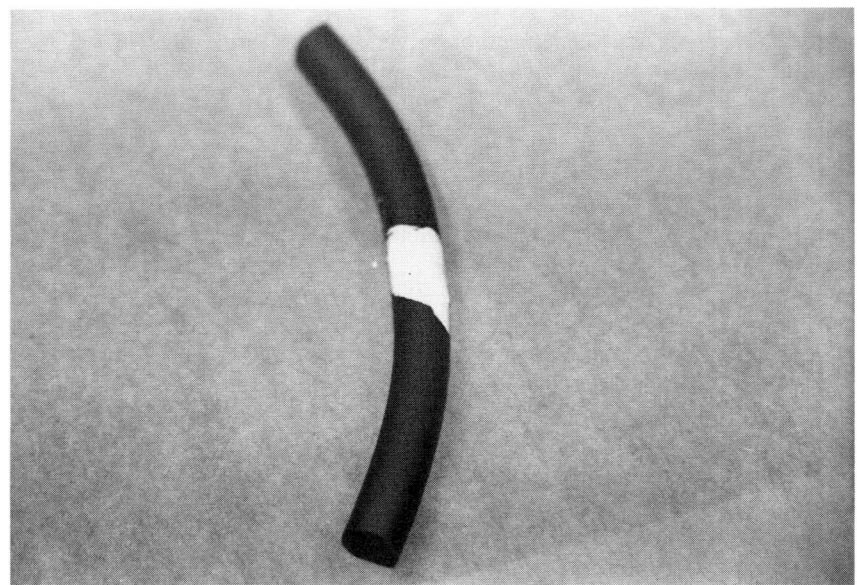

Figure 23.3. The sample in Figure 23.2 bonded to a mating part using a scarf butt splice

Figure 23.4. The spliced sample in Figure 23.3 is stretched in a tensile testing machine demonstrating the excellent adhesive and cohesive strength of the liquid fluoroelastomer bond. Note that the bond is also just as flexible, chemically resistant and heat resistant as the fluoropolymer being bonded. This is especially advantageous in the manufacture of fluoroelastomer gaskets from extrusions at much lower cost than integrally moulded gaskets

engine's fuel. Liquid fluoroelastomer is applied to the metal stem and around the interface with the fluoroelastomer hose. This keeps the hose and stem interface sealed from any fuel migration. The liquid fluoroelastomer acts as a protective coating for the metal and helps prolong the life of the fuel injection system.

9.3 DRY CLEANING

In the dry cleaning industry Perchloroethylene is one of the chemicals used to clean delicate fabrics. 'Perch', as it is commonly known, was recently identified as a potential hazard to ground water because of its ability to leach through concrete. The industry was forced to address the issue of spill containment inside a dry cleaning shop.

At first, the dry cleaners considered using epoxy-based coatings to resist the effect of Perch on concrete, but epoxies are not flexible materials. Epoxy coatings tend to crack under the stress of foot traffic and the movement of heavy machinery inside a dry cleaning shop. The industry needed something flexible and chemically resistant that could be applied as a coating. They also wanted a product that would cure at ambient temperatures.

Liquid Viton fluoroelastomers exhibit all of the properties that the dry cleaning industry needed for their specific application. Test patches of liquid Viton fluoroelastomer were applied to different floor areas in dry cleaning shops and the results were positive. The fluoroelastomer coating does not degrade in the presence of perchloroethylene, and it can handle the abrasive nature of foot traffic. Fluoroelastomer coatings are also easy to apply.

9.4 PRINTING

In the high-speed industrial printing industry, the need for a strong resilient rubber roller is ever present. Existing rubber rollers wear out prematurely due to excessive heat and the high abrasive nature of paper. Roller manufacturers today use Neoprene rubber as a roller material, but Neoprene does not have the abrasion resistance or chemical performance of a VDF–HFP fluoroelastomer. For this reason, some of the roller manufacturers are adopting a simple method of increasing the temperature and abrasion resistance of Neoprene rollers. They are overcoating existing Neoprene rollers with liquid fluoroelastomer. This gives the roller more abrasion resistance and does not cost as much as a solid fluoroelastomer roller.

9.5 ENVIRONMENTAL PROTECTION

A common application for liquid fluoroelastomers is found in industrial secondary containment areas. The area around hazardous chemical tanks must be carefully planned so that in the event of a tank failure, the chemical will be contained. This is usually done by building concrete dikes and earth berms around the tank. Joints must be designed into the concrete structure to allow for expansion and contraction caused by temperature changes.

In some concrete applications, the expansion joints can be left open or filled with tar, polysulfide, or some other flexible material. In the case of hazardous chemicals, the joint must be filled with material that can stand up to the chemical. As we have shown, fluoroelastomers will withstand exposure to chemicals that will dissolve polysulfide. Many applicators will apply a thin layer of the fluoroelastomer as a top coat on the polysulfide. This extra fluoroelastomer layer boosts the chemical resistance of the polysulfide without reducing flexibility.

Figure 23.5. A one-part fluoroelastomer caulk being applied in a concrete sample designed to simulate an expansion joint. Its excellent chemical resistance especially to HCl and chlorinated solvents makes fluoroelastomer caulks ideal for expansion gaps in concrete chemical containment barriers and for concrete flooring in dry-cleaning establishments. Furthermore, such caulks of fluoroelastomers have outstanding resistance to weathering

For those applicators who want the most in chemical resistance, a fluoroelastomer sealant is the material of choice for secondary containment joints. Fluoroelastomer caulks adhere very well to concrete and steel (Figure 23.5). They offer superior chemical resistance, remain flexible as the concrete moves and resist outdoor degradation from ozone attack and ultraviolet radiation. Fluoroelastomer sealants are growing as one of the most popular sealants for newly constructed containment areas. This is due in part to new environmental regulations. Corporations have to ensure that all containment areas meet these new guidelines, and those areas that do not must be improved with better sealants. Liquid fluoroelastomer caulks assure the integrity of a containment area and prevent any contamination of ground water below these areas.

10 CONCLUSION

Liquid fluoroelastomers have emerged as a distinct material, unique among fluoropolymers, that combine the advantages of solid fluoroelastomer with the versatility of a liquid. Their superior chemical resistance, coupled with their high-temperature service capabilities create more options for solving today's design problems that previously did not exist. There are no physical parameters, such as the dimensions of a molded fluoroelastomer part, to dictate, define and restrict its design for a special component. Instead, liquid fluoroelastomers behave like form-in-place gaskets and seals that give the engineer of today more flexibility in solving a problem.

REFERENCES

1. DuPont, Selection Guide for VITON Fluoroelastomers, 505 Blue Ball Road, PO Box 306, Elkton, MD 21911 (1981)
2. Robinson, R. A., and Thorn, A. D. Compound design, in *Rubber Products Manufacturing Technology* (Anil K. Bhowmick, Malcolm M. Hall and Henry A. Benary eds), Marcel Dekker, New York (1994) p. 68.
3. Crenshaw, L. E., and Tabb, D. L. Fluoroelastomers, in *The Vanderbilt Rubber Handbook* (Robert F. Ohm ed.), R. T. Vanderbilt Company, Norwalk, CT (1990) p. 217.
4. DuPont, The Volatile Products Evolved From Fluoroelastomer Compounds During Curing, 505 Blue Ball Road, PO Box 306, Elkton, MD 21911 (1976).
5. Schroeder, H. Fluorocarbon elastomers, in *Rubber Technology* (Maurice Morton ed.), Van Nostrand Reinhold, Melbourne, Victoria, Australia (1987) p. 410.
6. Landrock, Arthur H., *Adhesives Technology Handbook*, Noyes Publications, Park Ridge, NJ (1985) p. 54.
7. Hare, C. H. Protective coatings *Fundamentals of Chemistry and Composition*, Technology Publishing Company, Pittsburgh, PA (1994) p. 456.

24

Perfluoropolyethers (Synthesis, Characterization and Applications)

JOHN SCHEIRS
ExcelPlas Australia

1 INTRODUCTION

Perfluoropolyethers (PFPEs) are a class of low MW polymers (500–15 000 Dalton) that were originally developed [1–3] in the mid-1960s. Their molecular structure, comprising only carbon, fluorine and oxygen (Figure 24.1), makes these materials useful for applications under extreme conditions, in the presence of aggressive chemicals, and in oxidizing environments. They have approximately the same chemical stability as polytetrafluoroethylene in most cases.

PFPEs are liquids at room temperature with very low volatility and their viscosity shows little temperature dependence. In addition, they show almost no shear thinning even at very high shear rates. Furthermore, these truly amazing liquids are resistant to almost all oxidizing environments. Their exceptional stability is by virtue of the high bond strength of $C-O$ and $C-F$ bonds and the steric and electronic shielding provided by the fluorine atoms. Perhaps no other known compound can exhibit liquid-phase behaviour over such a wide temperature range ($-100\,°C$ to $+400\,°C$). The low dependence of viscosity on both temperature and shear rate as well as low losses by evaporation are the critical properties that make PFPEs excellent lubricants.

PFPEs are produced by a variety of different polymerization techniques. The basic repeat units are CF_2O, CF_2CF_2O, $CF_2CF_2CF_2O$ and $CF(CF_3)CF_2O$ while the terminal groups of the polymer chain can be CF_3O, C_2F_5O and C_3F_7O. The structure of the PFPE depends on the method used for the synthesis [3,4].

Modern Fluoropolymers. Edited by John Scheirs
© 1997 John Wiley & Sons Ltd

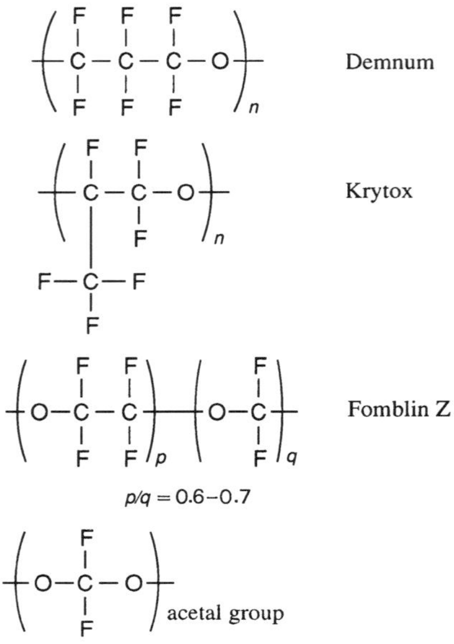

Demnum

Krytox

Fomblin Z

$p/q = 0.6-0.7$

acetal group

Figure 24.1. Structures of some commercially available PFPEs. Note: Fomblin® Z contains the 'acetal' structure

The outstanding chemical and thermal stabilities of PFPEs have led to their use in diverse fields of application such as lubricants for computer hard disks (where they minimize head/disk interfacial wear and friction), high-temperature greases, vacuum pump fluids and hydraulic oils. In addition, their hydrophobic and lyophobic properties makes them useful in colloids and cosmetics applications.

Yet despite their excellent combination of properties, PFPEs do suffer from some significant drawbacks [5]. For instance, while PFPEs are stable up to 410 °C in an inert, metal-free environment and up to 316 °C in an oxidizing, metal-free atmosphere, they are catalytically decomposed by metals at temperatures as low as 100 °C. As early as 1966, Sianesi identified that PFPEs could be decomposed by $AlCl_3$. In particular, the acetal linkage has now been identified as a site of instability in some PFPEs (Figure 24.1).

This chapter gives an overview of the manufacture, characterization and applications of PFPEs.

2 MANUFACTURE

2.1 NEUTRAL PFPEs

Neutral PFPEs are those with neutral or non-reactive end groups. These are distinct from functionalized PFPEs which have been modified to incorporate reactive end- groups.

2.1.1 Fomblins

PFPEs under the Fomblin® trade name are manufactured by Ausimont (Spinetta Marengo, Alessandria, Italy). Fomblin® PFPEs are prepared by reaction of fluoroolefins (such as TFE) with oxygen in a photochemical reactor under UV light ($\lambda < 300$ nm) at temperatures down to $-60\,°C$. A general schematic of the reactor is shown in Figure 24.2. Due to the nature of the polymerization process, which is

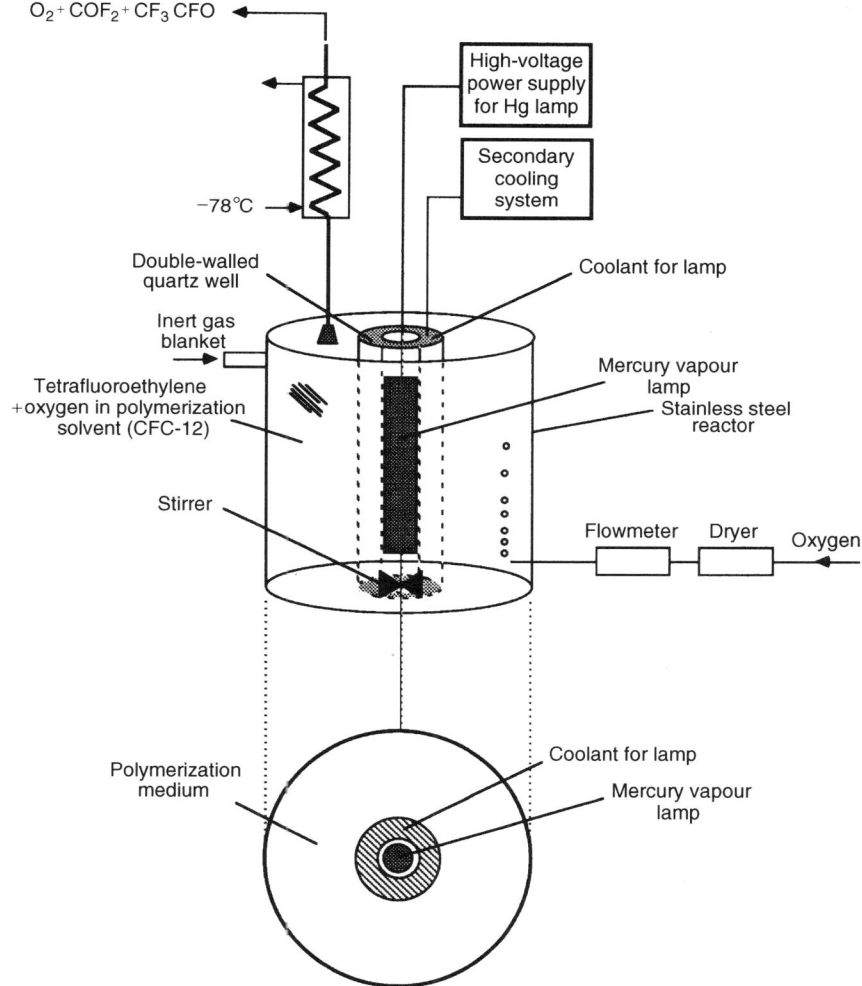

Figure 24.2. Schematic of a photochemical reactor for the preparation of Fomblin® PFPEs

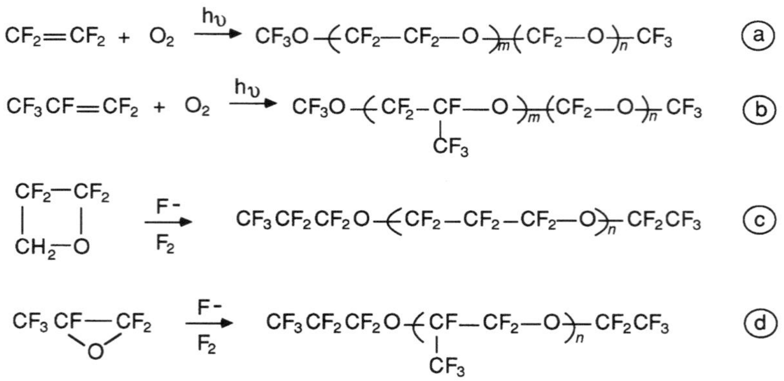

$$CF_2{=}CF_2 \; + \; O_2 \; \xrightarrow{h\upsilon} \; CF_3O\left(CF_2{-}CF_2{-}O\right)_m\left(CF_2{-}O\right)_n CF_3 \qquad \text{(a)}$$

$$CF_3CF{=}CF_2 \; + \; O_2 \; \xrightarrow{h\upsilon} \; CF_3O\left(CF_2{-}CF{-}O\right)_m\left(CF_2{-}O\right)_n CF_3 \qquad \text{(b)}$$
$$\underset{CF_3}{|}$$

$$\begin{matrix} CF_2{-}CF_2 \\ | \quad\;\; | \\ CH_2{-}O \end{matrix} \; \xrightarrow[F_2]{F\text{-}} \; CF_3CF_2CF_2O\left(CF_2{-}CF_2{-}CF_2{-}O\right)_n CF_2CF_3 \qquad \text{(c)}$$

$$CF_3CF{-}CF_2 \underset{O}{\diagdown\diagup} \; \xrightarrow[F_2]{F\text{-}} \; CF_3CF_2CF_2O\left(CF{-}CF_2{-}O\right)_n CF_2CF_3 \qquad \text{(d)}$$
$$\underset{CF_3}{|}$$

Figure 24.3. Synthesis reactions of: (a) Fomblin® Z by the photo-oxidation of TFE; (b) Fomblin® Y by the photooxidation HFP; (c) Demnum® by the ring-opening polymerization of tetrafluorooxetane; (d) Krytox® by the ring-opening, polymerization of hexafluoropropene oxide

essentially a photo-oxidation reaction, the first product of the reaction is a random peroxidic polyether known as the 'polyperoxide'. The explosive nature of this compound dictates that the reaction be performed at sub-zero temperatures. The need to conduct the polymerization of TFE and oxygen at low temperatures is highlighted by explosions which have occurred in the former Soviet Union while attempting to prepare PFPEs. Even at $0\,°C$ mixtures of TFE and oxygen are highly explosive. After polymerization, the peroxide bonds in the polyperoxide intermediate are broken down under controlled conditions to give the PFPE.

The two main types of Fomblin® fluids that have reached commercial significance are Fomblin® Y and Fomblin® Z. Fomblin® Y fluids are produced by the photo-oxidation of hexafluoropropylene [6] (C_3F_6). In this method of synthesis, head-to-head and tail-to-tail arrangements of the $-CF(CF_3)CF_2O-$ segment are present. The ratio of m : n for Fomblin® Y is 40 : 1 and the molecules possess trifluoromethyl branching (Figure 24.3). Fomblin® Z, on the other hand, has a linear perfluorinated structure and is prepared by the photo-oxidation of TFE at $-80\,°C$ [7]. It may be viewed as a random oxyethylene–oxymethylene copolymer with the ratio of CF_2CF_2O to CF_2O being approximately 0.5 to 2.0.

Figure 24.4 summarizes the steps involved in PFPE manufacture. The multi-step process increases the cost of manufacturing significantly and accounts for the high cost of commercial PFPEs.

2.1.1.1 Reactor Description

The reactor in which Fomblin® fluids are prepared consists of a cylindrical vessel filled with the polymerization medium. A chlorofluorocarbon (CFC-12) is used

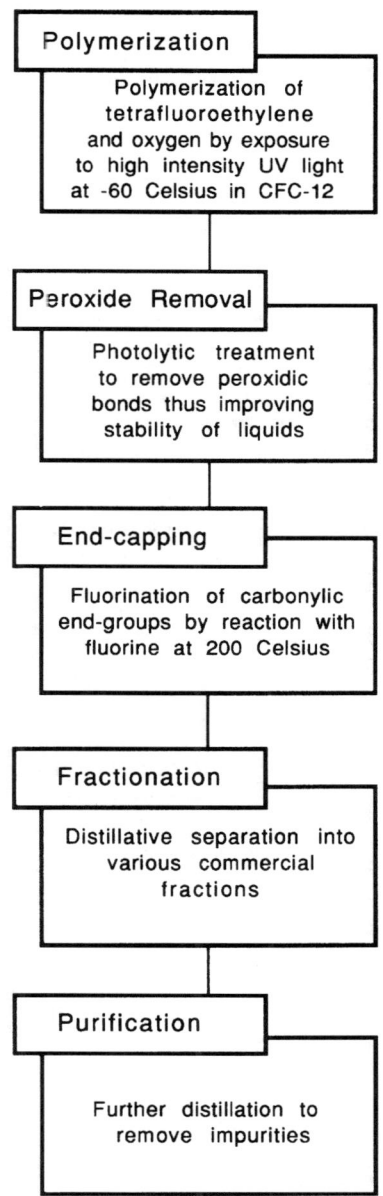

Figure 24.4. Basic steps involved in manufacture of PFPEs

as the polymerization medium. A high-intensity UV lamp ($\lambda = 238-354$ nm) is directly immersed in the polymerization liquid. The lamp is housed in a jacketed quartz well and is cooled by circulating coolants such as Galden D01 (Figure 24.2). The monomer and an excess of oxygen are continuously fed into the reactor via a gas dryer and flowmeter. The amount of oxygen entering the reactor is approximately three times the amount which reacts in one pass. The gaseous by-products (CF_2O, CF_3CFO and unreacted oxygen) escape through a reflux condenser operating at $-78\,^{\circ}C$ and a pressure regulating valve. Since TFE has a boiling point of $-76\,^{\circ}C$ it condenses at the condenser and returns to the reactor. The by-products CF_2O and CF_3CFO are scrubbed from the gas stream by KOH solution and the unreacted oxygen is then pumped and recycled back into the reactor [7].

2.1.1.2 Polymerization Initiation

The first step in the polymerization is the formation of a chromophore (light-absorbing compound). This is usually CF_3CFO or a peroxide which forms first. These compounds then undergo photolysis producing free radicals which initiate the polymerization. The main initiator of the photopolymerization is the trifluoromethyl radical ($CF_3\cdot$) which is formed by photolysis of CF_3COF in the reaction:

$$CF_3COF \longrightarrow CF_3 \cdot + \cdot COF \qquad (1)$$

The role of $CF_3\cdot$ as the initiation species was confirmed by the lack of an induction period upon adding CF_3COF to the monomer feed [8,9].

2.1.1.3 Photochemical Polymerization

The photochemical reaction is controlled by the concentration of reactants (TFE or HFP), temperature, residence time, removal of gaseous by-products (e.g. COF_2, CF_3COF), the light intensity and effective distance from the light source. Careful control of these parameters is important to avoid the occurence of a steep concentration gradient of polyperoxides across the reactor. The reaction can be monitored by using solution viscosity with suitable consideration of temperature and solvent dilution effects. The course of the polymerization is followed by measuring the oxygen consumed and the evolution of gaseous by-products (CF_2O and CF_3CFO). The composition of the exit gases can be continuously analysed by GC. The residence time in the reactor is regulated so that a compromise is attained between optimum monomer conversion and the viscosity of the system.

2.1.1.4 Peroxide Removal

In order to produce a stable (neutral) PFPE from the polyperoxide reaction product, the polymer must undergo a 'clean up' procedure to remove the thermolabile peroxide bonds in the polyperoxide. The peroxide linkages which represent weak links can be removed chemically or photochemically. In the production of neutral PFPEs, photochemical removal by photolysis is the preferred method since it causes a less drastic reduction of polymer MW than the chemical route. In the preparation of the difunctional PFPEs on the other hand, chemical peroxide removal is practised (see Figure 24.5).

2.1.1.5 Fluorination of End Groups

A fundamental step in the commercial manufacture of neutral PFPE fluids is the removal of their reactive end groups (acid fluoride and fluoroformate) [7]. These groups are usually ketonic or acyl fluoride groups and must be removed by end-capping treatment in order to obtain a stable, inert product such as neutral CF_3 groups. This end-capping operation is generally performed by a 'thermal process' in which the oligomeric PFPE is reacted with F_2 at temperatures higher than 200 °C [10]. A new end-capping process has recently been developed [10] based on photochemical fluorination process which is easier to control than the thermal process and which is much less aggressive. The photochemical process exposes the PFPE fluids to F_2 under the action of UV light ($\lambda = 238-354$ nm) at 24 °C for 8 h. In this process carboxylic, acylfluoride and trifluoromethylketone groups are quantitatively cleaved while fluoroformate groups are converted into stable perfluorinated groups.

2.1.1.6 Fractionation

After photochemical peroxide removal and end-capping by post-fluorination, the PFPE product is fractionated into various commercial fractions that have

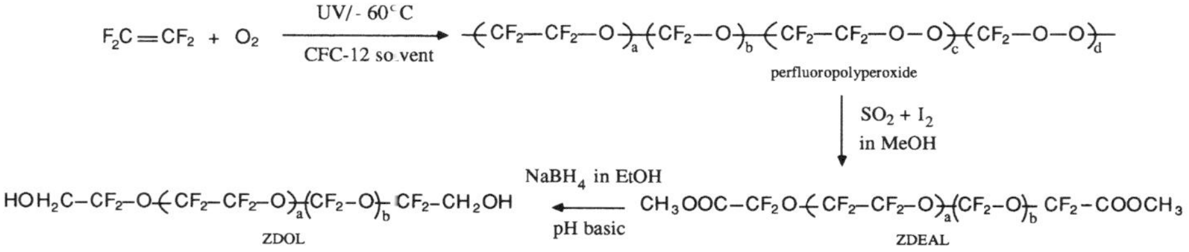

Figure 24.5. Synthesis scheme for functionalized perfluoropolyethers. The first step is the production of the perfluoropolyperoxide, this is then derivatized to ZDOL via ZDEAL

a near-continuous range of boiling points and viscosities for different application areas (Figure 24.6). Table 24.1 lists the physical properties of some representative PFPEs manufactured by Ausimont together with their application area.

2.1.2 Krytox®

Krytox® (E.I. DuPont de Nemours & Co.) (also known as K-lube) is synthesized by the base-catalysed (anionic) polymerization of hexafluoropropylene epoxide (also known as hexafluoropropene oxide) (HFPO) (Figure 24.3(d)) [11]. The acid

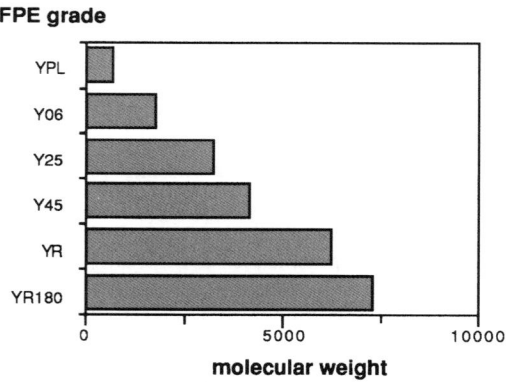

Figure 24.6. Range of molecular weights of Fomblin® Y PFPEs tailored for different application areas

Table 24.1. Typical properties of representative PFPEs manufactured by Ausimont

Properties	Galden D02 TS	Fomblin HC 04	Fomblin HV 25/9	Fomblin Y R
Average MW (Dalton)	750	1500	3200	6250
Kinematic viscosity (cSt, 20°C)	1.7	40	250	1300
Density (g/ml, 25°C)	1.77	1.87	1.90	1.91
Pour point (°C)	−97	−62	−35	−25
Vapour pressure (mm Hg, 20°C)	<1	10^{-3}	10^{-9}	10^{-7}
Surface tension (dyne/cm, 20°C)	17	21	22	24
Typical application area	Electronics	Cosmetics	Pump fluids	Lubricating oils

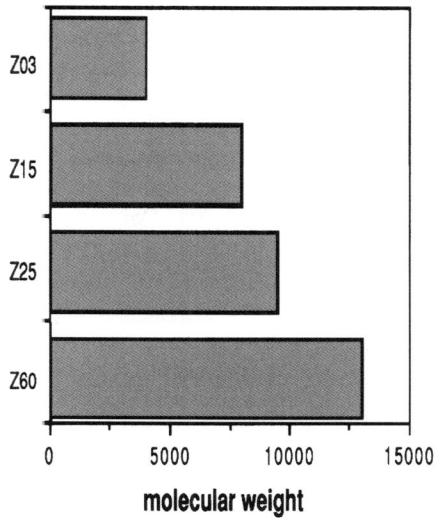

Figure 24.7. Range of molecular weights of Fomblin® Z PFPEs

fluoride end groups of this polymer are subsequently transformed to $C_x F_{2x+1}$ groups. This method of synthesis ensures almost exclusive head-to-tail arrangement and there are no structural irregularities other than a terminal hydrogen impurity. CsF is usually the basic catalyst used to polymerize HFPO. Krytox® fluids, like Fomblin® Y fluids are branched. Krytox® actually contains several polymer molecules which are all based on the same repeat unit in the backbone, but which have different end groups [12]. These different components result from the chemistry involved in the termination and capping during polymerization. A common terminating end group is $-CHF-CF_3$.

2.1.3 Demnum®

Demnum® fluids are produced by Daikin Industries (Osaka, Japan) by the Lewis acid-catalysed ring opening polymerization of 2,2,3,3-tetrafluorooxetane followed by direct fluorination under UV light, to convert the CH_2 groups to CF_2 moieties (Figure 24.3(c)) [13]. Demnum® is a linear homopolymer with monomer unit comprised of $CF_2CF_2CF_2O$ and terminal groups consisting of C_2F_5O and C_3F_7O in a 1 : 1 molar ratio.

2.1.4 Hostinert®

In the Hoechst process, polymerization of hexafluoropropylene epoxide gives a polyether acyl fluoride which is a fluorinated polyether. Hydrolysis and decarboxylation of this followed by treatment with aluminium chloride gives the Hostinert® PFPE [14]. Perfluorination can also be achieved by exposing the partially fluorinated polyether to fluorine at 50–250 °C [15].

Hoeschst have developed another method of producing PFPEs by the electrolytic decarboxylation of perfluorocarboxylic acids [16]. PFPE can be catalytically purified and stabilized by passing over a Cu−Mg silicate catalyst [16]. Ginzel [17] (from Hoechst) describes a similar method for the synthesis of PFPEs from oligomeric hexafluoropropenoxide by Kolbe electrolysis.

2.1.5 Direct Fluorination

Driven to lower manufacturing costs, the use of direct fluorination of hydrogen-containing polymers to produce PFPEs has been developed by Exfluor Research Corp. (Austin, Texas) [18,19]. This concept has been successfully used to make PFPEs in a single-step reaction with much cheaper starting materials and with a high degree of flexibility, given the range of hydrocarbon precursors available (e.g. polyethylene oxide, polydioxolane, and polytetrahydrofuran). The direct fluorination of both hydrocarbon ethers [20] and partially fluorinated vinyl ethers has been reported [21,22]. However, direct fluorination must be done very slowly under carefully controlled conditions to avoid decomposition of the precursors.

2.2 FUNCTIONALIZED PFPEs

Table 24.2 shows commercially available functionalized PFPEs together with their molecular weights. Functionalized PFPEs were developed for two major reasons: firstly to improve adhesion to magnetic computer disks and secondly to produce end-group functionalities so that the PFPE could be used as reactive intermediates in polymerization reactions.

Many of the functionalized Fomblin® Z products can be chemically derivatized via the functionalized intermediate known as ZDEAL which has methyl ester end groups. Chemical reduction of ZDEAL with sodium borohydride in ethanol produces the dihydroxy derivative (Fomblin® ZDOL) (Figure 24.5). The tetrafunctional derivative, ZTETRAOL, is prepared by the reaction of ZDOL with glycidol (3-hydroxypropylene oxide) at 70 °C in the presence of a potassium *tert*-butoxide catalyst (∼5% w/w). Table 24.3 lists typical properties of functionalized Fomblin® PFPEs.

The inertness of the PFPE backbone permits the use of strong reagents (e.g. $AlCl_3$, F_2) to effect modification of the reactive end groups only. This enables a range of chemical functionalities to be tailored through the use of specific

Table 24.2. Commercially available functionalized PFPEs

Trade name	Structure	M_n
Demnum SH	$F-(CF_2CF_2CF_2O)_nCF_2CF_2-COOH$	3800
Krytox 157 FSM	$CF_3CF_2O-(CF_2-CF(CF)_3-O)_nCF_2-COOH$	4500
Fomblin Z DIAC	$HOOC-CF_2O(CF_2CF_2O)_m(CF_2O)_nCF_2-COOH$	2000
Demnum SA	$F-(CF_2CF_2CF_2O)_nCF_2CF_2-CH_2OH$	4000
Fomblin Z DOL	$HOCH_2-CF_2O(CF_2CF_2O)_m(CF_2O)_nCF_2-CH_2OH$	2000
Fomblin Z DOLTX	$HO(CH_2CH_2O)_{1.5}CH_2\text{-}RF\text{-}CH_2(OCH_2CH_2)_{1.5}OH$	2200
Fomblin ZTETRAOL	$HOCH_2(OH)CHCH_2OCH_2-RF-CH_2OCH_2CH(OH)CH_2OH$	2000
Fomblin Z DEAL	$MeOOC-CF_2O(CF_2CF_2O)_m(CF_2O)_nCF_2-COOMe$	2000
Fomblin Z DISOC	$(OCN)(CH_3)C_6H_3NHC(O)-RF-C(O)NHC_6H_3(CH_3)(NCO)$	3000
Fomblin AM 2001	Piperonyl-RF-piperonyl	2100
Fomblin ACF	$CF_3OCOCH_2-RF-CH_2OCOCF_3$	—

$RF = -CF_2O(CF_2CF_2O)_m(CF_2O)_n CF_2-.$

Table 24.3. Typical properties of the functionalized Fomblin®
PFPE samples

Property	Z DOL	ZTETRAOL
Colour	Colourless	Colourless
Molecular weight (M_n) (Dalton)	950[a]	1100[a]
OH functionality (eq./mol)[b]	1.94	3.90
Viscosity at 20°C (cP)	40.1[c]	2800[c]
Density (g cm^{-3}) at 20°C	1.769	1.696
Polydispersity index (M_n/M_w)[a]	~1.1	~1.3
T_g (°C)[d]	−104	−88

[a] Determined by GPC with 500-10^6 A Styragel® columns and an azetropic mixture of CFC113 and acetone.
[b] Determined by NMR.
[c] Measured using a Cannon-Fenske capillary viscometer according to ASTM D445.
[d] Determined by DSC.

derivatization reactions. For instance, a carboxylic acid-terminated PFPE can be converted to an anhydride by reaction with P_2O_5 or to an acid chloride by reaction with thionyl chloride at temperatures in the range 100–200°C [23]. Similarly, oligomeric PFPE with fluoroformate end groups can be modified with fluorine in the presence of UV light [10].

3 CHARACTERIZATION AND PROPERTIES

PFPEs have been characterized by almost every analytical technique in existence including IR spectroscopy [24,25], liquid chromatography [26], mass spectroscopy [27,28], gel-permeation chromatography [29–34], electron paramagnetic resonance [35–42] and x-ray photoelectron spectroscopy [43–50].

Table 24.4. ^{19}F-NMR assignment table for Fomblin® Z

Signal	Chemical shift (ppm)	Assignment
a	−52.1	$-OCF_2CF_2\mathbf{OCF_2}OCF_2CF_2O-$
b	−53.7	$-OCF_2CF_2\mathbf{OCF_2}OCF_2OCF_2-$
c	−55.4	$-CF_2OCF_2\mathbf{OCF_2}OCF_2OCF_2-$
d	−89.1	$-OCF_2CF_2\mathbf{OCF_2}CF_2OCF_2-$
e	−90.7	$-CF_2OCF_2\mathbf{OCF_2}CF_2OCF_2-$
f	−83.8	$-OCF_2CF_2\mathbf{OCF_2}(CF_2)_nCF_2-$
g	−85.4	$-OCF_2\mathbf{OCF_2}(CF_2)_nCF_2-$
h	−125.8	$-OCF_2\mathbf{CF_2CF_2}CF_2O-$
i	−129.7	$-OCF_2\mathbf{CF_2}CF_2O-$
j	−56.3	$-OCF_2CF_2\mathbf{OCF_3}$
k	−58.0	$-OCF_2\mathbf{OCF_3}$
l	−27.7	$-OCF_2CF_2\mathbf{OCF_2}Cl$
m	−29.3	$-OCF_2\mathbf{OCF_2}Cl$
n	−74.5	$-OCF_2\mathbf{CF_2}Cl$
o	−78.0	$-OCF_2CF_2\mathbf{OCF_2}CH_2OR$
p	−80.0	$-OCF_2\mathbf{OCF_2}CH_2OR$

Note: $n = 1, 2$ and R = mono or bisadduct.

3.1 ^{19}F-NMR SPECTROMETRY

^{19}F-NMR is the most informative technique presently available for determining the structure of PFPEs since it can identify all aspects of the structure such as the ratio of monomer groups (m : n), end groups, branching, irregular monomer sequences and even number average MW. Table 24.4 shows the wide range of chemical shifts that fluorinated groups of PFPEs exhibit. ^{19}F-NMR exhibits a wide range of chemical shifts exceeding 400 ppm and this wide range often allows subtle environmental differences in the polymer chain to be discriminated. In general, the line width of the NMR signal ($\Delta\nu$) is inversely proportional to the relaxation time (T2) of a NMR nucleus, thus the narrower the line width, the larger the group mobility. The usefulness of ^{19}F-NMR in the structural characterization of PFPEs has been well demonstrated [7,51–57].

NMR can also be very useful for detecting the presence of residual hydrogen bonded to the chain ends of some Krytox® PFPEs. The structure of the hydrogen-bearing end group is $-O-CFH-CF_3$. Since the hydrogen is bonded directly to the chain it is easily distinguished from hydrogen-containing organic impurities by the fact that hydrogens bound to fluorinated chains resonate at lower fields and show characteristic multiplicity due to coupling effects.

3.2 INFRA-RED SPECTROSCOPY

Since PFPEs do not contain hydrogen they are relatively transparent over most of the IR spectrum (from 1400 to 4800 cm^{-1}). Figure 24.8 shows a transmission

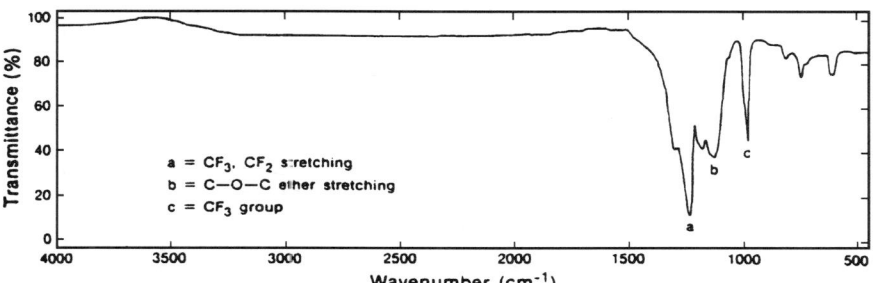

Figure 24.8. Infra-red spectrum of a HFP-based PFPE (Fomblin® YR) of thickness 0.35 μm on a silicon wafer. Note the absence of absorptions in the range 4000–1400 cm⁻¹. Reprinted from Pacansky, J., Wang, C. and Waltman, R. J., *J. Phys. Chem.,* **113**, 329 (1991) with kind permission from Elsevier Science SA, PO Box 564, 1001 Lausanne, Switzerland

FTIR spectrum of Fomblin® Z. The lack of peaks in the mid-IR is beneficial in space applications since the deposition of trace PFPE lubricants on optical components such as mirrors and windows will not appreciably distort sensitive measurements. The IR characteristics of PFPEs have been studied by Carre [58], Pakansky and Miller [59], Pianca [60], and Viswanathan [61].

While IR analysis of PFPEs is simple to perform by virtue of the fact they are liquids, the interpretation and attribution of the IR absorptions are not straightforward since the bands arise from various concerted stretching and bending modes (see Table 24.5). The major IR absorptions of PFPEs are CF_3 and CF_2 stretching at 1300 cm⁻¹, C–O–C stretching at 1100 cm⁻¹ and a relatively narrow absorption at 980 cm⁻¹. The paucity of absorptions in the frequency range 4000 to 1400 cm⁻¹ is advantageous in studying the degradation of PFPEs since products of degradation appear in areas where no peak overlap occurs, such as at 1885 cm⁻¹ (attributable to acid fluoride).

Table 24.5. Infra-red frequencies and assignments for Fomblin® Z

Wavenumber (cm⁻¹)	Assignment
1239	Cooperative motion involving ν (C–C) and ν (C–F) of the $-CF_2-$ groups
1185	ν (C–C) and symmetric ν (C–F) of the $-CF_3-$ groups
1133	Asymmetric ν (C–O) of the $-C–O–C-$ groups
985	Symmetric ν (C–F) of the $-CF_3$ groups
890	Vibrations associated with $-O–CF_3$ groups
865	Vibrations associated with $-O–CF_3$ groups
810	Symmetric ν (C–O) of the $-C–O–C-$ groups
687	Fundamental ν (C–F) band

ν denotes stretching band.

3.3   UV SPECTROSCOPY

The fluorination step in the manufacture of most PFPEs ensures that there are no hydrocarbon impurities present. However, since PFPEs are generally totally transparent down to 200 nm, any impurities can be readily detected by UV spectroscopy. Whilst neutral PFPEs are totally transparent to UV light, acid-functionalized PFPEs exhibit an absorption maximum at 240 nm due to the carboxylic acid functionality [62].

3.4   MASS SPECTROMETRY

PFPE samples can be a complex mixture of structurally different oligomers. As discussed, ^{19}F-NMR is the technique usually employed for determining the qualitative and quantitative composition of PFPEs, through the correlation of chemical shifts with structure and signal intensities with concentration [51,63]. NMR cannot, however, give structural information for submicrogram quantities of polymer coatings on surfaces due to sensitivity limitations.

Mass spectrometry (coupled with methods to desorb non-volatile species on surfaces) has more recently been evaluated to be a valuable tool for structural characterization of PFPE. An entire new field of characterization has opened up given the recent introduction of mass analysers capable of separating high-mass ions and methods of desorption ionization. This method involves desorbing/ionizing intact oligomers directly into the mass spectrometer without any chemical or physical treatment aimed at decreasing the molecular weight of the polymers [64].

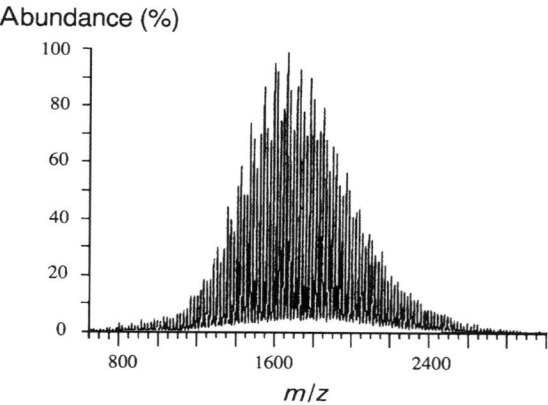

Figure 24.9.   Desorption chemical ionization mass spectra of a Fomblin® Z derivative. PFPEs exhibit very high sensitivity towards ionization mass spectroscopy making it an excellent technique for the structural characterization of PFPEs. Reprinted with permission from ref. 64. Copyright 1994 American Chemical Society

The mass spectra of PFPEs such as Fomblin® Y and Z are characterized by a regular series of peaks with a spacing of 16–18 Dalton (Figure 24.9). It has been demonstrated by Guarini et al. [64] that desorption chemical ionization mass spectrometry is an excellent method for obtaining structural information on PFPEs, such as the type and concentration of the structural repeat unit. The mass spectrum of PFPEs has been characterized by Cromwell [65].

3.4.1 Secondary Ion Mass Spectroscopy (SIMS)

The vapour pressure of PFPEs generally varies from 10^{-5} to $< 10^{-12}$ Torr at 20 °C which makes them suitable for measurement in the vacuum of the TOF–SIMS spectrometer. Furthermore, PFPEs are particularly well suited to TOF–SIMS analysis because of the highly reproducible nature of the relative intensities in the high mass fragmentation ion spectra [12]. Time-of-flight SIMS of PFPEs was pioneered by Benninghoven and coworkers [66,67]. They found that from the location of the peak and the spacings between consecutive peaks, the mass of the repeat unit could be determined. It has since been extensively applied to the structural characterization of PFPEs [12,52,66–72]. The information that is obtainable includes the identification of branching in PFPEs [73], the structure of the terminal groups of PFPEs [64] and quantitative data on their molecular weight and composition [12]. The fractional composition of the PFPE is obtained by measuring fragments related to different chain end group. Molecular weight information is obtained by determining the intensities of fragment species which are concluded to be representative of the monomer units and from these the number of repeating units within the polymer can be calculated. For example, in the case of Krytox® PFPEs: the end group $CF_3CF_2CF_2O-(CF(CF_3)-CF_2-O)_5-$ and the backbone fragment $(CF(CF_3)-CF_2-O)_6-$ have molecular mass values of 1015 and 996 amu respectively.

SIMS has thus become a favourite analytical technique because of the wealth of structural information that can be obtained. SIMS studies of PFPEs reveal an extensive series of high-mass ions whose distribution depends on the molecular weight and structure of the oligomers [12]. Figure 24.10 shows the negative secondary ion mass spectrum of Krytox® PFPE. TOF–SIMS gives excellent negative ion spectra of PFPEs. In contrast, the positive secondary ion spectra for these compounds have neither the intensity nor the abundance of high mass secondary ions that is characteristic in the negative ion spectra. The unusually high stability of the negative secondary ions is believed to be due to stabilization by the strong electronegative fluorine [68].

While ^{19}F NMR is an excellent technique for studying PFPEs in bulk, the analysis of ultra-thin PFPE films on substrates is more challenging. Recently, static secondary ion mass spectrometry (SSIMS) has proved useful in determining the number-average MW from a layer of PFPE just 3 nm thick using the intensities of peaks in the fragmentation pattern [74]. By measuring the intensity of

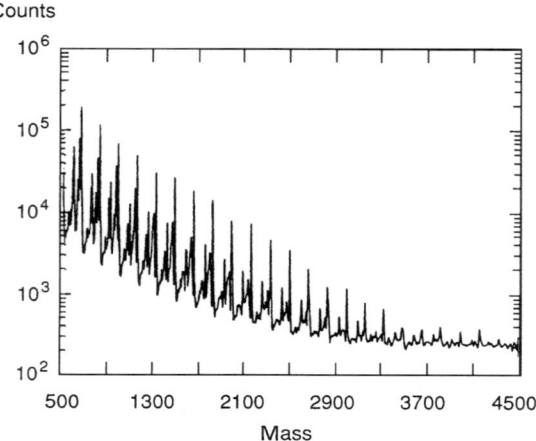

Counts

Figure 24.10. Negative secondary ion mass spectrum of Krytox® PFPE Reprinted with permission from ref. 33. Copyright 1990 American Chemical Society

certain characteristic secondary ion peaks, the thickness of PFPE lubricants on a magnetic disk has been estimated [75]. For instance, it was found that the SIMS intensity of the CF_3^+ ion of Fomblin® ZDEAL varies linearly with film thickness as measured by FTIR or ellipsometry, up to 3 nm. Beyond this thickness the signal becomes saturated as it is beyond the sampling depth of the SIMS technique.

3.5 X-RAY PHOTOELECTRON SPECTROSCOPY

XPS can be used to determine both the type of PFPE and its thickness on a surface. Since the carbon atoms in the PFPE molecular structure are in different environments they possess different (1s) binding energies, thus the XPS spectra can be used to characterize different PFPEs [45]. Figure 24.11 shows the XPS spectra for Krytox®, Fomblin® Z and Demnum® PFPEs. Demnum® has two types of carbons (C1 and C2) and the amount of C1 is twice that of C2 and this is reflected by the C1 peak having twice the intensity of C2. Krytox® on the other hand has three kinds of carbons (C5, C6 and C7) and the amount of each carbon is the same. Table 24.6 lists the XPS binding energies and relative intensities of the resolved peaks. It is apparent that the binding energy is affected by the electronegativity of the bonding atoms. Although XPS is a very informative technique when applied to PFPEs, care must be taken to prevent PFPE degradation by the x-ray irradiation during XPS measurements.

XPS has also revealed that the molecular orientation of groups at the surface of a liquid PFPE consists mainly of preferentially aligned terminal $-CF_3$ groups. While in the bulk of the liquid, the helical structure of PFPE presents more $-CF-$

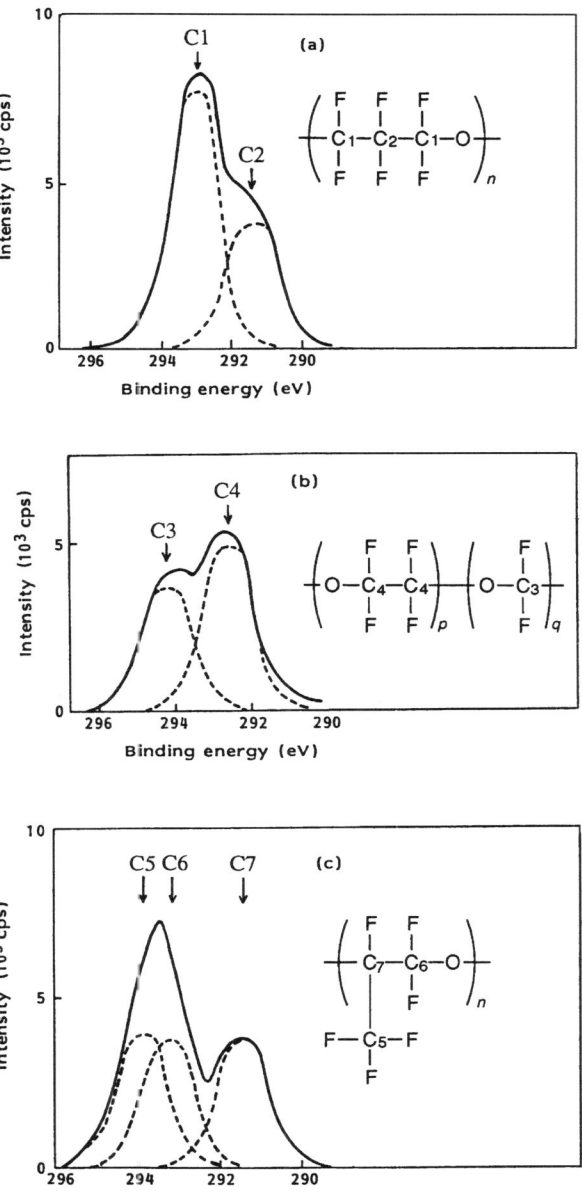

Figure 24.11. XPS spectra of C1s for (a) Demnum®, (b) Fomblin® Z and (c) Krytox® PFPEs (adapted from Mori and Morales [45])

Table 24.6. Binding energies of 1s carbon electrons and their intensity ratios for PFPEs (from Mori and Morales [45]

C1s	Demnum		Fomblin Z			Krytox	
Carbons	C1	C2	C3	C4	C5	C6	C7
Binding energy	292.9	291.4	294.2	292.6	293.8	293.2	291.4
Intensity ratio	2	1	1	1.3	1	1	1

than CF_3 groups. It was found that the surface of the PFPE liquid does not absorb any chemical species, implying that the only absorption site (the oxygen linkage) is beneath the surface [76].

3.6 MICROELLIPSOMETRY

The surface spreading properties of Fomblin® Z and Fomblin® ZDOL (hydroxy terminated) on silicon wafers has been investigated by microellipsometry [77]. ZDOL has a much lower mobility and a distinctly different thickness profile than Fomblin® Z which can be attributed to the stronger interaction of the hydroxyl end-groups with the substrate in the case of ZDOL.

3.7 ELECTRON SPIN RESONANCE

The characterization of PFPEs by electron spin resonance (ESR) has been extensively reported by Faucitano and coworkers [35–42] in a series of papers spanning the period 1981–1995. Among other things, ESR was used to study the radicals generated from peroxidic Fomblin® Y and Z (containing approximately 1% peroxy linkages) by γ- and UV-irradiation. It was found that the principal radicals formed were $R-O-CF_2\cdot$ in Fomblin® Z and $R-O-C\cdot F(CF_3)$ in the case of Fomblin® Y.

3.8 RHEOLOGY

The rheological behaviour of PFPEs at very high shear rates (10^9 s^{-1}) is a very important factor in their application as lubrication films in magnetic recording devices. In this extremely high shear rate regime, shear thinning (non-Newtonian behaviour) becomes extremely important. Workers at Ohio State University have constructed a special rheometer to study the behaviour of these PFPEs at high shear rates, since traditional rheometers are limited to moderate shear rates on account of viscous heating effects. The rheometer design is based upon the magnetic disk drive configuration in which an ultra-thin film of PFPE is sheared between a magnetic disk and a slider. By measuring the film thickness using a capacitance method it is possible to accurately measure the shear rate [78]. PFPEs behave as Newtonian liquids at shear rates up to $10\,000$ s^{-1} (Figure 24.12) [29].

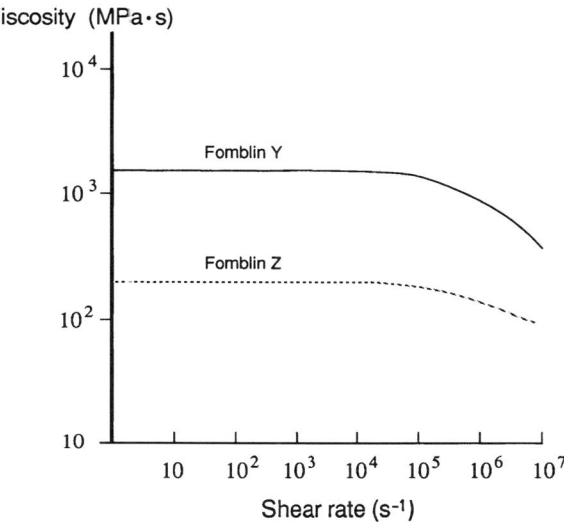

Figure 24.12. Shear rate dependence of the viscosities of Fomblin® Y (YR) and Fomblin® Z (Z25) at 25 °C. Note that both PFPEs start to exhibit non-Newtonian behaviour at shear rates higher than approximately 10 000 s⁻¹ (adapted from Cantow *et al.* [29]). There may also be a temperature effect superimposed at high shear rates

The temperature and pressure dependence of the viscosities of Fomblin® PFPEs have been studied by Cantow [30] using a modified Haake viscometer. It was found that Fomblin® Y because of the bulky side-chains branches has a higher flow activation energy and a more pronounced pressure dependence than Fomblin® Z. Figure 24.13 shows that at the same molecular weight Fomblin® Y exhibits a higher viscosity than Fomblin® Z. The higher viscosity profile of Fomblin® Y can also be related to its lower oxygen : carbon ratio compared with Fomblin® Z. The oxygen : carbon ratio dictates chain flexibility due to the fact that backbone oxygen atoms act as 'molecular hinges'.

3.8.1 Viscosity–temperature Properties

The viscosity–temperature behaviour of PFPEs is directly related to the oxygen ratio (C/O) in the polymer repeat unit. The high C/O ratio of the Krytox® fluids means large changes in viscosity occur with relatively small temperature changes. Furthermore, the branching of the Krytox® fluids leads to poor viscometric properties. In contrast, Fomblin® Z fluids have a low C/O ratio and exhibit good viscometric properties [79].

The number average MW of PFPEs can be easily determined by measuring the kinematic viscosity. This number average MW (M_n) can be related to the

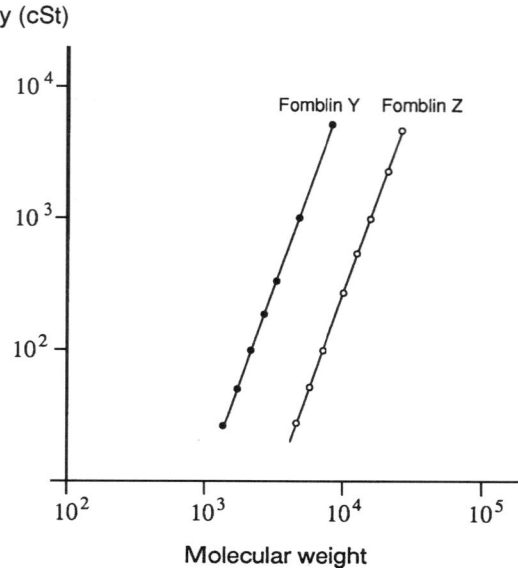

Figure 24.13. Viscosity–molecular weight relationship for PFPE fluids at 20 °C. Note: Fomblin® Y fluids have a higher viscosity than Fomblin® Z fluids of the same molecular weight

Table 24.7. Physical properties of commercial PFPEs

PFPE	Average MW (g/mol)	Kinematic viscosity (cSt at 20 °C)
Krytox 143 AY	3000	150
Krytox 143 AB	3700	230
Krytox 143 AX	4800	450
Krytox 143 AC	6250	800
Krytox 143 AD	8250	1600
Krytox 16256	11000	—
Fomblin Z-15	8000	148
Fomblin Z-25	9500	250
Fomblin Z-60	13000	600
Fomblin Y-25	3000	250
Fomblin Y-45	4100	450
Fomblin YR-1200	6000	1000
Demnum S-65	4500	150
Demnum S-100	5600	250
Demnum S-200	8400	500
Demnum D-700	14000	600
Demnum C-240	6500	630

kinematic viscosity (η) by the following equation: $M_n = 3.4 \times \eta^{0.4}$. Table 24.7 lists the MW and viscosity values of some commercial PFPEs.

3.9   THERMOGRAVIMETRY

Thermogravimetry has been extensively applied to PFPEs in order to determine their thermal stability [6,80,81].

3.10   GLASS TRANSITION TEMPERATURE

The glass transition (T_g) behaviour of PFPEs has been investigated by Danusso [82] and Marchionni [34,83]. It has been shown that the T_g of PFPEs increases with increasing MW as predicted by the Gibbs–Marzio equation (Figure 24.14). It is interesting to note that replacement of the majority of ($-OCF_2CF_2$) and ($-OCF_2$) groups with ($-OCF(CF_3)CF_2-$) as happens when going from Fomblin® Z to Fomblin® Y increases the T_g from -131 to $-57\,°C$. This increase is effectively due to the substitution with trifluoromethyl branching which 'stiffens' the chain and lowers chain mobility [83].

Figure 24.15 shows that increasing the oxygen : carbon ratio of PFPEs leads to a progressive decrease in the T_g. This is expected, as increasing the oxygen

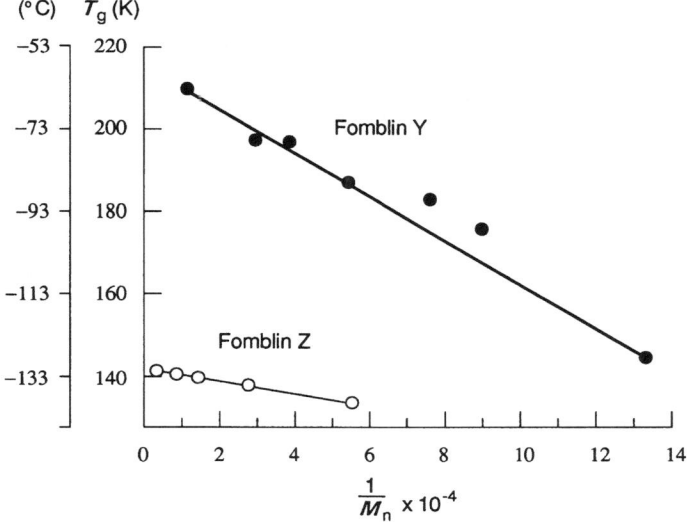

Figure 24.14. Dependence of glass transition temperature on the inverse of the molecular weight for FFPEs (adapted from Marchionni [83]). Reprinted from Marchionni, G., Ajroldi, E. with kind permission from Elsevier Science Ltd, The Boulevard, Langford Lane, Kidlington, OX5 1GB UK. and Pezzin, G. *Eur. Polym. J.*, **24**, 1211 (1988) Note: the branched PFPEs (Fomblin® Y) have stiffer molecular chains than the linear PFPEs (Fomblin® Z)

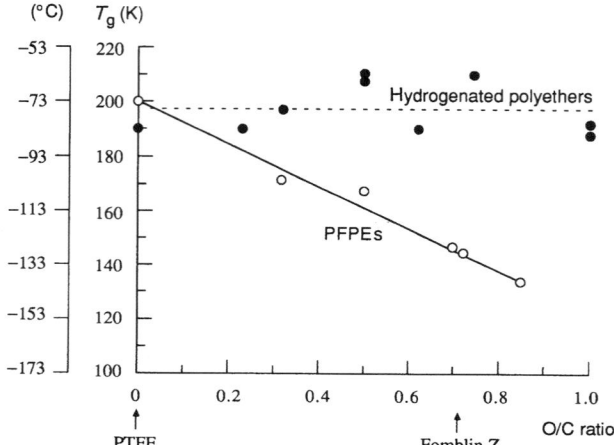

Figure 24.15. Relationship between the glass transition temperature and the oxygen/carbon ratio of both fluorinated and non-fluorinated polyethers (adapted from Marchionni *et al.* [79]). Reproduced by permission of American Chemical Society PTFE and Fomblin® Z are indicated on the x-axis for reference

content increases chain flexibility however, the T_g's of hydrogenated polyethers are rather insensitive to the content of oxygen. Since the oxygen : carbon ratio of the polymer can be conveniently varied by adjusting polymerization conditions this makes it easy to tailor the T_g of particular PFPEs.

3.11 SOLUTION PROPERTIES

The solution properties of PFPEs have been studied in CFC 113 and perfluoroheptane. CFC-113 also known as 1,1,2-trichlorotrifluoroethane ($CF_2Cl—CFCl_2$) is the best universal solvent for PFPEs. The solutions have been characterized by light scattering, vapour pressure osmometry and viscometry [84].

4   APPLICATIONS

PFPEs are used in a plethora of specialized and niche applications that exploit their unique performance properties. These properties include low vapour pressure, high thermal stability and low chemical reactivity. In addition, newer applications are emerging in which PFPEs functionalized at their chain ends are used as diols in the preparation of polymers, elastomers and resins where they confer hydrophobicity, enhanced chemical-, thermal- and oxidative stability, improved biocompatibility and increased gas permeability. In addition to those properties which typically characterize fluoropolymers, PFPEs exhibit a number of favourable attributes such as: good hydrodynamic film-forming capability, good boundary lubricating ability, wide liquid-phase temperature range and a small temperature dependence on viscosity.

4.1 MECHANICAL PUMP FLUIDS AND LUBRICANTS

Since PFPEs exhibit no significant alteration in their viscosity, even after protracted use at extreme temperatures (300 °C), they are well suited for use as pump fluids in critical applications. In particular, PFPEs are ideally suited for use as diffusion pump fluids [85–87]. PFPEs are also established as lubricants and working fluids in vacuum pumps used in the semiconductor manufacturing industry [83]. In fact, they are used as pump fluids in a variety of vacuum pumps from rotary to turbomolecular.

Figure 24.16. PFPEs are used as lubricants in bearing assemblies of satellites because of their low volatility in vacuum and their fluidity at extremely low temperatures

Their long-term retention of viscosity and their minor temperature dependence on viscosity has also made them the lubricant of choice in severe environments such as aerospace jet engines [18,88,89], high-temperature turbine engines [90] and satellite instrumentation (Figure 24.16). Their low vapour pressure, which is a mandatory for space applications, ensures that they are not lost easily to the vacuum of space. Figure 24.17 shows the relative evaporation rates of PFPEs compared to a mineral oil aerospace lubricant. On the basis of time to lose 1.0 ml of lubricant per 1 cm^2 outlet, PFPEs show excellent retention even at elevated temperatures. In particular, Fomblin® Z25 has been used to lubricate the ball-bearing assemblies of scanning instruments installed on several satellites. The vapour pressure of Fomblin® Z25 at 20 °C is less than 4×10^{-10} N m^{-2}. Its extremely low vapour pressure also ensures that there is no contamination of lenses and mirrors in the optics of on-board satellite instrumentation.

PFPEs are commonly used lubricants for compressors and valves which handle aggressive gases such as halogens and gaseous acids (e.g. HF, HCl and HBr). Such is their high stability and low chemical reactivity that PFPEs are the only fluids that have been approved as lubricants for ultracentrifuge bearings in uranium enrichment plants, where chemical resistance to UF$_6$ up to 130 °C is a stringent requirement [83,91]. They are also being considered for use in the next generation of military and supersonic aircraft. Presently they are used in many military aircraft including the Stealth bomber. In addition, the lubrication for the mechanism of prestige watches such as Rolex® is currently provided by PFPEs (Figure 24.18).

Despite the high thermal and chemical stability of PFPE lubricants, recent studies have found that under certain conditions (i.e. oxidizing conditions in the

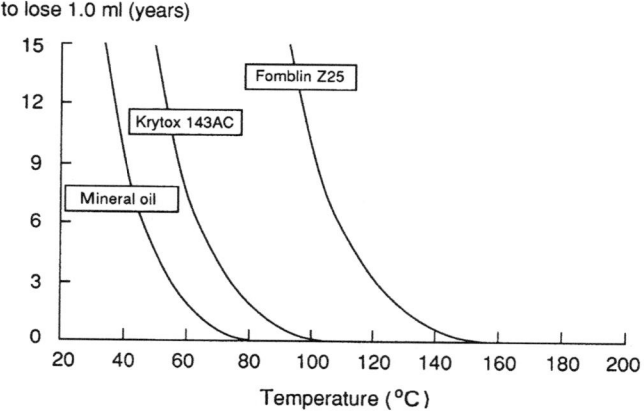

Figure 24.17. Relative volatility of PFPEs and mineral oil as a function of temperature. The low evaporation rate of Fomblin® Z makes it the lubricant of choice in the aerospace industry (adapted from Jones [140])

Figure 24.18. PFPEs are used as lubricants for the delicate and precise mechanisms of Rolex® watches

presence of ferrous or titanium alloys), PFPE lubricants can undergo degradation at relatively low temperature to yield corrosive degradation products [92]. In addition, since lubricants often contain a variety of additives for flow modification and performance enhancement, the overall stability of lubricant is determined by the effect of the additives on the PFPE. Thermal stability of PFPE lubricants containing various fillers has been assessed by measuring weight loss using thermogravimetry. The highest stabilizing effect on PFPE was found with fluorine-containing carbon black while the lowest stability was observed in the presence of Fe_2O_3 and PTFE thickening agents [93]. The stability of the lubricants was found to be increased by the addition of triazine containing additives [93]. Various functionalized PFPEs (e.g. PFPEs terminated with hydroxyl, carboxyl and phosphate groups) when added to base PFPE lubricants have been found to dramatically reduce the rate of wear under sliding contact in a vacuum [94].

4.2 GREASES

PFPEs can be thickened with PTFE powder to produce greases that find niche application for high-temperature use where other greases cannot perform adequately. However, greases prepared by thickening the PFPE fractions with a suspension of telomeric PTFE in trichlorotrifluoroethane are not suitable for high-vacuum application due to outgassing of the solvent. More recently, this problem has been overcome by formulating the greases with dry PTFE powder having an average particle size of 1 μm and MW of 5×10^5 Dalton [81].

PFPE-based greases are used extensively in aerospace applications for their chemical inertness and their resistance to high temperatures. While these properties are desirable for their intended application, they present difficulties when attempting to remove the grease for inspection and repair of military jet components. Traditionally, CFCs have been used to clean these surfaces; however, with their recent phase-out, alternatives have been used. Hydrofluorocarbons were found to be effective solvents particularly those with an eight-carbon chain length [95].

4.3 MAGNETIC RECORDING MEDIA LUBRICANTS

The exceptional lubricity of PFPEs makes them the specified fluid for use in computer hard-drives where their function is to protect the hard disk from the recording head and minimize head/disk interfacial wear and friction [43,77,96,97,98]. They are extensively used for the lubrication of magnetic recording media (MRM) coated on a substrate base which has an active coating of plated or vacuum deposited metal alloys.

In computer hard disks, the disks are composed of a metal alloy and the read/write head 'hovers' above the magnetic surface under the influence of a centrifugal air current generated over the spinning disk. The separation gap is of the order of 0.2 μm. Given such a fine tolerance it is critical that the disk does not possess any macroscopic unevenness. Since the read/write head cannot 'fly' when the disk is not spinning, effective lubrication is essential to deal with 'lift off and landing' situations (during start up/shut down of the disk) as well as to safeguard against the possibility of a 'crash' [96]. As for the requirements of lubricants for outer space, the low vapour pressure of PFPEs guarantees that they will not be lost from a hard disk as a consequence of volatilization over time.

Computer hard disks have a thin metal coating (eg. aluminium- or cobalt alloys) applied by vacuum sputtering. Good adhesion of the PFPE to the magnetic recording disk is necessary to maintain an integral film under the action of centrifugal loss during spinning of the disk. Problems with lubricant retention of hard disks has meant that certain neutral PFPEs have had to be functionalized with polar end-groups to improve adhesion to the metal-oxide disk surface. Two of these functionalized PFPEs are ZDOL and ZDIAC. Functionalized PFPEs, especially those with acid end-groups, show enhanced interaction with the metal oxide (e.g. ZrO_2) surface of the hard disk compared with the non-functionalized neutral oils. It has also been shown that PFPE acids (e.g. Demnum-SH) of shorter chain length are preferentially adsorbed compared to PFPEs with longer chains [99]. This is because, the low molecular weight chains (MW < 1300) possess a greater concentration of terminal acid end groups [99]. Ausimont have developed a functionalized Fomblin® with aromatic end-group functionalities (AM2001) that impart improved adhesion to the cobalt alloy surfaces of hard disks [96].

4.4 DIELECTRIC FLUIDS

Table 24.8 lists some electrical properties of linear and branched PFPEs. The excellent thermal stability of PFPE fluids combined with their outstanding dielectric properties makes them an ideal choice for dielectric fluids [100–102]. PFPEs find widespread use as lubricants in sliding electrical contacts and push-button switches where arc resistance and good lubrication are essential. While silicone oils are also renowned for their chemical stability, high-voltage arcing causes deposition of silicon dioxide on electrical contacts thus lowering conductivity of the contacts. PFPEs, on the other hand, break down to form only gaseous by-products and no solids deposition occurs. PFPE have been investigated by Pirelli Cables as dielectric fluids for use in electrical cables to prevent water ingress [103].

4.5 ELECTRONICS

The excellent electrical insulating properties of PFPEs can be graphically demonstrated by immersing a working television set in a bath of PFPE without

Table 24.8. Electrical properties of HFP and TFE-based PFPE's

Property	Standard	Fomblin Y^a (HFP–PFPE)	Fomblin Z^b (TFE–PFPE)
Dielectric strength (kV/100 mils)	ASTM D877	40	30
Dielectric constant (10^2–10^5 Hz)	ASTM D877	2.15	2.01
Resistivity (Ω.cm)	ASTM D257	10^{15}	4×10^{13}
Dissipation loss (10^2–10^5 Hz)	ASTM D150	4×10^{-4}	5×10^{-4}

[a] Y25.
[b] Z25.

any electrical shorting or fault occurring. PFPEs find widespread use in the electronics field for liquid burn-in testing fluids, thermal shock fluids and heat transfer/cooling media. With specific heat values around 0.23 cal $g^{-1} °C^{-1}$, PFPEs have good heat-transfer properties. The use of PFPEs as coolants has been described [17].

4.6 COATINGS

The low surface energy and high stability of PFPEs makes them eligible as protective materials for stone exposed to atmospheric attack [104,105]. Particularly limestone, since it is especially sensitive to attack by acidic rain. Various PFPEs have been investigated for preservation of masonry, where they can penetrate the pores of the masonry and prevent water permeation [106].

Fluorinated polyurethane coatings for protecting building materials such as brick, stones, masonry from weathering have also been made from crosslinking functionalized PFPEs such as Fluorolink D, cured with isocyanates [107]. ZDOL, which is capped with $-CH_2-OH$ endgroups, can be crosslinked with isocyanate or melamine hardeners to produce clear and durable coatings [108]. Despite the presence of the methylene group, this compound has excellent stability when exposed to UV irradiation as a consequence of the shielding of the $-CH_2-$ by the adjacent fluorine atoms. Functionalized PFPEs such as ZDOL can be crosslinked like conventional diols to produce surface coatings for masonry [109]. Pacansky [110] used a radiochemically cured diacrylate terminated ZDOL to give a low surface energy coating with H_2O contact angles of 107°. Polyurethane coatings have also been prepared from ZDOLTX and an isocyanate hardener [111]. The preparation and performance of such PFPE-based coatings has been reviewed by Scheirs [112–116].

4.7 PFPES AS INTERMEDIATES IN POLYMER SYNTHESIS
4.7.1 Polyurethanes

Segmented polyurethanes with PFPE blocks have been prepared by first functionalizing the liquid PFPE diol with isocyanate functionalities to give chain extension, then the functionalized PFPE (FPFPE) is reacted with short chain diols in the presence of a catalyst such as dibutyl tin dilaurate. These polymers are characterized by an unusually low-temperature elastomeric behaviour due to the low T_g of the fluorinated phase. In comparison to conventional hydrogenated polyurethanes, these fluorinated analogues exhibit enhanced thermal and chemical stability but at an increased cost [117]. They have been considered for biomedical applications such as tubes for use *in vivo*, because of their antithrombogenic properties.

4.7.2 Elastomers

FPFPE diols polymerized with MDI have been used to produce thermoplastic PU elastomers with good low-temperature flexibility and high-temperature mechanical property retention [118]. These elastomers have a tensile strength of 17.9 MPa and an elongation of 492% at 23 °C [118]. Applications include medical tubing.

4.7.3 Epoxy Resins

Chain extended PFPEs have been found to be miscible with bisphenol A (BPA) epoxy resins and can be cured through their terminal carboxylic acid functionalities. The addition of PFPE to a BPA epoxy imparts superior flexural strength, ductility and fracture toughness. By varying the proportion of PFPE and BPA, different morphologies are possible from clear castings to opaque systems comprising precipitated heterogeneous particles [119].

4.7.4 Polyester Resins

PFPEs can be used as comonomers in the preparation of poly(butylene terephthalate). While the functional groups of the FPFPE react completely with the other monomers, the distribution of the PFPE blocks is not homogeneous and phase separation between the PFPE and the PBT occurs. The PFPE imparts only a marginal improvement in fracture and wear resistance [120].

4.7.5 Stratifying Polymers and Paint Additives

The addition of PFPE during the polymerization of methyl methacrylate gives a polymer with an enhanced concentration of the fluoropolymer at the PMMA surface as shown by XPS [121]. The surface concentration of PFPE in this acrylic polymer varied little with the composition of the polymerization mixture indicating that the added PFPE migrates almost totally to the surface. A similar effect was observed when the PFPE was added to a commercial paint. This research is leading the way to the development of self-stratifying paints that segregate to give a fluorine-rich surface.

4.8 CHEMICAL ANALYSIS STANDARDS

PFPEs are ideal as mass calibration standards for mass spectroscopy [122,123]. The high response factor of fluorine in SIMS (negative ion mode) also makes PFPEs useful for leak testing high-vacuum systems. It has been shown that Krytox® gives repeating fragments of $CF(CF_3)CF_2O$ with mass of 166 amu, other peaks correspond to $CF + n(C_3F_6O)$ starting at 31. A characteristic peak at $m/z = 169$ is due to C_3F_7. In fact, from the authors own experience it has been found that PFPEs are persistent and easily detectable by TOF–SIMS at trace levels.

4.9 ULTRA-THIN LAYERS ON ACIDS

A novel application for PFPEs is as a thin layer (<1 μm) on the surface of concentrated mineral acids to impede water condensation, thus extending the lifetimes of acids [124]. PFPEs spread spontaneously on concentrated mineral acids to give stable and insoluble films. The characteristics of PFPEs which allow it to be used in this way are an outstanding chemical resistance to strong acids and the fact that they can spread into an ultra-thin layer on concentrated sulphuric acid, nitric acid and phosphoric acid. Since a correlation has been found between the Hammett activity of the acid and the degree of spreading, this spreading behaviour is believed to be driven by hydrogen bonding between the ether oxygen of the PFPE and the acid [124]. The weak attractive forces in neutral PFPE liquids also contribute to their propensity to spread into a layer [125].

4.10 COSMETICS

PFPEs represent ideal candidates for cosmetics since they are toxicologically innocuous and cause no dermal irritation. Furthermore, since PFPEs are both hydrophobic and lipophobic (i.e. incompatible with both water and oils) they form a third phase when added to a normal two-phase (water–oil) emulsion. This third phase creates a stable emulsion and prevents settling of the oil phase thus improving the shelf-life of cosmetics such as moisturizers. Also, as a result of their hydrophobic–lipophobic character, PFPEs can form thin protective films on human skin where they repel aqueous acids and caustic solutions as well as hydrocarbon oils and greases, thus they find application in barrier creams. Fomblin® HC is the particular grade used most often for cosmetic applications. Fomblin® PFPE polymers are used in cosmetic formulations at levels of up to 10% wt/wt. They also impart lubricity and ensure that the make-up spreads smoothly [126,127,128].

5 DEGRADATION AND STABILITY

5.1 INTRODUCTION

PFPEs are characterized by their exceptional stability and inertness. It is the absence of hydrogen atoms in their structure and the strong covalent character of C−F and C−O bonds that makes PFPEs remarkably non-reactive and stable as compared with their hydrogenated counterparts [83].

5.2 THERMAL STABILITY

PFPEs based on HFP, such as Krytox® have extremely high thermal stabilities with the onset of thermal decomposition occurring at temperatures as high as 410 °C [129]. Though, on examination of its structure, Krytox® would be antic-ipated to be thermally less stable than unbranched PFPEs such as Fomblin® Z, since tertiary carbon–fluorine bonds are normally less stable than those containing primary or secondary carbon atoms. In addition, Krytox® PFPEs can possess some hydrogen chain termination, nevertheless Fomblin® Z has lower thermal stability than Krytox®. The cause of this thermal instability has been attributed to the presence of acetal linkages (i.e. $[OCF_2O]_n$) in the main chain. Koch and Jantzen [130] examined the thermal oxidative stability of a number of PFPEs used as lubricating oils in high-temperature aircraft turbine engines and concluded that polymers without the acetal structure exhibit improved thermal stability.

Helmick and Jones [131] determined that the decomposition temperature of Demnum® ($378\,^{\circ}C \pm 2\,^{\circ}C$) is independent of molecular weight while the decomposition temperature of Krytox® was found to vary in the range 356–$376\,^{\circ}C$ depending on the polymer's molecular weight (with the longer chains giving lower decomposition temperatures).

The thermo-oxidative stability of commercial Krytox® PFPEs can be adversely affected by the presence of chain end impurities as a result of hydrogen termi-nated chains; e.g. $C_3F_7O[CF(CF_3)CF_2O]_XCF(CF_3)\mathbf{H}$. It is interesting to note that despite the presence of the hydrogen impurities these fluids show good stability in inert atmospheres even at temperatures of $343\,^{\circ}C$ [129]. It is also interesting that thermal treatment of Krytox® at $340\,^{\circ}C$ in an oxygen atmosphere leads to the removal of the hydrogen impurities without an appreciable decrease in MW. Other studies have established that the thermal stability of PFPEs is determined mainly by the stability of the end groups. Synder and Dolle [90] developed a micro-oxidation test for PFPE lubricants and determined that ZDOL decomposes at $240\,^{\circ}C$.

5.2.1 Gaseous Degradation Products

PFPEs generally undergo residue-free thermal degradation exclusively into gaseous products [6]. The nature of these volatiles has been investigated under degradative conditions. Sianesi found that during the thermal degradation of Fomblin® Y the major volatiles were C_3F_6, CF_3COF and COF_2 and the last two compounds were observed to hydrolyse to HF, CF_3COOH and HF and CO_2 respectively. However, when the decomposition is performed in air, COF_2 is a major decomposition product [6]. Gaseous products recovered after γ-irradiation under vacuum include COF_2 (major product), C_2F_4 (epoxide), and C_2F_6 in the case of Fomblin® Z, and C_2F_6, C_3F_8 in the case of Fomblin® Y and Krytox® [36]. Pan [49] studied the gaseous decomposition products of ZDOL during irradiation by x-ray photons and the primary products were COF_2, CO_2 and C_2F_6.

Electron beam irradiation studies show that unbranched PFPEs degrade more efficiently and produce more COF_2 than branched polymers when the monomer units contain CF_2CF_2O and/or COF_2 in the main chain. With a pendant perfluoromethyl (i.e. $CF(CF_3)CF_2O$) or an additional carbon atom in the main chain monomer unit (i.e. $CF_2CF_2CF_2O$), the electron beam induced degradation is comparatively less efficient, while in the branched PFPEs, CF_4 formation competes with COF_2 formation [132]. The gaseous products evolved during the wear-induced degradation of PFPEs (ZDOL) has been studied by Strom [98].

Since the degradation of PFPEs leads to the formation of a large amount of volatiles, the pressure rise in a closed system can be used to measure the extent of the degradation reaction. A device which has been used for many years to measure this pressure rise and thus the extent of thermal decomposition of PFPEs is the isoteniscope [131]. In fact, it is now a standard ASTM test procedure for measuring the initial thermal decomposition temperature of fluids (ASTM D-2879-75; 1980).

While PFPEs are toxicologically inert, their thermal decomposition products can be toxic. In particular, in the thermal degradation of Fomblin® Y, because of its branched structure, highly toxic perfluoroisobutene (PFIB) can be produced. In contrast, Fomblin® Z by virtue of its linear chain structure cannot form dangerous quantities of PFIB during thermal decomposition, although small quantities of COF_2 and CF_3COF are evolved [133].

5.3 DEGRADATION BY METALS

Thermodynamically, the thermal stability of PFPEs can be partially attributed to the fact that neither OF nor OF_2 are good leaving groups. Whereas, in fluorocarbon polymers, HF is a thermodynamically favoured species because it is an excellent leaving group. However, in the case of PFPEs, the formation of MF (where M = metal) is a thermodynamically favoured process and the high free energy of formation of M−F bonds renders PFPEs relatively unstable in the presence of metals, metal salts and oxides at elevated temperatures. This interaction represents the Achilles' heel of PFPEs and limits their awesome potential as lubricants.

PFPEs have been used as lubricants and greases for metal assemblies in space applications for over two decades [134]. At first these fluids performed satisfactorily, as early applications placed few demands on their performance. However, as other bearing components and materials became more advanced and applications more demanding, certain deficiencies of the PFPE lubricants have been exposed. The major shortcomings of PFPEs from a stability point of view are their instability in the presence of metals and Lewis acids.

It has been noted that PFPEs are susceptible to degradation by strong electron-transferring compounds such as Mg and Al at temperatures above $100\,°C$ and especially where shear produces freshly exposed metal surfaces. Combinations of PFPEs and metals that should be avoided are: (1) Fomblin® Y and titanium–aluminium alloys, and (2) Fomblin® Z and aluminium. Table 24.9 lists various engineering metals and their alloys which are reported to induce PFPE degradation.

Table 24.9. Various metals, alloys and salts that induce PFPE degradation

Metal/Alloy/Salt	PFPE	Tdec (°C)	Atm.	Ref.
Al	Fomblin Y	—	—	Sianesi [6]
Al	Fomblin Z	150	—	Herrera-Fierrro [47]
AlF_3, $AlCl_3$	Krytox	350	—	Carre [58]
$AlCl_3$	Fomblin Z	100	—	Bierschenk [18]
Fe alloys	Fomblin Z	315	—	Jones [135]
FeF_3	Krytox	300	—	Carre [58]
Fe_2O_3	Fomblin Z	185	—	Zehe [136]
M-50 steel	Krytox	343	Inert	Paciorek [129]
M-50 steel	Krytox	316	Oxidizing	Paciorek [129]
SiN/stainless steel	Krytox	—	—	Carre [44]
Stainless steel	Fomblin Z	—	—	Mori [142]
Steel	Fomblin Z	440	Vacuum	Mori [142]
Ti	Fomblin Z	288	Oxygen	Jones [5]
Ti	Fomblin Y	—	—	Sianesi [6]
Ti alloys	Krytox	316	Inert	Paciorek [129]
Ti(4Al,4Mn)	Krytox	288	—	Jones [135]
Ti(4Al,4Mn)	Krytox (H-free)	316	—	Paciorek [129]
Ti(6Al, 4V)	Krytox	—	—	Carre [58]
Ti(8Mn)	Krytox	—	—	Carre [58]
TiO_2	Fomblin Y	300	—	Kasai [57]
ZrF_4	Fomblin Y	250	—	Kasai [57]
ZrO_2	Fomblin Y	300	—	Kasai [57]

Tdec = decomposition temperature

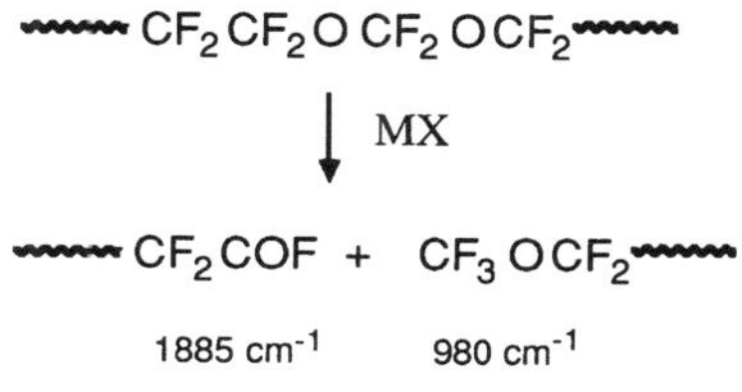

$$\text{CF}_2\text{CF}_2\text{O CF}_2\text{OCF}_2$$

$$\downarrow \text{MX}$$

$$\text{CF}_2\text{COF} + \text{CF}_3\text{OCF}_2$$

$$1885 \text{ cm}^{-1} \qquad 980 \text{ cm}^{-1}$$

Figure 24.19. Mechanism of metal-catalysed degradation of Fomblin® Z. MX represents a metal or alloy

5.3.1 Decomposition Reactions

It has been shown that exposing Krytox® to a Ti alloy (namely, Ti(4Al, 4Mn)) at elevated temperatures (e.g. 300 °C) leads to the production of a significant amount of volatiles followed by a drastic drop in MW of the PFPE [135]. The mechanism of this degradation is now well understood. The driving force for the reaction is the formation of metal fluorides (e.g. TiF_4 and AlF_3) which are formed by reaction of the primary degradation products of the PFPE (COF_2 and CF_3COF) with the metal oxides (Al_2O_3 and TiO_2). The metal fluorides once formed then act as catalysts for the chain scission reaction (Figure 24.19) which in turn produces more volatile degradation compounds. In the case of steel systems and PFPEs, FeF_3 has been identified as the primary reaction product and the main degradation catalyst [136]. It has been observed that Fomblin® Z degrades significantly more in the presence of steel than iron. This has been attributed to the fact that alloys and mixtures are more reactive than pure metals, since in the latter the lattice energy provides an extra reaction barrier.

Thus, the underlying cause of PFPE degradation in tribological environments is that the rubbing process removes the protective native oxide from the metal bearing surface and allows the PFPE to contact the atomically clean metal. The clean metal then degrades the PFPE even at room temperature to produce corrosive gases which subsequently react with the surface metal to produce metal fluorides which are strong Lewis acids. Since Lewis acids are electron acceptors they can readily attack and decompose PFPEs. The decomposition of PFPEs yields more reactive species which in turn produce more Lewis acids and thus the process becomes autocatalytic. This mechanism is consistent with the observation by Koch and Jantzen [130] that the interface between PFPEs and the metal surface is comprised of a mixed oxide-fluoride layer and that this layer exhibits catalytic activity in the presence of PFPEs. This mechanism of PFPE decomposition depends on the type and thickness of metal oxide, temperature, load and sliding speed [134]. Experience has shown that Krytox® fluids are more resistant to this mode of decomposition than Fomblin® Z fluids [134].

5.3.2 Kinetics of Metal Catalysed Degradation

Zehe and Faut [136] studied the degradation of Fomblin® Z in the presence of Fe_2O_3 and observed that the decomposition proceeds in two stages: first a slow catalytic reaction between the Fe_2O_3 and gaseous decomposition products of the PFPE leading to the formation of FeF_3, then in a distinct second stage the FeF_3 acts as a Lewis acid causing rapid catalytic decomposition of the PFPE. In an analogous study, Kasai [63] using thermogravimetry and NMR data examined the degradation of four PFPEs (Fomblin® Y, Fomblin® Z, Demnum® and Krytox®) in the presence of both Al_2O_3 and $AlCl_3$ at 200 °C (Figure 24.20). Interestingly, only Fomblin® Z decomposed significantly and it was concluded that it is the difluoroacetal groups in Fomblin® Z that are thermally weak links in the presence of the Lewis acid. Krytox® and Demnum® are stable because of the lack of the difluoroacetal group in their structure. As in the work by Zehe [136], the decomposition was again observed to proceed in two stages: the slow formation of AlF_3 followed by the rapid catalysed decomposition of the polymer. This reaction sequence is summarized in Figure 24.21.

5.3.3 Mechanism of Decomposition

Two mechanisms have been proposed for the initial step in the catalysed degradation of PFPEs. The Kasai mechanism [63] involves the donation of electron

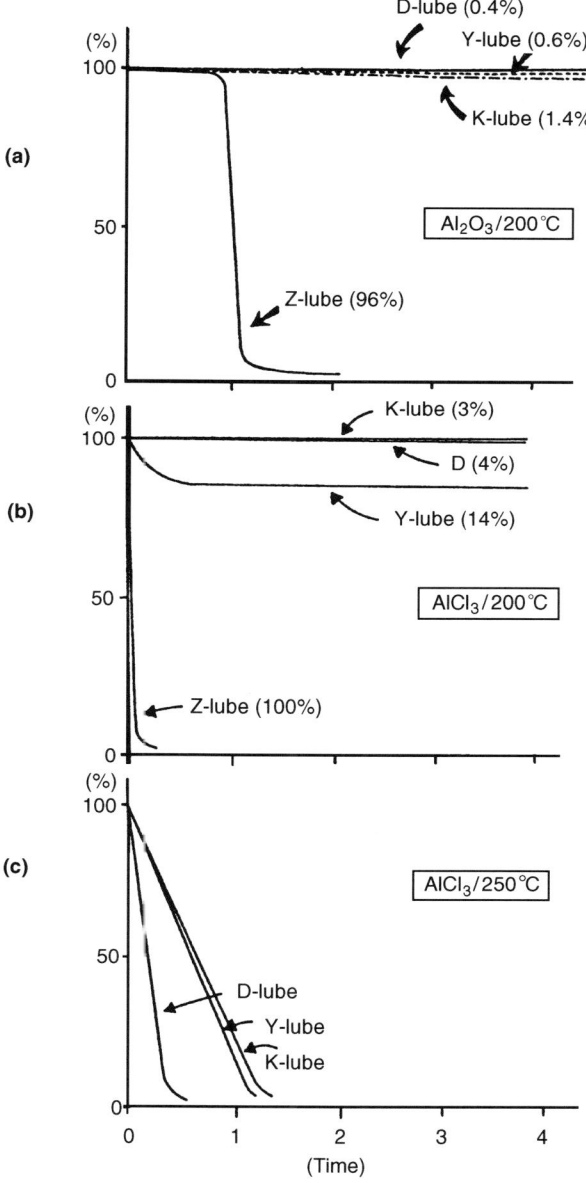

Figure 24.20. Thermogravimetry curves for PFPEs in contact with Al_2O_3 and $AlCl_3$ at elevated temperatures (from Kasai [57]). Reproduced by permission of American Chemical Society Note: D, K, Y and Z denote Demnum®, Krytox®, Fomblin® Y and Fomblin® Z respectively.

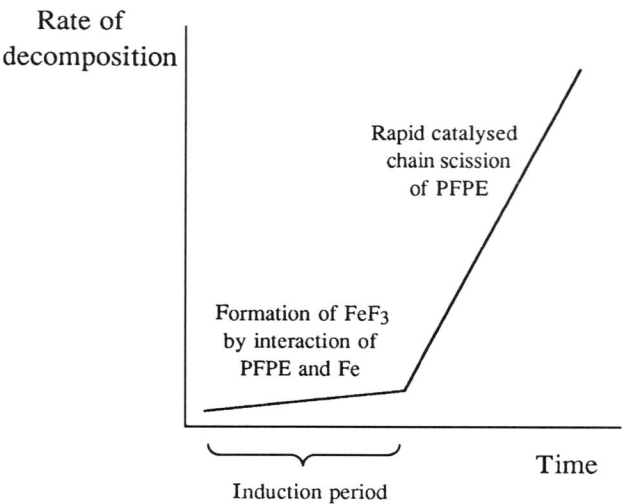

Rate of decomposition

Rapid catalysed
chain scission
of PFPE

Formation of FeF$_3$
by interaction of
PFPE and Fe

Time

Induction period

Figure 24.21. Schematic representation of the kinetic steps in the reaction between metals and PFPEs. During the initial stage, a slow breakdown by weakly acidic sites on the metal oxide surface results in the formation of COF$_2$ which converts the metal oxide to metal fluoride. The metal fluoride which is a strong Lewis acid then decomposes the PFPE

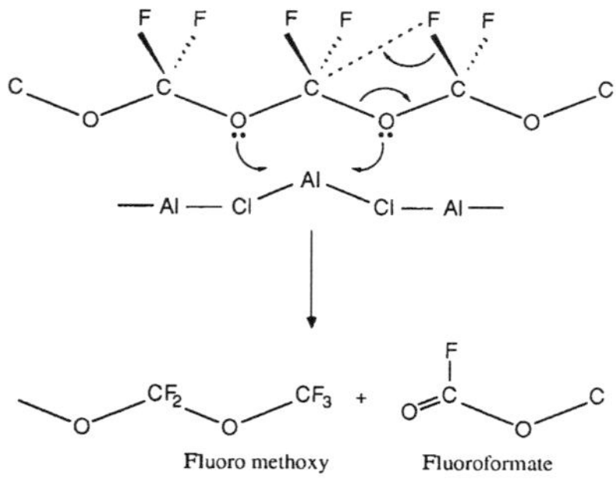

Fluoro methoxy   Fluoroformate

Figure 24.22. Mechanism of Lewis acid-catalysed degradation of Fomblin® Z. Note a partial positive charge on the acetal group induces a fluorine atom transfer from the adjacent CF$_2$ leading to scission of the PFPE chain. In the presence of Lewis acids (such as AlCl$_3$) Fomblin® Y is more stable than Fomblin® Z because of steric hindrance provided by the pendant trifluoromethyl branches suppress disproportionation by limiting access to catalytic sites (Kasai, Tang and Wheeler [63])

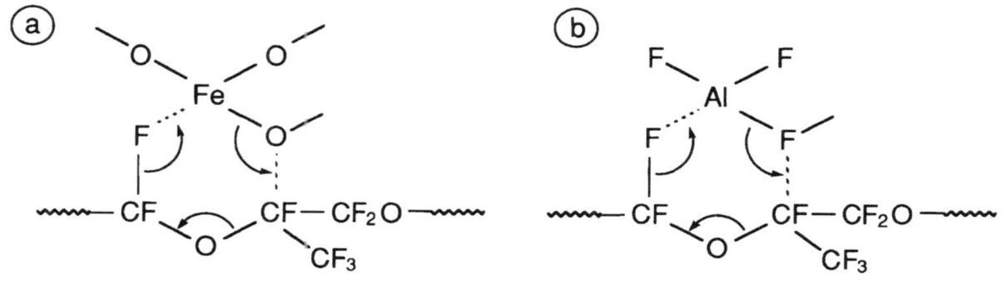

Figure 24.23. Mechanism involving acidic attack on the nonbonding electrons of the fluorine atoms, followed by fluorine abstraction to yield a stable carbenium species (after Zehe and Faut [136])

density from the oxygen atoms in the polymer chain to one or more acidic metal atoms on the Lewis acid surface (Figure 24.22), while, the Zehe mechanism [136] involves acidic attack on the non-bonding electrons of the fluorine atoms, followed by fluorine abstraction to yield a stable carbenium species (Figure 24.23).

5.3.4 Weak Links

Since the formation of the M−F bond is the overriding consideration when it comes to the thermal stability of PFPEs it follows that the weaker the C−F bond, the more readily the M−F bond will form. The question then remains 'Do all the linkages in PFPEs have the same stability, and which is the weakest?' Theoretical work by Smart and Dixon [137] has shown that the acetal linkages ($-OCF_2O-$) present in Fomblin® Y and Fomblin® Z have a lower thermo-oxidative stability than the various linkages present in Krytox® PFPEs. Furthermore, in Fomblin® Z (because of the higher concentration of both $-OCF_2O-$ and $-OCF_2OCF_2O-$ than in the Fomblin® Y fluids), these liquids show lower oxidative stability in the presence of metals, their oxides and alloys, at temperatures as low as 100 °C.

This is consistent with results reported in the literature where Fomblin® Z was found to be completely degraded in the presence of Fe_2O_3 at 185 °C while Krytox® and Demnum® fluids undergo less than 5% degradation under the same conditions. Table 24.9 shows that the degradation of Fomblin® Z can occur at temperatures as low as 100 °C in the presence of $AlCl_3$ (Bierschenk [18]). Through the application of XPS it has been shown that the degradation of Fomblin® Z is accompanied by the preferential depletion of $-OCF_2O-$ linkages which supports the arguments that these are the weak links in the PFPE chain.

Figure 24.24 shows 'space-filled' representations of the molecular structures of three commercial PFPEs. The oxygen atoms are fully rendered black. From these the accessibility and susceptibility of the two oxygen atoms of the acetal unit of Fomblin® Z is apparent. This explains the high reactivity of Fomblin® Z with both metals and Lewis acids. Although Krytox® PFPEs can also interact

Fomblin Z

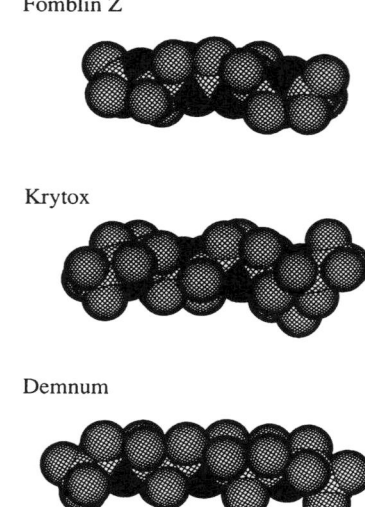

Krytox

Demnum

Figure 24.24. Space-filling representations of the molecular structures of commercial PFPEs. The oxygen atoms are fully rendered black. Note: the accessibility and susceptibility of the two oxygens of the acetal unit of Fomblin® Z (from Kasai [57]; reproduced by permission of American Chemical Society)

with metal atoms or Lewis acids this requires reorientation of the chain and thus presents a higher activation barrier.

5.3.5 Effect of Metal Type

From the literature it is apparent that titanium and its alloys exert one of the most powerful effects on accelerating the degradation of Fomblin® Z. Iron is also very reactive towards PFPEs because of the strength of the iron–fluoride bond and the strong electron-donating ability of metallic iron. Steel alloys are much less effective at promoting degradation while cobalt metal does not promote the decomposition of Fomblin® Z even in oxygen at 316 °C [129].

Different PFPEs exhibit different reactivities towards various metals. For example, Krytox® fluids are unaffected by Fe_2O_3, FeF_3, Al_2O_3 and AlF_3 under oxidizing conditions up to 316 °C. In contrast, Fomblin® Y and Z are degraded by these metal compounds. The reactivity of different PFPEs is very dependent on their structure and in the case of Fomblin® Z, as discussed earlier, on the presence of $-OCF_2O-$ linkages. Herrera-Fierro, Jones and Pepper [47,48] have shown that even at room temperature Fomblin® Z on aluminium oxide shows measurable degradation as determined by the depletion of $-OCF_2O-$ groups by XPS. In fact, the XPS technique proved central to determining that it is the acetal linkages in Fomblin® Z that make this PFPE susceptible to attack by metals and their salts. A decrease in the intensity of the high-energy C1s peak (that is the acetal carbon; $-O-CF_2-O$) was observed by XPS, suggesting the preferential consumption of this portion of the molecule in the presence of Lewis acids.

5.3.6 Al-catalysed Decomposition

PFPEs are strongly degraded by aluminium from freshly abraded metal surfaces (a strong electron transfer reducing agent). The decomposition of ZDOL in the presence of α and γ-alumina has been studied in detail by Morales [138]. It has been found that Al_2O_3 can catalyse the thermal degradation of ZDOL. A mechanism for the thermal decomposition of ZDOL in the presence of alumina involved an exothermic reaction and the production of AlF_3.

5.3.7 Degradation in Space

Metal-catalysed degradation of PFPE lubricants has caused failures in space equipment. This problem is particularly troublesome in space where because of the high vacuum present, stable metal oxides do not reform on metal after the original oxide layer is removed by tribological wear processes. Thus, the PFPE is constantly in contact with fresh metal surfaces. It is these fresh metal surfaces that show high reactivity with PFPEs. In addition to extensive degradation of the PFPEs in the presence of catalytic metals, severe corrosion problems on metallic surfaces have also been experienced. Problems have been encountered when using PFPEs as lubricants in ball-bearings on orbiting earth satellites — such as anomalous behaviour of a satellite scanner instrument [140]. The breakdown of PFPE lubricants leads to a sudden increase in instrument temperature and causes slip-stick behaviour of scanning mechanisms.

5.3.8 Effect of Atmosphere on Degradation

PFPEs generally have outstanding thermally stability up to $300\,^{\circ}C$ and, interestingly, oxygen exerts little effect on their thermal degradation. However, in the presence of certain catalysts (Lewis acids [e.g. $AlCl_3$], strong nucleophiles [e.g. amines] and electropositive metals [e.g. aluminium]), the degradation of PFPEs is strongly influenced by the presence of oxygen. Figure 24.25 shows the effects of both aluminium and atmosphere on the thermal degradation of Fomblin® Z.

The thermal decomposition of PFPE on titanium has been studied by reflectance spectroscopy [139], by measuring the conversion of Ti to TiF_3. It was found that the presence of oxygen accelerated the degradation.

The following order of lubricant lifetimes have been obtained using a wear apparatus comprising four stainless steel ball-bearings [140].

In air   Krytox® ∼ Fomblin®Z > Demnum®

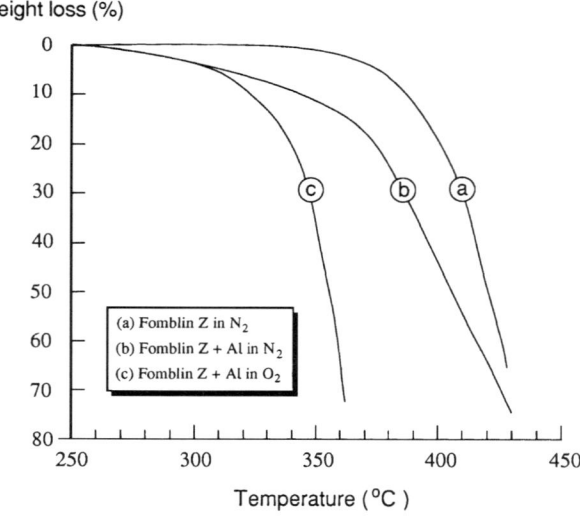

Figure 24.25. Thermogravimetric curves for Fomblin® Z. Note: the pronounced effect that both aluminium and oxygen have on the decomposition rate of the PFPE

A different order of lifetimes is found under oxygen starvation:

In vacuum   Krytox® ~ Demnum® ≫ Fomblin®Z

This trend is consistent with service experience of their performance in space environments where Krytox® has been found to have longer lifetimes than Fomblin® Z [140].

5.4   LEWIS ACIDS

The strong reactivity of PFPEs with Lewis acids is thermodynamically favoured because the reactions involve resonance stabilized carbo–cation intermediates,e.g.

$$\ddot{O}-C^+ \longrightarrow\ {}^+O=C,\ F-C^+ \longrightarrow\ {}^+F=C$$

The decomposition of PFPEs by Lewis acids has prompted research on model compounds, to establish the potential basic sites within the perfluoroalkyl ether structure. Work on the model PFPE perfluorodiethyl ether ($CF_3CF_2-O-CF_2CF_3$) has shown [141] that although protonation at the oxygen linkage is more energetically favoured, protonation at the fluorine is only marginally higher in energy and provides the formation of an excellent leaving group, namely HF.

The C−O bonds of the PFPE are susceptible to attack by Lewis acids because the lone pair of electrons of the oxygen can coordinate with the Lewis acid (Figure 24.22). This explains why Fomblin® Z is readily attacked by strong

Lewis acids at temperatures as low as 100 °C. However, in the case of Krytox® and Fomblin® Y the pendant CF_3 groups assist in sterically shielding the oxygen from attack.

5.5 BOUNDARY CONDITIONS

Boundary conditions occur in situations where relative motion of two surfaces that are contacting each another produce hot spots on a molecular level, for short durations. Such conditions have been encountered in bearings of spacecraft, aircraft engines and in computer hard drives. The temperatures obtained at the asperity (high points) contacts are sufficiently high (500 °C as estimated by Carre [44]) to cause thermal degradation of PFPE lubricants. Experiments with Fomblin® Z have shown that in the presence of steel under boundary conditions in vacuum, this PFPE degraded almost completely to yield volatile carbonyl fluoride [142]. In contrast, no volatiles were formed with Krytox® and Demnum® fluids under the same test conditions. The last two however yield metal fluorides, which also fortuitously function as effective lubricants. These metal fluorides though are also degradation catalysts. In the absence of oxygen, as in spacecraft orbital environments, the formation of FeF_3 and subsequent catalytic PFPE decomposition can be a significant pathway under boundary lubrication conditions [44]. In addition to thermal degradation, mechanochemical scission of PFPEs can occur under the extremely high shear rates (up to 10^{10} s^{-1}) encountered during sliding in boundary-lubricated systems.

5.6 THERMAL STABILIZATION

There has been a great deal of research performed on strategies to increase the thermo-oxidative stability of Fomblin® Z both by the addition of stabilizers and through modification of the chain structure. The use of oxidation inhibitors can offer significant improvements in PFPE stability by up to several orders of magnitude [5,135]. The most effective inhibitors developed thus far are perfluorophenyl phosphine and phosphatriazine. For instance, the stability of Fomblin® Z in the presence of titanium and ferrous alloys at temperatures up to 316 °C has been enhanced through end-capping the polymer with phospha-s-triazine groups [143].

5.7 UV DEGRADATION

The UV stability of PFPEs is exceptional as one would expect, given that the genesis of some types of PFPEs is from the UV-catalysed photo-oxidation of fluoro-olefins and where the photolabile groups (i.e. the peroxides) are subsequently removed. In addition, they absorb no radiation in the mid or near UV spectrum. However, if one uses incident light far enough into the UV spectrum, degradative reactions can be initiated. For instance, it has been demonstrated that PFPEs can be attached to a variety of substrates (e.g. amorphous carbon, silica, gold) by irradiation with 185 nm UV light. Interestingly, the PFPE does not wash off the surface, even with CFC-113 in which the PFPE is normally soluble [144]. It has been established that the mechanism operative here involves the UV light causing photoelectrons to be emitted from the substrate and these electrons then interact with the PFPE leading to the formation of F⁻ ions and PFPE radicals [144].

5.8 RADIATION RESISTANCE

Perfluorinated polymers generally exhibit high susceptibility to degradation by ionizing radiation and PFPEs are no exception. Although not extensively studied, it has been established that PFPEs are readily degraded by low-energy electrons, high-energy electrons, ion-beams and x-rays. The relative radiation resistance of Krytox®, Fomblin® Z and Demnum® has been studied by Mori and Morales [142] by exposing thin films of the PFPEs to x-rays in a commercial photoelectron spectrometer (XPS instrument) and measuring the change in chamber pressure as a function of time. It was found that Fomblin® Z demonstrates the highest rate of degradation followed by Krytox® (Figure 24.26). Demnum® exhibited the best resistance to radiation with its rate of degradation being about half that of the other two PFPEs. The mechanism of radiation degradation of Fomblin® Z is shown in Figure 24.27.

The superior radiation resistance of Demnum® compared with Fomblin® Z can be attributed to the fact that in Fomblin® Z, a radical is formed that propagates through the acetal linkages causing chain scission while in Demnum® a

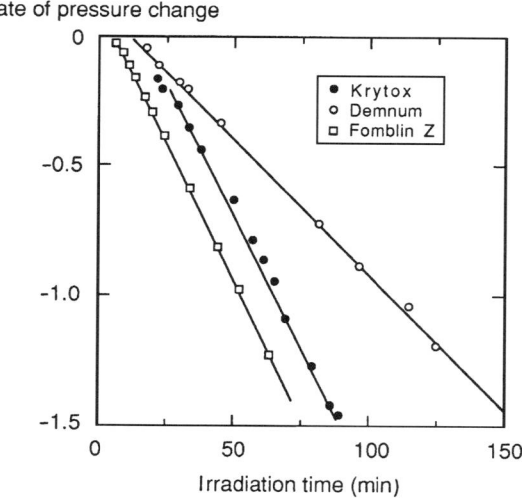

Figure 24.26. Pressure change during the radiation-induced degradation of PFPEs. Note: Fomblin® Z shows the fastest rate of degradation while Demnum® PFPE is the most stable (adapted from Mori and Morales [142])

$$CF_3\text{-}(OCF_2CF_2)_{\overline{p}}\text{-}(OCF_2)_{\overline{q}}\text{-}OCF_3 \xrightarrow{\text{X-rays}} F\cdot \qquad (1)$$

$$F\cdot \ + \ CF_3(OCF_2CF_2)_{\overline{p}}\text{-}(OCF_2)_{\overline{q}}\text{-}OCF_3 \longrightarrow \qquad (2)$$

$$CF_3\text{-}(OCF_2CF_2)_{\overline{p}}CF_3 \ + \ \cdot OCF_2OCF_2OCF_2\text{-}$$

$$\cdot OCF_2OCF_2OCF_2\text{-} \longrightarrow COF_2{\uparrow} \ + \ \cdot OCF_2OCF_2\text{-} \longrightarrow \qquad (3)$$

$$\text{Unzipping} \qquad COF_2{\uparrow} \ + \ \cdot OCF_2\text{-} \longrightarrow \text{etc.}$$

Figure 24.27. Reactions involved in the X-ray induced degradation of Fomblin® Z. Note the unzipping reaction (3) accounts for the copious quantities of carbonyl fluoride (COF_2) evolved during the decomposition of Fomblin® Z (from Mori and Morales [142])

stable radical is formed (namely $-OCF_2 \cdot CFCF_2O-$) which has a propensity to crosslink with other radicals rather than decompose [142].

5.9 ELECTRON BEAM IRRADIATION

The effects of electron beam irradiation on PFPEs have been extensively studied by Pacansky. Crosslinking of PFPEs has been observed in thin films on exposure to high-energy electron irradiation (25 keV) [25].

5.10 ATOMIC OXYGEN

PFPEs can be used aboard spacecraft as radiator fluids where they dissipate heat in order for the spacecraft to maintain an acceptable operating temperature [145]. In such radiators, liquid droplets of the PFPE are sprayed out into space to radiatively cool, they are then collected and recirculated [146].

In this application, the PFPEs are exposed to atomic oxygen, especially when the spacecraft is in low-Earth orbit. It has been calculated that atomic oxygen with an average energy of 4.25 eV will impact the spacecraft. PFPE lubricants have been exposed to atomic oxygen in order to simulate exposure environments in low-Earth orbits [147]. Tests have shown that PFPEs are more resistant to atomic oxygen attack than silicone-based liquids. However, PFPEs nevertheless show a weight loss when exposed to an oxygen plasma, with Krytox® PFPE being more resistant than Fomblin® Z [140].

CONCLUDING REMARKS

The versatile properties of PFPEs makes them excellent candidates for a host of applications. PFPEs combine all of the following novel properties into the one polymer:

- chemical inertness (except in the presence of reactive metals)
- high lubricity (due to their low intermolecular forces)
- high thermal stability
- low dependence of viscosity on temperature
- extremely low volatility (compared to other liquids)
- non-flammable
- the ability to spread into very thin layers (down to 20 Angstrom)
- hydrolytic stability in moist, hot environments
- low toxicity
- no odour or taste

PFPEs have very low Tg values (generally less than $-56°C$) because of the high mobility of the backbone chain segments. Their high chain flexibility is imparted by the oxygen atoms in the chains which essentially act as 'molecular hinges'. This allows them to exhibit liquid-phase behaviour over a wide temperature range, from cryogenic temperature to above 300°C. Obviously, as more demanding applications for polymers come into existence, PFPEs will play an increasingly important role, given their chemical stability and temperature resistance.

It is the absence of hydrogen atoms in the structure of PFPEs and the shielding provided by the electron clouds of the highly electronegative fluorine atoms which makes these liquids almost totally unreactive. For these reasons they are resistant to oxidation reactions that plague hydrocarbon polymers, since PFPEs have no hydrogen atoms that can be abstracted by free radicals.

The exceptional chemical resistance of PFPEs is highlighted by the fact that they can coexist with concentrated mineral acids without undergoing any measurable degradation. They can also survive intact, indefinitely in the presence of strong alkalis and halogens. This makes them ideal candidates for use as greases in the chemical processing industry.

Their unique rheological properties and high-temperature stability will continue to make PFPE oils the material of choice for demanding applications such as hydraulic fluids and high-temperature lubricants. PFPE lubricants can be successfully used to provide lubrication in environments where hydrocarbon lubricants would rapidly degrade; such as bearing assemblies in industrial ovens, furnaces and even in uranium enrichment plants. PFPE greases are used in cryogenic gas production since they do not pose a burning hazard in a pure oxygen atmosphere.

Since PFPEs are able to be functionalized with a range of end-groups, they can be customized to a myriad of applications. Thus functionalization is set to expand the opportunities for exploiting PFPEs to their maximum potential. Bifunctional derivatives of PFPEs can be used as reactive intermediates in polymer synthesis (e.g. fluorinated polyurethanes can be produced from PFPE diols) and also enable the production of lubricants whose polarity is tailored to that of the surface requiring lubrication. Such lubricants thus combine permanence with excellent thin-layer boundary lubrication characteristics. The solubility parameter of functionalized PFPEs is also modified by the incorporation of hydrogen-containing end groups into their structure thereby increasing compatibility and miscibility with polar organic solvents.

As a result of the abundance of oxygen atoms in their structure, PFPEs have an unusually high capability for dissolving gases. Similarly, PFPE-derived polymers and coatings exhibit high gas permeabilities. This property can be exploited in the production of 'breathable' coatings. Since PFPEs are also both oleophobic (oil-repellent) and hydrophobic (water-repellent) they can be used in the formulation of protective coatings.

A particularly interesting area that is currently emerging is that of microemulsions comprised of functionalized PFPEs in water. Aqueous microemulsion PFPE-based lubricants may allow the beneficial characteristics of PFPEs to be attained at a much reduced cost. Furthermore, ternary systems of functionalized PFPEs, water and alcohol can give monophasic systems over a wide composition range and these mixtures can be used as cleaning and degreasing agents. Growing fields of application for such PFPE emulsions are cosmetics, barrier creams and car polish/wax additives.

Perhaps the single most important factor limiting the wider application of PFPEs at present is their high cost. This is set to decrease as newer methods of PFPE production are realized. For instance, the direct fluorination of hydrogenated polyether precursors is a relatively low cost production route. However, degradation problems need to be overcome if high molecular weight PFPEs are to be commercially produced by this method.

As their cost is reduced through larger scale manufacture and cheaper synthesis methods, PFPEs may also be used in high-volume applications such as heat exchange fluids in the electrical industry owing to their excellent heat transfer and dielectric properties.

Functionalized PFPEs, by virtue of their non-flammable nature also have potential as fire extinguishants/fire suppressants and may indeed be used in the future as Halon® substitutes since PFPEs are note ozone depleting.

For more detailed information on the structure-property relationships of PFPEs the reader is referred to a recent review by Marchionni *et al.* from Ausimont, Italy [148].

ACKNOWLEDGEMENTS

Acknowledgement is given to the following persons for contributing information to this chapter: G. Nathanson (Dept. of Chemistry, University of Wisconsin), D. Fowler, P. Kasai and T. Karis (IBM Research, San Jose), B. Smart (Dupont R & D, Wilmington), M. Napier (Dept. of Chemistry, Harvard University), K. McGrath (Naval Laboratories, Washington), and W. R. Jones and W. Morales (NASA, Lewis Research Centre, Cleveland, Ohio). Appreciation is extended to

L. O'Toole (Laboratory for Surface & Interface Analysis, University of Sheffield, UK) for allowing access to their extensive SIMS facilities.

REFERENCES

1. Sianesi, D., Pasetti, A. and Corti, C., *Makromol. Chem.*, **86**, 308 (1965).
2. Sianesi, D. and Fontanelli, R., *Makromol. Chem.*, **102**, 115 (1967).
3. Sianesi, D., Pasetti, A., Fontanelli, R., Bernardi, G. C. and Caporiccio, G., *La Chimica e l'Industria*, **55**, 208 (1973).
4. Hill, J. T., *J. Macromol. Sci. Chem.*, **A8**, 499 (1974).
5. Jones, W. R., Paciorek, K. J. L., Ito, T. I. and Kratzer, R. H., *Ind. Eng. Chem. Prod. Res. Dev.*, **22**, 166 (1983).
6. Sianesi, D., Pasetti, A., Fontanelli, R. and Binaghi, M., *Wear*, **18**, 85 (1971).
7. Sianesi, D., Pasetti, A. and Belardinelli, G., US Patent 3 715 378 (1973) *Chem. Abs.*, **78**, 125179m.
8. Sianesi, D., Marchionni, G. and De Pasquale, R. J., in *Organofluorine Chemistry: Principles and Commercial Applications* (ed. R. E. Banks) Plenum Press, New York, 1994. p. 431.
9. Bargigia, G. A., Tonelli, C. and Tato, M., *J. Fluorine Chem.*, **36**, 449 (1987).
10. Marchionni, G., Staccione, A. and Gregorio, G., *J. Fluorine Chem.*, **47**, 515 (1990) *Chem. Abs.*, **113**, 98185p.
11. Gumprecht, W. H., *ASLE Trans.*, **9**, 24 (1966).
12. Fowler, D. E., Johnson, R. D., Van Leyen, D. and Benninghoven, A., *Anal. Chem.*, **62**, 2088 (1990).
13. Ohsaka, Y., *Petrotech (Tokyo)*, **8**, 840 (1985); Ohsaka, Y., *J. Jpn. Petrol. Inst.*, **8**, 1 (1985).
14. Meyer, M., European Patent EP 543 288 (to Hoechst A.-G) (1993) *Chem. Abs.*, **119**, 161082u; Meyer, M., Stapel, R., Kottmann, H. and Gries, T., European Patent EP 444 554 (1991) *Chem. Abs.*, **115**, 257001k.
15. Stapel, R., European Patent EP 437 844 (to Hoechst A.-G) (1991) *Chem. Abs.*, **115**, 115358h.
16. Ebmeyer, F., German Patent DE 4 213 642 (to Hoechst A.-G) (1992) *Chem. Abs.*, **120**, 220862u.
17. Ginzel, K. D., *DECHEMA Monogr.*, **125**, 631 (1992). *Chem. Abs.*, **117**, 139677d.
18. Bierschenk, T. R., Kawa, H., Juhlke, T. and Lagow, R. J., *NASA CR-182155* (1988).
19. Jones, W. R., Bierschenk, T. R., Juhlke, T. J., Kawa, H. and Lagow, R. J., *Ind. Eng. Chem. Prod. Res. Dev.*, **27**, 1497 (1988).
20. Persico, D. F., Gergardt, G. E., Lagow, R. J., *J. Am. Chem. Soc.*, **107**, 1197 (1985).
21. Hung, M. H., Farnham, W. B., Feiring, A. E. and Rozen, S., *J. Am. Chem. Soc.*, **115**, 8954 (1993). *Chem. Abs.*, **119**, 181350h.
22. Perisco, D. F. and Lagow, R. J., *J. Polym. Sci., A, Polym. Chem. Ed.*, **29**, 233 (1991).
23. Marchionni, G., Gavezotti, P. and Strepparola, E., South African Patent ZA 89/03, 233 (to Ausimont S.r.l) 1990. *Chem. Abs.*, **113**, 192231x.
24. Walder, F. T., Vidrine, D. W. and Hansen, G. C., *Appl. Spectrosc.*, **38**, 782 (1984).
25. Pacansky, J., Waltman, R. J. and Maier, M., *J. Phys. Chem.*, **91**, 1225 (1987).
26. Morales, W., 'A thin film degradation study of a fluorinated polyether liquid lubricant using a HPLC method', *NASA Technical Memorandum -87221* (1986) (available from NASA).

27. D'Anna, E., Leggieri, G., Luches, A. and Perrone, A., *J. Vac. Sci. Technol.*, **A5**, 3436 (1987).
28. Coburn, J. W. and Winters, H. F., *J. Appl. Phys.*, **60**, 3309 (1986).
29. Cantow, M. J. R., Larrabee, R. B., Barrall, E. M., Butner, R. S., Cotts, P., Levy, F. and Ting, T. W., *Makromol. Chem.*, **187**, 2475 (1986); Cantow, M. J. R., Ting, T. Y., Barrall, R. S., Porter, R. S. and George, E. R., *Rheo. Acta*, **25**, 69 (1986).
30. Cantow, M. J. R., Barrall, E. M., Wolf, B. A. and Geerissen, H., *J. Polym. Sci.*, **B25**, 603 (1987).
31. Wolf, B. A., Klimink, M. and Cantow, M. J. R., *J. Phys. Chem.*, **93**, 2672 (1988).
32. Gianotti, G., Levi, M. and Turri, S., *J. Appl. Polym. Sci.*, **51**, 973 (1994). *Chem. Abs.*, **120**, 192796e.
33. Hues, S. M., Wyatt, J. R., Colton, R. J. and Black, B. H., *Anal. Chem.*, **62**, 1074 (1990).
34. Marchionni, G., Ajroldi, G., Cinquina, P., Tampellini, E. and Pezzin, G., *Polym. Eng. Sci.*, **30**, 829 (1990).
35. Faucitano, A., Buttafava, A., Caporiccio, G. and Viola, C. T., *J. Am. Chem. Soc.*, **106**, 4172 (1984).
36. Faucitano, A., Buttafava, A., Guarda, P. A. and Marchionni, G., *J. Fluorine Chem.*, **64**, 189 (1993).
37. Faucitano, A., Buttafava, A., Martinotti, F. F., Caporiccio, G. and Corti, C., *J. Chem. Soc. Perkin Trans.*, **2**, 425 (1981).
38. Faucitano, A., Buttafava, A., Martinotti, F. F., Caporiccio, G., Corti, C., Maini, S. and Viola, C. T., *J. Fluorine Chem.*, **16**, 649 (1981).
39. Faucitano, A., Buttafava, A., Martinotti, F. F., Marchionni, G. and De Pasquale, R. J., *Tetrahedron Letters*, **29**, 5557 (1988).
40. Faucitano, A., Buttafava, A., Martinotti, F. F., Marchionni, G. and De Pasquale, R. J., *Radiat. Phys. Chem.*, **37**, 43 (1991).
41. Faucitano, A., Buttafava, A., Martinotti, F. F. and Marchionni, G., *Tetrahedron Letters*, **29**, 4611 (1988).
42. Faucitano, A., Buttafava, A., Patruno, V., Guadra, P. A. and Marchionni, G., *Radiat. Phys. Chem.*, **45**, 23 (1995).
43. Moulder, J. F., Hammond, J. S. and Smith, K. L., *Appl. Surf. Sci.*, **25**, 446 (1986).
44. Carre, D. J., *ASLE Trans.*, **29**, 121 (1986).
45. Mori, S. and Morales, W., *J. Vac. Sci. Technol.*, **A8**, 3354 (1990).
46. Napier, M. E. and Stair, P. C., *J. Vac. Sci. and Tech., A- Vac. Surf. and Films*, **10**, 2704 (1992); Napier, M. E. and Stair, P. C., *Surf. Sci.*, **298**, 201 (1993); Napier, M. E. and Stair, P. C., *Surf. Sci.*, **316**, 317 (1994).
47. Herrera-Fierro, P., Jones, W. R. and Pepper, S. V., *J. Vac. Sci. and Tech. A- Vac. Sci. and Films*, **10**, 2746 (1992).
48. Herrera-Fierro, P., Jones, W. R. and Pepper, S. V., *J. Vac. Sci. and Tech. A- Vac. Sci. and Films*, **11**, 354 (1993).
49. Pan, F. M., Lin, Y. L. and Horng, T., *Appl. Surf. Sci.*, **47**, 9 (1991). *Chem. Abs.*, **114**, 123232g.
50. Sherman, R. and Vossen, J., *J. Vac. Sci. Technol.*, **A8**, 3241 (1990). *Chem. Abs.*, **113**, 139084a.
51. Ciampelli, F., Venturi, M. T. and Sianesi, D., *Org. Mat. Res.*, **1**, 281 (1969).
52. Fowler, D. E., Johnson, R. D., Van Leyen, D. and Benninghoven, A., *Surf. Interface Anal.*, **17**, 125 (1991). *Chem. Abs.*, **114**, 165297n.
53. Pacansky, J., Wang, C. and Waltman, R. J., *J. Fluorine Chem.*, **32**, 283 (1986).
54. Pacansky, J. and Waltman, R. J., *J. Phys. Chem.*, **95**, 1512 (1991).

55. Karis, T. E., Novotny, V. J. and Johnson, R. D., *J. Appl. Polym. Sci.*, **50**, 1357 (1993).
56. Yanagisawa, M., *Tribol. Trans.*, **37**, 629 (1994). *Chem. Abs.*, **121**, 160477f.
57. Kasai, P. H., *Macromolecules*, **25**, 6791 (1992).
58. Carre, D. J. and Markowitz, J. A., *ASLE Trans.*, **28**, 40 (1985).
59. Pacansky, J., Miller, M., Hatton, W., Liu, B. and Scheiner, A., *J. Am. Chem. Soc.*, **113**, 329 (1991).
60. Pianca, M., Del Fanti, N., Barchiesi, E. and Marchionni, G., *Chem. Today*, Jan/Feb, 29 (1995).
61. Viswanathan, K. V. and Schulz, K. J., *Ceram. Trans.*, **19**, 197 (1991). *Chem. Abs.*, **117**, 192674b.
62. Kasai, P. H., *Chem. Mater.*, **6**, 1581 (1994).
63. Kasai, P. H., Tang, W. T. and Wheeler, P., *Appl. Surf. Sci.*, **51**, 201 (1991); Kasai, P. H. and Wheeler, P., *Appl. Surf. Sci.*, **52**, 91 (1991).
64. Guarini, A., Guglielmetti, G., Vincenti, M., Guarda, P. and Marchionni, G., *Anal. Chem.*, **65**, 970 (1993).
65. Cromwell, E. F., Reihs, K., de Vries, M. S. Ghaderi, S., Wendt, H. R. and Hunziker, H. E., *J. Phys. Chem.*, **97**, 4720 (1993). *Chem. Abs.*, **118**, 222327f; Anex, D. S., de Vries, M. S., Knebelkamp, A., Bargon, J., Wendt, H. R. and Hunziker, H. E., *Int. J. Mass Spectrom. Ion Processes*, **131**, 319 (1994). *Chem. Abs.*, **120**, 271695z.
66. Bletsos, I. V., Hercules, D. M., Magill, J. H., van Leyen, D., Niehuis, E. and Benninghoven, A., *Anal. Chem.*, **60**, 938 (1988). *Chem. Abs.*, **88**, 7219.
67. Bletsos, I. V., Hercules, D. M., Fowler, D., Van Leyen, D., and Benninghoven, A., *Anal. Chem.*, **62**, 1275 (1990). *Chem. Abs.*, **113**, 7152m.
68. Hues, S. M., Colton, R. J., Mowery, R. L., McGrath, K. J. and Wyatt, J. R., *Appl. Surf. Sci.*, **35**, 507 (1989).
69. McGrath, K. J., *J. Fluorine Chem.*, **49**, 171 (1990).
70. Steffens, P., Niehuis, E., Friese, T., Greifendorf, D. and Benninghoven, A., *J. Vac. Sci. Technol.*, **A3**, 1322 (1985).
71. Doering, R. and Schernau, U., *Farbe Lack*, **99**, 321 (1993). *Chem. Abs.*, **120**, 32897k.
72. Feld, H., Leute, A., Rading, D., Benninghoven, A., Chiarelli, M. P. and Hercules, D. M., *Anal. Chem.*, **65**, 1947 (1993).
73. Viswanathan, K. V., *J. Appl. Polym. Sci., Appl. Polym. Symp.*, **45**, 361 (1990). *Chem. Abs.*, **113**, 153411m.
74. Spool, A. M. and Kasai, P. H., *Macromolecules*, **29**, 1691 (1996).
75. Newman, J. G. and Viswanthan, K. V., *J. Vac. Sci. Technol.*, **A8**, 2388 (1990).
76. Ramasamy, S. and Pradeep, T., *J. Chem. Phys.*, **103**, 485 (1995).
77. O'Connor, T. M., Jhon, M. S., Bauer, C. L., Choi, J. W., Yoon, D. Y. and Karis, T. E., *Proc. 35th IUPAC Int. Symp. Macromol., MacroAkron '94*, paper 0-4.3-8 (1994); O'Connor, T. M. Jhon, M., Bauer, C. L., Min, B. G., Yoon, D. Y. and Karis, T. E., *Tribol. Lett.*, **1**, 219 (1995).
78. Jonsson, U. and Bhushan, B., *J. Appl. Phys.*, **78**, 3107 (1995).
79. Marchionni, G., Ajroldi, G., Righetti, M. C. and Pezzin, G., *Macromolecules*, **26**, 1751 (1993).
80. Corti, C., *Proc. 10th Int. Symp. on Fluorine Chem.*, Vancouver, 1982, p. 7.
81. Caporiccio, G., Flabbi, L., Marchionni, G. and Viola, G., *J. Synth. Lubr.*, **6**, 133 (1989).

82. Danusso, F., Levi, M., Gianotti, G. and Turri, S., *Polymer*, **34**, 3687 (1993); Danusso, F., Levi, M., Gianotti, G. and Turri, S., *Eur. Polym. J.*, **30**, 647 (1994); Danusso, F., Levi, M., Gianotti, G. and Turri, S., *Eur. Polym. J.*, **30**, 1449 (1994).
83. Marchionni, G., Ajroldi, G. and Pezzin, G., *Eur. Polym. J.*, **24**, 1211 (1988).
84. Sanguineti, A., Guarda, P. A., Marchionni, G. and Ajroldi, G., *Polymer*, **36**, 3697 (1995).
85. Henning, J. and Lotz, H., *Vacuum*, **27**, 171 (1977).
86. Laurenson, L., Dennis, N. T. M. and Newton, J., *Vacuum*, **29**, 433 (1979).
87. Hirsch, E. H. and McKay, T. J., *Vacuum*, **43**, 301 (1992). *Chem. Abs.*, **116**, 236970f
88. Johns, K., Corti, C., Montagna, L. and Srinivasan, P., *J. Physics D — Appl. Phys.*, **25**, 141 (1992).
89. Synder Jr., C. E., Gschwender, L. J. and Campbell, W. B., *Lubr. Eng.*, **38**, 41 (1988).
90. Synder Jr., C. E. and Dolle, R. E., *ASLE Trans.*, **19**, 171 (1976).
91. Ciancia, A., Ascensioni, A., Corti, C. and Caporiccio, G., *Nucl. Sci. Eng.*, **86**, 232 (1984).
92. Eapen, K. C., John, P. J. and Liang, J. C., *Macromol. Chem. Phys.*, **195**, 2887 (1994).
93. Skryabina, T. G., Nikonorov, E. M., Petrova, L. N. and Kobzova, R. I., *Khim. Tekhnol. Topl. Masel*, **7**, 33 (1988). *Chem. Abs.*, **109**, 130154h.
94. Masuko, M., Takeshita, N. and Okabe, H., *Tribology Trans.*, **38**, 679 (1995).
95. Thom, M. A., *Lubrication Engin.*, **51**, 726 (1995).
96. Gilson, R. and Grundy, P. J., *Proceedings of the 2nd Int. Conf. Fluorine in Coatings, Salford*, paper #20 (1994) (available from Paint Research Association, Middlesex TW11 8LD, UK).
97. Caporiccio, G., Strepparola, E. and Scarati, A. M., European Patent, 165 650 (1986) (to Montedison) *Chem. Abs.*, **104**, 186986.
98. Strom, B. D., Bogy, D. B., Walmsley, R. G., Brandt, J., Bhatia, C. S., *Wear*, **168**, 31 (1993). *Chem. Abs.*, **120**, 111257e.
99. Kasai, P. H., *J. Appl. Polym. Sci.*, **57**, 797 (1995).
100. Devins, J. C., *Natl. Acad. Sci. NRC Annu. Rep.*, 398 (1977).
101. Luches, A. and Provenzano, I., *J. Phys. D: Appl. Phys.*, **10**, 339 (1977).
102. Guglielmini, G., Misale, M. and Schenone, C., *Proc. 18th Actes Congr. Int. Froid*, **2**, 538 (1991). *Chem. Abs.*, **120**, 167671c.
103. Uccellio, B. and Basisio, C., British Patent 2 064 579 (to Pirelli Cables) (1982) *Chem. Abs.*, **96**, 70085.
104. Piacenti, F. and Camaiti, M., *J. Fluorine Chem.*, **68**, 227 (1994).
105. Frediani, P., Manganelli Del Fa, C., Matteoli, U. and Tiano, P., *Stud. Conserv.*, **27**, 31 (1982).
106. Moggi, G., *Proceedings of XVth Int. Conf. in Org. Coat. Sci. and Tech.*, Athens, pp. 283–298 (1989) (available from Angelos V. Patsis, Institute of Material Science, State University of New York, NY, USA); Moggi, G., *Proceedings of XVIth Int. Conf. in Org. Coat. Sci. and Tech.*, Athens, pp. 251–260 (1990) (available from Angelos V. Patsis, Institute of Material Science, State University of New York, NY, USA).
107. Lin, S-. C., Burks, S. J., Tonelli, C. and Lenti, D., European Patent, EP 689 908 (1996) *Chem. Abs.*, **124**, 205437y.
108. Simeone, G., Turri, S., Scicchitano, M. and Tonelli, G., *Angew. Makromol. Chem. (Appl. Macromol. Chem. Phys.)*, **236**, 111 (1996).

109. Visca, M. and Lenti, D., European Patent Application 106 149, 106 150 and 106 151 (to Ausimont) (1989); Visca, M. and Lenti, D., European Patent Application EP 337 311 (1990) (to Ausimont) *Chem. Abs.*, **112**, 11253s.
110. Pacansky, J. and Waltman, R. J., *Prog. Org. Coat.*, **18**, 79 (1990). *Chem. Abs.*, **113**, 174085y.
111. Tamaki, Y. and Murakawa, A., Japanese Patent JP 06 145 598 (1994). *Chem. Abs.*, **122**, 12215z.
112. Scheirs, J., 'Fluoropolymer coatings (new developments)', *The Polymeric Materials Encyclopedia* (ed. J. Salamone) CRC Press, Boca Raton, FL, USA 1995.
113. Scheirs, J., Burks, S. and Locaspi, A., *Trends in Polym. Sci.*, **3**, 74 (1995).
114. Scheirs, J., Costa, L., Camino, G., Tonelli, C., Scicchitano, M. and Turri, S., 'Photooxidation of functionalized perfluorinated polyethers: — I', unpublished work.
115. Scheirs, J., Costa, L., Camino, G., Tonelli, C., Scicchitano, M. and Turri, S., 'Photooxidation of Functionalized Perfluorinated Polyethers: — II', unpublished work.
116. Scheirs, J., Costa, L., Camino, G., Tonelli, C., Scicchitano, M. and Turri, S., 'Photooxidation of coatings based on functionalized perfluorinated polyethers: — III', unpublished work.
117. Tonelli, C., Trombetta, T., Scicchitano, M. and Castiglioni, G., *J. Appl. Polym. Sci.*, **57**, 1031 (1995).
118. Turri, S., Gianotti, G., Levi, M. and Tonelli, C., European Patent Application EP 621 298 (to Ausimont) (1994) *Chem. Abs.*, **122**, 242066f.
119. Mascia, L., Zitouni, F. and Tonelli, C., *Polym. Eng. Sci.*, **35**, 1069 (1995).
120. Toselli, M., Pilati, F., Fusari, M., Tonelli, C. and Castiglioni, C., *J. Appl. Polym. Sci.*, **54**, 2101 (1994).
121. Badyal, J. P. S., Chambers, R. D. and Joel, A. K., *J. Fluorine Chem.*, **60**, 297 (1993).
122. Warburton, G. A., McDowell, R. A., Taylor, K. T. and Chapman, J. R., *Adv. Mass Spectr.*, **8B**, 1953 (1980).
123. Bakhmendo, V. B., V'yunov, K. A., Ismagilov, N. G. and Sarkisov, Yu. S., *Izmer. Tekh.*, **8**, 56 (1988) *Chem. Abs.*, **109**, 239947v.
124. Klassen, J. K., Mitchell, M. B., Govini, S. T. and Nathanson, G. M., *J. Phys. Chem.*, **97**, 10166 (1993). *Chem. Abs.*, **119**, 211379u.
125. Saecker, M. E. and Nathanson, G. M., *J. Chem. Phys.*, **100**, 3999 (1994). *Chem. Abs.*, **120**, 192893j.
126. Naganuma, M. and Shigenori, K., Japanese Patent, JP 07 258 027 (1995) *Chem. Abs.* **124**, 37391h.
127. Brunetta, F., Guidolin, V. and Pantini, G., *Cosmetics and Toiletries*, Apr., 19 (1992).
128. Goedel, W. A., Wu, H., Friedenberg, M. C., Fuller, G. C., Foster, M. and Frank, C. W., *Langmuir*, **10**, 4209 (1994).
129. Paciorek, K. J. L. and Kratzer, R. H., *J. Fluorine Chem.*, **67**, 169 (1994).
130. Koch, B. and Jantzen, E., *Synth. Lubr.*, **12**, 191 (1995).
131. Helmick, L. S. and Jones, W. R., *NASA Tech. Memo: — 102493* (1990).
132. Pacansky, J. and Waltman, R. J., *Chem. Mater.*, **5**, 486 (1993). *Chem. Abs.*, **118**, 213706b.
133. Flabbi, L. and Briggs, R. S., *Brazing and Soldering*, **7**, 6 (1984).
134. Jones, W. R., *Tribology Trans.*, **38**, 557 (1995).
135. Jones, W. R., Paciorek, K. J. L., Harris, D. H., Smythe, M. E., Nakahara, J. H. and Kratzer, R. H., *Ind. Eng. Chem., Prod. Res. Dev.*, **24**, 417 (1985).

136. Zehe, M. J. and Faut, O. D., *Tribol. Trans.*, **33**, 634 (1990); Zehe, M. J. and Faut, O. D., *NASA Tech. Memo. — 101962* (1989).
137. Smart, B. E. and Dixon, D. A., *J. Fluorine Chem.*, **57**, 251 (1992).
138. Morales, W., *Tribol. Trans.*, **39**, 148 (1996).
139. Chandler, J. A., Lloyd, L. B., Farrow, M. F., Burnham, R. K. and Eyring, E. M., *Corrosion*, **36**, 152 (1980).
140. Jones, W. R., 'Properties of perfluoropolyethers for space applications', *NASA Technical Memorandum 106616* (1994) (available from NASA, Lewis Research Center, Cleveland, Ohio 44135).
141. Ball, D. W., *High Temp. Mater. Sci.*, **33**, 171 (1995).
142. Mori, S. and Morales, W., *Wear*, **132**, 111 (1989); Mori, S. and Morales, W., 'Degradation and crosslinking of perfluoroalkyl polyethers under x-ray irradiation in ultrahigh vacuum' *NASA Technical Publication 2910* (1989) (available from NASA, Lewis Research Center, Cleveland, Ohio 44135).
143. Jones, W. R., Paciorek, K. J. L., Nakahara, J. H., Smythe, M. E. and Kratzer, R. H., *Ind. Eng. Chem. Res.*, **29**, 1930 (1987).
144. Vurens, G. H., and Gudeman, C. S., Lin, L. J. and Foster, J. S., *Langmuir*, **8**, 1165 (1992); Vurens, G. H. and Mate, C. M., *Appl. Surf. Sci.*, **59**, 281 (1992).
145. Banks, B. A., 'The use of fluoropolymers in space applications' in *Modern Fluoropolymers* (ed. J. Scheirs), Wiley, London, 1997.
146. Gulino, D. and Coles, C., *J. Spacecraft and Rockets*, **25**, 99 (1988).
147. Ronghua, W., Wilbur, F. J., Buchholz, B. W. and Kustas, F. M., *Tribology Transactions*, **38**, 950 (1995).
148. Marchionni, G., Ajroldi, G. and Pezzin, G., in *Comprehensive Polymer Science*, (ed. G. Allen, S. L. Aggerwal and S. Russo), Pergamon Press, Oxford, 1996, p. 347.

25

PVDF in the Chemical Process Industry

DAVID A. SEILER
Elf Atochem North America Philadelphia, PA, USA

1 CHEMICAL STRUCTURE OF PVDF

Polyvinylidene fluoride (PVDF or PVF_2) is produced by the addition polymerization of 1,1-difluoroethene ($CH_2=CF_2$), also known as vinylidene fluoride (VDF or VF_2). The homopolymer is characterized by alternating carbon–hydrogen bonds with carbon–fluorine bonds:

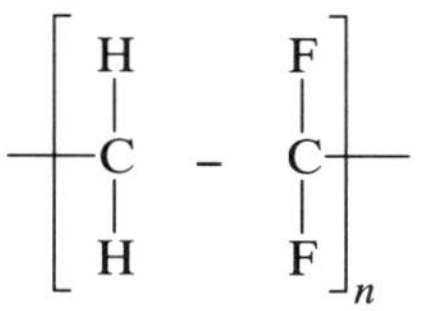

The structure of PVDF homopolymer is typically regular; however, some variability related to chain branching, head-to-head molecular formation and tail-to-tail molecular formation will exist depending on polymerization method and reactant products chosen. The polymer contains 59–60% fluorine and approximately 3% hydrogen by weight. PVDF homopolymers are partially crystalline ranging between 45 and 70% crystallinity depending on the above variables and the processing conditions and methods used.

Vinylidene fluoride-based copolymers, typically called PVDF copolymers, have become the products of choice in many applications that were formerly PVDF homopolymer [1]. These copolymers have also replaced other fluoropolymers or metals in the Chemical Process Industry (CPI) where PVDF homopolymer had limitations in impact strength and elongation that prevented its use.

Modern Fluoropolymers. Edited by John Scheirs
© 1997 John Wiley & Sons Ltd

The most common comonomers utilized with PVDF for CPI applications are hexafluoropropylene (HFP), chlorotrifluoroethylene (CTFE), and tetrafluoroethylene (TFE):

$$
\begin{array}{ccc}
\underset{\underset{F}{|}}{\overset{\overset{F}{|}}{C}} = \underset{\underset{CF_3}{|}}{\overset{\overset{F}{|}}{C}} \qquad & \underset{\underset{F}{|}}{\overset{\overset{Cl}{|}}{C}} = \underset{\underset{F}{|}}{\overset{\overset{F}{|}}{C}} \qquad & \underset{\underset{F}{|}}{\overset{\overset{F}{|}}{C}} = \underset{\underset{F}{|}}{\overset{\overset{F}{|}}{C}} \\
\text{HFP} & \text{CTFE} & \text{TFE}
\end{array}
$$

Other monomers exist that can be utilized to create VDF-based resins with interesting properties, but the above three are the most commonly used and readily available for current applications.

2 HISTORY OF PVDF

Polyvinylidene fluoride was first commercially introduced in 1961 by the Pennsalt Chemical Company (later known as Pennwalt Corporation now owned by Elf Atochem North America, Inc.). The original applications for the polymer still exist today. Early property evaluations of PVDF led to its use as piping and molded parts for plutonium recovery applications in nuclear facilities, abrasion-resistant insulation for computer back panel wire, and as a long life finish on metal panels used in architectural construction.

Since the original introduction by Pennsalt Chemical Co., and the first full-scale commercial plant built in Calvert City, Kentucky, USA, in 1965, several other plants were built by various companies to meet rising demand for PVDF and PVDF copolymers in many markets. Table 25.1 lists the current producers of PVDF for commercial applications worldwide.

Growth of PVDF has been fueled by several landmark applications. Heat-shrinkable tubing for electrical insulation in military applications and heat trace wiring were the original core markets in the 1960s. Strong growth in architectural finishes, plastic-lined steel for chemical plants, and fabrics for pulp and paper applications fueled large volume growth in the 1970s. In the early 1980s, the

Table 25.1. Producers of PVDF type products

Producer	Plant Location	Trademark
Ausimont	United States	Hylar
Daikin	Japan	Neoflon
Elf Atochem	United States	Kynar, Kynar Flex
	France	Kynar (Formerly Foraflon)
Kureha Chemical Industry Co., Ltd	Japan	KF
Solvay & Cie, SA	France	Solef

largest 'overnight' market for PVDF, especially PVDF copolymers, was created when the United States National Electrical Code (NEC) called for low-smoke, low-flame polymers to be used in plenum areas of buildings as an alternative to any polymer insulation placed in conduit. This plenum cable market for PVDF caused a worldwide shortage of PVDF-type polymers which brought on new PVDF plants and expansions of existing ones.

Also, in the 1980s, high-purity semiconductor chip manufacturers started to discover that plastics outperformed metals in the transport of high-purity washing chemicals in the manufacture of computer devices. By the mid-1980s, PVDF was becoming the material of choice compared to other cheaper polymers due to the fact that it could be easily processed without the need for any processing aids, stabilizers, fillers, or additives whatsoever [2–6].

In the 1990s, PVDF-type polymers maintained growth even during recession periods and they continue to be in high demand as materials for high-purity processing components (e.g. piping, tank linings, pumps, filtration products, fittings, flexible tubing, etc), architectural coatings, and new smaller markets such as use as an additive to other polymers to improve processing or performance, films for weathering protection of vinyls, and specialty hoses for fuel containment where permeation regulations exist. The market for PVDF products in plenum cable has declined since 1990 for various reasons, but still remains as a major area of application for this polymer family. New regulations promise renewed growth in electrical applications for the next few years.

The outlook for PVDF-type resins and other fluoropolymers is excellent. Expectations are for double digit growth up to and beyond the year 2000. Since PVDF is the least expensive fluoropolymer on a cost per volume ratio, it is often the first such material considered for new arising applications.

3 POLYMERIZATION

Commercial polymers based on vinylidene fluoride can be categorized into two types of polymerization methods. While there are other methods of producing PVDF homopolymers and copolymers, the most common methods are emulsion and suspension polymerization.

In emulsion polymerization, a number of reactant products are used such as fluorinated surfactants, initiators, and possibly chain terminators. Some manufacturers that supply very pure PVDF products for special applications employ high-purity rinsing of the emulsion latex before final drying [7]. This washing helps to eliminate any residual impurities such as polymerization initiator and surfactant before packaging as a free-flowing powder, or processing the powder into a pellet for extrusion or injection molding applications. The particles of the polymer isolated from the reaction vessel are agglomerated spherical particles ranging in diameter from 0.2 to 0.5 μm.

In suspension polymerizations, an aqueous recipe is used with initiators, colloidal dispersants (not always necessary), and chain transfer agents to control molecular weight. The final suspension consists of spherical particles which are approximately 100 μm in diameter. Suspension polymers are available either as a free-flowing powder or as a pellet for extrusion or injection molding applications.

Both emulsion and suspension PVDF in powder form can be milled into finer particle size with higher surface area for greater solubility to be used in coatings for metal components.

Comonomers such as HFP, CTFE, and TFE can be added at the start of polymerization or at other times in the reaction process to create VDF-based polymers with various crystallinities. These comonomers in existing commercial products typically impart flexibility, chemical resistance, elongation capability, solubility, impact resistance, clarity, and thermal stability in processing. On the downside, they often have lower melting points, higher permeation, lower tensile strength, and higher creep than PVDF hompolymers.

Only a small amount of comonomer is needed (less than 6%) in some cases to dramatically improve specific performance requirements if PVDF homopolymer is considered deficient. The maximum level of comonomer is limited by the phenomenon that the resin becomes an elastomer rather than a polymer if the comonomer ratio is over a certain percentage. For example, the addition of HFP at more than 20% would favor elastomeric tendencies.

PVDF for most applications does not contain additives. However, many compounded products are available for special applications. It has been found that small amounts of specially formulated PTFE can act as a process aid for PVDF-based resins. PTFE also acts as a surface lubricant that can lower the coefficient of friction of PVDF molded parts.

The addition of carbon black can be used to reduce mold shrinkage to nearly match polypropylene if the user can tolerate a black colored finished part. Glass spheres can also be compounded in PVDF for strength and shrinkage reduction, but component production must be done under a watchful eye to avoid decomposition created by the potential of silica reacting with PVDF at high processing temperatures. For products with reduced shrinkage and increased tensile strength, carbon fibers are sized with PVDF to create what is termed as carbon fiber reinforced PVDF. Commercial products exist with between 15 and 25% carbon fiber. The larger the percentage of carbon fiber, the lower the shrinkage and the higher the tensile strength and deflection temperature.

Specific smoke suppressants are available that can make certain PVDF-type resins virtually nonflammable. Typically such additives are added at levels well below 2% and do very little to change the mechanical properties of the resin. These resins that contain these additives can pass tests that few other plastics can pass. This has led to applications in institutional piping, duct work, and coatings for facilities where plastics offer an installation advantage, chemical advantage, or purity advantage over metals, but were previously not considered due to fire safety concerns.

PVDF accepts pigments and can be colored easily. Black, red, and blue resins are standard offerings and can be made from FDA listed pigments with very low percentage concentration. A whole spectrum of colors have been produced using concentrates should special colors be desired.

4 PROPERTIES OF PVDF

Polyvinylidene fluoride has a great balance of properties that make it suitable for many applications. The molecular structure with alternating CH_2 and CF_2 groups along the polymer chain forms a unique polymer with some of the best characteristics of polyethylene $(-CH_2-CH_2-)_n$ combined with performance approaching polytetrafluoroethylene $(-CF_2-CF_2-)_n$. The resultant polymer provides excellent properties in the following respects:

- Mechanical strength and toughness
- Resistant to fungi
- High abrasion resistance
- Low permeability to gases and liquids
- High thermal stability
- Low flame and smoke characteristics
- High dielectric strength
- Resistant to creep at elevated temperatures
- High purity
- Readily melt processable by many methods
- Resistant to most chemicals and solvents
- Rigid and flexible versions available
- Resistant to ultraviolet and nuclear radiation
- Impact resistant versions available
- Resistance to weathering
- Cold weather performance to $-40\,^\circ$C

The selection of PVDF resin for given applications has become more difficult due to the broad range of grades now available. Homopolymer properties have a much smaller variation between manufacturers than the difference range now available due to the commercially available copolymers that were developed in the 1980s and early 1990s.

In applications where PVDF homopolymer may not have been considered because of impact strength, elongation at break, chemical stress cracking, lack of clarity, or lack of flexibility, PVDF copolymers meet the need that may have been deficient, and at a similar cost. A set of property tables for PVDF homopolymers, VF_2/HFP polymers, and filled PVDF resins are shown in Tables 25.2–25.4.

Compared to other commercial fluoropolymers, PVDF has the lowest melting point, but actually has the highest heat deflection temperature under load.

Table 25.2. Properties of PVDF homopolymers

Properties	Test method	Units	Typical value Ranges[a]
Density/specific gravity		g/cm^3	1.75–1.80
Melting point	ASTM D3418	°C	160–178
Refractive index	ASTM D542	n25D	1.41–1.42
Water absorption	ASTM D570	%	0.01–0.04
Tensile strength @ yield	ASTM D638	MPa[b]	31–57
Tensile strength @ Break	ASTM D638	MPa[b]	27–52
Elongation @ break	ASTM D638	%	50–250
Tensile modulus	ASTM D882	MPa[b]	1030–2410
Flexural modulus	ASTM D790	MPa[b]	1130–2280
Izod Impact	ASTM D256 @ 25 °C	J/m[c]	
Notched			110–300
Unnotched			800–4500
Hardness	ASTM D2240	Shore D	75–80
Abrasion resistance	CS-17, 1000 g	mg/1000 cycles	5–9
Coefficient of sliding friction			
to steel			0.14–0.17
Heat deflection temperature	ASTM D648	°C	
0.46 MPa			119–142
1.82 MPa			84–118
Thermal conductivity	ASTM D433	W/K.m	0.17–0.19
Coefficient of linear expansion			
near room temperature	ASTM D696	°C^{-1}	0.7–1.5 ×10^{-4}
Limiting oxygen index	ASTM D2863	%	42–80

[a]Typical value ranges are broad due to the variability that can relate to sample preparation, molecular weight, molecular weight distribution, chain branching, polymerization method and data interpretation.
[b]To convert MPa to psi, multiply by 145.
[c]To convert J/m to ft-lb/in, divide by 53.38.

While some other fluoropolymers can be used above the 150 °C maximum continuous use temperature that PVDF can withstand, they tend to be softer and often must be used at very low pressure or act only as a lining bonded to metal that gives the necessary support structure. See Table 25.5 [8].

The combination of hardness and slipperyness imparts to PVDF homopolymer and some copolymers of VF$_2$/HFP the highest abrasion resistance of the fluoropolymer family. Ethylene chlorotrifluoroethylene (ECTFE) is a close second, but resins like Fluorinated Ethylene Propylene (FEP), polytetrafluoroethylene (PTFE), and perfluoroalkoxy (PFA) are as much as 25–100 times less abrasion resistant than PVDF. These properties often make PVDF very suitable for slurry applications involving corrosive chemicals.

The high crystallinity and surface tension properties of PVDF give it very low permeation values. In inert gas testing directly comparing fluoropolymers, PVDF is generally either the least permeable, or very close to the performance

Table 25.3. PVDF copolymer properties (commercial grades)

Properties	Test method	Units	Typical value ranges[a]
Density specific gravity		g/cm^3	1.76–1.79
Melting point	ASTM D3418	°C	125–170
Refractive index	ASTM D542	n25D	1.40–1.42
Water absorption	ASTM D570	%	0.03–0.05
Tensile strength @ yield	ASTM D638	MPa[b]	14–42
Tensile strength @ break	ASTM D638	MPa[b]	15–42
Elongation @ break	ASTM D638	%	200–500
Tensile modulus	ASTM D638	MPa[b]	480–1040
Flexural modulus	ASTM D790	MPa[b]	240–1180
Izod impact	ASTM D256 @ 25 °C	J/m[c]	
Notched			320-NB
Unnotched			3000-NB
Hardness	ASTM D2240	Shore D	55–75
Abrasion resistance	CS-17, 1000 g	mg/1000 cycles	6–20
Thermal conductivity	ASTM D433	W/K.m	0.16–0.18
Coefficient of linear expansion	ASTM D696	°C^{-1}	$1.0–1.5 \times 10^{-4}$
Limiting oxygen index	ASTM D2863	%	42–95

[a]Typical value ranges are broad due to variability that can be related to comonomer chosen, percentage of comonomer, method of addition of comonomer, sample preparation, molecular weight, molecular weight distribution, chain branching, polymerization method and data interpretation.
[b]To convert MPa to psi, multiply by 145.
[c]To convert J/m to ft-lb/in, divide by 53.38.

Table 25.4. Filled and reinforced PVDF properties

Properties	Test method	Units	Typical value Ranges[a]
Density specific gravity		g/cm^3	1.77–1.88
Melting point	ASTM D3418	°C	160–175
Tensile strength @ break	ASTM D638	MPa[i]	38–166
Flexural modulus	ASTM D790	MPa[j]	2410–8550
Izod impact	ASTM D256	J/m	
Notched			32–91
Unnotched			267–534

[a]Typical value ranges are broad due to variability that can be related to method of reinforcement, percentage of reinforcement, properties of base resin used, and sample preparation.
[b]To convert MPa to psi, multiply by 145.
[c]To convert J/m to ft-lb/in, divide by 53.38.

of the least permeable polymer. Degree of permeation has been linked to the polarity of the permeate and the substrate. In the case of highly corrosive chemicals such as chlorine gas and bromine, PVDF is unmatched in chemical and permeation resistance. Coupled with this, PVDF processing methods of extrusion and injection molding yield a more 'solid' part than a sintered resin such

Table 25.5. Comparison of Deflection Temperature and Melting Point of Fluoropolymers (ASTM D648)[a]

	Deflection temperature (°F)[b]		Melt point (°F)[c]
	66 psi[c]	264 psi[c]	
PVDF	298	235	352
PCTFE	258	167	424
PTFE	250	132	620
ECTFE	240	170	464
ETFE	220	165	518
PFA	164	118	590
FEP	158	124	554

[a]Taken from Table C in Ultrapure Water® July/August 1987, ref. 8 in this chapter.
[b]To convert °F to °C, Use °C = °F−32/1.8
[c]To convert psi to MPa, divide by 145.

as PTFE. See Table 25.6 for a comparison of published permeation values for fluoropolymers.

PVDF has excellent resistance to outdoor exposure and sunlight. Many years of outdoor exposure directly in sunlight seems to have little effect on the physical properties of the resin. Some increases in tensile strength and reduction in elongation occurs due to a small amount of crosslinking of the polymer over time. However, the ultraviolet light (UV) from sunlight tends to be a higher wavelength (>300 nm) than UV lamps that are used in pharmaceutical and semiconductor applications to kill bacteria. It has been found that accelerated UV light placed directly onto stressed (welded) PVDF piping below a wavelength of 257 nm causes discoloration and a noticeable amount of degradation of the polymer over a period of several years. UV light below 230 nm has a more detrimental effect if continuously exposed directly on PVDF. Within six months, the component can become noticeably brittle. If such a potential problem exists, it is recommended that for this limited area of a system either stainless steel or a longer performing fluoropolymer such as PFA be used.

Table 25.6. Gas permeability of fluoropolymers[a]

	PTFE	PFA	FEP	ETFE	PCTFE	ECTFE	PVDF	PVF
Water vapor g/m^2.d.bar	5	8	1	2	1	2	2	7
Air cm^3/m^2.d.bar	2 000	1 150	600	175	N/A	40	7	50
Oxygen cm^3/m^2.d.bar	1 500	N/A	2 900	350	60	100	20	12
Nitrogen cm^3/m^2.d.bar	500	N/A	1 200	120	10	40	30	1
Helium cm^3/m^2.d.bar	3 500	17 000	18 000	3 700	N/A	3 500	600	300
Carbon dioxide cm^3/m^2.d.bar	15 000	7 000	4 700	1 300	150	400	100	60

[a]Data published in 1980 Kunststoffe paper entitled *Fluorocarbon Films — Present Situation and Future Outlook*. Based on 100 μm film thickness at 23 °C. Gases method: ASTM D1434. Water vapor according to DIN 53122. N/A = not available.

Nuclear radiation that can be extremely detrimental to many polymers is not as aggressive to PVDF. PVDF preferentially favors crosslinking relative to chain scission when exposed to gamma radiation up to very high levels. Other than a color change and some loss of impact strength, many other properties actually improve up to 150 Mrad. Applications exist with PVDF in place after 1000 Mrad of exposure.

The fact that PVDF is partially soluble in ketones, amines and amides allows for the manufacture of dispersion coatings. Due to the high temperatures needed to bake out the solvents and fuse the polymer to the substrate, these coatings are typically applied to high-temperature substrates such as carbon steel, stainless steel, or aluminum. It has been found to be very difficult to apply PVDF liquid coatings to other polymers due to the fact that the temperatures needed to fuse the polymer will typically be high enough to distort the polymeric substrate.

Using similar solution technology and controlling the rate of solvent extraction can lead to the production of microporous membranes used in food, beverages, pure water, paint, and medical applications. Such membranes are able to withstand chemical attack and temperature for a long period of time.

5 PROCESSING AND FABRICATION

PVDF is a thermoplastic resin available as a pellet, powder, or liquid dispersion. A range of resins with various molecular weights offering alternatives in melt flow rate and viscosities are available. Unlike most other fluoropolymers, PVDF can be processed at sufficiently low temperatures to be fabricated on the same equipment and nearly the same conditions as polyethylene and polypropylene. Low molecular weight grades with high melt flow rates are used for injection molding. Higher molecular weight grades offer better melt strength and are used for extrusion of profiles. Even higher molecular weight resins are used in casting applications such as membranes.

Fluoropolymers exhibit relatively high shrinkage compared to most polymers. PVDF resins shrink about 3% (0.030 in/in) after molding. The resin can be filled with carbon (resulting in a black product) to bring the shrinkage very near polypropylene (around 2%). Carbon fiber reinforcement can be used to bring shrinkage to well below 1%.

Extrusion temperatures range from 210 to 290 °C depending on size of the profile. Molding temperatures range from 180 to 240 °C. In all cases, care must be taken not to exceed a melt temperature of 315 °C (600 °F) for extended periods of time. While PVDF has a wide processing window regardless of the processing method, the decomposition products are highly acidic and it is recommended that in the unlikely event of a decomposition, the processor have proper information on hand to aid in treating hydrofluoric acid (HF) burns and inhalation.

If PVDF is left static in processing equipment for any length of time, it should be turned down to 175 °C or below to avoid degradation. Before restarting the equipment, the temperatures should be turned up to the desired processing conditions.

Finally, PVDF can be sawed, machined, welded by many methods, and thermoformed to obtain a final component desirable for the task at hand. Many manufacturers of PVDF and PVDF stock shapes produce detailed descriptions of machining and welding guidelines to help companies interested in post fabrication.

6 CHEMICAL RESISTANCE

PVDF-based resins are totally chemically resistant to a wide group of chemicals at high temperatures. Most acids and acid mixtures, weak bases, halogens, halogenated solvents, hydrocarbons, alcohols, salts, and oxidants pose little problem for PVDF up to and above 90 °C for long-term use. In marginal applications for homopolymer, PVDF copolymers tend to have even better chemical resistance; however, this is sacrificed by the greater solubility in reagents known to swell or dissolve PVDF homopolymer.

On a short-term basis (less than 24 hours), not many chemicals are known to negatively affect PVDF. For longer term applications, PVDF is typically not the most logical choice for solutions containing a large percentage of strong bases (pH of solution above 12), ketones, primary amines, esters, organic acids, fuming acids, or ring chain compounds with a nitrogen link.

PVDF copolymers can be made to resist pH ranges easily to a pH of 13.5. These products, since they extend the useful range of PVDF with very little sacrifice of important properties, have become the polymer of choice for lined pipe and lined vessel applications.

In summary, typical PVDF-based resin chemical applications would be bromine, chlorine, methyl chloroform, chlorobenzene, chlorine dioxide, sodium hypochlorite, hydrogen bromine, hydrogen fluoride, hydrogen chloride, nitric acid, sulfuric acid, benzene, ozone (gaseous or in water), deionized water, chromic acid, ferric chloride, refrigerants, phosphoric acid, and combinations of all of the above.

7 APPLICATIONS

Any component available in PVC, CPVC, or polypropylene can be made available in PVDF. The most popular stock items in the CPI industry are piping, fittings, solid linings for metal pipe and metal vessels, flexible tubing, valves, instrumentation, strainers, pumps, stock shapes (rod, block, fabric-backed sheet), dual laminant lined tanks with fiberglass backing, monofilament, filter housings, fabrics, nozzles, mixers, dump tower packing, membranes, and porous products.

Various industries make use of the above products for special applications within their processes.

7.1 PULP AND PAPER

The bleaching operations in pulp mills utilize PVDF-based resins for piping chlorine, chlorine dioxide, sodium hypochlorite, ozone, and other whitening chemicals (Figure 25.1). Typically, for safety reasons related to concern about the potential of moving equipment banging into piping, pulp mills standardize on plastic-lined steel. These chemicals have an ability to permeate through many lining materials that otherwise provide the necessary chemical resistance. PVDF resins have been determined by lined piping suppliers to not need weepholes in exposure to chlorine. Other resins like PTFE and FEP require weepholes because so much chlorine permeates that it would otherwise collapse the liner.

Tanks, pumps, valves, and other components complement the PVDF-lined piping used in the bleaching area. In the bleach washer area, PVDF resin has become the material of choice for the wire for covering drums in all stages of

Figure 25.1. Fabric-coated PVDF copolymer sheet, welded inside this scrubber column, provides broad-range chemical resistance at elevated temperatures

bleaching. PVDF fabrics generally last two to three times longer than a similar fabric made from stainless steel [9].

Some PVDF piping is utilized in the paper manufacturing process for handling sulfuric acid which is used to decrease pH in the process.

7.2 SEMICONDUCTOR MANUFACTURING

Computer chip manufacturers that utilize both deionized water for washing and high-purity acids for etching silicon, have found that PVDF-based resins have the right balance of properties and the excellent purity required for making defect-free product. With circuits getting more complex and smaller, it is becoming increasingly important to minimize the amount of particulates that contact silicon wafers potentially causing short circuits and rejected product.

The major PVDF markets here are for piping of 18 MΩ deionized water, high purity water holding tanks, acid etch baths made from welded sheets, filter housings, acid transport piping, and other acid handling components. (Figure 25.2). Semiconductor manufacturers utilize hydrofluoric acid, sulfuric acid, nitric acid, and hydrochloric acid among other corrosive reagents. The materials of construction must not only stand up to chemical attack by these reagents, but must also avoid leaching any type of impurity. Since PVDF can be processed without any additives, it is an ideal material of construction for this industry. Several publications have been printed that expand on the use of PVDF in the semiconductor industry [10,11].

Figure 25.2.   High-purity PVDF components are readily injection moulded and machined

Table 25.7. Bulk analysis of elements contained within PVDF tested by neutron activation analysis

Element	Impurity detected (ppb)	Element	Impurity detected (ppb)
Sodium	44.60	Cadmium	<2.80
Potassium	<7.20	Indium	<5.10
Calcium	<4400.00	Tin	<190.00
Scandium	<0.02	Antimony	1.25
Titanium	<5600.00	Cesium	<0.35
Chromium	1.54	Barium	<42.00
Iron	<170.00	Lanthanum	<0.04
Cobalt	7.05	Cerium	<1.60
Nickel	<81.00	Europium	<0.14
Copper	4.51	Terbium	<0.16
Zinc	<13.00	Ytterbium	<0.09
Gallium	<0.12	Hafnium	<0.24
Arsenic	<0.10	Tantalum	<0.23
Selenium	<2.70	Tungsten	0.20
Bromine	16.60	Iridium	<0.01
Rubidium	<6.60	Platinum	<120.00
Strontium	<180.00	Gold	<0.01
Zirconium	<140.00	Mercury	<0.86
Molybdenum	<1.40	Thorium	<0.24
Silver	<1.90	Uranium	<0.21

Tables 25.7, 25.8, and 25.9 include common data available on purity analysis of PVDF. Metallic species, anionic species, and total organic carbon (TOC) are major concerns of semiconductor manufacturing facility engineers.

7.3 NUCLEAR INDUSTRY

Nuclear facilities that specialize in weapons research or other activity involving plutonium and uranium are utilizing PVDF for piping and vessel linings in purification of radioactive materials. To purify radioactive metals, work is done in a glove box in the presence of either nitric acid or hydrochloric acid (Figure 25.3) [12]. In choosing a handling material in the nuclear industry, two major considerations must be made: (1) the material must be chemically resistant to the fluids used for purification, and (2) the material must be resistant to nuclear radiation. This application requires the ultimate in life for the installation due to the fact that disposal of radioactive components is quite expensive. While PVDF would normally be considered overengineered for this service in a typical industrial application, the desire for the longest possible life makes it a popular choice in the nuclear industry.

Another application for PVDF in the nuclear industry is for handling tritium. PVDF is not very permeable to water and should these molecules become part

Table 25.8. Leachate detected in PVDF immersed in deionized water at 80°C. Results by ICP–MS

Element	PVDF leachate (ppb)	Element	PVDF leachate (ppb)
Lithium	<0.30	Tellurium	<0.25
Boron	<0.06	Cesium	<0.08
Lanthanum	<0.05	Barium	<0.08
Magnesium	<0.20	Cerium	<0.04
Aluminum	<0.10	Praseodymium	<0.10
Titanium	<0.10	Neodymium	<0.09
Vanadium	<0.11	Samarium	<0.06
Chromium	<0.26	Europium	<0.03
Manganese	<0.06	Gadolinium	<0.05
Copper	<0.08	Terbium	<0.10
Cobalt	<0.10	Dysprosium	<0.01
Nickel	<0.08	Holmium	<0.10
Zinc	<0.05	Erbium	<0.05
Gallium	<0.40	Thulium	<0.04
Germanium	<0.03	Ytterbium	<0.04
Arsenic	<0.03	Lutetium	<0.05
Rubidium	<0.01	Hafnium	<0.10
Strontium	<0.10	Tantalum	<0.08
Yttrium	<0.03	Tungsten	<0.02
Zirconium	<0.11	Rhenium	<0.10
Niobium	<0.03	Iridium	<0.02
Molybdenum	<0.02	Platinum	<0.13
Ruthenium	<0.12	Gold	<0.03
Palladium	<0.10	Mercury	<0.01
Silver	<0.03	Thallium	<0.08
Cadmium	<0.08	Lead	<0.07
Indium	<0.07	Bismuth	<0.15
Tin	<0.25	Thorium	<0.03
Antimony	<0.08	Uranium	<0.03

of the environmental water supply they would be deadly. PVDF can be buried, is resistant to weathering, and can be joined in continuous lengths via fusion. These properties make PVDF a superior choice for handling such chemicals.

7.4 MINING

Applications in the mining industry for PVDF are centered around chemical resistance, abrasion resistance, and long life in inaccessible places. Gold, silver, and copper go through an autoclaving process to separate them from other mining residue. Sulfuric acid and other etching chemicals are used in this process. As temperatures are elevated, the process becomes more effective. PVDF can be used to pipe these chemicals and as a liner on carbon steel to replace lead vessels, which are losing favor due to environmental concerns.

Table 25.9. Various extractible tests based on PVDF exposed for 24 hours to deionized water at 80 °C

Element	Impurity detected (ppb)	Test method
Calcium	3.40	Graphite furnace
Iron	0.60	Graphite furnace
Fluoride	21.70	Ion chromatography
Chloride	4.97	Ion chromatography
Bromide	0.13	Ion chromatography
Nitrate	5.79	Ion chromatography
Phosphate	<0.05	Ion chromatography
Sulfate	10.80	Ion chromatography
Lithium	<0.05	Ion chromatography
Sodium	3.40	Ion chromatography
Ammonium	4.19	Ion chromatography
Potassium	1.69	Ion chromatography
Silica	<3.00	Soluble silica
Organic carbon	8.00	TOC

Figure 25.3. Glove-box lined with PVDF for handling radioactive waste in the nuclear processing industry. PVDF crosslinks when exposed to nuclear radiation and maintains its mechanical properties at moderate doses

Just as in many applications for PVDF, mining applications often have a requirement for system longevity. While corrosion can be a concern, the most important aspect of a pump or other inaccessible system part is that it continues to perform over a long period of time and does not need to be replaced.

7.5 METAL PREPARATION

Metal finishing applications such as anodizing and plating use high-temperature acids to accomplish the goal. Large acid baths used here can be PVDF lined with welded sheet for long life. PVDF chamber pumps and piping are used to process these acids which can typically be at temperatures up to 105 °C.

7.6 PHARMACEUTICAL

Bromine and chlorine containing chemicals are commonly used in the manufacture of drugs. Halogens are notorious for corroding metal, so plastics are often used. Pharmaceutical companies tend to work very profitable processes and downtime is very costly. PVDF-based resins offer chemical resistance as well as excellent permeation resistance against halogenated chemicals. PVDF piping (solid and lined steel), vessels, pumps, and tower packing are all used extensively in the pharmaceutical manufacturing end as well as in the waste chemical and original reactant product handling end.

Recently, pharmaceutical companies have seen a greater need for higher purity water similar to that used in the semiconductor industry [13]. PVDF offers the purity and long-term resistance to disinfectants such as sodium hypochlorite and ozone that is required for a piping material (Figure 25.4). Also, sanitary tubing systems which can be autoclaved and/or reassembled are available to engineers concerned about system contamination from bacteria. PVDF membranes are used extensively in purification loops.

Finally, water for injection (WFI) systems that previously were limited to metal construction are now often made with PVDF components due to better purity obtained by a pure unfilled polymer that leaches a limited amount of metallic ions to the water. Many individual PVDF-based resins have been tested in compliance with *United States Pharmacopeia (USP)* Class 6 criteria.

7.7 FOOD AND BEVERAGE

PVDF is listed in several publications and/or letters which have been published explaining the suitability for its use in food and beverage applications. Such publications are: (1) Title 21, *Code of Federal Regulations,* Chapter 1, Part 177.2510; (2) *3-A Sanitary Standards for Multiple-Use Plastic Materials Used as Product Contact Surfaces for Dairy Equipment,* Serial No. 2000.

KYNAR® PVDF has obtained a letter from the United States Department of Agriculture (USDA) for use in storage areas in contact with meat and poultry and is also listed by the National Sanitation Foundation Standard 61 for potable water. There are most likely several worldwide listings for PVDF in food and beverage applications that have been left unmentioned here. There are many plastics that have application in the food and beverage industry. What is not always obvious to the average person is that certain juices, sauces, and other consumables are tremendously acidic and quickly corrode metals or polymers.

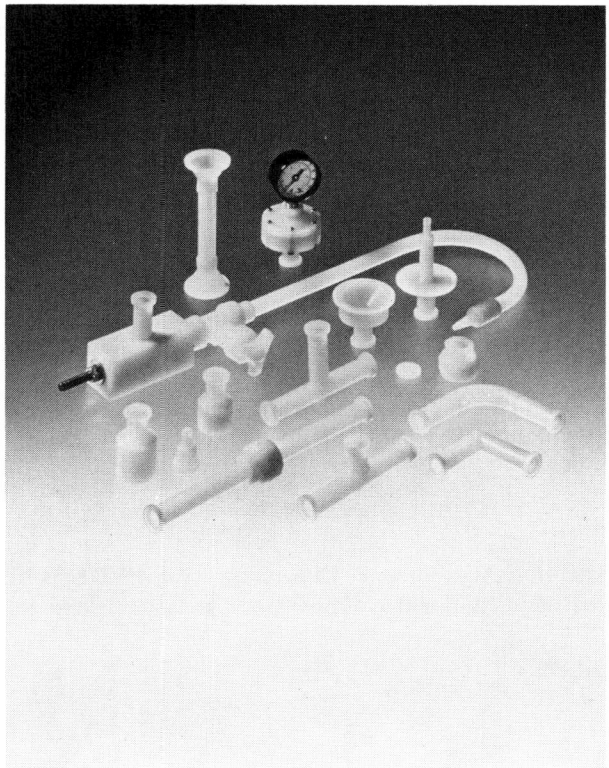

Figure 25.4. PVDF fittings and pipes are extensively used in the pharmaceutical industries where they offer purity and long-term resistance to disinfectants such as sodium hypochloride and ozone. Furthermore, since pharmaceutical operations are very profitable processes, downtime due to maintenance must be minimized

PVDF offers the necessary chemical resistance with the additional ability to withstand steam cleaning temperatures and certain cleaning agents that are used to kill bacteria that must be kept out of food processes. Also, the combination of smooth surface and abrasion resistance prevents food from sticking to the PVDF surface and from particles of PVDF getting into the food process. Tomato paste, cranberry juice, milk, lime juice, and various acidic sauces often require PVDF piping for the best in durability. PVDF does not stand up to continuous microwaving. Microwaves will degrade the polymer.

7.8 WASTEWATER

By its own nature, wastewater is typically a combination of many products that can potentially create interesting byproducts, exothermic reactions, and a wide pH range over time. The combination of properties attributed to PVDF makes it a potentially more universal material of construction compared to metals and other cheaper plastics.

Many acids become more reactive to metal at lower concentrations and cause quick oxidation. Usually with plastics, the performance is better the lower the concentration of acid. For example, if PVDF is known to last over 10 years in 93% sulfuric acid at 50 °C then it should last even longer at 60% sulfuric acid at 50 °C.

7.9 PESTICIDES

Pesticides, like some pharmaceutical products, are often chlorinated and/or brominated. Much of the earlier writing herein has been dedicated to PVDF-based resins and their excellent chemical and permeation resistance to halogens (Figure 25.5). This industry utilizes PVDF piping, pumps, vessels, and tower packing as a cost savings to the exotic metals which otherwise would be needed to handle these aggressive chemicals.

7.10 INSTITUTIONAL

The newest emerging high-volume CPI market for PVDF is for piping in air plenum areas of institutional buildings such as school laboratories, college

Figure 25.5. PVDF piping systems are commonly used in handling chlorinated solvents and pesticides. In this application outdoor weathering resistance is also required

research centers, chemical waste drainage lines in any building under government codes, and high-purity liquid handling. Smoke suppressed PVDF resins easily meet ASTM E-84 (also known as UL 723) and are being used for piping by many conscientious engineers in acid waste drainage where codes often call out metal or glass piping systems unless a recognized safe polymer is available.

In Canada, two grades of natural unfilled PVDF made by Elf Atochem pass ULC S102.2 for use directly in compliance with the Canadian Piping Code in plenum areas of buildings. No other unfilled polymer is known to have passed this stringent test, making PVDF the only viable choice for high-purity systems that need a purity higher than metal systems can offer.

REFERENCES

1. Robinson D. and Seiler D. A., 'Modifications of PVDF resins leading to new fabrication opportunities with improved service life', *Managing Corrosion With Plastics NACE Proceedings*, Vol. 10, No. 22, pp 1–14 (1991) October.
2. Craven R. A., Ackermann A. J., and Tremont P. L., 'High purity water technology for silicon-wafer cleaning', *Microcontamination*, August 1986.
3. Burggraaf P., 'Pick your plastic pipe carefully', *Semiconductor International*, July 1988.
4. Krause H. J., 'Modern plastics industrial piping systems', *Engineering Digest*, pp. 16–20 (November 1987).
5. Blume R., 'Preparing ultrapure water', *Chemical Engineering Progress*, pp. 55–57 (December 1987).
6. Zoccolante G., 'Innovations in water purification', *Semiconductor International*, pp. 86–89 (February 1987).
7. Locheed T., 'PVDF for ultrapure water', *Chemical Processing*, pp. 37–38 (January 1993).
8. Hanselka R., Williams R, and Bukay M., 'Materials of construction for water Systems - Part 1: Physical and chemical properties of plastics', *Ultrapure Water*, pp. 46–50 (July/August 1987).
9. Garner A., 'Chlorination washer corrosion resurveyed', *Pulp & Paper Canada*, pp. 77–84 (August 1988).
10. Governal R. A., 'Ultrapure Water: a battle every step of the way', *Semiconductor International*, pp. 176–180 (July 1994).
11. Henley M., 'Market review: semiconductor industry expansion should continue', *Ultrapure Water®*, pp 15–20 (May/June 1995).
12. Robinson D., Seiler D. A. 'Potential Applications For Kynar Flex® PVDF in the nuclear industry', *Proceedings, American Glovebox Society, 1993 National Conference, Linings, Coatings, and Materials Section 3C*, pp 10–14, Seattle, WA, August 16–19, 1993.
13. Keer D., 'Pharmaceuticals: upgrading purified water systems to today's standards', *Ultrapure Water®*, pp. 40–44, (July/August 1993).

26

Fluorinated Acrylic Ester Polymers

TETSUO SHIMIZU
Daikin Industries Ltd, Osaka, Japan

1 INTRODUCTION

Acrylic ester polymers in which some or nearly all hydrogens are replaced with fluorines have been prepared. Among these, polyacrylates and polymethacrylates having fluoroalkyl groups in the side chains, in particular, are used for a wide variety of industrial applications, such as textile finishes, protective coatings and surface modifiers. Meanwhile, the introduction of fluorines or a trifluoromethyl group in the acryloyl moieties of the esters expands the potential uses in the field of fluorinated acrylic ester polymers. For instance, poly(2-fluoroacrylate)s are commercially prepared as cladding materials for polymeric optical fibers. This chapter focuses on the preparation, structure and properties, and applications of the fluorinated acrylic ester polymers, whose monomeric unit contains at least one fluorine or fluorinated group.

2 PREPARATION

2.1 MONOMERS

Major monomers are acrylates and methacrylates of fluorinated alcohols, which are prepared by esterification of acrylic or methacrylic acids or the acid chlorides with fluorinated alcohols [1]. The major commercial processes for producing the fluoroalkyl intermediates for the alcohols are based on either electrochemical fluorination (ECF) of alkanoic or alkanesulfonyl chlorides, in anhydrous hydrogen fluoride [2], or telomerization of tetrafluoroethylene (TFE) with perfluoroalkyl iodides (R_FI) such as CF_3I, CF_3CF_2I and $(CF_3)_2CFI$ [3]. The ECF route from alkanesulfonyl chlorides and the latter TFE telomerization route to the fluorinated

Modern Fluoropolymers. Edited by John Scheirs
© 1997 John Wiley & Sons Ltd

alcohols are exemplified in equations (1) and (2), respectively.

$$C_8H_{17}SO_2Cl \xrightarrow{HF} C_8H_{17}SO_2F \xrightarrow{RNH_2} C_8H_{17}SO_2NHR \qquad (1)$$

$$\xrightarrow{ClCH_2CH_2OH} R_FSO_2N(R)CH_2CH_2OH \quad (R = CH_3, CH_3CH_2)$$

$$CF_3CF_2I \xrightarrow{TFE} CF_3CF_2(CF_2CF_2)_nI \xrightarrow{Ethylene} CF_3CF_2(CF_2CF_2)_nCH_2CH_2I \quad (2)$$
$$\textbf{a}$$

$$\textbf{a} \xrightarrow{oleum/H_2O} CF_3CF_2(CF_2CF_2)_nCH_2CH_2OH \quad \text{or}$$

$$\textbf{a} \xrightarrow{KOH} CF_3CF_2(CF_2CF_2)_nCH_2{=}CH_2$$

$$\xrightarrow{[O]} CF_3CF_2(CF_2CF_2)_nCOOH \xrightarrow{[H]} CF_3CF_2(CF_2CF_2)_nCH_2OH$$

Moreover, although the production scale is far smaller than for the above, a wide variety of fluorinated alcohols are prepared for the esters. For instance, ω-hydroperfluoroalkyl carbinols $H(CF_2CF_2)_nCH_2OH$, branched lower alcohols such as $(CF_3)_2CHOH$ and $(CF_3)_2CFOH$, perfluoropolyether alcohols such as

$$CF_3O(C_3F_6O)_m(CF_2O)_nCF_2CH_2OH,$$

and fluoroalkylene diols such as $HOCH_2(CF_2)_nCH_2OH$. ω-Hydroperfluoroalkyl carbinols are obtained directly from TFE telomerization with CH_3OH.

Several methods have been suggested for the synthesis of 2-fluoroacrylic acid or the derivatives for the intermediates of 2-fluoroacrylate monomers [4]. For instance, Ohmori, Takaki and Kitahara reported a method to obtain $F(O)CCF{=}CH_2$ in high yield using tetrafluorooxetane (equation 3) [5].

$$CF_2{=}CF_2 \xrightarrow{CH_2O/HF} \begin{array}{c} CF_2{-}CF_2 \\ | \qquad | \\ CH_2{-}O \end{array} \xrightarrow[DMF]{NaI, Zn} F(O)CCF{=}CH_2 \quad (3)$$

For the synthesis of 2-trifluoromethylacrylic acid or its derivatives, several methods are known [6]. More recently, Fuchigami and Suzuki reported carbonylation of $CF_3C(Br){=}CH_2$ with CO and H_2O using I^-, base and palladium catalysts to get the acid in a high yield of more than 80% (equation 4) [7].

$$CF_3C(Br){=}CH_2 + CO + H_2O \xrightarrow[(Ethyl)_3 N/THF]{PdCl_2(PPh_3)_2, KI} HO(O)C(CF_3){=}CH_2 \quad (4)$$

Other fluorinated 2-substituted acrylate monomers, such as fluoroalkyl 2-cyanoacrylates $R_fOC(O)C(CN){=}CH_2$ [8] and fluoroalkyl 2-chloroacrylates $R_fOC(O)C(Cl){=}CH_2$, have been prepared, where R_f is a fluoroalkyl or occasionally a fluoroalkoxyalkyl group. $R_fOC(O)C(Cl){=}CH_2$ are easily prepared using the corresponding acrylates as starting materials (equation 5).

$$R_fO(O)CCH=CH_2 \xrightarrow{\text{Cl}_2} R_fO(O)CCHClCH_2Cl \xrightarrow{-HCl} R_fO(O)CC(Cl)=CH_2$$

(5)

In addition, 2,3-difluoroacrylates $R^1OC(O)CF=CHF$ [9] and 2,3,3-trifluoro-acrylates $R^1OC(O)CF=CF_2$ [10], where R^1 is an alkyl (R) or R_f, are also synthesized.

2.2 POLYMERIZATIONS

Fluoroalkyl acrylates and methacrylates

$$(R_fOC(O)CH=CH_2, R_fOC(O)C(CH_3)=CH_2)$$

can be easily polymerized using free-radical initiators such as peroxides or azo compounds, in bulk, solution and emulsion methods, which is almost similar to the polymerizations of nonfluorinated analogs. Most of the polymers are commercially prepared as copolymers to tailor molecules to specific applications. The monomers copolymerize readily with each other or with other nonfluorinated vinyl monomers such as alkyl acrylates and methacrylates, styrene, acrylonitrile, vinyl acetate and vinyl chloride.

2-Fluoroacrylate monomers $R^1OC(O)CF=CH_2$ can also be easily polymerized or copolymerized with a variety of vinyl monomers under radical conditions, being similar to acrylates and methacrylates [11–13]. The esters usually show large homo-polymerizability, giving high molecular-weight polymers, compared with the corresponding acrylates and methacrylates. According to the report of Theis et al. [13], the polymerization rate constants $(k_p/k_t^{0.5})$ of $CH_3OC(O)CF=CH_2$ and $(CF_3)_2CHOC(O)CF=CH_2$ are 10–19 times larger than those of the corresponding methacrylates. The absolute values of propagation and termination rate constants (k_p, k_t) are given for ethyl 2-fluoroacrylate [12]. In addition, it is noteworthy that 2-fluoroacrylate polymerizations are less inhibited by oxygen than acrylates and methacrylates.

On the other hand, 2-trifluoromethylacrylates $R^1OC(O)C(CF_3)=CH_2$ are barely homopolymerizable under radical conditions because of an electron-drawing effect and steric hindrance effect of the CF_3-substituent [14]. However, radical copolymerizations with comonomers having electron-rich double bonds, such as vinyl ethers and α-olefins, are reported, giving alternative copolymers, which is explained by a charge transfer polymerization mechanism [15,16].

$R^1OC(O)CF=CF_2$ is also not readily homopolymerizable. Polymerization of $R^1OC(O)CF=CHF$ is possible but very slow [13]. For $R_fOC(O)C(Cl)=CH_2$ or $R_fOC(O)C(CN)=CH_2$, both are known to be easily polymerized in the patent literature [8,17]. Especially, the polymerization of $R_fOC(O)C(CN)=CH_2$ is initiated by moisture in air [8].

Table 26.1 lists Q–e values for Alfrey-Price's scheme to estimate the effect of the fluorosubstituents on the reactivity of acrylic ester monomers. Fluoroalkyl

Table 26.1. Q and e-values of acrylic ester monomers, $R^1O(O)CC(R^2)=CH_2$

R^1	R^2	Q	e	Ref.
CH_3-	$-H$	0.42	0.60	
CF_3CH_2-	$-H$	0.97	1.13	19
$(CF_3)_2CH-$	$-H$	0.79	1.36	19
$CF_3CF_2CF_2CH_2-$	$-H$	0.78	1.15	20
CH_3-	$-CH_3$	0.74	0.40	
CF_3CH_2-	$-CH_3$	1.13	0.98	19
$(CF_3)_2CH-$	$-CH_3$	1.38	1.30	19
$CF_3CFHCF_2CH_2-$	$-CH_3$	2.24	1.2	21
$H(CF_2)_4CH_2-$	$-CH_3$	1.51	0.94	21
CH_3-	$-F$	0.47	0.73	11
CH_3CH_2-	$-F$	0.49	0.68	12
CH_3-	$-CF_3$	0.8	2.9	22

acrylates and methacrylates exhibit large e-values compared with their nonfluorinated counterparts, with $CH_3OC(O)C(CF_3)=CH_2$ exhibiting the largest e-value, because of an electron-drawing effect of their fluorosubstituents. However, there are little differences in $Q-e$ values between $ROC(O)CF=CH_2$ (R; CH_3, CH_3CH_2) and the corresponding acrylate and methacrylate. This is due to the characteristic of fluorine directly attached to a double bond. Taking advantage of the large e-values, several anionic polymerizations have been investigated [18].

3 STRUCTURE AND PROPERTIES

3.1 STRUCTURE

Fluorinated acrylic ester polymers radically polymerized are known to be generally amorphous. However, when the esters of the higher normal fluorinated alcohols, typically $F(CF_2)_n(CH_2)_mOH$ with $n = 7 - 11$ and $m = 1$ or 2, are polymerized, side-chain crystallization occurs in the fluoroalkyl groups and imparts crystalline properties to the polymers [23,24]. The linear side chains crystallize to form lamellar packings having single- and/or double-layered structures at room temperature, in which the fluoroalkyl chains are perpendicular to the main chains [25,26]. Figure 26.1 shows a schematic representation of the lamellar structure and the X-ray diffraction pattern of poly[2-(perfluorooctyl)ethyl acrylate]. The presence of four inner rings in the pattern indicates the high degree of perfection in the double-layered structure of the $C_8F_{17}CH_2CH_2-$ fluoroalkyl side chains. Side-chain crystallization behaviors play an important role to the surface properties of the polymers, as described later.

Moreover, it is to be noted that radically polymerized poly(fluoroalkyl 2-fluoroacrylate)s show crystalline properties even if the fluoroalkyl side chains are short ($n < 7$) [27], while the corresponding polyacrylates and polymethacrylates

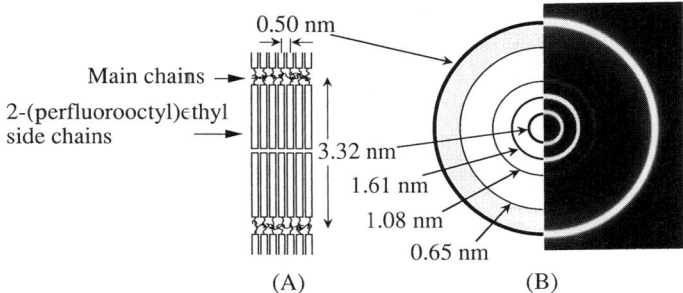

Figure 26.1. Schematic representation of the lamellar structure in poly[2-(per-fluorooctyl)ethyl acrylate] (A), and its X-ray diffraction pattern and Bragg-spacings (B)

are mostly amorphous as reported [23,24]. The crystallites found in the short fluoroalkyl poly (2-fluoroacrylate)s are considered to be caused by increased intermolecular interaction between the main chains, rather than the stereoregularity. However, the crystalline properties are not found when the fluoroalkyl side chains are replaced by branched ones like $(CF_3)_2CH-$, or by alkyl groups.

3.2 THERMAL AND MECHANICAL PROPERTIES

Table 26.2 lists glass-transition temperatures (T_g) measured by differential scanning calorimetry DSC and dynamic mechanical measurement for representative fluorinated acrylic ester polymers. In comparison, among the poly(2-substituted acrylate)s having identical side groups, the T_g's tend to be in the order H < CH_3 < F < Cl for the respective substituents. It is characteristic of poly(2-fluoroacrylate)s to exhibit T_g's 10–30 °C higher than their polymethacrylate analogues. As listed in Table 26.2, melting points (T_m) are observed, not only for polymethacrylate and polyacrylates having crystallizable longer fluoroalkyl groups, but also poly(2-fluoroacrylate)s having linear fluoroalkyl side chains.

Table 26.3 shows a comparison between poly(2-substituted acrylate)s, of several mechanical properties. In comparison among the poly(2-substituted acrylate)s having identical side groups, dynamic moduli (E' at 25 °C) are in the order H << F < CH_3 < Cl for the respective substituents. However, at high temperatures, the moduli of the poly(2-fluoroacrylate)s are reported to become relatively larger than those of the polymethacrylates [28]. This is likely to be due to the increased intermolecular interaction of the formers. Comparing poly(2-fluoroacrylate)s with polymethacrylates for the values of stress–strain tests and the hardness values, the former group show tougher behavior than their corresponding polymethacrylates [28].

For thermal degradation of the polymers in air, the temperatures at 5 wt% loss $(T_d$ in Table 26.2) are in the order Cl < CH_3 < H < F for the respective substituents. Poly(2-fluoroacrylate)s possess outstanding thermal stability.

Table 26.2. Thermal properties of fluorinated acrylic ester polymers

Polymer[a]		T_g/°C		T_m[b]	T_d[d]
R_1	R_2	DSC[b]	(DMA[c])	/°C	/°C
CF_3CH_2-	$-CH_3$	81	(93)	ND[e]	276
$H(CF_2)_2CH_2-$	$-CH_3$	80	(91)	ND	291
$F(CF_2)_2CH_2-$	$-CH_3$	73	(85)	ND	283
$(CF_3)_2CH-$	$-CH_3$	56[f]			
$H(CF_2)_4CH_2-$	$-CH_3$	46	(59)	ND	281
$F(CF_2)_8CH_2-$	$-CH_3$	56		89	
$F(CF_2)_8CH_2CH_2-$	$-CH_3$	57	(61)	89	264
$(CH_3-$	$-CH_3)$[h]	99		ND	311
CF_3CH_2-	$-H$	5		ND	334
$F(CF_2)_2CH_2-$	$-H$	-13		ND	311
$F(CF_2)_8CH_2CH_2-$	$-H$	52	(58)	74	274
CF_3CH_2-	$-F$	110	(130)	214	362
$H(CF_2)_2CH_2-$	$-F$	86	(110)	176	361
$F(CF_2)_2CH_2-$	$-F$	97	(113)	220	352
$(CF_3)_2CH-$	$-F$	102		ND	350
$H(CF_2)_4CH_2-$	$-F$	55	(99)	141	351
$F(CF_2)_8CH_2-$	$-F$		(127)	185	
$F(CF_2)_8CH_2CH_2-$	$-F$	60	(66)	126	313
CH_3-	$-F$	129		ND	345
CH_3CH_2-	$-F$	81		ND	323
CF_3CH_2-	$-Cl$	133[g]		ND	
$F(CF_2)_2CH_2-$	$-Cl$	131		ND	229
$F(CF_2)_8CH_2CH_2-$	$-Cl$	61	(65)	143	254
$(CH_3-$	$-Cl)$[h]	152		ND	

[a] $-(CR^2-CH_2)_p-$
      |
      $COOR^1$

[b] Differential scanning calorimetry (heating rate; 10 °C min^{-1})
[c] Damping peak temperatures by dynamic mechanical measurement at 1 Hz.
[d] Degradation temperatures at 5 wt% loss in air.
[e] Not detected.
[f] Ref. 29.
[g] Ref. 30.
[h] Nonfluorinated polymers.

3.3 SURFACE PROPERTIES

Surface properties of fluorinated acrylic ester polymers have been widely studied, because of their usefulness to industrial applications which take advantage of the extremly low surface energy. Table 26.4 lists several representative poly(2-substituted acrylate)s and the critical surface tensions of wetting γ_c defined by Zisman, which are the extrapolated values of 0 contact angle or $\cos\theta = 1$. In general, the γ_c value depends on the length, composition, branching and terminal

Table 26.3. Comparison of several mechanical properties between poly (2-substituted acrylate)s

Polymer[a]		Dynamic modulus (E') /GPa	Tensile strength /MPa	Elongation[b] /%	Vickers hardness[b] /kg mm^{-2}	Rockwell hardness[b] /M scale
R^1	R^2					
CH$_3-$	$-$CH$_3$	5.2[c]	61.8	7.1	20.5	90
CF$_3$CH$_2-$	$-$CH$_3$	1.7[d]	27.8	6.3	10.7	47
CF$_3$CF$_2$CH$_2-$	$-$CH$_3$	1.2[d]	24.9	4.8	8.7	31
CH$_3-$	$-$F	4.0[c]	70.3	7.4	29.7	95
CF$_3$CH$_2-$	$-$F	1.5[d]	35.1	16.6	15.7	62
CF$_3$CF$_2$CH$_2-$	$-$F	0.9[d]	28.0	16.9	9.5	75
CF$_3$CF$_2$CH$_2-$	$-$H	0.002[d]				
CF$_3$CF$_2$CH$_2-$	$-$Cl	1.6[d]				

[a] $-\!(CR^2\!-\!CH_2)_{\overline{p}}\!-$
　　　|
　　COOR^1
[b] Ref. 32. Reproduced by permission of Nikkan Kogyo Shinbunsha.
[c] Ref. 28 (at 25 °C, 35 Hz)
[d] Ref. 31 (at 25 °C, 1 Hz)

Table 26.4. Critical surface tensions (γ_c/mNm^{-1}) of fluorinated acrylic ester polymers

Alkyl or fluoroalkyl side group	2-Substituent			Refs
	H	CH$_3$	F	
CH$_3-$		38.0	20.0	28,28
CF$_3$CH$_2-$		19.0	15.0	36,36
C$_2$F$_5$CH$_2-$	15.5	15.5	12.0	34,36,36
H(CF$_2$)$_2$CH$_2-$		20.4	19.0	36,36
n-C$_3$F$_7$CH$_2-$	15.2			35
n-C$_4$F$_9$CH$_2$CH$_2-$	10.2			33
H(CF$_2$)$_4$CH$_2-$	17.0	16.6	15.8	35,36,28
H(CF$_2$)$_6$CH$_2-$	18.5			35
n-C$_7$F$_{15}$CH$_2-$	10.4	10.6		35,35
n-C$_8$F$_{17}$CH$_2-$	10.4			35
iso-C$_7$F$_{15}$CH$_2$CH$_2-$	12.6			34
n-C$_8$F$_{17}$CH$_2$CH$_2-$	11.0	10.6		33,28
H(CF$_2$)$_8$CH$_2-$	13.0			35
n-C$_{10}$F$_{21}$CH$_2$CH$_2-$	11.0			33
H(CF$_2$)$_{10}$CH$_2-$	14.5–15.0			35
(CF$_3$)$_2$CF$-$	14.1			35
(CF$_3$)$_2$CH$-$	15.0–15.4	14.8–15.4		35,35
n-C$_{12}$F$_{25}$CH$_2$CH$_2-$	10.5			33

groups of fluoroalkyl side chains, and the crystallinity and tacticity of polymers
[35]. When the carbon number (n) of CF_2 moieties in normal fluoroalkyl side
groups [n-$C_nF_{2n+1}(CH_2)_m-$, $m = 1$ or 2], is larger than or equal to 4, the
γ_c values of polyacrylates and polymethacrylates drop to the lowest value of
$10-11$ mNm^{-1}. However, the polymers whose terminal groups of the side chains
that are branched or contain HCF_2- groups show values a few millinewton per
meter larger than those of CF_3- terminated linear side chains. In addition, the
differences of 2-substituents in γ_c are observed. Poly(2-fluoroacrylate)s exhibit
lower critical surface tensions than the corresponding polymethacrylates [36].

Meanwhile, with respect to the practical use of oil- and water-repellency,
it is pointed out that dynamic contact angles are more important than static
ones. That is, the larger the receding contact angle (θ_r), the more repellent
surface we can obtain [33]. Figure 26.2 shows variations of advancing contact
angles (θ_a) and θ_r obtained by the Wilhelmy plate method for the surfaces
of poly[2-(perfluoroalkyl)ethyl acrylate]s as a function of the carbon number
(n) of CF_2 moieties in the side chains. Interestingly, θ_r is very low when n
is below 7. This result indicates that θ_r strongly reflects crystallinity in the
fluoroalkyl side chains, since the side-chain crystallization occurs when $n \geqslant 7$.
For the quenched polymers with lower crystallinity, accordingly, the θ_r values
decrease distinctly [37].

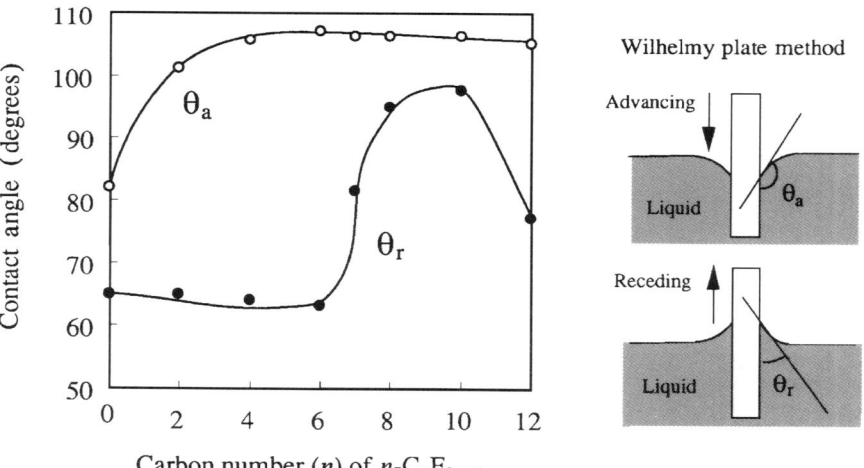

Figure 26.2. Variations of advancing contact angles (Θ_a) and receeding contact
angles (Θ_r) for the surfaces of poly[2-(perfluoroalkyl)ethyl acrylate)s as a function
of the carbon number (n) in the $C_nF_{2n+1}CH_2CH_2-$ side chains by the Wilhelmy
plate method. (Reproduced with permission from reference 33. Copyright 1994 CMC
Co. Ltd.)

3.4   OTHER PROPERTIES

3.4.1   Optical Properties

Replacement of C—H bonding by C—F in general decreases the refractive index of the polymer by an amount proportional to the density of fluorine, since a fluorine atom or a C—F bonding has small polarizability [29].

3.4.2   Solubility

Acrylic ester polymers of the lower fluorinated alcohols are commonly soluble in conventional polar solvents such as DMF, THF and ethyl acetate. However, the analogues with crystallizable longer n-fluoroalkyl side chains tend to be insoluble. Several fluorinated solvents such as perfluoro(2-butyltetrahydrofuran) (FC-75), 1,3-bis(trifluoromethyl)benzene and HCFC-225 ($C_3HF_5Cl_2$) are employed for their coating applications.

4   APPLICATIONS

4.1   TEXTILE FINISHES

Applications of fluorinated acrylic ester polymers to impart oil- and water-repellency to textiles such as apparel, soft furnishings and carpets, are in wide commercial use today (Scotchgard® by 3M, Teflon® by DuPont, Asahigard® by Asahi Glass, Unidyne® by Daikin, etc.). According to the patents in this area, most of the finishes are based on methacrylate or polyacrylate monomers having structures of either $R_FCH_2CH_2OC(O)CR^1{=}CH_2$ (R_F = per- fluoroalkyl, R^1 = H, CH_3) or sulfonamides of the type

$$R_FSO_2N(R)CH_2CH_2OC(O)CH{=}CH_2 \quad (R = CH_3, CH_3CH_2).$$

   These monomers are usually emulsion-copolymerized in the presence of surfactants, together with a range of nonfluorinated vinyl monomers, and then formulated with other ingredients to adjust the required performance. The comonomers are typically classified into two kinds, one is alkyl acrylates or methacrylates to afford enhanced durability, and/or modification of the "feel" of the treated fabric. The other is smaller amounts of cross-linking monomers having functional groups such as OH and glycidyl to improve a resistance of the finishes to laundering and dry cleaning. In addition, hydrophilic components are in some cases incorporated into the polymers to improve wettability in water whilst retaining stain-repellency in air, resulting in a good soil-release effect during laundering. The finishes formulated with these amphiphilic type of polymers are called 'soil-release agents'. Representative examples of the amphiphilic polymers are block copolymers composed of polyacrylate segments having fluoroalkyl pendant groups and poly(ethylene oxide) segments [38], and copolymers of fluoroalkyl acrylates with acrylates having poly(ethylene oxide) groups [39]. Details on the subject are reviewed by Kissa [40].

4.2 ELECTRONICS SECTOR

4.2.1 Resists

In the production of high-density electronic integrated circuits, polymeric resists for microlithography are indispensable materials. PMMA has been widely used as a positive resist because of its high resolution; however, it has low sensitivity for electron beam and X-rays. In order to improve the radiation sensitivity, a great number of acrylic polymers including the fluorinated ones have been investigated. Among them, Kakuchi *et al.* found that poly(2,2,3,4,4-hexafluorobutyl methacrylate) is 46 times more sensitive and higher in contrast than PMMA, and has the resolution limit of 0.3 μm [41]. Meanwhile, Tada found that poly(2,2,2-trifluoroethyl 2-chloroacrylate) can be a highly sensitive positive resist with improved thermal properties $(T_g;\ 133\,^\circ C)$ [30]. The latter resist is commercialized by Toray Industries under the name of EBR®-9 (Figure 26.3). Table 26.5 lists properties for the positive resist polymers compared with PMMA.

4.2.2 Protective Coatings

Printed circuit boards (PCBs) on which electronic components are mounted, are frequently coated with a protective coating to preserve their electrical properties from moisture. Polyacrylates and polymethacrylates having longer n-perfluoroalkyl side chains, and sometimes the copolymers with non-fluorinated

Figure 26.3. EBR®-9 resist pattern produced from a polyfluoroalkyl acrylate polymer

Table 26.5. Properties of fluorinated acrylic ester polymers for positive resists[a]

Polymer[b]		T_g	Resolution	Contrast	Sensitivity		
R^1	R^2	(°C)	(μm)	(γ value[c])	E-beam (μC cm^{-2})	X-ray (mJ cm^{-2})	Deep UV (mJ cm^{-2})
CH$_3$CHFCF$_2$CH$_2$–	–CH$_3$	50	0.3	4.5	0.4	52	90
HCF$_2$CF$_2$(CH$_3$)$_2$C–	–CH$_3$	101	0.3	2.6	1	160	320
CF$_3$CH$_2$–	–Cl	133	0.2–0.3	4.5	1	52	90
(PMMA)		105	0.1	2.5	500	4000	700

[a] Reproduced with permission from Ref. 42. Copyright 1982 Nikkan Kogyo Shinbunsha
[b] –[R^{1}OC(O)C(R^2)–CH$_2$]$_p$–
[c] Ref. 42.

vinyl monomers, are used for the coating. The fluorinated coatings provide an efficient moisture protection with thinner coating than that of conventional ones such as acrylics and urethanes [43]. In addition, interestingly, their coatings on PCBs do not prevent solder from wetting and adhering to the circuits, giving the ease to repair electronic components [44]. Fluorinated acrylic polymers are also applied as a protective coating when soldering electric parts. The soldering flux is prevented from penetrating to or depositing on the peripheral portion around the soldering portion [45].

4.2.3 Charge Control Agents

In the Xerographic process, the electrostatic image on the photoreceptor is developed by dry toner powders. For a two-component developing system, charging of the developer is accomplished by agitating a mixture of toner and carrier beads. In the triboelectrification, charge control agents are frequently used, by adding them to the toner or coating them on the carrier beads. For negative charge control, fluorinated acrylate polymers are often used [46], since fluorine-containing polymers generally have a tribo characteristic of negative charge, which is a unique functionality of fluorine. For the carrier coating, not only negative chargeability, but the ease of coating, strong adhesion to the core materials of the carrier beads, and tribo-durability are required. Fluorinated acrylate polymers or their copolymers well meet the requirements.

4.3 OPTICS SECTOR
4.3.1 Optical Fibers

At present, fluorinated acrylic ester polymers are commercially used as cladding materials for polymeric optical fibers (POF) whose core is typically made of poly(methyl methacrylate), and in some cases, for silica optical fibres (PCF: polymer coated fibres). These are the largest applications taking advantage of the low refractive index which is unique to fluoropolymers. For the cladding of POF, polymethacrylates or poly(2-fluoroacrylate)s having rather short fluoroalkyl side

TETSUO SHIMIZU

groups such as CF_3CH_2-, $HCF_2CF_2CH_2-$, $CF_3CF_2CH_2-$ and $(CF_3)_2CH-$ are usually used [47]. In addition, they are often copolymerized with other acrylic ester monomers to adjust the required properties. Although vinylidene fluoride (VDF)-based resins are also used for claddings, fluorinated acrylic ester polymers exhibit better performances in transparency and attenuation loss. The photo-curable coatings specially formulated with oligomers of fluorinated acrylic esters are also employed, but are limited for PCF [48]. (These types of coatings are also being applied for various displays to impart antireflection.)

Present uses of POF or PCF are mainly lightings (Figure 26.4), optical sensors, short-length data links, etc. However, there has been investigations of new POF composed of fluoropolymeric core materials for longer distance optical communications such as LAN (local area network), to replace silica fibres which are inferior to POF in handling and connecting [49]. For this, intrinsic attenuation losses in the present PMMA core fibres, which are dominated by $C-H$ overtone absorptions in a spectral region between 600 and 900 nm, must be reduced. So, acrylic ester polymers which are substituted by heavier elements than hydrogen, such as deuterium and fluorine, have been proposed. Figure 26.5 exemplifies dramatic decreases in attenuation for the deuterated or fluorinated POFs compared with a PMMA-based one. Comparing deuterated and fluorinated polymers, the latter has much advantage economically. Moreover, it has low water absorption because of its hydrophobicity (water absorption seriously increases losses at the near-infrared region).

Among a number of candidates for core materials, fluoroalkyl 2-fluoroacrylate polymers such as poly(hexafluoroisopropyl 2-fluoroacrylate) are considered to be well suited [13,50], offering not only low attenuation losses but also better thermal resistance than PMMA. Meanwhile, amorphous perfluoroplastics such as Teflon® AF and Cytop® are also being evaluated for this purpose. Their suitability will depend on their balance of cost and performance.

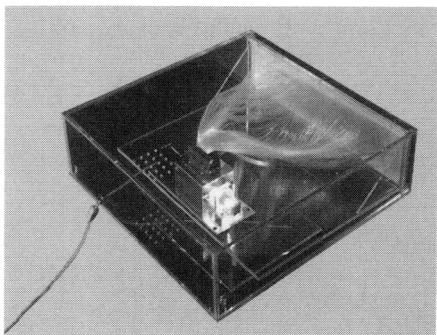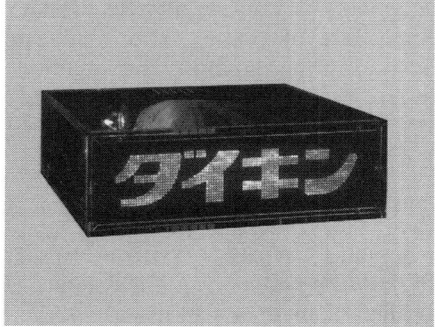

Figure 26.4. Illumination using polymeric optical fibers made from polyfluoroalkyl acrylate

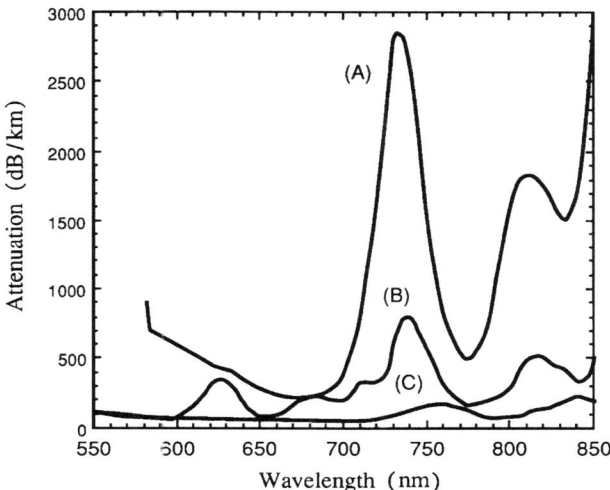

Figure 26.5. Attenuation losses of graded index POFs as a function of wavelength. (A) PMMA-base; (B) fluorinated-base; (C) deuterated-base. (Reproduced with permission from reference 49. Copyright 1995 the European Institute for Communications and Networks.)

4.3.2 Optical Adhesives

Some fluorinated adhesives with photo-curability are used for fabricating optical communication devices. The base resins are designed to match with the reflactive indices of various kinds of optical glass components by introducing fluorines. They mainly consist of fluorinated epoxyacrylates or epoxymethacrylates derived from fluorinated diols ($HO-R_f-OH$) such as bisphenol-AF and bis(1,1,1,3,3,3-hexafluoro-2-hydroxyisopropyl)cyclohexane (equation 6) [51].

$$CH_2=C(R)COCH_2CHCH_2O-R_f-OCH_2-CHCH_2OCC(R)=CH_2$$
$$\underset{O}{\|}\ \ \underset{OH}{|}\ \ \ \ \ \ \ \ \ \underset{OH}{|}\ \ \underset{O}{\|}\ \ \ \ (R = H, CH_3)$$

(6)

4.3.3 Contact Lenses

It is well known that the introduction of fluorine or, more often, siloxane units into polymers increases oxygen permeability. Therefore, for the polymeric materials for hard type contact lenses, fluorinated moieties, typically fluoroalkyl methacrylates are frequently incorporated as comonomers, together with siloxanyl methacrylates. Figure 26.6 illustrates the variation of oxygen permeabilities for terpolymers composed for methyl methacrylate (MMA), 2,2,3,3,4,4,4-heptafluorobutyl methacrylate (HFBuMA) and dimethacrylate macromonomers having poly(dimethyl siloxane) units (**A**) [52]. The P values increase with

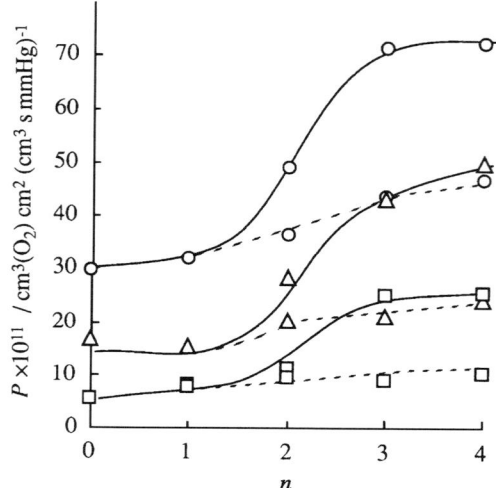

Figure 26.6. Oxygen permeability P for terpolymers consisting of methyl methacrylate (MMA), 2,2,3,3,4,4,4-heptafluorobutyl methacrylate (HFBuMA) and dimethacrylate macromonomers (**A**). Here n is the number of Si(CH3)2O–units in **A**. The solid and dashed lines indicate 10 mol% and 5 mol% of **A** in the terpolymers respectively. The contents of HFBuMA are 25 mol% (□), 50 mol% (△) and 75 mol% (○) respectively. (Reproduced with permission from reference 51. Copyright 1992 Hüthig & Wepf Verlag.)
A: $H_2C=C(CH_3)-CO-O-[Si(CH_3)_2O]_n-Si(CH_3)_2O-OC-(CH_3)C=CH_2$

increasing HFBuMA content as well as with an increasing number of units (n) and macromonomer content in the terpolymers.

Diacrylic esters having perfluoropolyether units (equation 7) are also expected to be a functional comonomer for contact lenses, probably giving higher permeability and some other improved properties [53].

$$H_2C=C(R)\underset{\underset{O}{\|}}{C}OCH_2CF_2O(CF_2CF_2O)_m(CF_2O)_nCF_2CH_2O\underset{\underset{O}{\|}}{C}C(R)=CH_2 \qquad (7)$$
$$(R = H, CH_3)$$

4.4 SURFACE MODIFIERS

Fluorinated acrylic ester polymers are used as modifiers to promote blending instead of coating, imparting functionality of the fluoroalkyl groups to the surfaces of other resins or paints. Since they tend to accumulate on the surface of the substrates faced to air, the surface is easily modified. The modifiers are usually added to paints to enhance leveling or dispersing pigments, and sometimes to improve moisture-proof qualities, and are blended to resins to give oil- and water-repellency, or low friction properties. In addition, they are used as anti-blocking

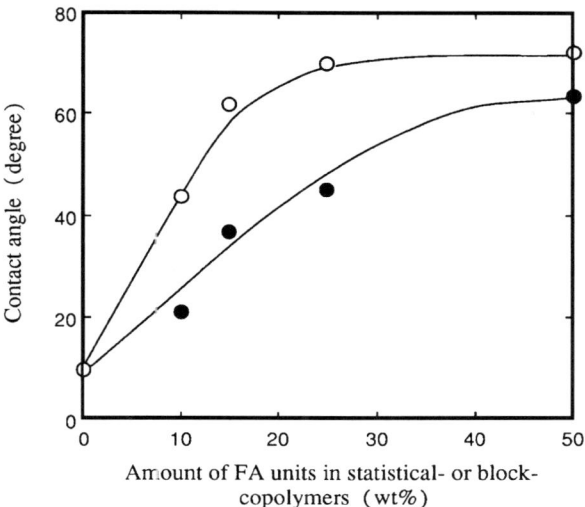

Figure 26.7. Contact angles with dodecane on the surfaces of poly[methyl methacrylate (MMA)-*stat*-butyl methacrylate (BMA)] films, into which are mixed 1% of statistical-or block-copolymers containing $C_8F_{17}CH_2CH_2OOCCH = CH_2$ (FA) monomeric units. (○) Poly(MMA-*stat*-BMA)-*block*-poly(FA); (●) poly(MMA-*stat*-BMA-*stat*-FA). (Reproduced with permission from reference 55. Copyright 1989 the Society of Polymer Science, Japan.)

agents. Commercially available modifiers are copolymers with rather low molecular weights of approximately 3000–10 000, consisting of fluoroalkyl acrylic esters, and nonfluorinated monomers which give affinity to the substrates to be modified, such as acrylic esters of the higher alcohols and the esters containing poly(ethylene oxide) groups.

Other promising techniques to obtain modified surfaces make use of block-or graft-copolymers with amphiphile characteristics [54,55]. Figure 26.7 illustrates one such modification. Shown are the modification of the contact angle of dodecane on the surface of polyacrylic films into which are mixed 1% of another copolymer containing 2-(perfluorooctyl)ethyl acrylate units. The top curve represents the effect of adding block-copolymers, the bottom curve represents the effect of adding statistical-copolymers. The data indicate that the former are more effective than the latter [55].

REFERENCES

1. Codding, D. W., Reid, T S., Ahlbrecht, A. H. *et al.*, *J. Polym. Sci.*, **15**, 515–19 (1955).
2. Burdon, J. and Tatlow, J. C., *Advances in Fluorine Chemistry*, Vol. 1, Butterworths, London, 1960, pp. 129–65.

3. Fielding, H. C., *Organofluorine Chemicals and Their Industrial Applications*, Banks, R. E. ed., Ellis Horwood, London, 1979, Chapter 11.
4. Boguslavskaya, L. S. and Chuvatkin, N. N., *Macromol. Symp.*, **82**, 51–6 (1994).
5. Ohmori, A., Takaki, S. and Kitahara, T., US Patent 4 604 482 (Aug. 5, 1986).
6. Wakselman, C., *Macromol. Symp.*, **82**, 77–87 (1994).
7. Fuchigami, T. and Suzuki, Y., Kokai Tokkyo Koho JP 60-94933 (May 28, 1985).
8. Chang, R. W. H., Banitt, E. H. and Joos, R. W., US Patent 3 540 126 (Nov. 17, 1970).
9. Wirners, G., Veter, H., Ludoerf, H. *et al.*, PCT Int. Appl. WO 90/12040 (Oct. 18, 1990).
10. Samejima, S. and Kaneko, I., Kokai Tokkyo Koho JP 59-88445 (May 22, 1984).
11. Pittman, Jr. C. U., Ueda, M., Iri, K. *et al.*, *Macromolecules*, **13**, 1031–6 (1980).
12. Yamada, B., Kotani, T., Yoshioka, M. *et al.*, *J. Polym. Sci., Polym. Chem. Ed.*, **22**, 2381–93 (1984).
13. Theis, J., Groh, W., Shütze, G. *et al.*, *Annual Technical Conference (ANTEC), Society of Plastics Engineers*, 893–6 (1990).
14. Iwatsuki, S., Itoh, T. and Iida, T., *J. Polym. Sci., Part A, Polym. Chem.*, **32**, 1389–92 (1994).
15. Koishi, T., Tanaka, I., Yasumura, T. *et al.*, Kokai Tokkyo Koho JP 60-42411 (Mar. 6, 1985).
16. Aglietto, M., Passaglia, E., Mirabello, L. M. *et al.*, *Macromol. Chem. Phys.*, **196**, 2843–53 (1995).
17. Kataoka, M., Kokai Tokkyo Koho JP 61-170735 (Aug. 1, 1986).
18. Narita, T., Hagiwara, T. and Hamana, H., *Makromol. Chem., Rapid Commun.*, **6**, 175–78 (1985). Narita, T., Hagiwara, T. and Hamana, *Makromol. Chem., Rapid Commun.*, **6**, 301–4 (1985). Narita, T., Hagiwara, T., Hamana, H. *et al.*, *Polymer Journal*, **20**, 519–23. (1988).
19. Narita, T., Hagiwara, T. and Hamana, H., *Makromol. Chem., Rapid Commun.*, **6**, 5–7 (1985).
20. Sandberg, C. L. and Bovey, F. A., *J. Polym. Sci.*, **15**, 553–7 (1955).
21. Kárpátyová, A., Barto, J. and Paleta, O., *Makromol. Chem.*, **191**, 2901–3 (1990).
22. Iwatsuki, S., Kondo, A. and Harashima, H., *Macromolecules*, **17**, 2473–9 (1984).
23. Okawara, A., Maekawa, T., Ishida, Y. *et al.*, *Polymer Preprints, Japan*, **40**, 3898–3900 (1991).
24. Budovskaya, L. D., Ivanova, V. N., Oskar, L. N. *et al.*, *Vysokomol. Soedin., Ser. A*, **32**, 561–5, 1990.
25. Volkov, V. V., Platé, N. A., Takahara, A. *et al.*, *Polymer*, **33**, 1316–20 (1992).
26. Shimizu, T., Tanaka, Y., Kutsumizu, S. *et al.*, *Macromol. Symp.*, **82**, 173–84 (1994).
27. Shimizu, T., Tanaka, Y., Kutsumizu, S. *et al.*, *Macromolecules*, **26**, 6694–6 (1993). Shimizu, T., Tanaka, Y., Kutsumizu, S. *et al.*, *Macromolecules*, **29**, 156–64 (1996). Shimizu, T., Tanaka, Y., Ohkawa, M. *et al.*, *Macromolecules*, **29**, 3540–4 (1996).
28. Ishiwari, K., Ohmori, A. and Koizumi, S., *Nippon Kagaku Kaishi*, **10**, 1924–8 (1985).
29. Gaynor, J., Schueneman, G., Schuman, P. *et al.*, *J. Appl. Polym. Sci.*, **50**, 1645–53 (1993).
30. Tada, T., *J. Electrochem. Soc.*, **126**, 1829–30 (1979).
31. Shimizu, T., unpublished results.
32. Ohmori, A., '*Fusso Jushi Handbook (Japanese)*,' Satokawa, T., ed., Nikkan Kogyo Shinbunsha, Tokyo, 1990, p. 733.
33. Kamata, S., *Newest Aspect of Fluoro Functional Materials*, Yamabe, M. and Matsuo, M. eds., CMC Co., Tokyo, 1994, pp. 151–65.

34. Maekawa, T., Okawara, A., Yoshioka, R. *et al.*, *Proc. 16th Fusso Kagaku Toronkai* (Japan), pp. 25-6, Oct. 1991.
35. Pittman, A. G., '*Fluoropolymers*,' Wall, L. A. ed., Wiley-Interscience, New York, 1972, pp. 419-49.
36. Koizumi, S., Ohmori, A. and Shimizu, T., *J. Surface Sci. Soc. Jpn*, **13**, 428-33 (1992).
37. Katano, Y., Tomono, H. and Nakajima, T., *Macromolecules*, **27**, 2342-4 (1994).
38. Sherman, P. O., Smith, S. and Johannessen, B., *Textile Res. J.*, **39**, 449-59 (1969).
39. Kirimoto, K. and Hayashi, T., Kokai Tokkyo Koho JP 49-75472 (July 20, 1974). Midori, K., Saito, Y., Tsujimoto, A. *et al.*, Kokai Tokkyo Koho JP 6-1116340 (Apr. 26, 1994).
40. Kissa, E. '*Handbook of Fiber Science and Technology: Vol. II, Chemical Processing of Fibers and Fabrics, Functional Finishes, Part B*,' Lewin, M. and Sello, S. B. eds, Marcel Dekker, New York, 1984, Chapters 2-3.
41. Kakuchi, M., Sugawara, S., Murase, K. *et al.*, *J. Electrochem. Soc.*, **124**, 1648-51 (1977).
42. Satokawa, T., *Kinosei Ganfusso Kobunshi (Japanese)*, Nikkan Kogyo Shinbunsha, Tokyo, 1982, p. 126.
43. Yamauchi, M. and Ohtoshi, S., *Reports Res. Lab. Asahi Glass Co., Ltd.*, **38**, 167-76 (1988).
44. Sprengling, G. R., US Patent 3 931 454 (Jan. 6, 1976).
45. Ohtoshi, S. and Hase, F., US Patent 4 615 479 (Oct. 7, 1986).
46. Nomura, Y., Aoki, M. and Nemoto, S., Kokai Tokkyo Koho JP 53-97435 (Aug. 25, 1978). Shigeta, K., Takahashi, J., Ohmori, A. *et al.*, Kokai Tokkyo Koho JP 61-12069 (June 9, 1986). Yabuuchi, N. and Aoki, T., Kokai Tokkyo Koho JP 62-39878 (Feb. 20, 1987).
47. Ohmori, A., Tomihashi, N. and Kitahara, T., US Patent 4 720 166 (Jan. 19, 1988). Yamamoto, T. and Nishida, K., Kokai Tokkyo Koho JP 61-103107 (May 21, 1986).
48. Hashimoto, Y., Kamei, M. and Baba, T., Kokai Tokkyo Koho JP 62-199643 (Sept. 3, 1987). Hashimoto, Y., Kamei, M. and Baba, T., Kokai Tokkyo Koho JP 62-250047 (Oct. 30, 1987).
49. Koike, Y., *Proc. 3rd Inter. Conf. on Plastic Optical Fibres & Applications*, Yokohama, Oct. 26-8, 1994, pp. 16-20.
50. Boutevin, B., Rousseau, A. and Bosc, D., *J. Polym. Sci., Part A, Polym. Chem. Ed.*, **30**, 1279-86 (1992). Takezawa, T., Tanno, S., Taketani, N. *et al.*, *J. Appl. Polym. Sci.*, **42**, 3195-3203 (1991).
51. Nakamura, K., Murata, N. and Maruno, T., *Review of the Electrical Communications Laboratories, NTT*, **37**, 127-32 (1989). Maruo, T., Ishibashi, S. and Nakamura, K., *J. Polym. Sci., Part A: Polym. Chem.*, **23**, 3211-14 (1994).
52. Koßmehl, G., Fluthwedel, A. and Schäfer, H., *Makromol. Chem.*, **193**, 157-66 (1992).
53. Rice, D. E. and Ihlenfeld, J. V., US Patent 4 818 801 (Apr. 4, 1989).
54. Yamashita, Y., Tsukahara, Y. and Ito, H., *Polymer Bull.*, **7**, 289-94 (1982).
55. Oshibe, Y., Ishigaki, H., Ohmura, H. *et al.*, *Kobunshi Ronbunshu*, **46**, 89-94 (1989).

27

ECTFE Copolymers

GARY STANITIS
Ausimont, USA, Thorofare, NJ, USA

1 INTRODUCTION

Chlorotrifluoroethylene–ethylene copolymers (commonly referred to as ECTFE) were first polymerized by workers at DuPont in 1946 [1]. The structure and properties of ECTFE copolymers have been well studied and documented in the literature (see References and Further Reading). Allied Chemical first commercialized the production of ECTFE in 1974, under the trade name of HALAR®, at their Elizabeth, New Jersey, production site [1]. The product was initially marketed for use in the chemical process industry because of its excellent chemical resistance properties, as well as strength and toughness at elevated temperatures (see Tables 27.1 –27.3 for physical properties of typical commercial 1 : 1 alternation copolymers). Due to increased demand in the early 1980s, Allied Chemical built a second plant to produce ECTFE in Orange, Texas. In 1986 Allied Corporation sold the ECTFE business, including both the Elizabeth, New Jersey, and Orange, Texas, ECTFE production units, to Ausimont USA, Inc., a wholly owned subsidiary of Montedison SpA of Milan, Italy. In 1988 the Elizabeth, New Jersey, production unit was sold back to Allied Corporation for the production of PCTFE resins. Allied has not restarted ECTFE production at Elizabeth. Today, the Ausimont USA Orange, Texas, plant is the only manufacturing unit in the world producing significant volumes of 1 : 1 alternating ECTFE copolymers [2].

2 MANUFACTURING

CTFE (chlorotrifluoroethylene) monomer is produced commercially in the United States by AlliedSignal Corporation at its Baton Rouge, Louisiana, plant.

Modern Fluoropolymers. Edited by John Scheirs
© 1997 John Wiley & Sons Ltd

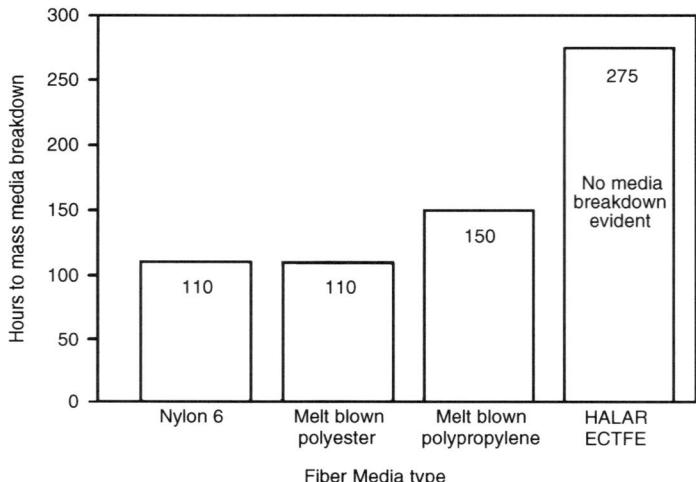

Figure 27.1. Media degradation by ozone (residual ozone concentration of 0.50 ppm except polypropylene at 0.30 ppm). Most commonly available melt blown filter medias break down after a short exposure to ozone. HALAR fluoropolymer, however, shows no deterioration even after 275 hours of exposure. Reproduced by permission of the Filterite Division of Memtec America Corporation

AlliedSignal's capacity for CTFE monomer is estimated to be several million pounds per year. This is based on AlliedSignal's internal usage for CTFE–VDF polymers [2] and Ausimont USA's usage for ECTFE [2]. The process used is zinc dechlorination of CFC-113 (1, 1, 2-trichloro-1, 2, 2-trifluoroethane). CTFE is also produced by Daikin in Japan using similar process technology. CTFE monomer will not auto-polymerize at ambient temperatures and so is transported without inhibitor. Like all fully and partially fluorinated ethylenes, CTFE monomer can undergo a disproportionation reaction, and must be handled properly. The presence of oxygen greatly increases the sensitivity of the mixture. Oxygen and CTFE also react to form high molecular weight peroxides which can precipitate from solution. Oxygen concentrations of commercial CTFE are maintained below 50 ppm, and more typically between 5 and 20 ppm.

An aqueous free radical suspension process is used to polymerize ECTFE commercially. Water, CTFE, and in some instances additives to adjust interfacial surface tension, are charged to a reactor vessel. The vessel is agitated to give an oil in water suspension. Polymerization is carried out at low temperatures. This is done to reduce the amount of ethylene blocks in the polymer backbone which are susceptible to thermal degradation. Commercial polymer with an overall CTFE : ethylene molar ratio of 1 : 1 contains ethylene blocks and CTFE blocks of less than 10 mol% each [3]. Reaction pressure is adjusted to give the desired copolymer ratio [4]. Typically moderate pressures of less than 500 psig

Table 27.1. Typical properties of ECTFE fluoropolymer (data supplied courtesy of Ausimont USA, Inc.)

Property	Units	1–3 melt flow	15–20 melt flow
Mechanical properties			
Tensile strength			
at yield	MPa (psi)	30 (4300)	29 (4200)
at break	MPa (psi)	54 (7800)	46 (6600)
Elongation at break	%	250	260
Flexural modulus	MPa	1690	1670
	(psi)	(2.45×10^5)	(2.45×10^5)
Impact resistance			
Notched Izod @ 23 °C	J/m	No break	No break
Notched Izod @ −40°C	J/m	122	64
Electrical properties			
Dielectric strength			
25 micron thick	kV/mm	80	80
3.18 millimeter (1/8 in.)	kV/mm	14	13
Dielectric constant			
at 1000 Hz		2.5	2.47
at 1 MHz		2.59	2.57
Flammability			
Oxygen index		52 minimum[a]	52 minimum[a]
UL 94 vertical, 0.007 in.		94 V-0	94 V-0
UL 910 (up to 200 pair)		Pass	Pass
Thermal properties			
Melting point	°C	240	240
Brittleness temperature	°C	< −76	< −76
Maximum service temperature	°C	150	150
Heat distortion temperature			
Under load			
0.45 millimeter/meter^2	°C	90	90
1.8 millimeter/meter^2	°C	63	63
Other properties			
Specific gravity		1.63 to 1.73	1.63 to 1.73
Moisture absorption		<0.1%	<0.1%
Taber abrasion			
ASTM D 1044			
500 revs	Volume loss, cc	0.002	—
1000 revs	Volume loss, cc	0.005	—
Critical surface tension	Dyn/cm	32	—
Crystallinity (X-ray measurements)	%	50	—

[a]Actual measured values typically range from 60 to 64

Table 27.2. Coefficient of linear thermal expansion of ECTFE fluoropolymer (data supplied courtesy of Ausimont USA, Inc.)

Temperature range, (°C)	Coefficient of linear expansion
−30−+50	8×10^{-5}
50–85	10×10^{-5}
85–125	13.5×10^{-5}
125–180	16.5×10^{-5}

Table 27.3. Radiation resistance of ECTFE fluoropolymer (data supplied courtesy of Ausimont USA, Inc.)

Cobalt-60 dosage (Mrad)	Tensile strength at break, (MPa)	% Elongation
0	48	210
50	32	105
100	30	65
500	28	20
1000	19	10

(3.45 MPa) are used. Pressure is controlled by ethylene addition during polymerization. A proprietary, low-temperature organic peroxide is used to initiate the reaction. Chain transfer agents are used to control the molecular weight which is typically measured by melt flow index [5]. After polymerization unreacted monomers and volatile reactor additives are recovered and purified for reuse.

Modified copolymers have also been produced commercially to reduce high-temperature stress cracking. Typically a modifying monomer is added to the polymerization reactor prior to reaction. Addition may be continuous during the reaction if reactivity ratios dictate. Typically modified polymers are less crystalline and have reduced melting points [3].

Because ECTFE is insoluble in either CTFE or ethylene monomer, polymer precipitates from monomer droplets as a fine powder, typically less than 20 microns in major dimension. Over the course of polymerization the powder can agglomerate into roughly spherical beads. The reactor product consists of a broad particle size mixture of beads and powder. The reactor bead is dewatered and dried. Low concentrations of stabilizer additives are typically used to improve the thermal stability of the material. The stabilized powder is then either extruded and pelletized, or ground and screened into powder coating grades.

3 PRODUCT FORMS

ECTFE is available in a variety of forms and molecular weights. The most common form is hot cut pellets. The pellets can be utilized in all thermoplastic processes such as extrusion, injection molded, blow molded, compression molded, fiber spinning, etc. Because ECTFE is corrosive in the melt, the wetted surfaces of molding machines and extruders used to process it are typically lined with a high nickel content alloy, such as Hastelloy C-276. Recent additive technology by Ausimont USA has led to a family of products that can be processed on equipment with conventional materials of construction. This is achieved by reduced thermal instability and improved acid scavenging. The wide range of molecular weights allow melt processing of a variety of shapes and sizes. High molecular weight material (melt index of 1.0) is used to extrude sheet for lining. Medium melt flow materials are used for pipe extrusion and cable jacketing. Very high melt flow material is used to injection mold complex parts for use in ambient temperature non-load bearing applications (see Figure 27.2) and for fiber spinning.

Modified ECTFE polymers are produced for coating grades. The addition of the modifying monomer greatly reduces high temperature stress cracking [3]. The first commercial grade was modified with hexafluoroisobutylene (HFIB) and was introduced by Allied chemical. Improved modifiers have since been identified. These include perfluorohexylethylene [3], perfluoroisoalkoxy perfluoroalkyl ethylenes [6], and perfluoropropylvinylether (see Table 27.4) [7]. A product modified with perfluoropropylvinylether was commercialized in 1996 by Ausimont USA for demanding powder coating applications.

Figure 27.2. Injection-molded microtitration plates from ECTFE

Table 27.4. Effects of polymer modifiers on stress cracking properties of ECTFE (data supplied courtesy of Ausimont USA, Inc.)

Polymer type	Tensile creep, (218 psi applied stress), time to break (h)	
	135 °C	150 °C
HFIB modified ECTFE	40	0.2
PFPVE modified ECTFE	>1700	12.0

Coating grades are made in either powder form for electrostatic powder coating (see Figure 27.3) or very fine pellets for rotomolding and lining. Formulated primers are used to improve adhesion and moisture permeability for powder-coated metal systems. One, or several, top coats, depending on the thickness required, are then applied. Typically no primer is used for roto-lining. ECTFE has excellent chemical and thermal resistance compared with other common coating materials [8].

The fine crystalline structure of ECTFE [9] allows the fabrication of fibers using a variety of processes. In 1973 Allied Chemical Corporation demonstrated that fine denier multi-filament fibers with good strength per denier could be produced [10]. Multi-filament fibers are not currently commercial, although prototypes have been made for evaluation.

In the late 1980s Ausimont USA, in conjunction with the University of Tennessee, developed a process for melt blown non-woven fiber from

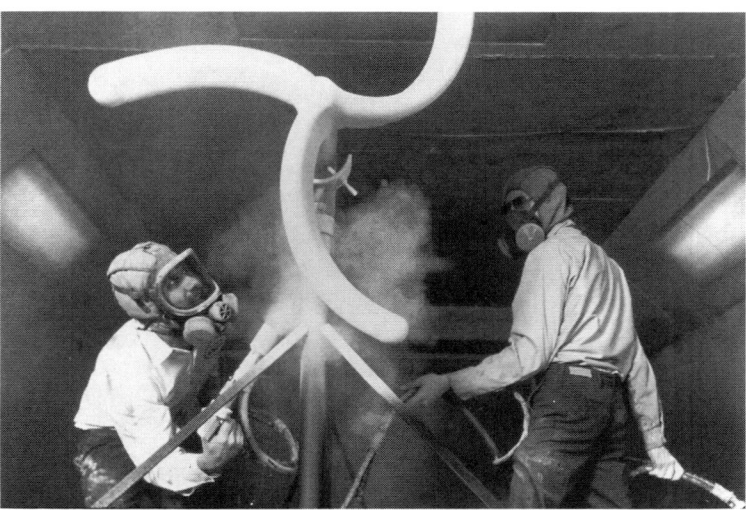

Figure 27.3. Powder spray coating of a tank agitator with ECTFE

ECTFE [11,12]. Melt blown fiber webs in a variety of weights are commercially available. Further downstream processes are used to impart additional benefits to the fabric. Ausimont USA obtained a patent in 1994 for a method to improve fabric strength by hydroentanglement [13]. In 1993 Cobale Company, L. L. C. received a patent for using ECTFE melt blown fiber as a surfacing veil in fiberglass reinforced plastic (FRP) composites [14]. The chemical resistance of ECTFE is much greater than glass or polyester veils that are commonly used. Its presence as a surfacing veil limits the attack on the composite from aggressive chemicals. Perforated veils have been developed to improve resin penetration during the composite fabrication process.

ECTFE sheet is available in both polyester backed and unbacked versions. Backed sheets are used for lining tanks, scrubbers, etc. The polyester fabric is partially incorporated into the melt during sheet extrusion to give a strong bond. Adhesives are applied to the exposed fabric for adhesion to surfaces. Once sheets are adhered the seams are welded using conventional plastic welding techniques to give an uninterrupted lining. ECTFE sheet can be adhered to metals or fiberglass. Unbacked sheet is used for loose inserts and can be fabricated and/or machined for stand alone items of ECTFE.

4 APPLICATIONS

The original market for ECTFE was the chemical process industry. ECTFE is a partially fluorinated polymer and has very good room temperature physical properties. Although the yield strength is lower than PVDF, ECTFE is tough and less prone to brittle failure. The high crystallinity and fluorine content of ECTFE makes it resistant to most industrial chemicals. It works particularly well in caustic environments which are known to cause stress cracking in PVDF. ECTFE also has very low permeability to many gases such as water vapor, chlorine and hydrogen chloride. Permeability can be a concern with fully fluorinated materials such as FEP and PTFE that can require weep holes, which is a source of fugitive emissions [15].

Due to its low permeability and good chemical resistance ECTFE is often used in chlorine and caustic environments. It has found application in chlorine/caustic production processes. Patents have been granted to PPG industries [16] and OxyTech Systems [17] for the manufacture of asbestos diaphragms using ECTFE as a binder. ECTFE lined tanks and pipes have also been used in chlorine/caustic plants [18,19]. Cell covers, outlet boxes, and piping made from FRP composites using ECTFE surfacing veil are undergoing long-term testing in plants in Texas and Louisiana. Pulp and paper technologies that rely on chlorine and sodium hypochlorite bleaching agents have utilized pipes and scrubbers lined with ECTFE. ECTFE lined steel and lined fiber glass have replaced materials such as PVDF and unlined fiber glass in many applications.

There are a variety of other CPI applications, other than chlorine and caustic environments, where ECTFE has proven to be the material of choice. Some examples which have been cited in Ausimont USA news releases are: ECTFE lined FRP tanks successful use for hot sulfuric acid storage used in plating processes for the production of alkaline batteries. An ECTFE lined FRP scrubbing tower used to separate methanol from HCl at 300 °F. M. E. Loosbergh of Ausimont, Belgium, gives the following additional examples: large (10 m × 6 m × 4 m and 9 m × 8 m × 7 m) tanks have been fabricated aboard ships for ocean transport of H_2SO_4, 98% HCl, $AlCl_3$, $FeCl_3$, and caustic. Large pickling tanks for stainless steel production operate at 84 °C with sulfuric acid, nitric acid and hydrofluoric acid. Large ECTFE lined organic decomposing tanks have been fabricated from ECTFE lined FRP for use in Germany.

One producer of FRP piping is promoting the use of ECTFE veil, in conjunction with improved resin technology, to improve performance in hot sulfuric acid. A unique application patented by Pall Corporation is the use of ECTFE melt blown fiber web to make a coalescing element for separating immiscible liquids. Liquids with interfacial surface tensions differing by as little as 0.6 dyne/cm^2

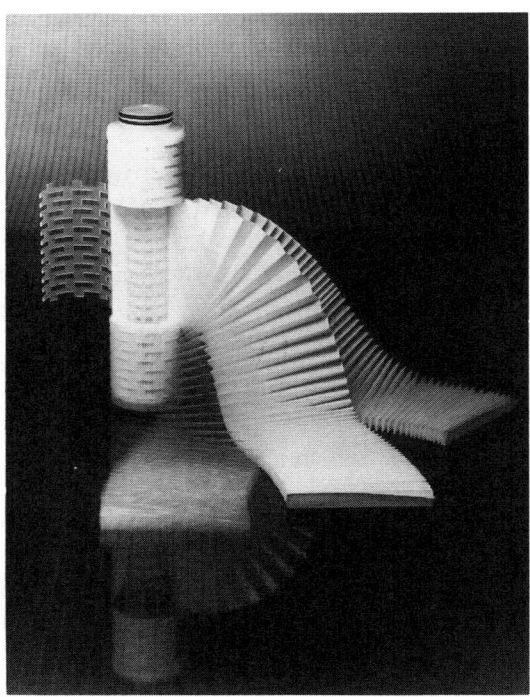

Figure 27.4.   ECTFE filter element (courtesy of Memtec America Corporation)

can be separated efficiently in their coalescer design for which 'the preferred embodiment ... employs ... ECTFE non-woven web ...' [20]. Melt blown fiber webs are also used to make cartridge filters for particulate removal. Filtration media, media supports, and filter housings can all be manufactured from ECTFE (see Figure 27.4).

The single largest market for ECTFE has been as an insulation material, both primary and jacketing, in voice and data copper cables used in building plenums [2] (see Figure 27.5). This market was created in 1975 when NFPA (National Fire Protection Agency) subcommittee 90A allowed the use of plenum cables without protective conduit if fire performance was equivalent to cable in conduits. This exception was included in the National Electric Code (NEC), section 300–22. DuPont and AT & T corporation demonstrated that cables with fluoropolymer jackets, specifically FEP, could meet this criteria. A test protocol was developed at Underwriters Laboratories in 1981 to measure the fire performance of cables [21]. The test, referred to as UL 910, has become the industry standard in the USA for rating plenum cables. This test is also identified in NFPA, section 262. Although the use of fluoropolymers represented a large increase in material cost, the installed cost of the system was much lower than cables within conduit. Soon after the work with FEP, Allied and Pennwalt corporations demonstrated that other fluoropolymers, namely ECTFE and PVDF, could also meet the strict fire performance requirements. ECTFE proved to be an ideal material for plenum wire and cable. Its low dielectric constant of 2.6 at 1 MHz allow it to be

Figure 27.5.   Plenum cable jacketed with ECTFE fluoropolymer

used for both primary and jacketing insulation. The high oxygen index, 60–64, and its tendency to char when exposed to flame temperatures allow it to pass the difficult UL 910 test protocol. Other properties which contribute to ECTFE's successful use as jacketing and primary insulation in unshielded twisted pair (UTP) voice, data, and fire alarm cables are toughness over a broad temperature range, low moisture absorption, lack of brittle failure, and low cost compared with fully fluorinated materials. During the late 1980s and early 1990s heavily formulated PVC compounds have eroded the market share of fluoropolymers in this application [2].

With the development of very high speed data transmission cables in the early 1990s (referred to as category 5 cables) FEP, with a low dielectric of 2.1, became the primary insulation of choice. Due to increased demand on FEP and double digit yearly growth rates for plenum data cable, a shortage of FEP ensued. Cable manufacturers were able to design hybrid cables that utilized FEP combined with flame-retardant polyolefins. A flexible grade of ECTFE based on a plasticized CTFE rich copolymer (56 mol% CTFE) was introduced in 1995 as the jacket for these hybrid cables. Technologies to make primary insulation from foamed alloys of ECTFE and low dielectric materials were developed by some manufacturers, but never fully commercialized. Raychem Corporation received a patent for one of these compounds [22]. Chemically foamed ECTFE is used in some cable constructions. Foamed material gives a better dielectric constant and lower raw material price.

Due to its high-temperature properties ECTFE is also used in specialty cable designs. In automotive applications ECTFE jacketed cables are used inside fuel tanks for level control, for hook-up wires with a 150 °C thermal rating of 3000 hours, and in heating cables for car seats. ECTFE jacketed cables are also used in aerospace due to their excellent cut-through resistance at 140 –150 °C.

Fluoropolymers have gained wide acceptance in the semiconductor industry due to their excellent resistance to aggressive chemicals and deionized water, as well as their high level of cleanliness. ECTFE powder-coated stainless steel exhaust fume ducting is selected frequently for use in computer chip fabrication plants (see Figure 27.6). The excellent chemical resistance of ECTFE prevents corrosion due to aggressive chemicals in the exhaust air. The low flammability and smoke generation allows installation without a sprinkler system, which is required when FRP ducting is used.

Water used in the manufacture of semiconductor chips must be extremely clean to prevent damage to the very fine chip architecture. Seamless ECTFE powder coated ultra-pure water (UPW) treatment and storage tanks are commonly used (see Figure 27.7). The lack of seams reduce the trapping and release of dirt and micro-organisms. Data developed by Gary Husted of MicroTechco Research, Inc. for Ausimont USA has demonstrated that the very smooth surface of ECTFE compared with stainless steel and other fluoropolymers (see Figure 27.8) reduces

Figure 27.6. ECTFE-lined exhaust ducting

Figure 27.7. Seamless ECTFE powder coated mixed deionizer resin bed tank

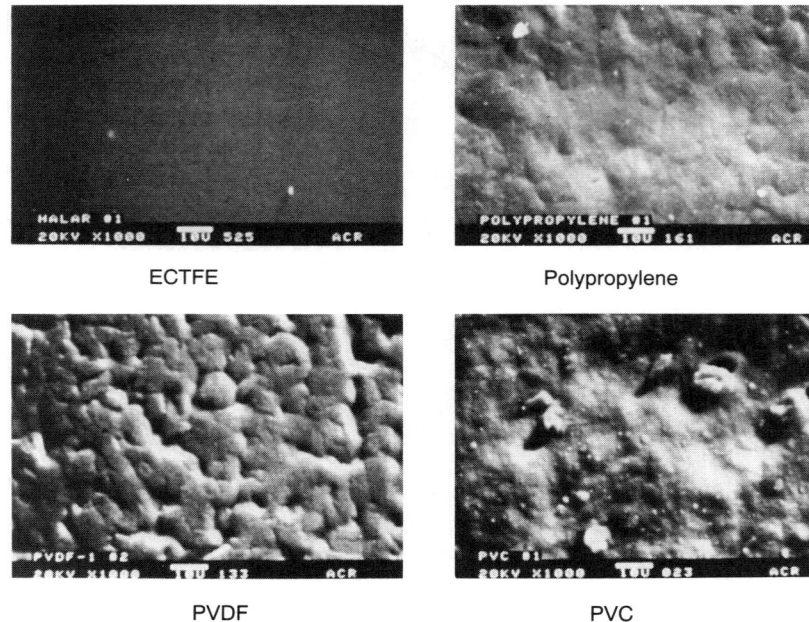

ECTFE   Polypropylene

PVDF   PVC

Figure 27.8.   Surface roughness of molded plastics (1000 X magnification)

dramatically the tendency for adhesion and growth of bacteria. The surface smoothness of ECTFE can be attributed to its very fine crystal dimensions [9].

PVDF is typically used for ambient temperature deionized water delivery piping. The semiconductor industry is shifting use to higher temperature water to reduce the growth of bacteria and improve wafer rinsing efficiency. At higher temperatures PVDF can discolor and is prone to thermal stress cracking. ECTFE powder coated stainless steel piping is currently being evaluated by the industry for larger diameter UPW piping. For smaller diameter piping self-supported extruded pipe can be used (see Figure 27.9). ECTFE is also fabricated into filter media (melt blown fiber) and filter housings. ECTFE's excellent ozone resistance (see Figure 27.1 supplied courtesy of Memtec America Corporation) makes it particularly suitable for use in ozonated water streams. Ozone treatment is growing in popularity as a method to eliminate bacteria in UPW. Traditional filter materials, such as polypropylene, suffer dramatic media breakdown after only a few days.

Although PFA has a dominant role, machined, molded, and coated ECTFE parts are also used for wafer handling systems in wet chemical processes (see Figure 27.10).

Offshore oil and gas production technology relies on flexible pipes for crude oil transport. Polymer barrier layers are incorporated into the flexible piping to

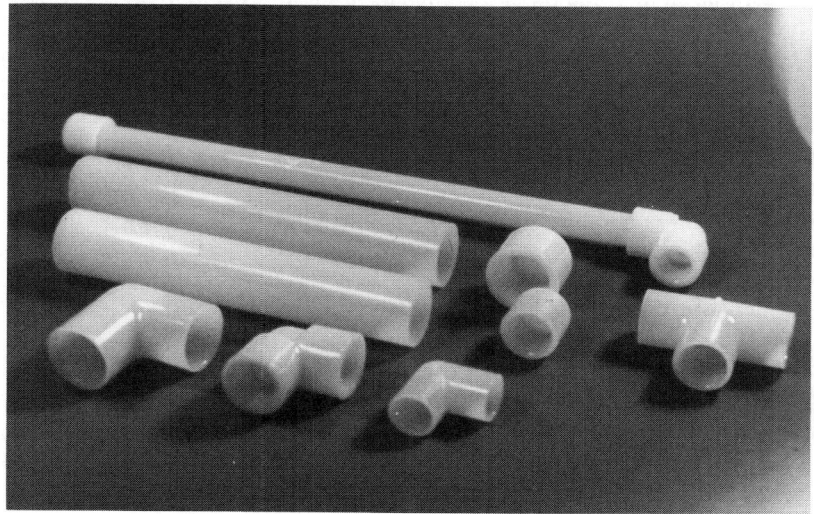

Figure 27.9.   Self-supported ECTFE pipes and fittings

Figure 27.10.   ECTFE-lined semiconductor wafer baskets

reduce permeation of crude oil and methanol (used to prevent hydrate formation) into the surrounding sea water. For high-temperature production fields plasticized PVDF lining is often used. Recent industry publications include ECTFE as a potential material of choice in these applications, particularly at temperatures over 120°C [23]. Advantages of ECTFE over PVDF are better physical properties at high temperature (120–180 °C), less tendency for notch-induced brittle failure, and lower coefficient of thermal expansion. The lower elongation at yield for ECTFE will require larger diameter spools. ECTFE has also been used to encapsulate stainless steel tubing used to inject chemicals into oil-producing wells. ECTFE is chosen for its toughness, chemical resistance, and excellent high-temperature properties. It is also used as a jacket for data logging cables used in oil production for the same reasons.

ECTFE is used in a variety of very special applications. Wire ties used in the space shuttle and the Hubble Space Telescope are made from ECTFE due to its very low out-gassing properties. Braided cable jackets made from ECTFE monofilament strands are used in military and commercial aircraft as a protective sleeve over cables due to its toughness and high-temperature properties. ECTFE film is used as release sheet in the fabrication of high-temperature composites for aerospace applications.

REFERENCES

1. Miller, W. A., 'Chlorotrifluoroethylene–ethylene copolymers' in Mark, H. F., Bikales, N. M., Overberger, C. G., and Menges, G. (ed), *Encyclopedia of Polymer Science and Engineering*, 2nd edn, pp. 480–491 (1986).
2. Haley, Michael J., with Leder, A., and Sakuma, Y., 'CEH marketing research report: fluoropolymers Draft-1' *SRI International Chemical Economics Handbook*, pp. 69–70, March 29, 1995.
3. Reimschuesel, H.K., Marti, J., and Murthy, N.S., 'Ethylene–chlorotrifluoro copolymers. II. Effects of structure on resistance to thermal stress cracking', *Journal of Polymer Science: Part A: Polymer Chemistry*, **26**, 43–59 (1988).
4. Schulze, Steven R. United States Patent 4 053 445 (October 11, 1977), to Allied Chemical Corporation.
5. 'Standard specification for E-CTFE-fluoroplastic molding, extrusion, and coating materials: designation D 3275-91a' *Annual Book of ASTM Standards*, Volume 08.02, Plastics (II) (1995).
6. Reimschuessel, Herbert K., Rahl, Forrest J., and Ulmer Harry E., United States Patent 4 736 006 (April 5, 1988), to Ausimont USA, Incorporated.
7. Coates, M. and Kent, B., 'Modification of fluoropolymers', SPI Fluoropolymer Division Spring Conference (San Diego). March 12, 1996.
8. Blackwell, J. P., Brady, D. G., and Hill, H. W. Jr., 'Comparison of electrostatic spray-applied powder coatings', *Journal of Coatings Technology*, **50**, (643), 62–66, August 1978.
9. Rabolt, John F., 'Molecular dynamics of a crystal–crystal phase transition in an alternating copolymer of ethylene and chlorotrifluoroethylene using the Raman active longitudinal acoustical mode (LAM)', *Polymer*, **22**, 890–895, July 1981.

10. Robertson, A. Bruce, 'Poly (ethylene-chlorotrifluoroethylene) fibers'. *Applied Polymer Symposium*, No. 21, 89–100 (1973).
11. Fagan, J. P. and Wadsworth, Larry C., 'Melt blown processing and characterization of HALAR® fluoropolymers' Textiles and Nonwovens Development Center (TANDEC), The University of Tennessee, 1–15, February 28, 1991.
12. Wadsworth, Larry C. and Khan, Ahamad Y. A. Finer fibered melt blown webs from poly (ethylene–chlorotrifluoroethylene) [ECTFE] copolymer', Textiles and Nonwovens Development Center (TANDEC), The University of Tennessee, 1–9, February 28, 1991.
13. Fagan, Joseph P. United States Patent 5 422 159 (June 6, 1995), to Ausimont USA, Incorporated.
14. Bailey, Edward D. Jr United States Patent 5 534 337 (July 9, 1996), to Cobale Company, LLC.
15. Miller, W. A., 'How to choose a fluoropolymer', *Chemical Engineering*, 163–167, April 1993.
16. Rechlicz, Thomas A. and Maloney, Bernard A. United States Patent 4 065 534 (December 27, 1977), to PPG Industries, Incorporated.
17. Schulz, Arthur C., Bommaraju, Tilak V., Kiszewski, Robert, and Keller, Ursula I. United States Patent 4 810 345 (March 7, 1989), to OxyTech Systems, Incorporated.
18. Loosbergh, M. E., 'E-CTFE helps amalgam cells produce better quality caustic soda in Europe', *Managing Corrosion with Plastics*, No. 6, 711–718 (1987).
19. Fisher, G. and Lund, R. 'Performance of plastic materials in chlorine and caustic service', *Managing Corrosion with Plastics*, No. 6, 611–714 (1987).
20. Williamson, Kenneth M., Whitney, Scott A., and Rausch, Alan R. United States Patent 5 480 547 (January 2, 1996), to Pall Corporation.
21. Glew, Charles A. 'Evolution of materials for communications wiring', *Wire Technology International*, 30–32 (September 1995).
22. Mehan, Ashok, United States Patent 5 468 782 (November 21, 1995), to Raychem Corporation.
23. Kalman, Mark, Belcher, John, Chen, Bin, Fraser, Dana, Ethridge, Andrew, and Loper, Cobie, 'Development and testing of non-bonded flexible pipe for high temperature/high pressure/deep water/dynamic Sour Service Applications' Offshore Technology Conference, Houston Texas, May 6–9, 1996.

FURTHER READING

1. Y. P. Khanna and T. J. Taylor, 'Dielectric properties of HALAR, an alternating copolymer of ethylene and chlorotrifluoroethylene', *Journal of Applied Polymer Science*, **38**, 135–145 (1989).
2. John P. Sibilia and Robert J. Schaffhauser, 'Molecular transitions in an alternating copolymer of ethylene and chlorotrifluoroethylene', *Journal of Polymer Science: Polymer Physics Edition*, **14**, 1021–1028 (1976).

28

Perfluoropolymers Obtained by Cyclopolymerization and Their Applications

N. SUGIYAMA

Research Centre, Asahi Glass Co. Ltd, Kanagawa-ku, Yokohama, Japan

1 INTRODUCTION

Since PTFE was discovered in 1938 and commercialized by DuPont, various fluoroplastics and fluoroelastomers have been developed [1]. They have been widely used not only in fundamental industry but also in leading edge technology because of their excellent properties, e.g. heat resistance, chemical stability, electrical and surface properties. Fluoropolymers can be classified into categories as shown in Table 28.1, on the basis of their crystallinity and degree of fluorine substitution. Most recently developed polymers are CYTOP® and Teflon AF®, which are classified in the amorphous perfluorinated resin category. These novel fluoroplastics were commercialized by Asahi Glass and DuPont respectively.

CYTOP® and Teflon® AF (Figure 28.1) have the same excellent chemical, thermal and electrical properties as conventional perfluorinated resins such as PTFE and PFA. In addition, they possess other favourable characteristics such as high optical transparency and good solubility in selected fluorinated solvents. Their excellent clarity and good solubility are a consequence of their amorphous morphology and is attributable to the cyclic structure in their backbone. Teflon® AF is obtained by copolymerization of TFE and perfluoro -2,2-dimethyl dioxole (PDD) and its properties change with PDD content as mentioned in some reports [2]. Other experimental amorphous perfluoropolymers obtained by polymerization of dioxole analogs [3], methylene dioxorane [4] and dioxine [5] are also reported. This chapter, however refers to preparation, properties and some applications of CYTOP®.

Modern Fluoropolymers. Edited by John Scheirs
© 1997 John Wiley & Sons Ltd

Table 28.1.   Categories of fluoropolymers

		Partially fluorinated	Perfluorinated
Crystalline	Resin	ETFE	PTFE
		PVDF	PFA
		PVF	FEP
		PCTFE	
Amorphous	Resin	LUMIFLON®	CYTOP®
			TEFLON® AF
	Elastomer	FKM	KALREZ®
		AFLAS®	

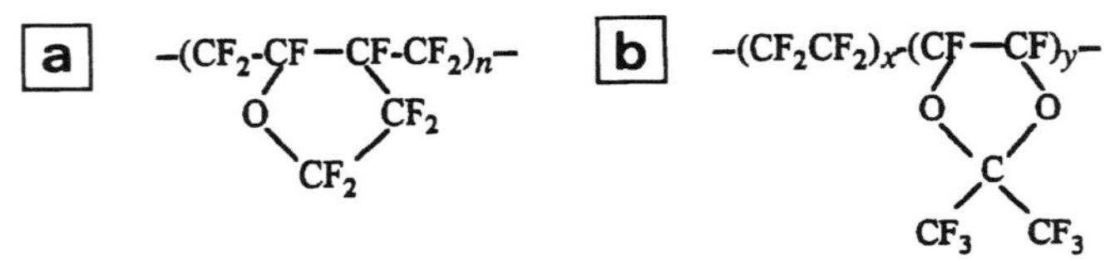

Figure 28.1.   Structures of amorphous perfluoropolymers: (a) CYTOP® developed by Asahi Glass (1988) and (b) Teflon® AF developed by DuPont (1989)

2   CYCLOPOLYMERIZATION OF FLUORINE-CONTAINING DIENES

CYTOP® is synthesized by cyclopolymerization of perfluorodiene, which is a very unique preparation. Cyclopolymerization of hydrocarbonic bifunctional monomer is well known [6]. But only a few attempts for perfluorodienes have been reported [7]. Chlorofluorohexadiene and perfluorohexadiene have been polymerized by γ-ray irradiation under extremely high pressure (13 000 atm) to give high molecular weight polymers. The polymerization of perfluorohexadiene is shown in Scheme 1. The polymerization conditions, however, are unsuitable for commercial production.

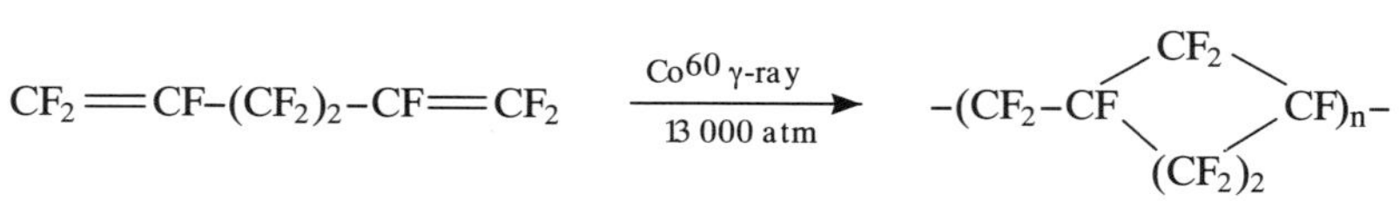

Scheme 1.   Cyclopolymerization reaction of perfluorohexadiene by γ-ray irradiation under extremely high pressure

On the other hand, solution polymerization of perfluoro-divinylether has been reported [8] (Scheme 2). This monomer was cyclopolymerized under mild conditions, but gel formation occurred, when the monomer concentration exceeded 12%. This indicates that intermolecular linear propagation and crosslinking occurred at the same time during cyclopolymerization.

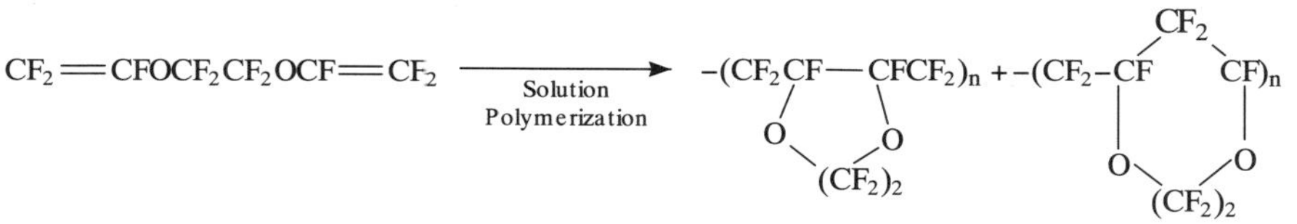

$$CF_2=CFOCF_2CF_2OCF=CF_2 \xrightarrow[\text{Polymerization}]{\text{Solution}}$$

-(CF_2CF—CFCF_2)_n + -(CF_2-CF   CF)_n

Scheme 2.   Solution polymerization of perfluoro-divinylether under mild conditions

A generalized schematic of cyclopolymerization is shown in Scheme 3. In order to obtain a soluble polymer, it is necessary that the cyclization predominates over linear propagation ($k_c \gg k_1$). The formation of pendant group by linear propagation results in intermolecular crosslinking.

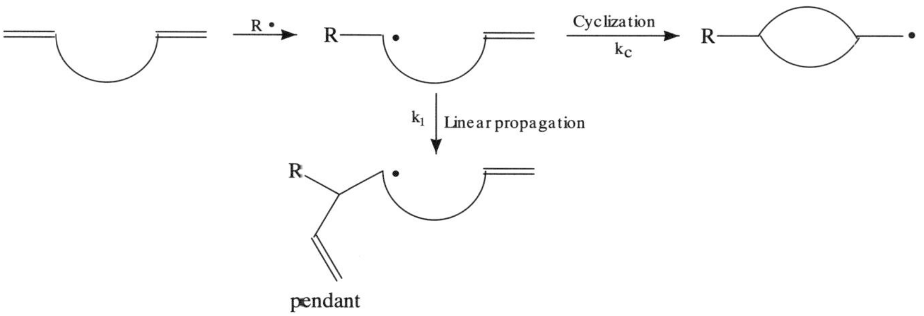

Scheme 3.   Schematic of the cyclopolymerization reaction showing that cyclization competes with linear propagation. (Note: k_1 and k_c represent the rate constants for the cyclization and linear propagation reactions respectively)

Some fluorinated, nonconjugated diene monomers have been prepared and the polymerization has been investigated [9] as shown in Table 28.2. Monomers 1,2,8 and 9 respectively, gave soluble, tough, high molecular weight polymers with 100% cyclization ($k_1 = 0$) despite being a bulk polymerization process. In particular, the polymerization rate for monomers 8 and 9 was faster than for the other monomers, which was explained by alternative polymerization ability between CF_2=CFO— and CH_2=CH—. On the other hand, monomers 10 and 11 without an ether oxygen in their structure gave an insoluble polymer due to both cyclization and linear propagation reactions occurring. Commercially attractive monomers are 1 (perfluoro-allyl vinyl ether: AVE) and 2 (perfluoro-butenyl vinyl ether: BVE) [10]. It was thought that a five-membered ring was formed through the cyclopolymerization in both monomers 1 and 2, though two mechanisms of the ring formation are possible (Schemes 4 and 5).

Table 28.2. Polymerization of fluorinated dienes

	Monomer	Initiator	Temperature (°C)	Time	Yield (%)	$[\eta]$ (dl/g)	T_g (°C)	Remarks
1	$CF_2=CFOCF_2CF=CF_2$	$(C_3F_7COO)_2$	25	24 h	90	0.5	69	Tough
2	$CF_2=CFO(CF_2)_2CF=CF_2$	IPP	40	20 h	85	0.5	108	Tough
3	$CF_2=CFO(CF_2)_3CF=CF_2$	IPP	65	20 h	4	< 0.1	84[a]	Brittle
4	$CF_2=CFOCF_2CF=CFCF_3$	PBIB	60	1 day	13	—	—	Grease
5	$CF_2=CFOCFCF=CF_2$ with CF_3	—	—	—	—	—	—	Unstable Monomer
6	$CF_2=CFO(CF_2)_2CF=CFCF_3$	IPP, PBIB	25–70	2 days	0	—	—	
7	$CF_2=CFOCF_2CFCF=CF_2$ with CF_3	IPP	40	2 days	40	< 0.1	118[a]	Brittle
8	$CF_2=CFO(CF_2)_2CH=CH_2$	$(C_3F_7COO)_2$	20	1 h	> 90	1	90	Tough
9	$CF_2=CFOCF_2CFCH=CH_2$ with CF_3	$(C_3F_7COO)_2$	25	15 h	81	> 1.0	108[a]	Tough
10	$CF_2=CF(CF_2)_2CH=CH_2$	IPP	60	4 h	40	—	120[a]	Crosslinked
11	$CF_2=CF(CF_2)_2CH=CF_2$	IPP	40	48 h	5	—	27[a]	Crosslinked

[a] measured by DSC (others : DMA)
IPP: $((CH_3)_2CHOCOO)_2$, PBIB: $(CH_3)_3COOCOCH(CH_3)_2$.

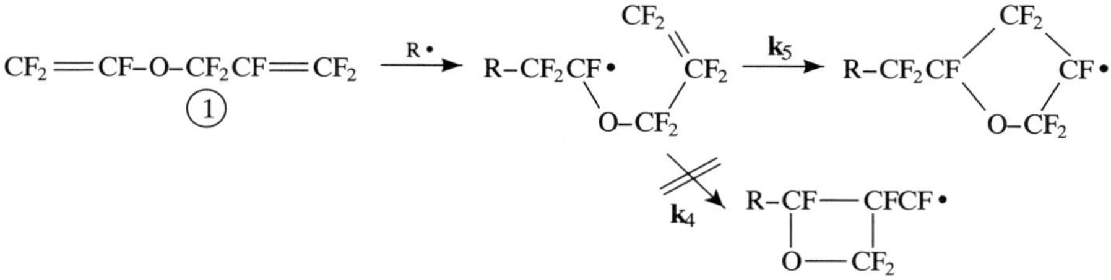

$$CF_2\!=\!CF\!-\!O\!-\!CF_2CF\!=\!CF_2 \quad \xrightarrow{R\,\bullet} \quad R\!-\!CF_2CF\bullet \; \begin{array}{c} CF_2 \\ \| \\ CF_2 \\ | \\ O\!-\!CF_2 \end{array} \quad \xrightarrow{k_5} \quad R\!-\!CF_2CF \underset{O-CF_2}{\overset{CF_2}{\diagup\diagdown}} CF\bullet$$

(1)

$$\xdashrightarrow{k_4} \quad \begin{array}{c} R\!-\!CF\!-\!CFCF\bullet \\ | \qquad | \\ O\!-\!CF_2 \end{array}$$

Scheme 4. Cyclopolymerization mechanism of perfluoro-allyl vinyl ether (AVE). Two mechanisms of the ring formation are possible though five-membered ring formation predominates over 4-membered ring formation. (Note: k_4 and k_5 represent the rate constants for the 4-membered ring formation and 5-membered ring formation respectively)

$$CF_2\!=\!CF\!-\!O\!-\!(CF_2)_2CF\!=\!CF_2 \quad \xrightarrow{R\,\bullet} \quad R\!-\!CF_2CF\bullet \quad \begin{array}{c} CF_2 \\ \backslash \\ CF \\ | \\ CF_2 \\ CF_2 \end{array} \quad \xrightarrow{k_5} \quad R\!-\!CF_2CF\!-\!CFCF_2\bullet$$

(2)

$$\xdashrightarrow{k_6} \quad R\!-\!CF_2CF \quad \begin{array}{c} CF_2 \\ CF\bullet \\ CF_2 \\ CF_2 \end{array}$$

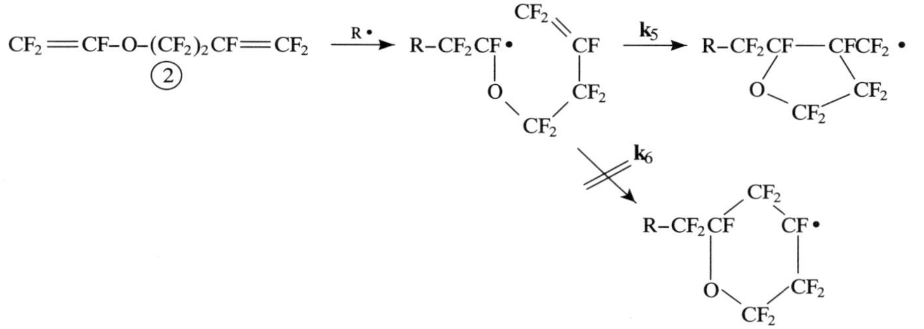

Scheme 5. Cyclopolymerization mechanism of perfluoro-butenyl vinyl ether (BVE). As in the above case for AVE, two mechanisms of the ring formation are possible though 5-membered ring formation predominates over 6-membered ring formation. (Note: k_5 and k_6 represent the rate constants for the 5-membered ring formation and 6-membered ring formation respectively)

The polymer structure was investigated by ^{13}C-NMR spectroscopy [11]. It was suggested that five-membered ring formation was dominant in both polymerization reactions for monomers 1 and 2. In addition, thermal decomposition chromatography (pyrolysis GC) of the two polymers was performed at 600 °C. From the results of poly(AVE), monomer was detected, which can indicate depolymerization, but C_2F_4 units were not detected. On the other hand, C_5F_xO and C_2F_4 units were detected in the case of poly(BVE). Figure 28.2 shows the sites where chain scission would occur during thermal decomposition of poly(AVE) and poly(BVE).

Chlorination of BVE monomer was carried out to give one molar adduct [9(a)]. From the ^{19}F-NMR spectrum, the compound was assumed to be a five-membered ring structure. Furthermore, thermodynamic parameters of a model compound at the initial, transition and product state (R=F on Scheme 5) was calculated by means of semi-empirical molecular orbital (MO) method — Parametric Method 3

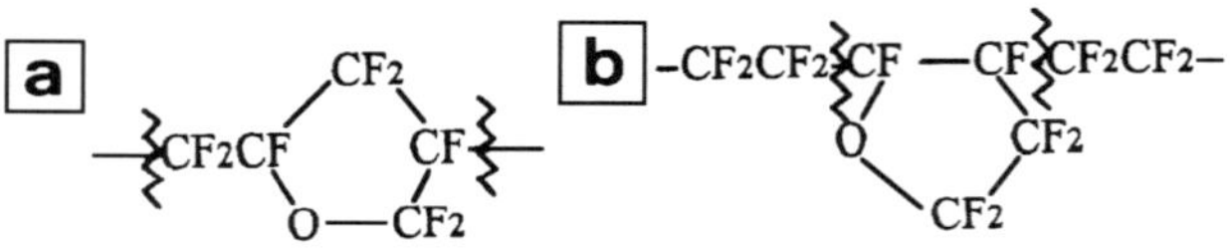

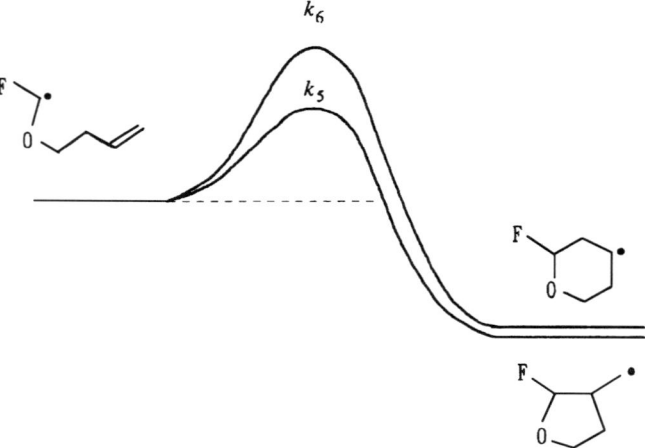

Figure 28.2. Sites where chain scission would occur during thermal decomposition of poly(AVE) and poly(BVE)

Figure 28.3. Representation of the thermodynamic barrier and reaction rate for cyclization of the BVE monomer calculated using a semi-empirical molecular orbital method (Parametric Method 3). It can be seen from the change in Gibbs free energy (ΔG) that the formation of the five-membered ring is favoured over that of the six-membered ring

(PM3) for both cases of five and six-membered ring formation respectively [11] (Figure 28.3). Gibbs free energy changes (ΔG) from initial to the transition state showed that the five-membered formation possesses a lower activation barrier than that of the six-membered ring. In addition, calculations of cyclization rate constant based on the theory of Eyring absolute reaction rate were performed. The rate constant for the five-membered form was 128 times greater than for the six-membered one, at 300 K.

3 PROPERTIES

Various properties of CYTOP® compared with crystalline fluoropolymers and PMMA (a representative polymer with good optical properties) are presented in Tables 28.3 and 28.4 [12].

Table 28.3. Optical and electrical properties of CYTOP® compared with competitive polymers

Property	CYTOP®	PTFE	PFA	PMMA	Remarks
Light transmittance (%)	95	Opaque	Opaque	93	Visible region
Refractive index	1.34	1.35	1.35	1.49	Abbe's refractometer
Abbe's number	90	—	—	55	Abbe's refractometer
Dielectric constant	2.1~2.2	< 2.1	2.1	4	60 Hz~1 MHz
Dissipation factor	0.0007	< 0.0002	< 0.0002	0.04	60 Hz
Volume resistivity (Ω cm)	> 10^{17}	> 10^{18}	> 10^{18}	> 10^{16}	At RT in air
Dielectric strength (KV/0.1 mm)	11	13	12	2	At RT in air

Table 28.4. Physical and mechanical properties of CYTOP® compared with competitive polymers

Property	CYTOP®	PTFE	PFA	PMMA	Remarks
Glass transition temperature (°C)	108	130	75	105~120	By DSC
Melting point (°C)	Not observed	327	310	160 (isotactic)	By DSC
Density (g cm^{-3})	2.03	2.14~2.20	2.12~2.17	1.20	At 25 °C
Contact angle of water (degrees)	110	114	115	80	At 25 °C
Critical surface tension (dyne cm^{-1})	19	18	18	39	At 25 °C
Water absorption (%)	< 0.01	< 0.01	< 0.01	0.3	60 °C, H_2O
Tensile strength (kg cm^{-2})	390	140~350	280~320	650~730	
Elongation at break (%)	150	200~400	280~300	3~5	
Yield strength (kg cm^{-2})	400	110~160	110~150	(650)	
Tensile modulus (kg cm^{-2})	12 000	4000	5800	30 000	

CYTOP® has similar physical and chemical properties to PTFE and PFA. Moreover, its tensile strength and yield strength are higher than that of PTFE and PFA, thus overall the mechanical properties are superior (Figure 28.4). Its optical properties are also quite unique. Figure 28.5 shows light transparency of a 200 μm film in the range from 200 to 700 nm, and its clarity in particular

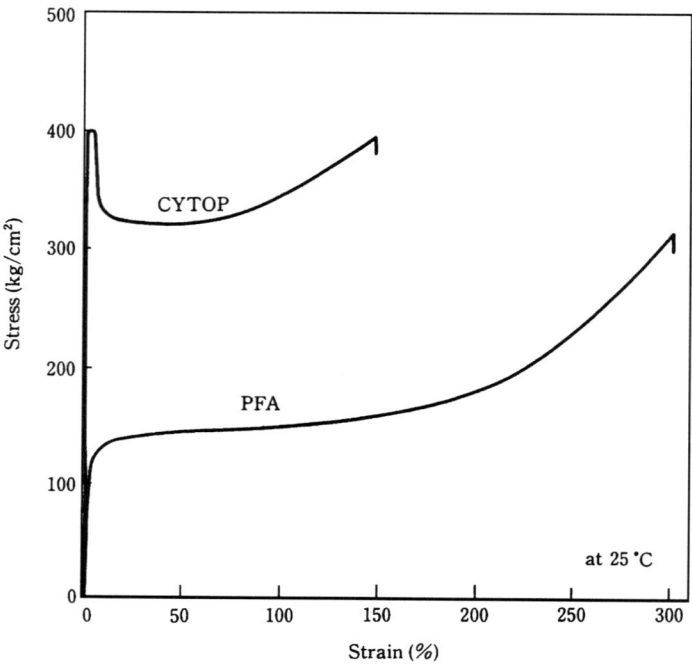

Figure 28.4.   Stress–strain behaviour for CYTOP® as compared with PFA

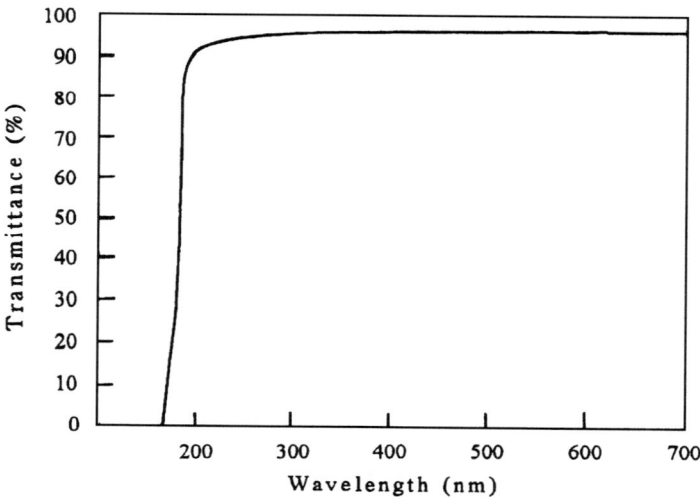

Figure 28.5.   UV-visible light transmission spectrum of a 200 μm film of CYTOP®

is very high even in the UV region because absorption due to electronic transitions do not exist. Loss of transmittance by a few percent is due to surface reflection that occurs. Furthermore, near-IR can also be transmitted as shown in Figure 28.6 (rod-shaped sample of 30 cm length was used). This optical property, in particular, is expected to increase the applications potential of such new fluoropolymers including Teflon® AF [13].

Another important property of CYTOP® is its solubility in selected fluorinated solvents. Since fluorinated solvents are completely inert toward many substrate materials like metals, ceramics, plastics and rubbers, CYTOP® coatings can be applied without any damage or any undesirable change to the substrate. For example, silicon wafers, glass plates and plastic plates, which have a flat surface, can be coated by the method of spinning, dipping and die coating. Films are obtained that are free from pin holes and having uniform thickness. The low surface tension of the solution also enables it to spread into porous material and cover the entire surface.

In practical applications, good adhesion to a variety of substrates is a very important property. It is difficult for conventional perfluoropolymers to adhere to many substrates due to their inert nature and low surface energies. But, carboxylic acid or alkoxysilane moieties can be introduced into the polymer molecule of CYTOP® [14], so that the coating is able to adhere to various materials. This

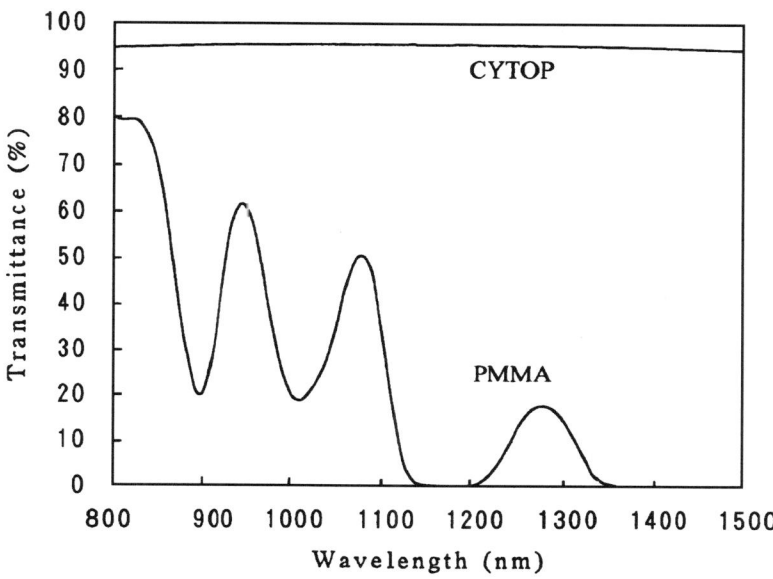

Figure 28.6. NIR transmission spectrum of CYTOP® as compared with PMMA. Sample pathlength = 30 cm

adhesion is promoted by the functional groups as shown in Figure 28.7. If the substrate surface can be pretreated by a chemical or physical method, strong adhesion can be achieved. Adhesion to the various substrates was summarized in Table 28.5.

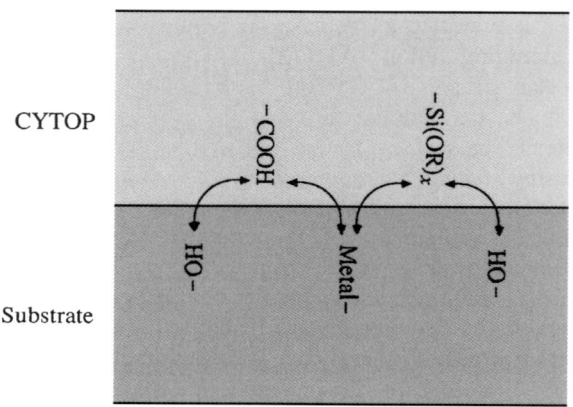

Figure 28.7. Schematic of the adhesion mechanism between CYTOP® and substrate

Table 28.5. Methods of promoting adhesion of CYTOP® to various substrates

Substrate	Pretreatment
Glass Quartz Silicon wafers	Silane coupling agent
Fe, stainless steel Aluminium Silver	None
Poly(methyl methacrylate) Polycarbonate Polystyrene	1. UV/ozone, plasma 2. Silane coupling agent 3. Epoxide primer

4 APPLICATIONS

CYTOP® is useful for production of 'pellicles', which are photomask covers for dust protection and to prevent contamination during the microlithography process in semiconductor production (Figure 28.8). With the increasing miniaturization of integrated circuitry, the wavelength of the light source used in resist manufacture has gone from near- and mid-UV (300–450 nm) to far-UV (< 300 nm). In particular, KrF excimer lasers (248 nm) are used for 256 MB RAM fabrication. The commonly used pellicle materials based on nitrocellulose derivatives cannot

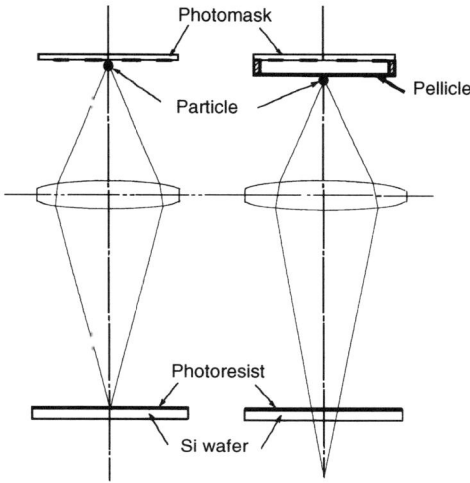

Figure 28.8. The use of CYTOP® as pellicle for microlithography is semi-conductor production

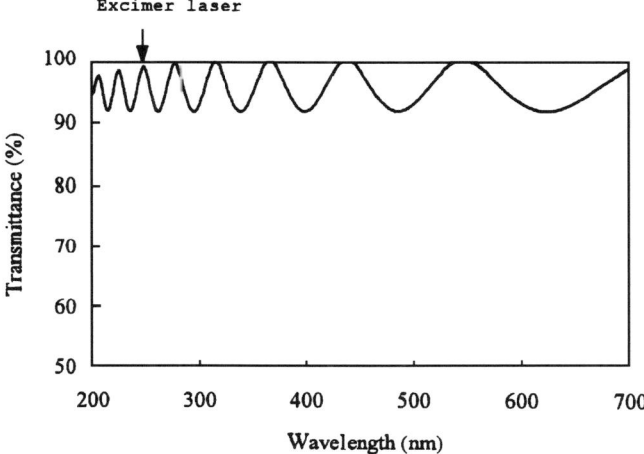

Figure 28.9. Interference fringes of a thin film of CYTOP

be used due to far-UV absorption and subsequent degradation. CYTOP®, on the other hand, can be used because of its transmission characteristics in the far-UV (Figure 28.5). The pellicle is constructed by using a 0.8–0.9 μm thin film supported on an aluminum frame. At the optimum thickness, transmission is maximized by the effect of light interference (Figure 28.9).

CYTOP® can also be used in the manufacture of anti-reflective coatings due to its low refractive index. A product for this application has been commercialized under the ARCTOP® tradename [15]. It has a multilayer structure composed of a top layer of CYTOP® and a middle layer of a high refractive index polymer, layered on urethane base film (Figure 28.10). Magnesium fluoride is mainly used for anti-reflective coatings on display units, e.g. CRT (cathode ray tubes) and LCD (liquid crystal displays). However, ARCTOP® is produced at lower cost by means of a continuous wet coating process and exhibits less reflection than the inorganic coatings which are produced by vapor deposition.

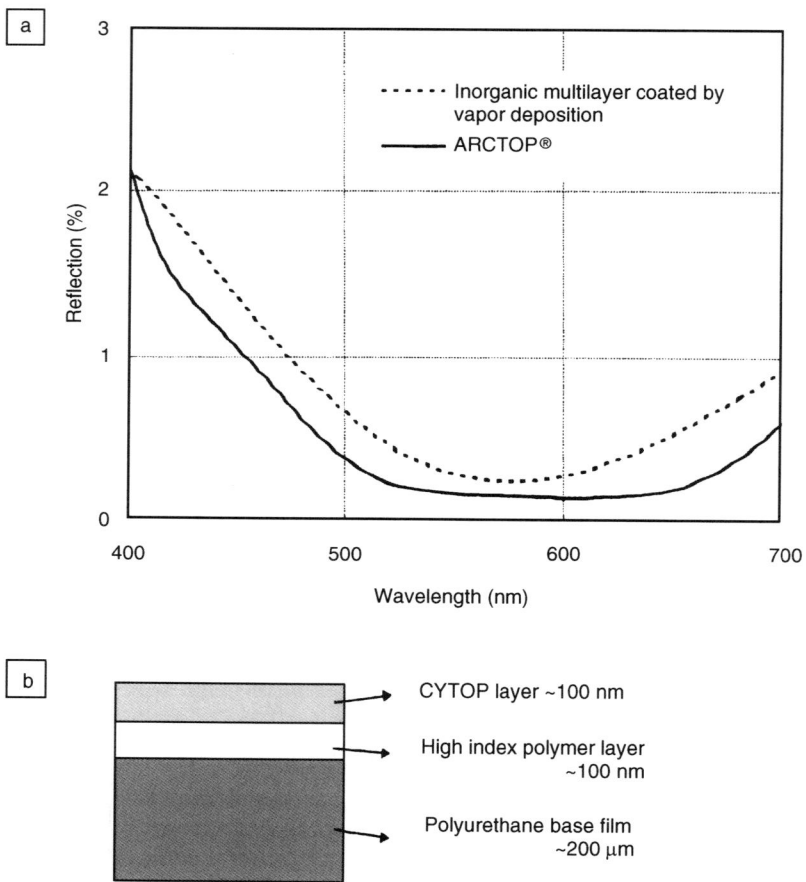

Figure 28.10. An anti-reflective film (ARCTOP®) which comprises an upper layer of CYTOP®. (a) reflection profile and (b) cross-section of anti-reflective sandwich structure

The use of amorphous fluoropolymers in the electronic industry is expected to increase dramatically as new applications are realized. Its low dielectric constant and low water uptake are suitable attributes not only for protecting material for underlying devices, but also for inter-layer dielectrics replacing silicon-based materials such as silicon dioxide or silicon nitride in fabricating integrated circuits [16]. A dry etching process for CYTOP® using a novel photoresist of low surface energy has been developed (Figure 28.11) and the resulting micropattern is shown in Figure 28.12.

Recently, the development of plastic optical fibres for short-range communication has become an active research area [17] because plastic optical fibres offer several advantages in comparison to glass optical fibre, such as flexibility, ease of connection and low cost light-emitting diode (LED) system. PMMA is generally used for this application, but its maximum transmission wavelength is only about 650 nm because of the existence of a vibrational overtone absorption attributed to C–H bonds. In order to improve the transmission properties at higher wavelengths, deuterated PMMA and PS have been investigated [18], but by calculation of the overtone absorption it was found that fully halogenated polymers show negligible loss contribution and attain the lowest loss for organic polymers [19]. For this reason perfluoropolymers like CYTOP® are expected to play an important role in plastic optical fibres as communication devices.

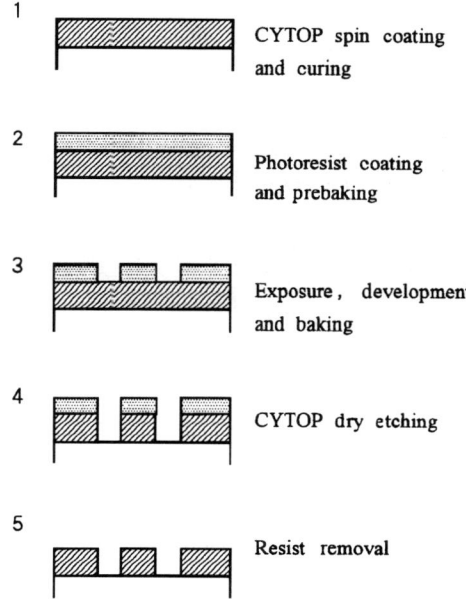

1   CYTOP spin coating and curing

2   Photoresist coating and prebaking

3   Exposure, development and baking

4   CYTOP dry etching

5   Resist removal

Figure 28.11. Schematic of the CYTOP® patterning process for manufacturing interlayer dielectrics by dry etching

Figure 28.12. CYTOP® micropattern for semiconductor fabrication produced by laser microlithography

REFERENCES

1. (a) Carlson, D. P. and Schmiegel, W., 'Fluoropolymers', *Ullmann's Encyclopedia of Industrial Chemistry*, Vol. A11, Verlag Chem., Weinheim, p. 39 (1988); (b) Feiring, A. E., 'Fluoroplastics', *Organofluorine Chemistry: Principles and Commercial Applications* (Banks, R. E., Smart, B. E. and Tatlow, J. C. eds), Plenum, New York, p. 339 (1994); (c) Chandrasekaran, S., 'Chlorotrifluoroethylene homopolymer', *Encyclopedia of Polymer Science and Engineering* (Mark,

H. F., Bikales, N. M., Overbarger, C. G. and Menges, G. eds), Wiley, New York, Vol. 3, 463 (1989); (e) Brasure, D. and Ebnesajjad, S., 'Vinyl fluoride polymers', *ibid.*, Vol. 17, 468 (1989); (f) Dohany, J. E. and Humphrey, J. S., 'Vinylidene fluoride polymers', *ibid.*, 532 (1989).

2. (a) Resnick, P. R., 'The preparation and properties of a new family of amorphous fluoropolymers: Teflon® AF, *Polym. Prepr., Am. Chem. Soc., Div. Polym. Chem.*, **31**, 312 (1990); (b) Buck, W. H. and Resnick, P. R., 'Properties of amorphous fluoropolymer based on 2,2-bistrifluoromethyl-4, 5-difluoro-1,3-dioxole', *Electrochemical Soc., Spring Meeting, Honolulu, Hawaii, Extended Abstracts*, **93**-1, p. 548 (1993); Buck, W. H. and Resnick, P. R., *Teflon® AF Technical Information Bulletin*, DuPont, Wilmington, DE (1993).

3. (a) Hung, M. H., *Macromolecules*, **24**, 6660 (1991); (b) Sugiyama, N. and Murobusi, H., Japanese Patent Kokai 05–09224 (1993).

4. Selman, S. and Squire, E. N., US Patent 3 308 107 (1967).

5. Krespan, C. G. and Dixon, D. A., *J. Org. Chem.*, **56**, 3915 (1991).

6. Gibbs, W. E. and Barton, J. M., 'The mechanism of cyclopolymerization of nonconjugated diolefins' in *Vinyl Polymerization, Kinetics and Mechanism of Polymerization Series*, Vol. 1, Part 1, Marcel Dekker, New York, p. 59 (1967).

7. Wall, L. A., 'High pressure polymerization', *Fluoropolymers* (L. A. Wall ed.), Wiley, New York, p. 127 (1972).

8. Darby, R. A., Perfluorocyclic Ether Polymers, US Patent 3 418 302 (1968).

9. (a) Oharu, K., Sugiyama, N., Nakamura, M. and Kaneko, I., 'Preparation and reaction of perfluoro (alkenyl vinyl ether)', *Reports Res. Lab. Asahi Glass Co. Ltd.*, **41**, 51 (1991); (b) Oharu, K. Sugiyama, N. *et al.*, 'New fluoropolymers obtained by cyclopolymerization', 3, *Abstracts of 16th Symposium on Fluorine Chemistry (JPN)*, (1991).

10. (a) Nakamura, M. *et al.*, US Patent 4 897 457 (1997); (b) Nakamura, M., Kawasaki, T., Unoki, M., Oharu, K., Sugiyama, N. and Kojima, G., 'Synthesis and properties of a new fluoropolymer obtained by cyclopolymerization', *Progress in Pacific Polymer Science*, Springer-Verlag, p. 369 (1991).

11. Sugiyama, N. and Nakamura, M., 'New fluoropolymers obtained by cyclopolymerization', 5, *Abstracts of 18th Symposium on Fluorine Chemistry (JPN)*, 37 (1993).

12. *CYTOP Technical Bulletin*, Asahi Glass Co. Ltd, Japan (1990).

13. Lowry, J. H., Mendlowitz, J. S., Mo, S. L. and Subramanian, N. S., 'Optical characteristics of Teflon AF fluoroplastic materials', *SPIE. Invited Paper*, **1330**, 142 (1990).

14. Matuo, M. *et al.*, Japanese Patent Kokai 04–226177 (1992).

15. (a) Aosaki, K., 'Anti-reflective film by continuous coating process', *Sinsozai (JPN)*, p. 55 (1995.2); (b) Hasegawa, T. *et al.*, Japanese Patent Kokai 07–168005 (1995).

16. (a) Kuramochi, T., Kiyokawa, H., Ono, T. Miyasaka, K., 'Multi-chip-module substrate decreasing signal delay and improving thermal conductivity', *Proceedings IEEE/CHMT'91 IEMT Symposium*, 255 (1991); (b) Chang, C., Richards, C., Vu, Q., Mack, A. S., Fraser, D. B. Yokotuka, S. and Nakamura, M., 'The use of CPFP — a new fluoropolymer as an interlayer dielectric for ULSI application,' *Extended Abstracts, ECS Spring Meeting*, 93–1, 445 (1993).

17. Koike, Y., 'High-speed multimedia POF network', *Conference Proceedings, POF'94*, Yokohama, p. 16 (1994).

18. Kaino, T., *J. Polym. Sci., Part A*, **25**, 37 (1987).

19. Groh, W., 'Overtone absorption in macromolecules for polymer optical fibres,' *Macromol. Chem.*, **189**, 1261 (1988).

29

CTFE/Vinyl Ether Copolymers

TERUO TAKAKURA

Chemicals General Division, ASAHI GLASS Co., Ltd., Yokohama, Japan

1 INTRODUCTION

Polymerization of chlorotrifluoroethylene (CTFE) was reported in 1937 [1], prior to the discovery of poly(tetrafluoroethylene) (PTFE) [2], but copolymerization of CTFE with other vinyl monomers had received less attention than would have been expected. The radical copolymerization of CTFE with vinyl acetate, styrene and methyl methacrylate was first reported in detail by Thomas and O'Shaughnessy [3]. Ragazzeni *et al.* reported that the copolymerization of CTFE with ethylene or propylene proceeded in an alternating manner [4]. Pioneering work in this area involved the radiation copolymerization of CTFE with ethyl vinyl ether by Tabata in 1971 [5]. Meanwhile, a new type of paint resin consisting of fluoroethylene (TFE or CTFE)/vinyl ether copolymer (PFEVE) was developed and commercialized by Asahi Glass Co. Ltd, under a trade name of LUMIFLON® [6]. This stimulated the development of new fluoropolymers for paint applications [7,8]. This chapter refers to copolymerization chemistry, properties and applications of CTFE/vinyl ethers copolymer, together with recent developments.

2 COPOLYMERIZATION OF CTFE WITH VINYL ETHERS

Q and e values [9] of CTFE, vinyl ethers (VE) and vinyl acetate are listed in Table 29.1, together with calculated monomer reactivity ratios (r_1, r_2 and $r_1 \times r_2$) according to Alfrey and Price equations [10]. CTFE is an acceptor monomer, whereas VE and vinyl acetate are donor monomers. As the values of $r_1 \times r_2$ show clearly, alternating copolymers are expected with CTFE/VE systems. Figure 29.1 shows the copolymerization curves of CTFE with some VEs [6(a)]. In the case of CTFE/vinyl acetate, the monomer reactivity ratios show that the alternating

Modern Fluoropolymers. Edited by John Scheirs

© 1997 John Wiley & Sons Ltd

Table 29.1.　Q and e values of chlorotrifluoroethylene and other comonomers [9] and calculated monomer reactivity ratio

	Q	e	Reactivity ratio (1 : CTFE, 2 : comonomer)		
			r_1	r_2	$r_1 \times r_2$
Chlorotrifluoroethylene	0.026	1.56	—	—	—
Ethyl vinyl ether	0.018	−1.80	7.6×10^{-3}	1.6×10^{-3}	1.2×10^{-5}
Butyl vinyl ether	0.038	−1.50	5.8×10^{-3}	1.5×10^{-2}	8.7×10^{-5}
Vinyl acetate	0.026	−0.88	1.8×10^{-2}	1.2×10^{-1}	2.1×10^{-3}

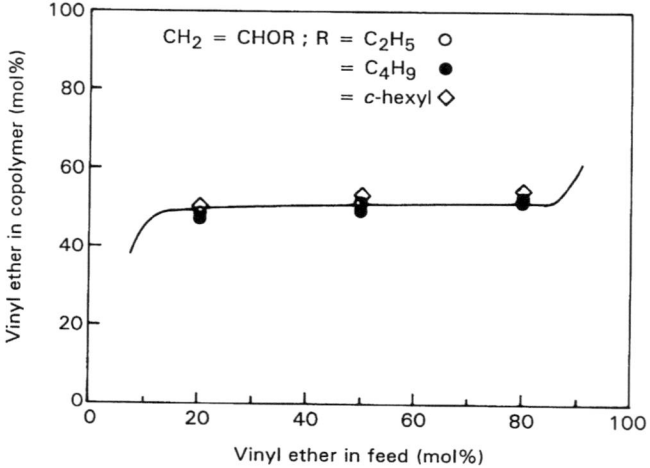

Figure 29.1.　Copolymerization composition curves for the copolymerization of CTFE and vinyl ethers [6(a)]

structure of the copolymer is not so strictly controlled [11] as that of CTFE/VE copolymers.

The detailed copolymerization kinetics of CTFE/VE were investigated by Boutevin *et al.* [12]. They have concluded that the polymerization proceeds by the propagation of free monomers as shown in Scheme 1, and not by the propagation of donor–acceptor charge transfer complexes (Scheme 2) as previously thought.

3　APPLICATIONS OF CTFE/VE COPOLYMER AS A COATING MATERIAL

Although fluoropolymers are expected to have several advantages as coatings, poor solubility in organic solvents and the need for high baking temperature have

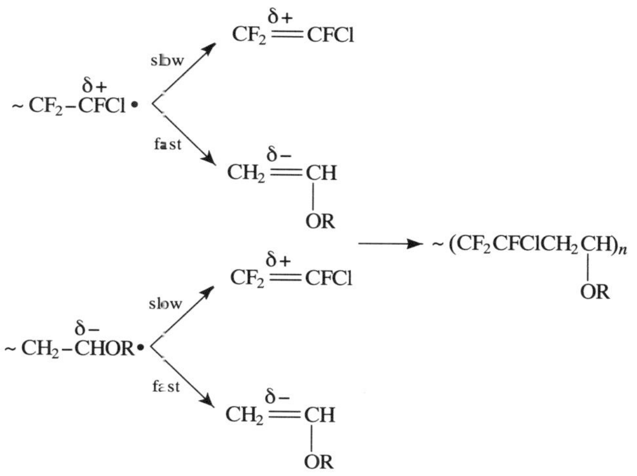

Scheme 1

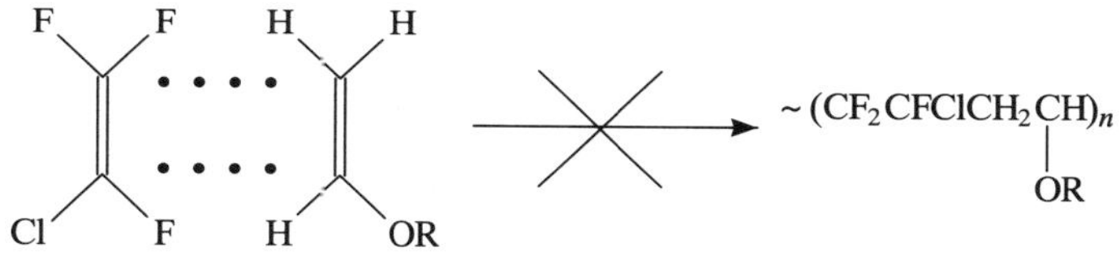

Scheme 2

restricted their application fields. PTFE, tetrafluoroethylene-hexafluoropropylene copolymer (FEP), tetrafluoroethylene-perfluoro(alkyl vinyl ether) copolymer (PFA), tetrafluoroethylene-ethylene copolymer (ETFE) and poly(vinylidene fluoride) (PVDF) are presently applied as anti-stick and anticorrosive powder coatings [13]. Some fluoropolymers are also applied in a water or organic solvent-borne dispersion form (Table 29.2), and among them PVDF is the most widely used fluoropolymer as a dispersion-type weather-resistant paint. Film lamination of melt-processable fluoroplastics is an alternative technology in these fields.

LUMIFLON® is a specially designed fluoropolymer for paint applications based on PFEVE (mainly CTFE/VE) copolymers. PFEVE, an amorphous fluoropolymer, has various characteristics that make it suitable as a base polymer for paints, such as solubility in organic solvents, film-forming ability at an ambient temperature and transparency of the resultant film [6(c)]. CTFE is a more suitable fluoromonomer than TFE with regard to solubility in organic solvents and compatibility with hardeners or pigments. It is likely that the chlorine atom plays an important role in improving these properties. As shown in Figure 29.2, selection and combinations of different types of VE could add further characteristics

Table 29.2. Conventional fluoropolymers for coating applications (except powder coatings)

Fluoropolymer	Dispersion media	Baking temperature (°C)	Application
PTFE	Water	>330 °C	Non-sticking coating Wire coating
FEP PFA	Water	>260 °C	Non-sticking coating
PCTFE[a]	Water	>210 °C	Wire coating
PVDF PVF[b]	Solvent	>170 °C >200 °C	Weather-resistant paint

[a] Polychlorotrifluoroethylene.
[b] Poly(vinyl fluoride).

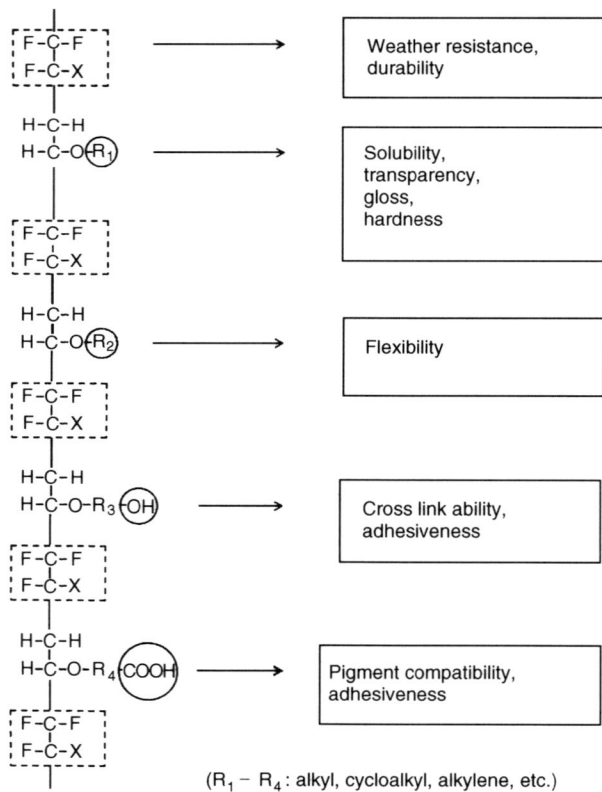

(R₁ – R₄: alkyl, cycloalkyl, alkylene, etc.)

Figure 29.2. Molecular design of LUMIFLON® (note X = Cl or H)

Table 29.3. Comparison of PFEVE-based paints with PVDF-based ones

	PFEVE	PVDF
Type	Solution	Dispersion
Cure temperature	Room temperature ($\sim$23 °C)	High temperatures >250 °C
Pigment compatibility	Good	Poor
Flexibility of coated film	Good	Good
Weatherability	Excellent	Excellent
Stain resistance	Excellent	Excellent
Gloss (60° $\sim$ 60° reflection)	$\sim$80%	25$\sim$35%
Recoatability	Excellent	Poor

required for a base polymer of paints. The highly alternating structure of CTFE and VE units are responsible for the excellent outdoor weather resistance of LUMIFLON®. Chemically stable fluoroorefin units protect the rather unstable VE units. Varying the structure of R_1 and R_2, flexibility and solubility can be modified; for example, the glass transition temperature (T_g) of the polymer can be varied between 20 and 70 °C. Introducing a hydroxyl group in the side chain enables the copolymer to be cured by polyisocyanate or melamine resins. Excellent compatibility with pigments is attained by partial carboxylation of the pendant hydroxyl group with acid anhydride. The copolymer with a higher COOH content is soluble in an aqueous medium by neutralizing it with organic amines.

PFEVE-based paint is superior to the dispersion-type PVDF-based paints for the properties shown in Table 29.3. As previously mentioned, the former has good film-forming properties and can be chemically cured even at room temperature. Whereas PVDF has no curable sites and needs higher temperature, greater than 250 °C, to form a continuous film. The lower pigment compatibility and the higher baking temperatures restrict the range of pigments available for PVDF. Both fluoropolymer coatings are flexible and have excellent weather and stain resistance compared with the conventional non-fluorinated paints. PFEVE-based coatings have higher gloss and better recoatability than PVDF-based ones coatings. Figure 29.3 shows the gloss retention of PFEVE coatings compared with PVDF and non-fluorinated acrylic-based paints by the accelerated weathering test (sunshine weather-o-meter). LUMIFLON®-based coatings exhibit excellent weathering resistance without any destabilizing effect by the hardeners. The weather resistance has also been confirmed by the EMMAQUA test carried out in the Arizona desert [12].

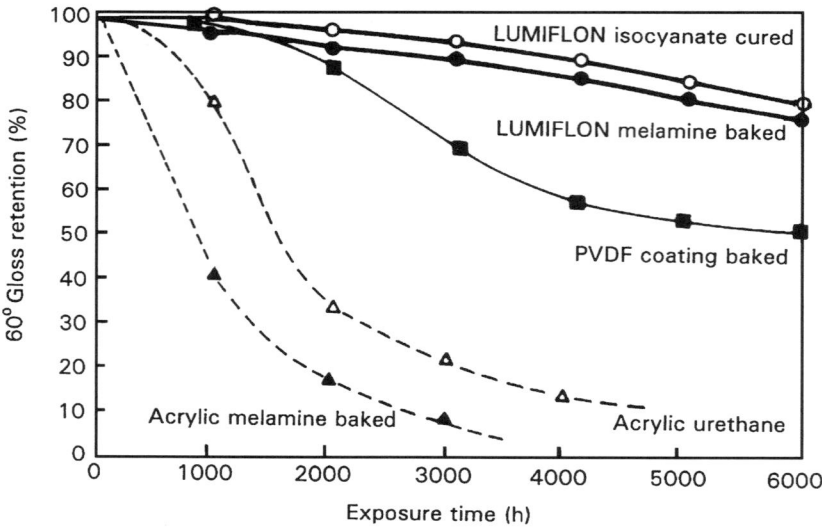

Figure 29.3. Accelerated weathering test of LUMIFLON®, PVDF and acrylic-based coatings (using white enamels)

Some fundamental properties of LUMIFLON® are shown in Table 29.4. The molecular weight of the polymer was determined by gel-permeation chromatography and it is a major factor in determining the painting methods. Several grades with different molecular weight are now available. Over the past decade, since LUMIFLON® was launched into the market, it has been applied in many fields. Its excellent weathering resistance and curability at room temperature have opened up a large market for large architectural structure such as office buildings and bridges where on-site coating is required (Figure 29.4). Giving satisfactory results in this field, it is now being applied to transport like automobiles, trains and ships, and other specialty industrial coatings like signs and solar panels.

Table 29.4. Fundamental properties of LUMIFLON PFEVE

Fluorine content (wt%)	$25 \sim 30$
OH value (mgKOH/g)	$47 \sim 52$
COOH value (mgKOH/g)	$0 \sim 5$
Molecular weight M_n	$0.8 \times 10^4 \sim 6 \times 10^4$
M_w	$1.0 \times 10^4 \sim 15 \times 10^4$
Specific gravity (g/cm³)	$1.4 \sim 1.5$
Glass transition temperature (°C)	$20 \sim 70$
Decomposition temperature (°C)	$240 \sim 250$
Solubility parameter	8.8 (calcd.)

Figure 29.4. LUMIFLON® top coated 'Rainbow Bridge', Tokyo, Japan

4 RECENT DEVELOPMENTS

In view of the need to protect the environment, new grades of LUMIFLON® have been developed [14] to meet the requirements for water-borne, electrodeposition and powder coatings. Substantial progress in the emulsion polymerization technology of PFEVE has enabled the production of these new grades. Fifty per cent polymer concentration, with high storage stability of the emulsion has been attained by introduction of hydrophilic poly(ethylene oxide) units in the side chain of the polymer as well as by optimization of emulsifiers. LUMIFLON® emulsions have film-forming ability at ambient temperature and show almost the same weather resistance independent of the solvent type. The new products are widely used for application on exterior walls as an environmentally friendly paint. Crosslinkable PFEVE emulsions have also been developed with water-based polyisocyanates. This technology may open up further new applications for these emulsion-type paints.

Asahi Glass has recently developed and commercialized a base polymer for sealants. LUMISEAL®, a trade name of the new product, is also a FEVE-based copolymer. Introducing a long flexible graft chain in R_2 of Figure 29.2, both weather resistance and flexibility required for a sealing material are successfully attained. Silicone sealants are well known as high-performance sealing materials with excellent weathering resistance, but oligomeric silicone, fractions which are difficult to eliminate from these compounds, diffuses to the surface and stains the substrate to be sealed. Whilst modified silicone sealants do not cause such a stain they lack sufficient weather resistance. LUMISEAL®-based sealants show high weatherability like silicones but without the staining problem. It is expected they can maintain their sealing property for at least 20 years with no maintenance.

REFERENCES

1. Brit. Pat. 465 520 (to I. G. Farbenindustrie) [CA 31, 7145(1937)].
2. Plunkett, R. J., US Pat. 2 230 654 (to Kinetic Chemicals) [CA 35, 3365(1941)].
3. Thomas, W. M. and O'Shaughnessy, T., Kinetics of chlorotrifluoroethylene polymerization, *J. Polym. Sci.*, **11**, 455 (1953).
4. Ragazzini, M. *et al.*, Copolymerization of propylene and chlorotrifluoroethylene. estimation of monomer reactivity ratios, *Europ. Polym. J.* **3**, 129, 137 (1967); **6**, 763 (1970).
5. Tabata, Y. and Du Plessis, T. A., Radiation-induced copolymerization of chlorotrifluoroethylene with ethyl vinyl ether, *J. Polym. Sci: Part A-1*, **9**, 3425 (1971).
6. (a) Kojima, G. and Yamabe, M., A Solvent-soluble fluororesin for paint, *J. Synth. Org. Chem., Jpn.*, **42**, 841 (1984); (b) Yamabe, M. *et al.*, New fluoropolymer coatings, *Organic Coatings: Science and Technology*, Vol. 7 (Parfitt, G. and Patsis, A. eds), Marcel Dekker, New York and Basel, p. 25 (1984); (c) Munekata, S., Fluoropolymers as coating materials, *Progress in Organic Coatings*, **16**, 113 (1988).
7. Yamabe, M., Fluoropolymer coatings, *Organofluorine Chemistry; Principles and Commercial Applications* (Banks, R. E. *et al.* eds), Plenum, New York, p. 397 (1994).
8. Scheirs, J. *et al.*, Developments in fluoropolymer coatings, *Trends Polym. Sci. (TRIPS)* **3**, 74 (1995).
9. Greenly, R. Z., *Polymer Handbook, 3rd edn*, (Brandup, J. *et al* eds), Wiley, New York, II/267 (1989).
10. Alfrey, T. and Price, C. C., Relative reactivities in vinyl copolymerization, *J. Polym. Sci.*, **2**, 101 (1947).
11. Murray, D. L. *et al.*, The use of sequence distributions to determine monomer feed compositions in the emulsion copolymerization of chlorotrifluoroethylene with vinyl acetate and vinyl propionate, *Polymer*, **36**, 3841 (1995).
12. Boutevin, B. *et al.*, Studies of the alternating copolymerization of vinyl ethers with chlorotrifluoroethylene, *Macromolecules*, **1992**, 25, 2842.
13. Khaladkar, P. R., A Comparison of fluoropolymer linings, *Material Performance (MP)*, **33**, 35 (1994).
14. Yamauchi, M. *et al.*, Fluoropolymer emulsions, *European Coatings J.*, 124 (1996).

30

Fluorinated Thermoplastic Elastomers

MASAYOSHI TATEMOTO and TETSUO SHIMIZU
Daikin Industries, Osaka, Japan

1 INTRODUCTION

Since the serendipitous discovery of 'Kraton' block-copolymers of styrene and butadiene by Shell Oil Co [1], a number of thermoplastic elastomers (TPE) have been developed to afford elastomers the ease of processing when compared with the troublesome and expensive procedures required for their vulcanization. Non-fluorinated TPE has been a commercial success. What happened to fluorinated thermoplastic elastomers (FTPE)?

Two kinds of FTPE are industrially produced in Japan. One is the block-copolymer type which is composed of a central fluoroelastomer soft segment and multiple terminal fluoroplastics hard segments. This has been accomplished as a result of developing 'iodine-transfer polymerization' which behaves like a 'living' radical polymerization [2]. This type of FTPE was first introduced commercially in 1982 from Daikin Industries under the trade name of Dai-el® Thermoplastic [3]. Another is the graft-copolymer type composed of main-chain fluoroelastomers and side-chain fluoroplastics. In the preparation, initially there is an introduction of monomeric units having peroxide groups as grafting sites in the main chains, followed by graft polymerization. The graft-copolymer type of FTPE was introduced commercially in 1987 from Central Glass Co. (Cefral Soft®) [4]. Current production of Dai-el® Thermoplastic and Cefral Soft® is estimated to be a total of about 100 tons in 1995.

Several other attempts for example by dynamic vulcanization [5,6], have also been made to prepare FTPE. Moreover, soft fluoroplastics melt processable at lower temperatures have recently developed at 3M Company* (THV®) [7].

---

* Note: 3M Company is now called Dyneon

Modern Fluoropolymers. Edited by John Scheirs
© 1997 John Wiley & Sons Ltd

These would compete with the present FTPE in the 'plastomeric' field (THV® is described in a different chapter). In answer to the question posed in the first paragraph of this chapter, we should answer that, presently, the technology and applications of FTPE (or soft fluoroplastic) are now still under development.

2 PREPARATION

2.1 IODINE-TRANSFER POLYMERIZATION

The first TPE, Kraton, was realized by the living anionic polymerization technique. The same technique has not been successfully applied to the fluorinated analogues in spite of enormous efforts by a number of chemists so far. On the other hand, iodine-transfer polymerization makes it possible to perform a 'living' polymerization in radical conditions. In this process (equation 1)

$$
\begin{array}{c}
\text{I-R}_F\text{-I} \\
\downarrow \text{Monomer A} \\
\text{I-(A)}_n\text{-R}_F\text{-(A)}_n\text{-I} \qquad (1) \\
\downarrow \text{Monomer B} \\
\text{I-(B)}_m\text{-(A)}_n\text{-R}_F\text{-(A)}_n\text{-(B)}_m\text{-I}
\end{array}
$$

typically, monomer (A) is polymerized with a small amount of peroxides in the presence of perfluoroalkyl diiodides (IR_FI; usually, low molecular weight compounds) to form soft segments. This is followed by block polymerization with the second monomer (B) to form hard segments. Terminal iodines in each reaction step are stable but labile to active polymeric radicals under the polymerization conditions to be removed by chain-transfer reaction. The resultant polymeric radicals continue propagation reactions with added monomers and chain transfer reactions with other terminal iodines. Accordingly, the tri-block-type copolymers are easily obtained using two kinds of monomers different from each other. It is also possible to prepare multi-block copolymers by this process. Not only a variety of fluoromonomers but some hydrocarbon monomers can be applied to this reaction as monomer (B). Copolymers of vinylidene fluoride (VDF) with hexafluoropropylene (HFP) and optionally tetrafluoroethylene (TFE) are the most conventional selection as soft segments. Poly (TFE-*alt*-ethylene) (ETFE) or poly(vinylidene fluoride) (PVDF) are feasible selections for hard segments. The commercial products are prepared in these combinations.

2.2 GRAFT-COPOLYMERIZATION USING POLYMERIC PEROXIDES

The graft-copolymer type of FTPE is reported to be prepared basically by the following two steps, according to the literature (equation 2) [4,8]. In the first step,

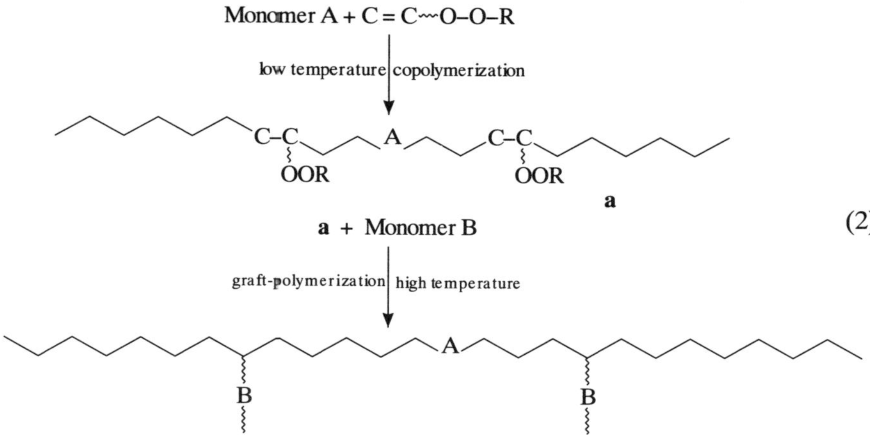

Monomer A + C = C~O–O–R

low temperature | copolymerization

a + Monomer B (2)

graft-polymerization | high temperature

fluoroelastomers are prepared at a low temperature, using unsaturated peroxides such as $[CH_2=CHCH_2OC(O)-O-O-tert\text{-butyl}]$ as well as conventional fluoromonomers such as a combination of VDF and chlorotrifluoroethylene (CTFE). The peroxides having C=C bonds must be selected with consideration for the appropriate copolymerizability with fluoromonomers and the higher decomposition temperature of the peroxide moieties. In the second step, post-polymerizations to form crystalline segments are repeatedly performed using mainly VDF, successively raising the reaction temperature. This graft-copolymerization process in principle affords a wide selection of fluoromonomers to tailor the copolymers to the required properties.

2.3 OTHERS

Taking advantage of the functionality of fluorinated polyether diols, polyurethane-based FTPE can be synthesized by reacting the diols with aromatic diisocyanates, giving block copolymers with fluorinated polyether soft segments [9]. Dynamic vulcanization could also be a general method to prepare FTPE. It is widely applied to non-fluorinated TPE, where thermoplastics and elastomers are dynamically vulcanized during the melt-blending stage using cure agents. For instance, a blend of a perfluoroplastic and a perfluoroelastomer containing cure sites, or a combination of VDF-based fluoroelastomers and high-performance thermoplastics such as polyamides, poly(butylene terephthalate) and poly(phenylene sulfide) are reported [5,6].

3 STRUCTURE AND PROPERTIES

3.1 BLOCK-COPOLYMERS

Figure 30.1 shows a schematic representation of the microstructure for the block copolymer (T-530). The hard segments form crystalline domains whose size is

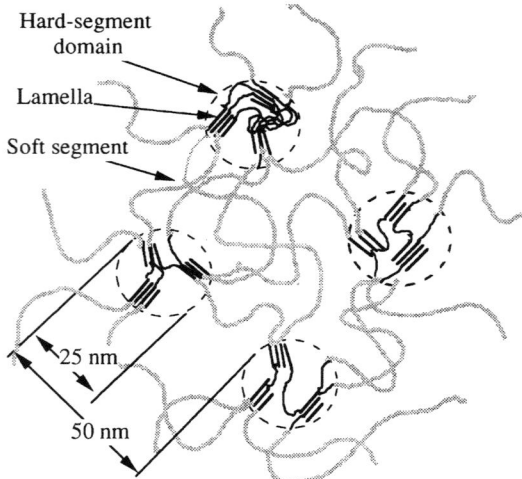

Hard-segment domain

Lamella

Soft segment

25 nm

50 nm

Figure 30.1. Schematic representation of the microstructure for the block-copoly-mer type of FTPE. (Dai-el® Thermoplastic T-530)

estimated to be less than 100 nm was found to be approximately 25 nm on the average in diameter by small-angle X-ray scattering measurements (SAXS) [10]. Since the size of dispersed phase is very small, the block copolymers show an excellent transparency. In Table 30.1, the general properties of Dai-el® Thermo-plastic are listed [11]. They are primarily designed to have similar flexibility to conventional vulcanized fluoroelastomers, in addition to thermoplasticity. Both T-530 and T-550, having ETFE-based hard segments, exhibit higher melting points than T-630 whose hard segments are based on PVDF, and show good melt flowability. Tear strengths and tensile strengths are fairly good for T-530 and T-550. Elongations are very large for the three grades. This characteristic permits their fabrication into shrinkable tubing and films.

Moreover, the FTPEs show a good resistance to most chemicals and show excellent weathering durability. However, they swell or are soluble in polar solvents such as methyl ethyl ketone (MEK) and N,N-dimethylformamide (DMF). T-630 is suitable, in particular, for use in coatings and for fabric impregnation.

Dynamic mechanical properties of the block copolymers are shown in Figure 30.2. Two relaxations, which correspond to those of fluoroelastomeric soft segments, are observed. The larger ones correspond to the glass transitions. The profiles are similar to those of crosslinked fluoroelastomers. However, dynamic moduli (E') curves indicate that the block copolymers tend to be brittle or lose toughness around 140°C for T-530 and 110°C for T-630 respectively, although the temperatures are far below the respective melting points. Improving mechanical properties at increased temperatures is a problem yet to be solved, as it is for

Table 30.1. General properties of Dai-el® Thermoplastic

Property	Unit	Dai-el thermoplastic			Cured fluoroelastomer
		T-530	T-550	T-630	
Specific gravity	d_{25}	1.89	1.88	1.89	1.8 ~ 2.1
Hardness	(JIS A)	67	73	61	55 ~ 90
Melting point	°C	c 220	c 220	c 160	None
Melt flow index	g/10 min, 250°C, 10 kg	8 ~ 20	5 ~ 8	2 ~ 5	—
Decomp. start. temp.	°C (in air)	380	380	400	>400
Thermal conductivity	cal cm^{-1} s^{-1} °C^{-1}	3.6×10^{-4}	3.6×10^{-4}	3.6×10^{-4}	6.0×10^{-4}
Specific heat	cal g^{-1} °C^{-1}	0.3	0.3	0.3	0.3
Low temp. torsion test	(Gehman) T-50, °C	−9	−9	−10	−20 ~ −8
Tensile strength	MPa	12	17	4	7 ~ 22
Elongation at break	%	650	600	>1000	600 ~ 150
Tear strength	kN m^{-1}	28	30	21	17 ~ 25
Resilience	% (20 °C)	10	10	10	10 ~ 15
Friction coefficient		0.6	0.5	0.4	0.6 ~ 0.7
Tabor wear	mg/1000 rev				
CS-17	(1000 g)	2	2	2	5 ~ 70
H-22	(1000 g)	10	14	2	46 ~ 113
Compression set	% (50°C × 24 h)	11	13	80	5 ~ 27
	% (100°C × 24 h)	—	—	89	4 ~ 25
Electrical properties					
Volume resistivity	Ω cm	5×10^{13}	6×10^{14}	1×10^{15}	1×10^{13}
Breakdown voltage	kV mm^{-1}	14	14	16	9.3
Dielectric constant	(23°C, 103 Hz)	6.6	6.2	7.7	13.8
Critical surface tension	mN m^{-1}	20.5	—	19.6	—
Refractive index	n_D^{20}	1.357	—	—	—
Total luminous transmittance	% / 1 mm	87	—	84	—
Flammability	(Oxygen index)	66	100	75 ~ 100	—
Gas permeability N$_2$	cm^3 mm m^{-2} day^{-1} atm^{-1}	82	—	119	48
O$_2$		136	—	174	118
CO$_2$		111	—	211	109
He		1715	—	2120	1820

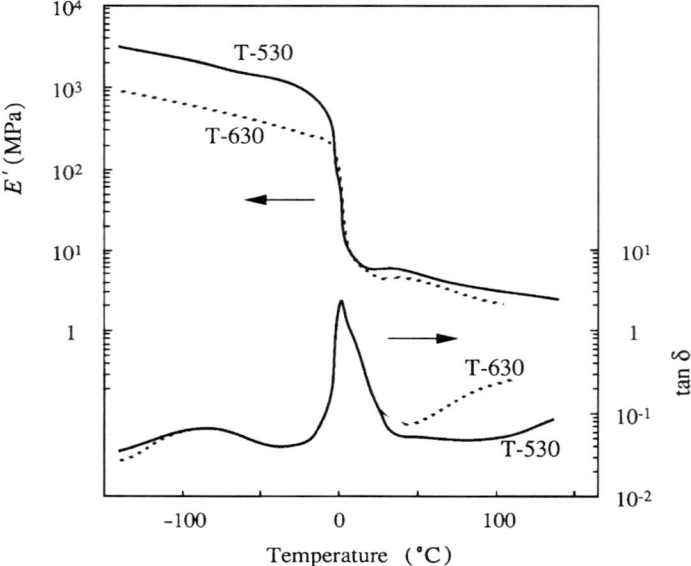

Figure 30.2. Dynamic mechanical properties of Dai-el® Thermoplastic

conventional TPE. The lower toughness of the FTPEs is probably due to the lower intermolecular cohesive energy in the crystalline domains of fluoroplastics, and also to their lower content in the block copolymers.

To improve toughness, irradiation treatments of mouldings are occasionally performed under inert atmosphere, rather than increasing the content of hard segments in the block copolymers. As illustrated in Table 30.2, irradiation increases their tensile strengths and significantly reduces their compression

Table 30.2. Gamma irradiation of Dai-el® thermoplastic

Dose (kGy)	T-530				T-550				T-630			
	M_{100}[a] (MPa)	TS[b] (MPa)	E[c] (%)	CS[d] (%)	M_{100} (MPa)	TS (MPa)	E (%)	CS (%)	M_{100} (MPa)	TS (MPa)	E (%)	CS (%)
0	1.9	10.8	650	—	2.2	16.7	600	—	1.2	2.0	>1000	100
10	1.9	12.7	680	65	2.2	17.7	610	98	1.4	5.3	934	94
30	1.9	14.7	620	57	2.2	19.6	630	95	1.1	12.2	690	80
100	1.8	19.6	580	34	2.2	19.6	500	50	1.2	12.9	450	39
150	1.8	17.7	500	23	2.2	19.6	450	40	1.7	9.5	240	26

[a]M_{100}: modulus at 100% elongation.
[b]TS: tensile strength.
[c]E: elongation at break.
[d]CS: compression set, P-24 O-ring, 25% compression, 70 h at 150°C.

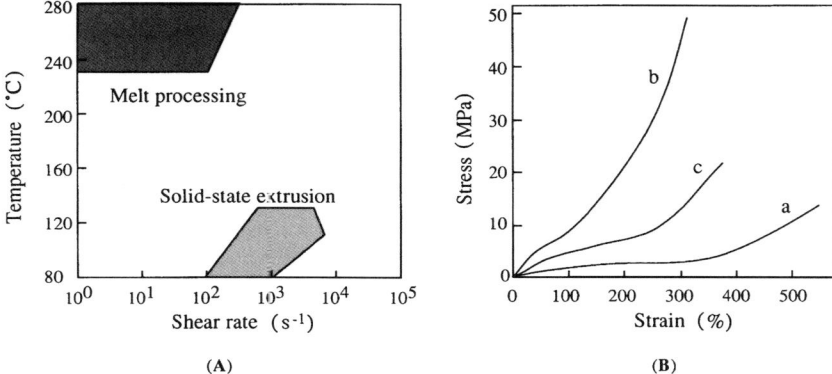

Figure 30.3. (A) Processing conditions of melt-processing and 'solid-state' extrusion for Dai-el® Thermoplastic T-530. (Reproduced with permission from reference 12. Copyright 1994 the Society of Polymer Science, Japan.) (B) Stress and strain curves of the original FTPE (T-530) (a), solid-state extruded T-530 (b), conventional cured fluoroelastomer filled with carbon black (c), at an elongation rate of 500% min⁻¹ and a temperature of 25 °C. (Reproduced with permission from reference 13. Copyright 1991 Kogyo Chosakai Publishing Co.) [13]

sets at 150°C. The appropriate irradiation accordingly expands their service temperatures substantially without impairing transparency and flexibility. 'Solid-state moulding', is also proposed [12] as another method of toughening. Small amounts of peroxides and cure agents such as triallylisocyanurate are blended with the block copolymer. The compound is then moulded far below the melting point. Figure 30.3 exemplifies extrusion conditions of this technique for T-530, compared with those of ordinary melt extrusion. Extrusion is preferably done at higher shear rates. The moulds are successively cured by heating above the decomposition temperatures of the peroxides. This technique is in principle adaptable to versatile processing methods such as compression, transfer, extrusion, injection and calendering. Characteristic features of cured mouldings by this technique are their excellent tensile properties as demonstrated in Figure 30.3.

3.2 GRAFT-COPOLYMERS

In contrast, in Cefral Soft®, the side chains of hard segments (PVDF) aggregate together to form crystalline domains, as illustrated in Figure 30.4. The amount of PVDF hard segments in the graft copolymers seems to be more than that of Dai-el® Thermoplastic T-630 having a similar hard segments, although this has not been described in detail. As a result, they show medium values in flexibility between fluoroelastomers and PTFE, but decreased transparencies. In Table 30.3, the general properties of Cefral Soft® are listed [14]. The melting points are

Table 30.3. General properties of CEFRAL SOFT® [Reproduced with permission from ref. 14]

Properties		CEFRAL SOFT	Fluoroelastomer	PTFE	PVDF	Method
Specific weight		1.78	1.80 ~ 1.85	2.13 ~ 2.19	1.76	ASTM D 792
Tensile strength	MPa	23 ~ 31.5	9 ~ 15	28 ~ 42	32 ~ 60	ASTM D 638
Elongation at break	%	460 ~ 500	190 ~ 420	300 ~ 450	50 ~ 200	ASTM D 638
Tensile modulus	MPa	80 ~ 99	—	410	860	ASTM D 638
Tear strength	$kN\ m^{-1}$	110 ~ 120	30 ~ 40	—	—	JIS K 6301
Izod impact strength with notch	$N\ m\ m^{-1}$	>600	—	160	110	ASTM D 256
Hardness	Shore D	40 ~ 50	~ 30	50 ~ 65	~ 80	ASTM D 2240
Impact resilience	%	28 ~ 35	5 ~ 15	—	—	JIS K 6301
Compression set		Fair	Good	Poor	—	
Gas permeability $\times 10^{-9}$ ($m^3\ m\ m^{-2}\ day^{-1}\ atm^{-1}$)	O_2	27	118	~ 400	5.4	JIS Z 1707
	N_2	9	48	~ 150	3.5	JIS Z 1707
Oxygen index		54	75 ~ 100	>95	43	JIS K 7201
Melting point	°C	162 ~ 165	—	327	165 ~ 170	DSC
Glass transition temperature	°C	−25	−19.5	—	−38	DSC
Decomposition starting temperature	°C	380	>400	>400	400	TGA
Maximum service temperature	°C	120 ~ 150	230	290	130 ~ 150	—
Brittleness temperature	°C	−55	−45	−270	−40	JIS K 6301
Water absorption	%	0.04	—	0	0.04	ASTM D 570
Chemical resistance Acids		Excellent	Good	Excellent	Excellent	—
Alkalies		Excellent	Excellent	Excellent	Excellent	—
Ketone, ester and the like		Poor	Poor	Excellent	Good	—
Other oils and solvents		Excellent	Excellent	Excellent	Excellent	—
Material which is easy to elute		Not included	Included	Not included	Not included	—

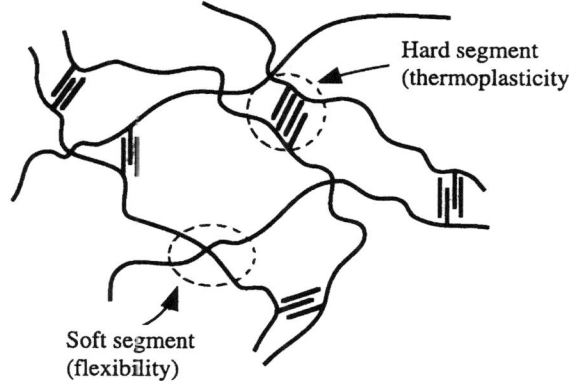

Figure 30.4. Schematic representation of the microstructure for the graft-copolymer type of FTPE (Cefral Soft®). (Reproduced with permission from reference 14.)

almost the same as that for PVDF. Chemical resistance is excellent, particularly to concentrated inorganic acids such as HCl, H_2SO_4, SO_3, HNO_3 and H_3PO_4. Gas permeability is much better than that of a conventional cured fluoroelastomer. In addition, it is soluble in polar solvents, since it is based on PVDF.

With respect to dynamically vulcanized FTPE composed of a fluoroelastomer and a polyamide, it has been reported that the maximum continuous use temperature is very high, between 177 and 200°C, however, there is a possible limitation in the resistance to methanol, gasohol and hot brake fluid [6].

4 APPLICATIONS

Major uses for FTPEs (Dai-el® Thermoplastic and Sefral Soft®) are for sealing materials in the chemical and semiconductor industries. These are mostly O-rings, V-rings, gaskets and diaphrams, which reflect their excellent chemical resistance and high purity (Figure 30.5). These fabricated articles are often cured by actinic radiation without adding any other components [15]. Tubing and inner linings of multi-layer hoses for corrosive gases or a ultra-pure water, and vessel liners for inorganic acids such as HF are also applications in these sectors [16].

In the electrical and wire/cable industries, FTPEs are used for wire coating and wire/cable sheathing, taking advantage of their flexibility and low flammability in addition to oil, fuel and chemical resistance [17,18]

In architecture, the flexibility and excellent weatherability of FTPEs make them suitable for tents and greenhouses. Laminates of Cefral Soft® films (Figure 30.6) on polyester fiber-reinforced poly(vinyl chloride) sheets have been described [16,19].

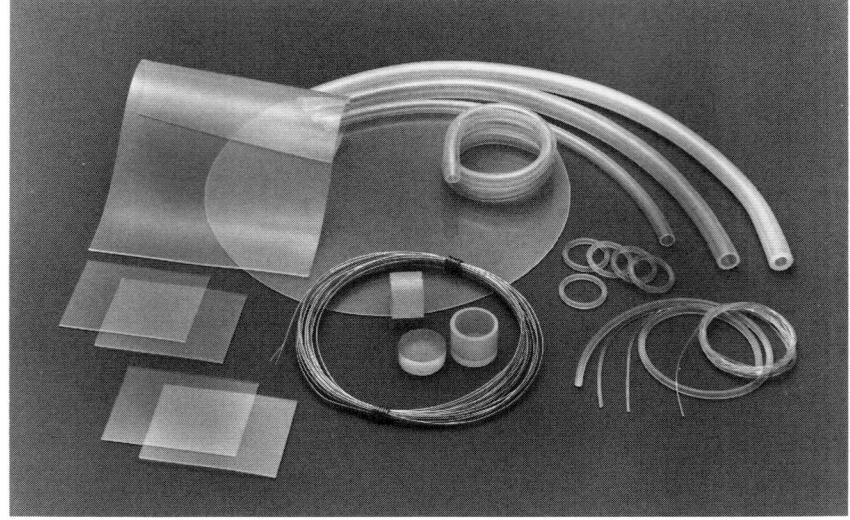

Figure 30.5. Various products of Dai-el® Thermoplastic. (Reprinted from reference 11.)

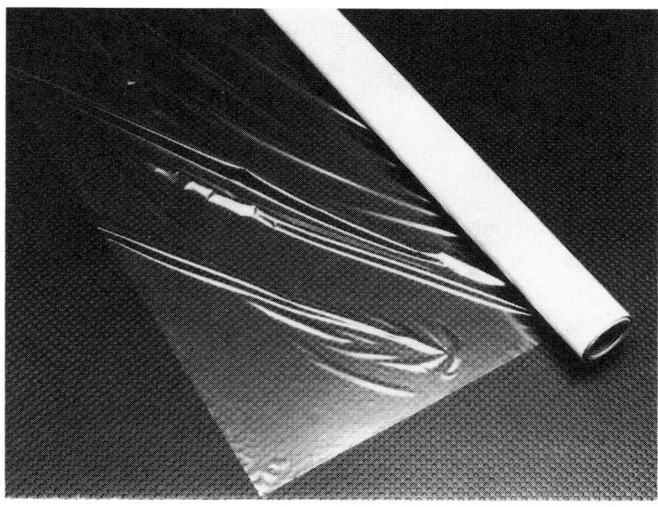

Figure 30.6. Cefral Soft® Film. (Reproduced with permission from reference 14.)

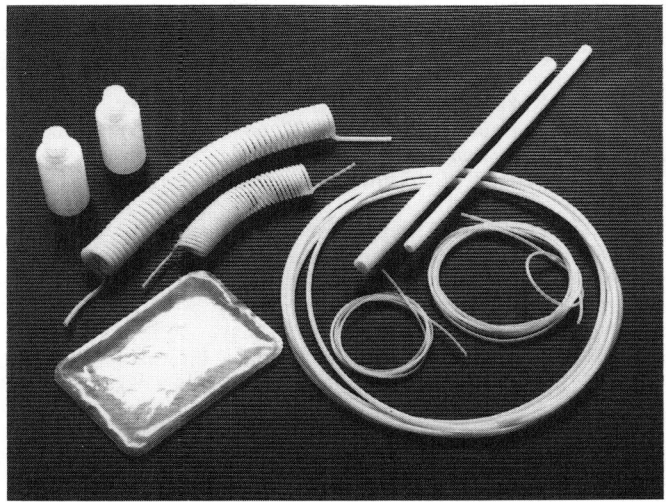

Figure 30.7. Various products of Cefral Soft® for food processing industry and for sanitary goods. (Reproduced with permission from reference 14.)

Since contamination resulting from extraction and dilution is minimized, FTPEs are safe to be used in the food processing industry and for sanitary goods. Figure 30.7 shows applications of Cefral Soft® for bottles, a package, tubing and pipes in these fields.

Other hopeful applications of FTPEs include use as modifiers for elastomers or plastics to improve their processibility or hardness [20]. The use of FTPE solution in DMF for preparing artificial leather is found in ref. 21. The resistance to fuels or gasohols may lead to the use of these materials in automotive fuel hoses. A method for fabrication of three-layered fuel hoses are the subject of ref. 22, where FTPE is used for an intermediate layer between polyamides outer and fluoroplastics inner layers.

REFERENCES

1. Snider, A. V., *Rubber World*, **152**, 90-1 (1965).
2. Tatemoto, M., *Kobunshi Ronbunshu*, **49**, 765-83 (1992). Oka, M. and Tatemoto, M., *Contemporary Topics in Polym. Sci.*, Vol. 4, Plenum Press, New York, 1985, pp. 763–77.
3. Tatemoto, M., *Intern. Polymer Sci. and Technol.*, **12**, 4, T/85-T/91 (1985), translated from Tatemoto, M., *Nippon Gomu Kyokaishi*, **57**, 761–7 (1984).
4. Kawashima, C., *Fusso Jushi Handbook*, Satokawa, T. ed., Nikkan Kogyo Shinbunsya, Tokyo, 1990, pp. 671–86.
5. Logothetis, A. L. and Stewart, C. W., US Patent 4 713 418 (Dec. 15, 1987).

6. Goebel, K.D. and Nam, S., *Proc. 3rd Int. Conf. Thermoplast. Elastomer, Mark. Prod.*, 55–81 (1990). Goebel, K.D., US Patent 5 371 143 (Dec. 6, 1994).
7. Minnesota Mining and Manufacturing Co., THV® Fluoroplastic Technical Information.
8. Kawashima, C., Kokai Tokkyo Koho JP 58-206615 (Dec. 1, 1983). Katoh, E., Kawashima, C. and Ando, I., *Polymer J.*, **27**, 645–50 (1995). Kawashima, C., *Proc. Symp. New Fluorine-Containing Materials*, pp. 52–9, Tokyo, May 9, 1985.
9. Tonelli, C., Trombetta, T., Scicchitano, M. *et al.*, *J. Appl. Polym. Sci.*, **59**, 311–27 (1996).
10. Tatemoto, M. and Ishiwari, K., unpublished results.
11. Daikin Industries, Ltd., Dai-el® Thermoplastic Product Information.
12. Tatemoto, M. and Sakaguchi, K., *5th SPSJ Intern. Polym. Conf., Prep.*, p. 250, Osaka, 1994.
13. Tatemoto, M., *Japan Plastics*, **42**, 71–6 (1991).
14. Central Glass Co., Ltd, CEFRAL SOFT® Technical Information and Technical Data.
15. Tatemoto, M., Tomoda, M., Kawachi, M. *et al.*, Kokai Tokkyo Koho JP 59-62635 (Apr. 10, 1984).
16. Kawashima, C. and Koga, S., *Japan Plastics*, **39**, 98–106 (1988).
17. Cheng, T. C., Kaduk, B. A., Mehan, A. K. *et al.*, US Patent 4 935 467 (June 19, 1988).
18. Kawamura, K., Kawashima, C. and Koga, S., US Patent 5 294 669 (Mar. 15, 1994).
19. Katsuragawa, S., Kawashima, C. and Masaki, T., US Patent 4 749 610 (June 7, 1988).
20. Kawashima, C. and Yasumura, T., Kokai Tokkyo Koho JP 59-30847 (Feb. 18, 1984). Ueda, Y., Kawachi, M. and Hosokawa, K., Kokai Tokkyo Koho JP 59-68363 (Apr. 18, 1984). Kawashima, C., Minegishi, S., Ogasawara, S. *et al.*, US Patent 4 748 204 (May 31, 1988).
21. Katsuragawa, S., Kawashima, C., Shiga, Y. *et al.*, Kokai Tokkyo Koho JP 1-22547 (Jan. 25, 1989).
22. Kawashima, C., Koga, S., Nakahata, S. *et al.*, US Patent 5 441 782 (Aug. 15, 1995).

31

Thermoplastic Copolymers of Vinylidene Fluoride

CLAUDE TOURNUT
Elf Atochem Centre de Recherche Rhône-Alpes, Pierre-Benite, France

1 BACKGROUND

Among the fluoromonomers, vinylidene fluoride (VDF) is one of the most important and has been offered for many years as homopolymers and copolymers essentially with other fluoromonomers. These are known as elastomers or thermoplastics according to the nature and the content of the comonomers. The PVDF homopolymer is extensively described in Chapter 25. The present chapter is dedicated to the copolymers containing at least 50 mol % of vinylidene fluoride and which are thermoplastics. The fluoroelastomers containing VDF are described in Chapters 2 and 32.

The first copolymers of vinylidene fluoride were described in a patent as early as 1944 [1], but the first commercial copolymers were elastomers obtained by copolymerisation of VDF and approximately 30–50 mol% of chlorotrifluoroethylene (CTFE) [2] or hexafluoropropene (HFP) [3] followed by terpolymer including tetrafluoroethylene (TFE) [4].

The VDF/CTFE or VDF/HFP copolymers containing less than 15 mol% of comonomer have thermoplastic properties [5] and are known as flexible PVDF. They were launched in the early 1980s and represent today the most important production among the thermoplastic copolymers of VDF.

Other important copolymers of VDF are produced by copolymerisation of VDF with tetrafluoroethylene (TFE) or with TFE plus HFP. With these three monomers, properties and applications depend on the composition. Only the copolymers containing at least 70% in weight of VDF are presented in this chapter.

The use of copolymers of vinylidene fluoride and trifluoroethylene is rather limited, but has the advantage of a high added value for their piezoelectric

Modern Fluoropolymers. Edited by John Scheirs
© 1997 John Wiley & Sons Ltd

and pyroelectric properties. While the copolymerisation of VDF with monomers containing hydroxy or carboxylic groups gives functionalized copolymers of VDF for coatings applications.

2 COPOLYMERS VDF–HFP AND VDF–CTFE: FLEXIBLE PVDF

This family is the most important among VDF thermoplastic copolymers. They were first described in 1955 [5] but were developed commercially in the early 1980s when the plenum market was open to fluoropolymers in the USA (see section 2.3.1). PVDF was the cheapest material for this application, but its high flexural modulus (an advantage for several applications) was a disadvantage in this area.

Therefore, PVDF producers began to copolymerise VDF with HFP or CTFE to provide flexible PVDF. Today, at least 25 grades are offered on the market depending on the nature of the comonomer, its content and its distribution, molecular weight, form (powder or pellets) or the presence of additives to improve specific properties. Table 31.1 provides a list of the most important copolymers [6–8].

2.1 PREPARATION

VDF can copolymerise with HFP or CTFE in a similar polymerization process to that used to prepare the homopolymer (see Chapter 25 and ref. 9). Both emulsion and suspension processes can provide these copolymers. A random copolymer (called hereafter 'homogeneous copolymer') can be obtained by feeding into the polymerisation reactor a mixture of the two comonomers instead of VDF alone. To obtain a more constant composition of the copolymer, it is useful to adjust the composition of the monomers in the polymerisation vessel according to their reactivity ratios. In fact from this point of view, there is an important difference between HFP and CTFE, the reactivity ratios for the two systems being.

$$\text{VDF/CTFE} \quad r_1 = 0.13 \quad r_2 = 3.73$$
$$\text{VDF/HFP} \quad r_1 = 2.45 \quad r_2 = 0.00$$

Therefore, for a given composition of the monomers, with HFP, VDF disappears quicker than HFP and with CTFE, this comonomer disappears quicker than VDF.

This property is used to prepare heterogeneous or 'core-shell' copolymers. Thus, with CTFE, it is possible to produce copolymers which present a very good compromise between mechanical and thermal properties (shown in section 2.2) by introducing CTFE only during the first part of the polymerisation [10] or only in the first load of monomers [11]. In the first part of the polymerisation, and according to the composition of the mixture of monomers, the copolymer produced could be an elastomeric one. When the introduction of CTFE is stopped,

Table 31.1.   Main Commercial VDF Thermoplastic copolymers

VDF–HFP random copolymers			
Producer and trade mark	Grades	Melting point (°C)	Flexural modulus (MPa)
Elf Atochem			
Kynar Flex®	2850	155–160	1100–1200
Kynar Flex®	2800, 2820 2821, 2822[a,b]	141–145	450–630
Kynar Flex®	2750[b]	132–138	340–400
Solvay SA			
Solef®	11008, 11010 11012[a,b]	158–160	850–1000
Solef®	21508, 21510[a]	132–133	350–440
Kureha Chemical Industry Co., Ltd			
KF Polymer®	1200	172	1320
	2000	158	540
	2300	151	600
VDF–CTFE heterogeneous copolymers			
Elf Atochem			
Kynar®	5050	167	800
Solvay SA			
Solef®	31508	168	425
Kureha Chemical Industry Co., Ltd			
KF Polymer®	1500	167	910

[a]The different grades have the same composition but different molecular weights.
[b]Grades with improved limiting, oxygen, index. (LOI) are available.

the composition of the copolymer moves from an elastomeric to a thermoplastic one and finally to an almost PVDF homopolymer. Thus by adjusting the average composition of the copolymer and the ratio of elastomeric to thermoplastic part, it is possible to tailor the properties of the final copolymer.

With HFP, such copolymers can also be produced but, in this case, the polymerisation is started with VDF alone (or with a small amount of HFP), with a greater amount of HFP being introduced quickly in the second part of the polymerisation [12].

2.2   PROPERTIES

Keep in mind that these copolymers were developed first to provide a flexible PVDF and the first target was to reduce the flexural modulus. But, at the same

Table 31.2. Main Properties of VDF/HFP Copolymers Versus HFP Content

Properties	Standard	Unit	Comparative values for PVDF homopolymer	Average values for VDF/HFP copolymers with HFP weight content		
				5%	10%	15%
Specific gravity	ISO R 1183 D		1.77–1.79	1.78	1.78	1.78
Refractive index			1.42	1.42	1.41	1.41
Mechanical properties at 23°C						
Tensile properties	ISO R 527-2 and ISO 12086					
Tensile stress at yield		MPa	49–52	32–39	23–27	15–21
Elongation at yield		%	10	10–12	10–12	12–15
Tensile stress at break		MPa	35–45	23–26	24–42	28–42
Elongation at break		%	20–100	200–300	200–450	250–500
Flexural modulus	ISO 178	MPa	1800	1100	700	400
Izod notched impact	ISO 180	J/m	160/100	230	680	1400
Shore hardness	ISO 868	D	76–80	70	65–70	60–65
Thermal properties						
Crystalline melting point	ISO 12086	°C	169–174	155–160	140–145	132–138
Heat of fusion		J/g	55–65	35	24	17
Thermal conductivity		W/m K	0.17	0.17	0.17	0.17
Specific heat		J/g K	1.2	1.2	1.2	1.2
Vicat softening point (B50)	ISO 306	°C	140	90	70	60
Heat deflection temperature	ISO 75	°C	104–108	55	40–50	40
Glass transition temperature		°C	−38	−38	−36	
Electrical properties at 23°C						
Dielectric constant for 1 kHz			8–10	7.6	7.9–10	10.6
tg delta loss factor for 1 kHz			0.005–0.019	0.02	0.02–0.04	0.05
Fire resistance						
Limit oxygen index	ISO 4589	%	44	44	44	44
Inflammability rating (0.8 mm)	UL 94	Class	VO	VO	VO	VO

time, some properties are altered, whereas others are not modified in comparison with the homopolymer, as we can see in Table 31.2.

In terms of standards, a designation system and basis for specification is provided by ISO 12086-95 Parts 1 and 2 [13,14].

Thus for a VDF/HFP random copolymer, a standard grade for extrusion, sold as natural granules, and having:

- a melting point between 130 and 140 °C
- a MFR between 2 and 5 when tested at 230 °C and 5 kg load
- a tensile yield stress between 15 and 20 MPa
- elongation at break between 600 and 800%
- tensile modulus less than 500 MPa
- density between 1.7 and 1.8

the designation is

ISO 12086 - VDF/HFP-R, EGN, M.5E5.B.K.A.C,,,

The significance of the different letters or numbers used in the designation is given in the various tables of Part 1, whereas the methods used for the determination of the properties are given in Part 2 of the standard.

2.2.1 Properties Relatively Independent of Polymer Composition

As the comonomers are fluoromonomers, the thermoplastic copolymers of VDF with HFP or CTFE maintain the very good UV resistance and weatherability of the fluoropolymers. Their low flammability is also unmodified (UL 94 V0 for 0.8 mm) as well as the smoke emission. Nevertheless, these properties which are key requirements for the plenum application can be further improved with additives like calcium molybdate [15] or aluminium silicates [16].

The molecular weight and molecular weight distribution are not greatly modified by the copolymerisation and are more tailored by the polymerisation parameters (eg. pressure, temperature, initiator, chain transfer agent). Therefore, the rheological properties of a given copolymer are very close to those of the homopolymer produced with the same polymerisation parameters.

The thermal stability of VDF copolymers is also similar to that of the PVDF, even with CTFE which contains chlorine. The thermal decomposition of the VDF copolymers begins at over 350 °C, far higher than the melting point (less than 170 °C) allowing a wide processing window.

The electrical properties of VDF copolymers can also be considered as being independent of composition with these polymers having a relatively high dielectric constant and loss factor in comparison with other fluoropolymers.

2.2.2 Properties Dependent on the Composition

The introduction of a bulky atom (such as Cl) or a group (such as CF_3) in the chain of the copolymer hinders their crystallisation which can be measured by the heat of fusion. The crystallinity decreases when the HFP comonomer content increases (see Table 31.2). At the same time, the mechanical and thermal properties are affected as well and to a certain extent, the chemical resistance.

2.2.2.1 Mechanical Properties

The decrease in crystallinity with increasing comonomer leads to the required decrease in the flexural properties and also to a decrease of the tensile properties. For a given comonomer (for example CTFE), the flexural modulus decreases regularly when the CTFE content increases, as we can see in Table 31.3 and Figure 31.1. It is amazing that the structure of the copolymer has almost no influence on the modulus, i.e. a core-shell (heterogeneous) copolymer has the same modulus as a random (homogeneous) copolymer having the same comonomer content [17].

As expected, the tensile properties and the compression modulus vary like the flexural modulus. Concerning the tensile properties, the tensile strength at yield decreases with the comonomer content, whereas the elongation at break increases and the elongation at yield are almost constant.

The impact resistance at room temperature is improved by copolymerisation, particularly in the case of heterogeneous copolymers.

2.2.2.2 Thermal Properties

The crystalline melting point of VDF copolymers with HFP or CTFE depends greatly on their composition and structure. Thus, for random copolymers, the melting point decreases with the comonomer content (Table 31.3 and Figure 31.2), whereas in copolymers with a core-shell structure, the melting point does not differ greatly from the homopolymer, even at high comonomer contents (Figure 31.2). Consequently, with the core-shell process, it is possible

Table 31.3.   Melting point and flexural modulus of VDF/CTFE copolymers

% CTFE	Homogeneous copolymers		Heterogeneous copolymers	
	Melting point (°C)	Flexural modulus (MPa)	Melting point (°C)	Flexural modulus (MPa)
0	169	1800		
5	158	1240		
8	150	900	168	850
12			164	560
14	142	435	163	440

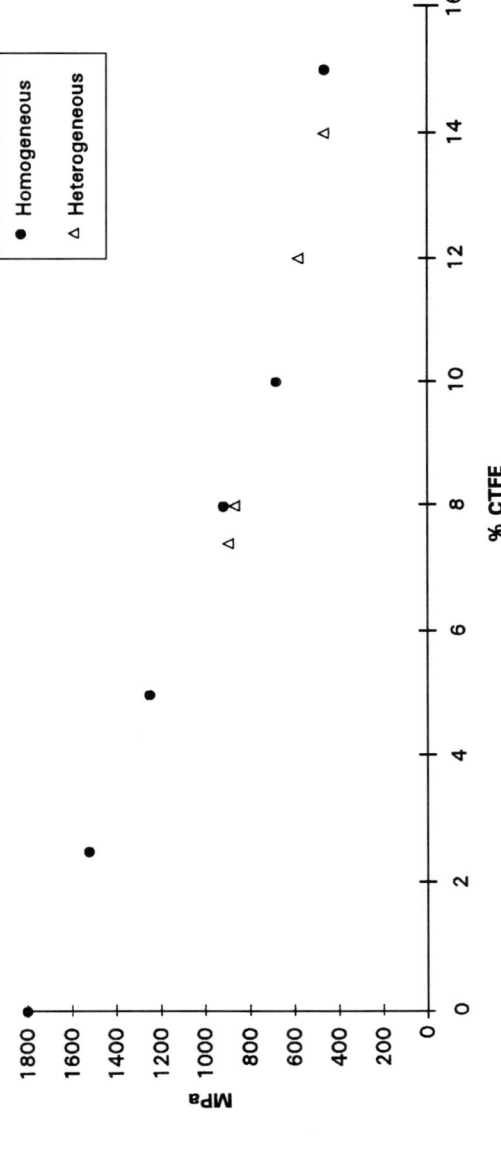

Figure 31.1. Flexural modulus vs CTFE content

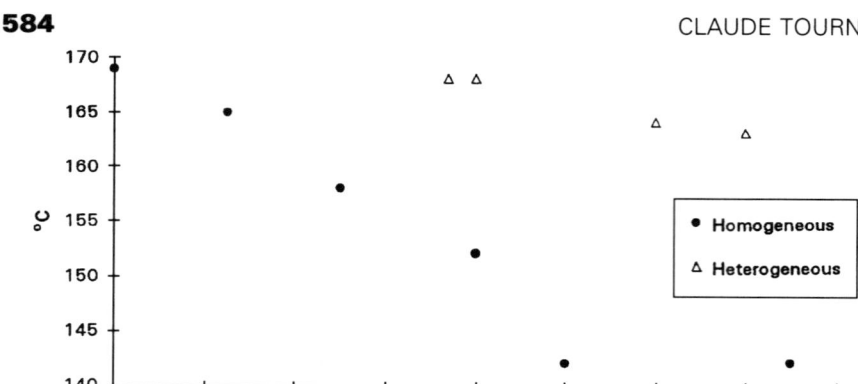

Figure 31.2. Melting point vs CTFE content

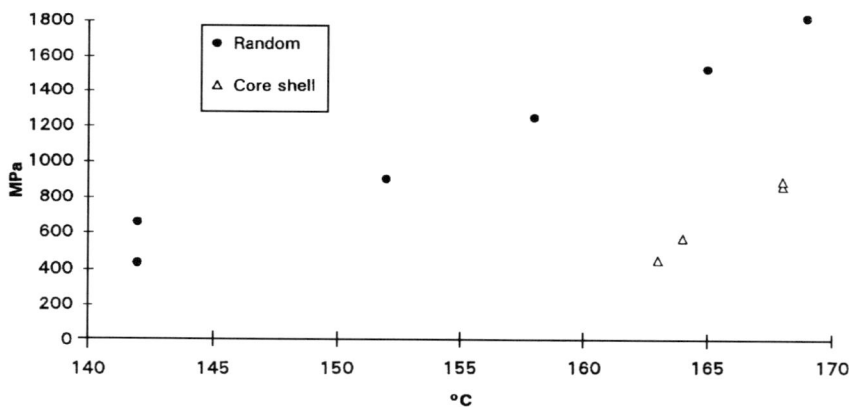

Figure 31.3. Flexural vs melting point for VDF–CTFE copolymers

to get copolymers with low flexural modulus and high melting point, as we can see on Figure 31.3. Such copolymers comply with the 150 °C rating from UL (Underwriters Laboratories) for the plenum applications (see below).

Other thermal properties like Vicat softening point or heat deflection temperature vary more with the comonomer content rather than with structure. The glass transition temperature, however, is almost invariant with comonomer content.

2.2.2.3 Chemical Resistance

On average, VDF–HFP and VDF–CTFE have the same excellent chemical resistance as the homopolymer PVDF for a variety of chemical reagents like inorganic acids, salt solutions, and hydrocarbons. Nevertheless their decreased crystallinity leads to some important differences. Generally speaking, the copolymers are less sensitive to stress cracking than the homopolymer, particularly in alkaline solutions or in highly concentrated sulphuric acid. Table 31.4 provides a comparison between a homopolymer (Kynar® 1000 HD) and a heterogeneous VDF–CTFE copolymer (Kynar® 5050 HD) for some critical reagents. At the same time, copolymers show higher swelling and higher permeability in organic solvents.

2.2.2.4 Crosslinking

By electron-beam irradiation it is possible to easily crosslink VDF–HFP copolymers. Crosslinking improves both the high-temperature performance properties and the creep resistance.

Table 31.4. Comparative chemical resistance of Kynar® 1000 HD (homopolymer) and Kynar® 5050 (VDF/CTFE copolymer)

Reagent	Temperature (°C)	5050	1000	Comments
Acids				
HCl 37%	90	+	+	
HF 70%	50	+	+	
HNO$_3$ 97%	90	–	–	Very high swelling especially, with copolymer
H$_2$SO$_4$ 68%	90	+	+	
H$_2$SO$_4$ 90%	90	+	+	
H$_2$SO$_4$ 99.2%	50	+	–	Stress-cracking with homopolymer
H$_2$SO$_4$ 99.2%	90	–	–	
Acetic acid	90	–	–	
Halogens				
Bromine	50	+	+	
Solvents				
Cyclohexanone	90	–	–	
Trichloroethylene	90	0	+	High swelling with copolymer
Perchloroethane	90	0	+	High swelling with copolymer
Tetrachloroethane	90	+	+	
Ethanol	90	+	+	
Propanol	90	+	+	
Bases				
NaOH pH 13	90	+	–	Stress cracking with homopolymer
NaOH pH 14	90	+	–	Stress cracking with homopolymer
NaOH 10%	90	+	–	Stress cracking with homopolymer
NaOH 20%	90	+	–	Stress cracking with homopolymer
NaOH 40%	50	+	+	
Bleach + base	40	+	–	Stress cracking with homopolymer
Bleach (150 g/l)	90	+	–	Stress cracking with homopolymer

Note: + Means that the material is resistant to the reagent; – means that it is not resistant to the reagent; 0 means that its resistance is limited and must not be subject to stress.

2.3 PROCESSING AND APPLICATIONS

Because of their good thermal stability and rheological properties, VDF–HFP and VDF–CTFE copolymers can be processed like PVDF homopolymers with conventional equipment by compression or injection moulding, and especially by extrusion. Special alloys are not needed as is the case with other thermoplastic fluoropolymers. Also, the semi-finished products made of VDF copolymers can be welded like the homopolymer and also to the homopolymer if necessary.

The main applications of PVDF copolymers result directly from their major properties: flexibility and resistance to stress cracking. Given this, electrical applications and chemical process industry (CPI) are the primary areas of use.

Typically VDF–HFP and VDF–CTFE can be extruded by using an extruder with a length at least 20 times the diameter (20 L/D) and a compression ratio of 3. The typical temperature profiles vary from about 200 °C at the feed zone, 225–240 °C at the transition zone and metering zone and 250 °C at the die.

2.3.1 Electrical Applications

Due to their good capability to crosslink, the VDF–HFP copolymers are used in the wire and cable industry to provide crosslinked heat-shrinkable items. Nevertheless, the most important application of the flexible PVDF is the so-called 'plenum' market which appeared in the USA at the end of the 1970s. According to a regulation published in 1974 by the NEC (National Electric Code), it was mandatory to install in the plenum of the buildings only wire and cable protected by a metallic pipe to prevent propagation of fire and smoke in case of emergency. One exception was made for cables which could demonstrate that, by burning in standard conditions described in UL 910 (known as the modified Steiner tunnel test), the fire propagation and the smoke emission were low enough to comply. At the beginning of the 1980s, only some fluoropolymers (PVDF, ECTFE, FEP) were able to pass the UL 910 test. Among these, the PVDF was the cheapest, but it was considered too rigid, and therefore it opened the door (of the market) to the flexible copolymers. Infact, due to their high dielectric constant and loss factor, the VDF copolymers are generally used only for the secondary insulation (jacketing), whereas the primary insulation is provided by FEP or ECTFE. In the 1990s, some PVC compounds were able to meet the UL 910 requirements and they are now used on this market at least for the less sophisticated cables. Therefore, the market for the flexible PVDF is decreasing. Nevertheless, it remains very important for this family of copolymers.

2.3.2 Chemical Applications (CPI)

VDF/HFP and VDF/CTFE copolymers are presently used in CPI where they successfully complete the numerous applications of the PVDF homopolymer because of their better flexibility and better resistance to stress cracking in specific reagents like sodium hydroxide, chlorine (bleaching agents), sodium chlorate, and high concentrated sulphuric acid. Even though other fluoropolymers (like ECTFE or FEP) can be used for such applications, VDF copolymers are far cheaper than their competitors.

The flexibility of VDF copolymers is particularly advantageous for the lining of metallic tanks, for example road tankers. Constructions made with Kynar Flex® VDF copolymers can be reinforced by fibre-reinforced polyester. Such constructions are shown on Figures 31.4 and 31.5. Figure 31.4 shows a chlorine neutralising column, diameter 2 m, height 18 m, made with 4 mm thick sheets in Kynar Flex® 2850, reinforced by FRP and used to absorb chlorine in a 20% solution of sodium hydroxide to produce sodium hypochlorite at a temperature of 60 °C.

Figure 31.4.   Chlorine neutralising column made in Kynar Flex® 2850 reinforced with with FRP. Diameter 2 m; height 18 m. (Reproduced with permission)

Figure 31.5. Inside view of a 110 m³ container for sodium chlorate solution made from Kynar® 5050. (Reproduced with permission)

Figure 31.5 shows the inside of a 110 m³ container made from Kynar® 5050 reinforced with FRP used to store sodium chlorate solution. In this picture one can see the welding lines.

In the CPI, VDF copolymers are also used as injection-moulded items (tower packing), rotomoulded containers with high impact resistance and flexible pipes.

2.3.3 Other Applications

2.3.3.1 Processing Aid for PE

The VDF/HFP thermoplastic copolymers can be used like VDF–HFP elastomers as processing aids in the extrusion of linear low-density polyethylene (LLDPE).

The copolymer is first compounded with PE to produce a masterbatch and this is then added to the LLDPE at a level of 200–800 ppm. The use of VDF copolymers improves the quality of the blown films and allows then to be extruded at high speeds without melt fracture. The clarity of the film is also improved.

2.3.3.2 Li Batteries

To provide rechargeable batteries for the numerous new portable electric or electronic tools, battery producers are developing a new generation of Li ions batteries. The electricity is produced by exchange of Li^+ ions between $LiMO_2$ salts (M being Ni, Co, Mn) at the positive electrode and carbon (graphite) at the negative electrode. In such constructions, VDF/HFP copolymers are used as a binder for the Li salts as well as a separator in some constructions [18]. This market emerged only in 1995.

3   VDF–TFE AND VDF–TFE–HFP COPOLYMERS

The two most important fluoromonomers : tetrafluoroethylene (TFE) and vinylidene fluoride (VDF) are copolymerizable across the entire composition range. The reactivity ratios are very close for the system VDF/TFE: $r_1 = 0.73$; $r_2 = 0.75$ respectively. These copolymers can be made by using emulsion or suspension processes developed for VDF homopolymerization. Solution polymerization of VDF and TFE is also possible.

When the content of TFE is increased in the copolymers, the melting point decreases from about 170 °C for the PVDF homopolymer to about 120 °C for a copolymer containing roughly 20 mol% of TFE and increases after this composition.

As a matter of fact, at present, only two compositions have found industrial applications. A copolymer with 5% by weight of TFE is produced in Russia and is used in place of the PVDF homopolymer despite limited service temperature performance in relation to the lower melting point. The most important copolymer has a 20% TFE content and is marketed under the trade names Kynar® SL or Kynar® 7200. The main properties are given in Table 31.5 in comparison with the PVDF homopolymer. An important difference between these copolymers and PVDF is the solubility which is greatly improved in several solvents shown in Table 31.6. This solubility and the relatively low melting temperature allow applications in the paint industry to prepare coatings having roughly the same properties as PVDF coatings, but with the possibility of curing at lower temperature, near the melting point of the copolymer.

Kynar® 7200 can also be easily melt processed by standard extrusion and moulding techniques into film, sheet, tube, pipe, cable jackets and other standard or special shapes; the processing temperatures are normally within the 190–260 °C range. By adding HFP or CTFE to the mixture of VDF and TFE monomers, the crystallinity of the copolymer is reduced and solubility in solvents

Table 31.5. Main Properties of VDF/TFE and VDF/TFE/HFP Copolymers

Properties	Standard	Unit	Comparative values for PVDF homopolymer	Average values for copolymers	
				VDF/TFE Kynar® 7200	VDF/TFE/HFP Kynar® 9300
Specific gravity	ISO R 1183 D		1.77–1.79	1.88	1.85
Refractive index			1.42	1.4	1.384
Mechanical properties at 23°C					
Tensile properties	ISO R 527-2				
Tensile stress at yield	and	MPa	49–52	15–19	7–8
Elongation at yield	ISO 12086	%	10		29
Tensile stress at break		MPa	35–45	32–45	15–16
Elongation at break		%	20–100	500–800	1000
Flexural modulus	ISO 178	MPa	1800	600–700	130
Izod notched impact	ISO 180	J/m	160/100	1100	
Shore hardness	ISO 868	D	76–80		
Thermal properties					
Crystalline melting point	ISO 12086	°C	169–174	122–126	87–93
Heat of fusion		J/g	55–65	13–21	16–19
Heat deflection temperature	ISO 75	°C	104–108	39–43	
Electrical properties at 23°C					
Dielectric constant for 1 kHz			8–10	7.2	
tg delta loss factor for 1 kHz			0.005–0.019	0.013	
Fire resistance					
Limit oxygen index	ISO 4589	%	44	44	44
Inflammability rating (0.8 mm)	UL 94	class	VO	VO	VO

Table 31.6. Solubility of PVDF and its Copolymers in organic solvents at room temperature [25]

Solvent	Kynar® 500	Kynar® SL	Kynar® ADS
Hydrocarbon	0	0	0
Alcohol	0	0	0
Acetone	0	++	++
Methyl ethyl ketone	0	++	++
Methyl isobutyl ketone	0	+	+
Cyclohexanone	0	+	++
Isophorone	0	++	++
Ethyl acetate	0	++	+
Tetrahydrofuran	0	+	++
1-4-dioxane	0	+	++
Dimethyl acetamide	++	++	++
Dimethyl formamide	++	++	++

Note: ++ Good solubility; + partially soluble; 0 : no solubility.

Table 31.7. Paint formulation for Kynar® 500 and Kynar® ADS [25]

Components	Homopolymer Kynar® 500	Copolymer Kynar® ADS
Kynar® 500[a]	23.8	
Kynar® ADS[a]		13.7
Paraloid® B 44[b]	21.7	14.6
Isophorone	28.1	
Dimethyl phthalate	4.8	
Methyl isobutyl ketone		39.8
Cyclohexanone		15.3
Diacetone alcohol		6.1
TiO$_2$–Ti pure R960®[c]	21.6	10.5
Solid content (%)	54	30
Processing coil coating on aluminium		
Metal temperature (°C)	240/255	23
Duration of curing	45–60 s	8 h/one week[d]
Primer thickness (μm)	5/7	
Finish thickness (μm)	20/25	

[a]Elf Atochem.
[b]Rohm & Haas: PMMA in solution at 40%.
[c]DuPont de Nemours.
[d]Dry after 8 h, complete hardness after one week at room temperature.

is increased. (see Table 31.6). The reduction of the melting point also enables the preparation of paints that dry at room temperature. The commercial product for this application is the Kynar® ADS, which is also available in the form of pellets under the trade name Kynar® 9300. Its composition is roughly 72/18/10 of VDF/TFE/HFP respectively. Some properties are given in Table 31.5.

In the coatings industry, the copolymers VDF/TFE or VDF/TFE/HFP are generally formulated with acrylic resin as is the case with PVDF homopolymer (see Chapter 14) and with solvents capable of dissolving Kynar® ADS. In fact, PVDF is only dispersed in the solvents, whereas the copolymer is actually dissolved.

Table 31.7 provides a comparison of the formulations and application made with PVDF homopolymer (Kynar® 500) and with the VDF/TFE/HFP copolymer (Kynar® ADS). The use of the copolymer allows repair and refurbishment of items already coated with PVDF whilst maintaining excellent weathering resistance.

By changing the composition of the VDF/TFE/HFP terpolymers some other commercial products are obtained which, according to their composition, are elastomers or thermoplastics like THV described in the Chapter 13.

4 VDF-TRIFLUOROETHYLENE COPOLYMERS

Trifluoroethylene (TrFE) polymerises easily and can copolymerise in all proportions with VDF. In fact, at present, only the VDF–TrFE copolymers containing between 17 and 50 mol% of TrFE have a small commercial development limited to a couple of tons per year. These copolymers are very expensive because of the high price of the monomer. Nevertheless, they have found a market for their piezo and pyroelectric properties [19]. It is known that the PVDF crystallised in its so called beta form presents such piezoelectric properties. Usually, PVDF crystallises in its alpha form and can only give the beta form by stretching at moderate temperature. By introducing enough of TrFE (17%) in the PVDF chain, the copolymer directly gives the beta crystalline form from the molten state. Therefore it is possible with these copolymers to fabricate piezo electric items in any required form, by injection moulding for example.

5 VDF BLOCK COPOLYMERS

In 1984 Daïkin Industries Ltd presented new fluorinated block copolymers comprised of an elastomeric VDF/HFP copolymer and a PVDF homopolymer or an ethylene-tetrafluoroethylene copolymer. The combination of a soft segment (the elastomer) and a hard segment (the thermoplastic) gives the thermoplastic elastomers. This copolymer is marketed under the trade name Dai El® Thermoplastics [20] (see Chapter 30).

5.1 PREPARATION

The fluorinated thermoplastic elastomers are prepared by a two-step process using iodine transfer polymerisation [21]. An alkyl diiodine derivative initiates the copolymerisation of the VDF and HFP to give the first elastomeric block.

VDF is then introduced alone to produce the hard segment according to the following reactions:

$$IRI + \text{monomer A} \longrightarrow I(A) \times R(A) \times I$$

$$I(A) \times R(A) \times I + \text{monomer B} \longrightarrow I(B)y(A) \times R(A) \times (B)y$$

where A is a mixture of VDF and HFP and B is VDF in the case of Dai El® T-630 or ETFE in the case of Dai El® T-530 or 550.

5.2   PROPERTIES

Dai El® T-630 has a specific gravity of 1.89, a good transparency, a melting point of about 160 °C and its decomposition starts at 400 °C. The tensile strength for the uncured polymer is low (2 MPa) with a high elongation at break (>1000%). These properties can be improved by curing. Its chemical resistance is similar to those of the PVDF homopolymer, but it swells in several solvents and is soluble in ketones as well as in dimethylformamide. The dielectric constant is 7.7 for 1 kHz at 23 °C and decreases when the frequency increases.

5.3   PROCESSING AND APPLICATIONS

The thermoplastic elastomer can be processed by any conventional process without curing or can also be cured like the elastomers. Compression moulding can be made at 200 °C and extrusion from 200 to 230 °C.

Any curing method used by fluoroelastomers can be used by Dai El® thermoplastics (see Chapter 5). Peroxide cures and polyol cures are recommended. The curing is made in two steps : 15–20 min under press at 160–170 °C followed by heating in an oven at 180 °C (peroxide curing) or 230 °C (polyol curing). Dai El® thermoplastics can be blended with conventional fluoroelastomers. Coatings and impregnation processes can be made with the polymer in solvents.

The main applications are: tubes, seals, coatings, diaphragm in chemical industry, biochemistry, food industry, electronics, semiconductors, engineering construction.

6   FUNCTIONALIZED COPOLYMERS OF VDF

Thanks to their combination of high weathering resistance and low surface energy, PVDF fluoropolymers are well suited for very high quality coatings. In this field the PVDF Kynar® 500 is the benchmark for the top level of quality in coil and spray coatings. The drawback of these coatings is the necessity to heat at a high temperature (240 °C) during the coil coating or spray coating process. The copolymers VDF/TFE or VDF/TFE/HFP on the other hand, allow applications at low temperature (see Table 31.3), but these coatings are more sensitive to solvents. Recently, new crosslinkable copolymers of vinylidene fluoride were presented by Elf Atochem [22–25]. The novelty of these copolymers is that they have a functional group, like hydroxy or other group, able to react to the conventional reagents used in the paints industry such as isocyanates or melamine to give coatings with the same ageing properties as the non-functionalized copolymers, but with a higher hardness and better solvent resistance.

6.1 PREPARATION OF FUNCTIONALIZED COPOLYMERS OF VDF

Such copolymers can be prepared by copolymerisation of VDF (optionally with another fluoromonomer like TFE or CTFE) with a monomer containing a functional group. Allylic monomers with hydroxy or epoxy group like 3-allyloxyethanol, 3-allyloxypropanediol (AOPD), allyl glycidylether are all available at a reasonable price and do not inhibit the polymerisation of the fluoromonomers.

Generally speaking, these monomers are soluble in water and the polymerisations are conducted in solution using one solvent (or a mixture of solvents) like alcohols, esters or ketones and with a percarbonate or perpivalate as initiator.

Depending on the solvent and the solids content, the final solution can be concentrated or used directly for the preparation of the paint formulation. Optionally, the copolymer can be transferred into another solvent which is better adapted to the desired paint or varnish.

6.2 PROPERTIES AND APPLICATIONS

The solvents used for the polymerisation act as a chain transfer agent giving a copolymer with a relatively low molecular weight ($\overline{M}_n$ from 5000 to 10 000) and allowing a highly concentrated solution with low viscosity. But with such molecular weights the copolymers have poor mechanical properties by themselves and have to be crosslinked to realize their full potential. Crosslinking is achieved through conventional crosslinking agents used in the paints industry such as isocyanates or melamines. Because of their reactivity, the functionalized copolymers are used as two-component systems: resin and curing agents. For ambient temperature coatings, it is necessary to use an isocyanate as hardener and the paint is supplied as a two-pack composition. Complete curing is achieved after one week at room temperature.

When a stoving finish is preferred, a blocked isocyanate or a suitable melamine can be used as the curing agent and the paint may be delivered as a single-pack product with a curing time of 30 minutes at 80 °C with an isocyanate and at 130 °C with a melamine. Examples of formulations for the copolymer VDF/TFE/AOPD having a molar composition of the fluoromonomers of 64/36 and a hydroxy content of 1.65 milliequivalent per gram is given in Table 31.8.

These formulations generally have good adhesion on different materials like metals (aluminium, galvanised steel), but also thermoplastics like PVC or thermosetting resins like phenolic resins or glass fibre reinforced polyesters. Coating on concrete is also possible.

Table 31.8. Paint formulation for functionalized copolymers of VDF [25]

Solvent	Ethyl acetate	Butyl acetate
Concentration (%)	30	50
Quantity (g)	72.2	56.1
TiO$_2$ R960[a] (g)	21.8	25.9
Tolonate HDT[b] (g)		8.8
Cymel 370[c] (g)	6	
Butyl acetate (g)		9.1
DBTL[d] (g)		0.1
Baking (time, temperature) (min, °C)	30 min, 130 °C	30 min, 80 °C or 7 days, 20 °C
Film thickness (μm)	15	25
MEK resistance (double rubs)	>100	>100
Gloss 60° (%)	60	60
Hardness Persoz (s)	240	240
Impact resistance (kg*cm)	>50	>50

[a]Pigment from DuPont de Nemours.
[b]Isocyanate from Rhône-Poulenc.
[c]Polymeric melamine from Cyanamid (90% in isobutanol).
[d]Dibutyltin dilaurate (isocyanate catalyst).

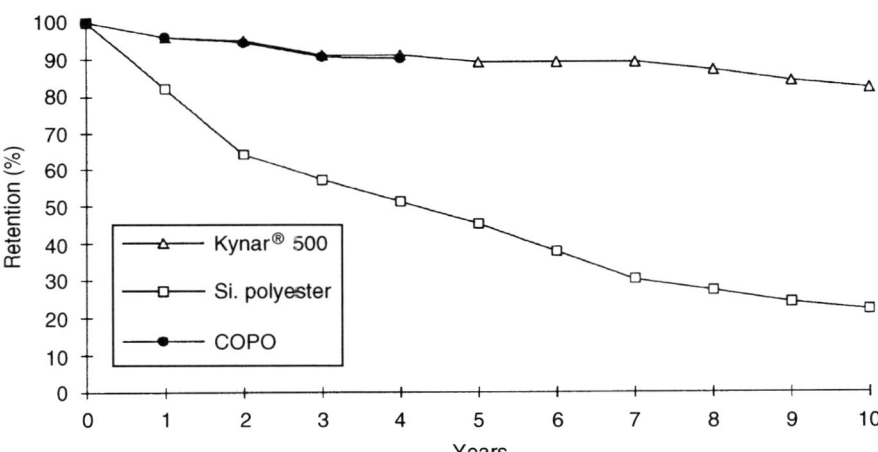

Figure 31.6. Gloss after exposure in Florida [25]

Such coatings have excellent accelerated and outdoor ageing characteristics as can be seen in Figure 31.6 giving a comparison of the gloss retention after exposure in Florida of a paint made with PVDF homopolymer (Kynar® 500), a VDF/TFE/AOPD copolymer (referred to as COPO) and a siliconized polyester.

The corrosion resistance of these paints is also very good, particularly against acids. Conclusion, these new functionalized copolymers of VDF are excellent binders for high-performance paints for exterior applications.

REFERENCES

1. Ford T. A., E. I. DuPont de Nemours, US Pat. 2468054 (1944).
2. Dittman A. *et al.*, M. W. Kellog, US Pat. 2738343 and US Pat. 2752331 (1956).
3. Rexford D. R., E. I. DuPont de Nemours, US Pat. 3051677 (1964).
4. Pailthorp J. R., Schroeder H. E., E. I. DuPont de Nemours, US Pat. 2968649 (1961).
5. Lo E. S., 3M Co, US Pat. 3178399 (1965).
6. Elf Atochem technical Brochures KYNAR® and KYNAR FLEX®, KYNAR FLEX® Fluoropolymer Resin
7. Solvay S. A., Technical Brochure SOLEF®.
8. Kureha Chemical Industry Co. Ltd, Technical Brochure KF POLYMER®.
9. Dohany J. E., Humphrey J. S., 'Vinylidene fluoride polymers', *Encyclopedia of Polymer Science and Technology*, Wiley, New York, 1989, vol. 17, p. 532.
10. Blaise J., Kappler P., Elf Atochem, Eur Pat. 280591, US Pat 4851479 (1987).
11. Metz J. Y., Plissart P., Solvay S. A., Eur Pat. 554931 (1993).
12. Barber L. A., Elf Atochem North America Inc., Eur Pat. 456019, US Pat. 5093427 (1992).
13. ISO 12086-1, *Plastics — Fluoropolymer Dispersions and Moulding and Extrusion Materials — Part 1: Designation System and Basis for Specification* (1995).
14. ISO 12086-2, *Plastics — Fluoropolymer Dispersions and Moulding and Extrusion Materials — Part 2: Preparation of Test Specimens and Determination of Properties* (1995).
15. Hannecart E., Solvay S. A., Eur Pat. 107220, US Pat. 4898906 (1989).
16. Bartoszek E. J., Pennwalt Corporation, USP 4804702 (1989), USP 4881794 (1989).
17. Tournut C., 3rd Chemical Congress of North America and 195th ACS meeting, Toronto 1988, poster and abstract 21.
18. Gozdz A. S. *et al.*, Bell Communications Research Inc., USP 5456000 (1995).
19. Higashihata Y., Sako J. and Yagi T., Piezoelectricity and vinylidene fluoride — trifluoroethylene copolymers, *Ferroelectrics*, **32**, 85 (1981).
20. Tatemoto M., *Nippon Gomu Kyokaishi*, **57** (11), 761 (1984).
21. Tatemoto M., *Kobunshi Ronbunshu*, **49** (10), 765 (1992).
22. Kappler P., Perillon J. L., Elf Atochem, Eur Pat. 396444, US Pat. 5082911 (1989).
23. Kappler P., Perillon J. L., Elf Atochem, Eur Pat. 396445, US Pat. 5037922 (1989).
24. Kappler P., Perillon J. L., Elf Atochem, Eur Pat. 433106, US Pat. 5079320 (1989).
25. Tournut C., Kappler P., Perillon J. L., *Surface Coatings International 1995* (3), 99.

32

Fluoroelastomers

ALBERT VAN CLEEFF
DuPont Dow Elastomers LLC, Wilmington, DE, USA

1 INTRODUCTION

Fluoroelastomers are well known for their excellent chemical and oil resistance, and can be used at much higher temperatures than any other known elastomer, a performance attributed to the strength of the carbon–fluorine bond, to steric hindrance and to strong van der Waals' forces [1–9]. Although polytetrafluoroethylene was already known in 1938, the synthesis of fluoroelastomers was not achieved until the mid-1950s when fluoroelastomers based on vinylidene fluoride (VF_2) and chlorotrifluoroethylene (CTFE) became commercially available. They were replaced shortly thereafter by elastomers made with VF_2 and hexafluoropropylene (HFP). The scope of these polymers was further broadened by the incorporation of a third monomer, tetrafluoroethylene (TFE). These VF_2-based polymers were first vulcanized with polyamines, and later with polyhydroxy compounds (polyols). Further improvements in chemical resistance were obtained through peroxide cures, requiring the presence of bromine and/or iodine cure sites. Peroxide cures were first developed to allow the vulcanization of a new class of fluoroelastomers, based on TFE, VF_2 and perfluoromethylvinyl ether (PMVE), characterized by much improved low-temperature characteristics. Fully fluorinated elastomers, based on TFE and PMVE only, have the best chemical resistance and high-temperature performance of all. A somewhat different class of fluoroelastomers is based on TFE and propylene, providing good base resistance. There are also a number of specialty fluoroelastomers commercially available including polymers containing ethylene and thermoplastic fluoroelastomers.

Fluoroelastomers started out as expensive curiosities, suitable only for military applications. However, environmental concerns and ever increasing demands with regard to low- and high-temperature sealing performance and chemical resistance continued to impose new standards. Not only basic VF_2/HFP and TFE/VF_2/HFP fluoroelastomers but also more costly PMVE- and other

Modern Fluoropolymers. Edited by John Scheirs
© 1997 John Wiley & Sons Ltd

perfluoroalkylvinylether-containing fluoroelastomers and perfluoroelastomers are now commonly used in industrial as well as automotive applications. The fluoroelastomer industry is rising to the new challenges as demonstrated by patent activities concurrent with the introduction of new products.

Patent and technical literature is covered in this review through 1995. There are separate chapters elsewhere in this book on 'Processing of fluoroelastomers' (Chapter 5), on 'Perfluoroelastomers' (Chapter 10,19) and on 'Liquid fluoroelastomers' (Chapter 23).

2 MONOMER COMBINATIONS

Table 32.1 lists the main monomers used in commercially available polymers.

Elastomers require low levels of crystallinity and low glass transition temperatures. Polymers prepared from TFE and/or VF_2 monomers have low glass transition temperatures but are crystalline. Amorphous elastomers are obtained by copolymerizing with monomers possessing a bulky side group attached to the vinyl group (Figure 32.1) [4,10].

3 GENERAL PROPERTIES OF FLUOROELASTOMERS

Fluoroelastomers withstand high temperatures better than all other elastomers with the exception of perfluoroelastomers. Their vulcanizates remain elastic

Table 32.1. Monomers

$CH_2{=}CF_2$	VF_2	Vinylidene fluoride
$CH_3CF{=}CF_2$	HFP	Hexafluoropropylene
$CF_2{=}CF_2$	TFE	Tetrafluoroethylene
$CF_3OCF{=}CF_2$	PMVE	Perfluoromethylvinyl ether
$CF_3(CF_2)_nOCF{=}CF_2$	PAVE	Perfluoroalkylvinyl ether
$ClCF{=}CF_2$	CTFE	Chlorotrifluoroethylene
$CH_2{=}CH_2$	E	Ethylene
$CH_3CH_2{=}CH_2$	P	Propylene

	$\overset{\displaystyle CF_3}{\underset{\displaystyle CF_2{=}CF}{\vert}}$	$\overset{\displaystyle OCF_3}{\underset{\displaystyle CF_2{=}CF}{\vert}}$	$\overset{\displaystyle Cl}{\underset{\displaystyle CF_2{=}CF}{\vert}}$	$\overset{\displaystyle CH_3}{\underset{\displaystyle CH_2{=}CH}{\vert}}$
$CF_2{=}CH_2$	X	X (+ TFE)	X	
$CF_2{=}CF_2$	X (+ VF_2)	X		X (+ VF_2)

Figure 32.1. Main monomer combinations used in the manufacture of fluoroelastomers. *Note:* X indicates a monomer combination which gives commercially available fluoroelastomers

indefinitely when exposed to laboratory air aging at temperatures up to 204 °C (400 °F) [11]. Continuous service limits are generally considered to be:

$$3000 \text{ h at } 232\,^\circ\text{C } (450\,^\circ\text{F})$$
$$1000 \text{ h at } 260\,^\circ\text{C } (500\,^\circ\text{F})$$
$$240 \text{ h at } 288\,^\circ\text{C } (550\,^\circ\text{F})$$
$$48 \text{ h at } 316\,^\circ\text{C } (600\,^\circ\text{F})$$

Fluoroelastomer seals retain their sealing force, after compression for long periods in severe environments. After 100 h in air at 150 °C, fluoroelastomers retain more than 90% of their sealing force, while seals of fluorosilicone, polyacrylate and nitrile retain only 70, 60 and 40% respectively. Dynamic seal applications have been successful down to −40 °C, static applications down to −160 °C.

Fluoroelastomers show excellent resistance to oils, fuels, lubricants and mineral acids. They resist aliphatic and aromatic hydrocarbons which act as solvents for other elastomers. For instance, a standard VF2/HFP dipolymer, when exposed to ASTM oil no. 3 for 168 hours at 150 °C, swells only 1.7 vol%, compared to up to 20 vol% for HNBR elastomers, up to 30 vol% for VMQ silicone rubber, up to 14 vol% for ACM acrylic elastomers (Figure 32.2). They also show extremely low permeability to a broad range of substances including

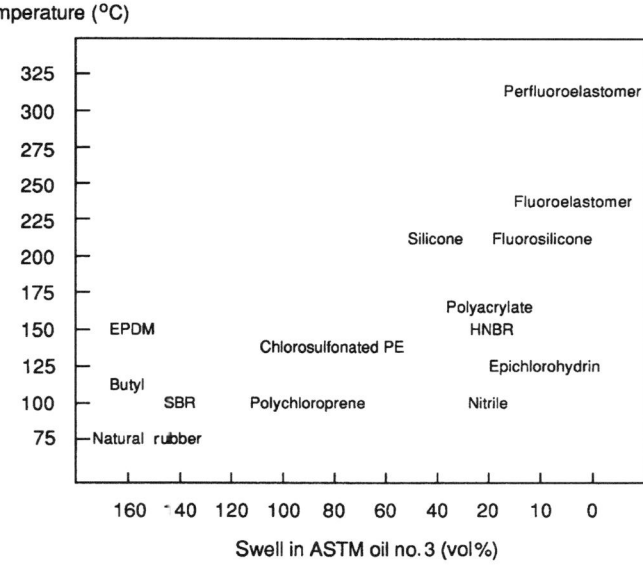

Figure 32.2. Heat and oil resistance grid for commercial elastomers. Perfluoroelastomers are clearly superior as they occupy the upper position on both property axes. SBR (styrene–butadiene rubber), EPDM (ethylene–propylene diene rubber), HNBR (hydrogenated nitrile rubber)

oxygenated automotive fuels. They have excellent resistance to atmospheric oxidation, sunlight and ozone, showing no sign of degradation even after 20 years. They are also more resistant to burning than hydrocarbon rubbers.

4 FLUOROELASTOMERS BASED ON VINYLIDENE FLUORIDE

Dipolymers of VF_2 with CTFE were developed by M. W. Kellog under contract from the Army Quartermaster Corps. They were introduced in 1955 by 3M under the trade name Kel-F [12]. Their vulcanizates showed a significant improvement in heat and chemical resistance versus hydrocarbon elastomers. DuPont introduced in 1957 the first copolymers of VF_2 and HFP under the trade name Viton®, which showed even better thermal stability and resistance to chemicals, less swell in fuels, oils and hydraulic fluids [13,14].

Commercial acceptance of this VF_2/HFP copolymer rapidly surpassed that of the earlier, chlorine-containing products which at the moment are commercially available only from the CIS. Similar VF_2/HFP elastomers were introduced in 1958 by 3M under the trade name Fluorel. Ausimont (Tecnoflon) and Daikin (Dai-El) are other producers. Ausimont introduced earlier fluoroelastomers based on VF_2 and 1-hydro-pentafluoropropylene. These polymers showed shortcomings, especially in heat resistance, and are no longer produced. Asahi Chemical (Miraflon) participated in this market for about 10 years, but stopped production in 1995. Small amounts of Russian and Chinese fluoroelastomers are being traded in some markets. Fluoroelastomers produced by Nippon Mektron Co. are not available on the open market.

VF_2/HFP elastomers contain about 60 wt% VF_2 (roughly 4 moles of VF_2 for each mole of HFP). Such polymers are commercially produced over a wide range of molecular weights. They can be cured with polyamines and polyols, yielding products with good resistance to chemicals and high temperatures. They are used in compression, transfer and injection molding for the production of O-rings, valve stem seals and shaft seals, in extrusion for fuel hose and tubing, and in solution coatings for fabrics, tanks and chemical containers.

The presence of VF_2 can lead to considerable swell in some highly polar solvents such as low molecular weight ketones and esters. Solvent resistance is improved by lowering the VF_2 content (hydrogen level), which is synonymous with raising the fluorine level in the polymer. Since it is difficult to incorporate high levels of HFP into these polymers, TFE is added instead. This not only lowers the swell in polar solvents but also improves their high-temperature performance [15,16].

Up to 30 wt% TFE is incorporated, raising the fluorine level (lowering the hydrogen level) from 66 wt% for the dipolymers (1.9 wt% H), to 68 wt% for standard terpolymers (1.4 wt% H), to 70 wt% for the high-fluorine types

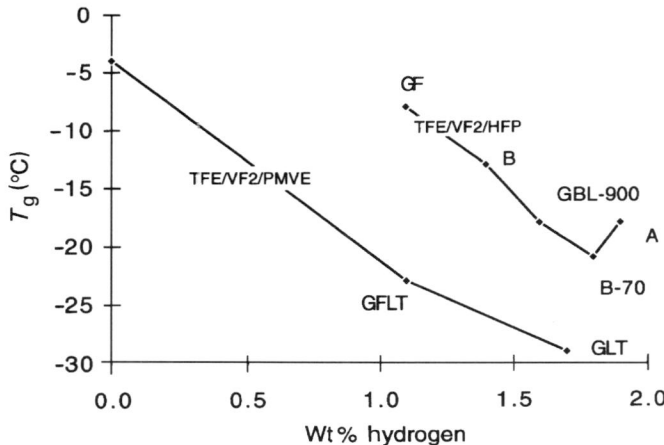

Figure 32.3. Glass transition temperature values for Viton® fluoroelastomers as a function of their hydrogen content. Note: A, B, B-70, GF, GLT, GFLT denote the position where these respective Viton grades lie in relation to one another

(1.1 wt% H). For reference, polytetrafluoroethylene contains 76 wt% fluorine, and of course no hydrogen.

Low-temperature performance is adversely affected by increased fluorine levels, as expressed by glass transition temperature (Figure 32.3). The same is true for room temperature compression set resistance [17–19]. However, even high-fluorine elastomers continue to show adequate vulcanizate properties. Applications include molded goods such as O-rings and shaft seals, and extruded tubing and hose. High-fluorine fluoroelastomers show extremely low permeability towards automotive fuels, containing oxygenated additives such as methanol and MTBE and meet the toughest environmental regulations. For example, the permeability of high fluorine types in 85% Fuel C / 15% methanol is 20 times less than for nylon, 200 times less than for fluorosilicones and HNBR, hence their use in fuel hose and other automotive applications. Calendered goods include flue duct expansion joints.

5 COMPOUNDING AND CURING OF FLUOROELASTOMERS

Fluoroelastomers are normally cured under pressure at elevated temperatures. The press cure is followed by further heat treatment in circulating air ovens, preferably at temperatures higher than the maximum application temperature. Pressureless cures have been developed where the use of pressure is not practical or possible.

6 AMINE CURES

Early attempts of curing fluoroelastomers were only moderately successful and are only of historic interest. Polyamine cures were the first cure systems of commercial importance [20]. Polyamines first act as a base and form double bonds in the polymer chain by removing HF. These unsaturated sites will then

Table 32.2. Properties of typical fluoroelastomers in dihydroxy cures

Viton®	A-500	B-600	GF
Monomers	VF$_2$/HFP	TFE/VF$_2$/HFP	TFE/VF$_2$/HFP
Wt% fluorine	66	68.5	70
Wt% hydrogen	1.9	1.4	1.1
Viscosity, ML $1 + 10/121\,°C$	50	65	60
Compounds			
Polymer	100	100	100
Viton® Curative no. 50	2.5	—	—
Viton® Curative no. 20	—	3.0	4
Viton® Curative no. 30	—	3.8	6
MT Black (N-990)	30	30	30
Maglite D	3	3	3
Calcium hydroxide	6	6	6
VPA no. 1	0.5	—	—
VPA no. 3	0.5	1	—
Stock properties			
Mooney Scorch MS at 121 °C			
Minimum viscosity, units	47	65	70
Time to 10 unit rise	>30	>30	>30
ODR at 177 °C, micro die, 3 ° arc, 12 min			
ML, N.M	2.0	2.9	3.7
t$_s$2 (min)	1.5	1.5	1.2
t'90 (min)	2.9	3.2	4.3
M_H, N.M	14.2	10.7	9.2
Vulcanizate properties			
Press cure: 10 min at 177 °C; post cure: 24 h at 232 °C			
Stress/strain at 23 °C — original			
100% Modulus, (MPa)	7.2	5.2	7.4
Tensile strength (MPa)	15.6	12.5	13.0
Elongation at break (%)	197	260	200
Hardness durometer A pts	78	72	80
Compression set, method B			
70 h at 200 °C	15	28	44
Fluid resistance, volume swell (%)			
Fuel C, 70 h/23 °C	—	2.9	3
Methanol, 70 h/23 °C	94	19	4

react readily with additional polyamine to form crosslinks. Metal oxides such as MgO, CaO and PbO are added to neutralize the HF formed. Polyamines are very reactive and produce very scorchy compounds. In order to improve scorch resistance, blocked diamines are preferred. Examples are hexamethylene diamine, used as its carbamate (Diak* 1) and N,N-dicinnamylidene-1,6-hexanediamine (Diak* 3), both introduced by DuPont in the late 1950s. Diamine crosslinks are not stable at high temperatures, and tend to revert in the presence of steam. They are used mostly in less demanding applications.

7 POLYHYDROXY CURES

The introduction of polyhydroxy curatives constituted a major breakthrough. They give rapid, yet scorch-safe, cures, forming stable crosslinks, yielding vulcanizates with excellent tensile properties and compression set resistance. The cure mechanism was elucidated by W. W. Schmiegel of DuPont [22]. The mechanism involves dehydrofluorination of the main chain by strong bases, such as calcium hydroxide, magnesium oxide, in the presence of phase transfer catalysts (often called accelerators). The polyol then adds readily across the polymer chains to form crosslinks. Bisphenol AF is commonly used, in combination with accelerators such as benzyltriphenylphosphonium chloride (BTPPC) [23–29]. Prereacted curatives, formed by reacting for instance bisphenol AF with accelerators such as BTPPC to form the BTPP$^+$–bisphenol AF$^-$ salt, have been described [30].

Curing conditions depend on the formulation and application but are typically 5–10 min at 180 °C. The cure is followed by a post-cure, normally 24 hours at 200–230 °C. During the post-cure cycle not only is the network further stabilized, resulting in much reduced compression set values, but by-products from the cure such as triphenylphosphine oxide are removed.

Typical properties of fluoroelastomers are displayed in Table 32.2. Figure 32.4 displays the excellent resistance of fluoroelastomers towards sulfuric acid at elevated temperatures, as encountered in flue duct expansion joints.

The main drawback of bisphenol cures is the presence of some unsaturation in the vulcanized parts. They are more readily attacked by amines than vulcanizates from peroxide cures, especially under demanding conditions.

Fluoroelastomers containing PMVE are not normally cured with bisphenols. They would lose trifluoromethoxide during the dehydrofluorination step. Such polymers are therefore always cured with peroxides, as described below. Patent literature describes some degree of success when using bisphenol cures with tetrapolymers, containing the four major monomers, TFE, VF$_2$, HFP and PMVE [31–34].

8 PEROXIDE CURES

Bromine-containing peroxide curable fluoroelastomers were first introduced by DuPont in 1977 [35,36,76]. The bromine level is typically between 0.5 and

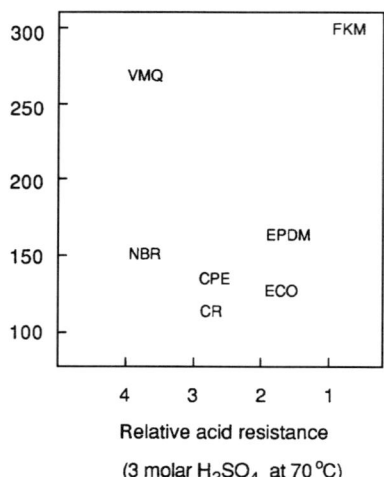

Figure 32.4. Comparative rating of heat and acid resistance of various synthetic elastomers. Relative heat resistance is based on the criteria that samples retain >100% elongation at break after 70 h at the quoted temperature. Relative acid resistance is in accordance with the Fluid Sealing Association rating system where 1 is satisfactory and 4 is unsatisfactory. Note: CPE (chlorinated polyethylene), CR (chloroprene rubber), ECO (epichlorohydrin rubber), EPDM (ethylene–propylene diene rubber), FKM (fluoroelastomer), NBR (nitrile rubber), VMQ (silicone rubber)

1.0 wt%, randomly distributed in the main chain. Examples include polymers based on TFE/VF$_2$/PMVE as well as on TFE/VF$_2$/HFP, available from DuPont under the trade name Viton®. Ausimont and 3M (now Dyneon Co.) also have bromine-containing fluoroelastomers in their product line, under their Tecnoflon and Fluorel trade names. In a later development iodine, in addition to bromine, was incorporated in polymers, produced by DuPont and Ausimont [37–40]. Iodine, being very reactive, not only gives faster cures but also less mold fouling, important for injection molding operations. Polymers containing iodine normally have low Mooney viscosities.

Daikin is the main supplier of iodine-capped fluoroelastomers. They are produced under special polymerization conditions, using I(CF$_2$)$_n$I as the chain transfer agent. These polymers show efficient cures in the presence of peroxides [41]. They are available under the trade name Dai-El.

Peroxides used in the peroxide cures of fluoroelastomers are by necessity quite stable. Otherwise it would be too difficult to prevent scorch during compounding and further processing. Examples are 2,5-dimethyl-2,5-di(*t*-butylperoxy)- hexane and dicumylperoxide. They are almost always used in the presence of polyfunctional coagents, such as triallylisocyanurate (TAIC) and trimethallylisocyanurate

(TMAIC); triallylcyanurate (TAC), a standard coagent for other elastomers, is also sometimes used.

Finlay of DuPont proposed the following cure mechanism [42–44]: Methyl radicals are formed by the decomposition of peroxides. These radicals react first with the coagent to form coagent radicals. The coagent radicals abstract bromine and/or iodine from the polymer chain, forming polymeric radicals which in turn react with the coagent to form cross-links. Very little methylbromide or methyliodide is formed in these reactions.

Table 32.3. Properties cf typical fluoroelastomers in peroxide cures

Viton*	GBL-200	GF	GFLT
Monomers	TFE/VF$_2$/HFP	TFE/VF$_2$/HFP	TFE/VF$_2$/PMVE
Cure sites	I + Br	Br	Br
Wt% fluorine	67	70	67
Wt% hydrogen	1.6	1.1	1.1
Viscosity, ML $1 + 10/121\,^\circ$C		65	
Compounds			
Polymer	100	100	100
MT Black (N-990)	30	30	30
Zinc oxide	3	—	3
Sublimed Litharge	—	3	—
Diak* no. 7	3	3	2.5
Luperco 101-XL	3	3	2.5
Stock properties			
Mooney Scorch MS at 121 $^\circ$C			
Minimum viscosity, units	19	47	25
Time to 10 unit rise	>30	>30	25
ODR at 177 $^\circ$C, Micro die, 1 $^\circ$ arc, 12 min			
t_s2 (min)	0.9	1.5	1.1
t'90 (min)	3.5	6.0	5.3
Vulcanizate properties			
Press cure: 10 min at 177 $^\circ$C; post cure: 24 h at 232 $^\circ$C			
Stress/strain at 23 $^\circ$C — original			
100% modulus (MPa)	5.4	5.4	8.2
Tensile strength (MPa)	21.5	17.6	16.3
Elongation at break (%)	246	260	149
Hardness, durometer A pts	71	76	73
Compression set, method B			
70 h at 200 $^\circ$C	46	38	36
Low-temperature properties			
Brittle point ($^\circ$C)	−47	−45	−54
TR$_{10}$ ($^\circ$C)	−15	−6	−24
T_g, DSC ($^\circ$C)	−16	−8	−24

Peroxide cures do not require the presence of strong bases. Suitable acid acceptors are magnesium oxide, lead oxide and especially zinc oxide. Their presence also prevents the homolytic decomposition of peroxides, which can occur under acidic circumstances. No double bonds are formed, the polymer chains remain saturated.

Peroxide-cured vulcanizates, therefore, react less readily and maintain elastomeric properties longer when exposed to typical, amine-containing, corrosion inhibitors, present in modern motor oils. Typical properties of peroxide-cured fluoroelastomers are displayed in Table 32.3.

9 FILLERS

Fluoroelastomers can be compounded with a variety of fillers. The incorporation of fillers results in higher compound viscosity, lower elongation at break, higher hardness, modulus and tensile strength. Most common are MT carbon black (N-990), Blanc Fixe (barium sulfate) and Nyad® 400 (fibrous calcium silicate). Austin black, SRF black (N-774), titanium dioxide, red iron oxide, diatomaceous silica and Teflon powders are also used.

10 PROCESS AIDS

Process aids are commonly used to improve release from molds and calenders, and to provide a smooth finish for extrudates. Examples include Carnauba Wax, VPA no. 1, VPA no. 2 and VPA no. 3, used at levels between 0.5 and 2.0 phr. Armeen 18D and Proton Sponge are used in peroxide cures.

For a discussion of processing of fluoroelastomers, consult Chapter 5 in the book.

11 FLUOROELASTOMERS BASED ON TFE AND PROPYLENE

TFE polymerizes with propylene to form almost strictly alternating polymers, as described in patents by DuPont [46] and Asahi Glass [47–51]. Asahi Glass introduced such polymers commercially in 1975 under the trade name Aflas. They show much better base resistance than standard fluoroelastomers, in addition to excellent resistance towards acids and polar solvents. They are used in applications such as shaft seals, exposed at elevated temperatures to amine-based corrosion inhibitors present in modern motor oils. The high level of hydrocarbon monomer is reflected in higher swell in aromatic solvents, but less swell in certain polar solvents. The swell of a typical TFE/P polymer in toluene for instance is 40 vol%. Low-temperature performance is mediocre, as judged by a glass transition temperature of about 0 °C. Polymers based on TFE, propylene and VF$_2$ show improved low-temperature properties, but give up some chemical resistance and resistance towards polar solvents.

TFE/P fluoroelastomers are normally cured with peroxides in the presence of coagents [52]. Curability would be mediocre but the commercially available dipolymers contain an appropriate level of unsaturation in the polymer backbone. This is achieved by exposing the polymer to inorganic bases in combination with solvents. Such polymers show severe discoloration as a result. TFE/P polymers do not respond to bisphenol cures. TFE/P/VF$_2$ fluoroelastomers respond to nucleophilic cures to a degree, depending on VF2 level [53–54]. Prereacted bisphenol curatives have been described for use in TFE/propylene/VF$_2$ terpolymers [30].

12 FLUOROELASTOMERS CONTAINING ETHYLENE

12.1   TFE/ETHYLENE/PMVE FLUOROELASTOMERS

DuPont introduced in 1986 fluoroelastomers based on TFE, ethylene and PMVE. A bromine cure site is provided to allow peroxide curability [55,56]. They are suitable for long-time service in severe environments, particularly oilfield applications, where the seals are exposed to high temperatures in the presence of wet hydrocarbons containing H$_2$S and amine based corrosion inhibitors. They have better low-temperature properties than TFE/PMVE and TFE/P elastomers and moreover do not show the high swell of TFE/P polymers in hydrocarbons. Properties of a typical TFE/E/PMVE polymer, displayed in Tables 32.4 and 32.5, show its superior performance over other hydrogen-containing fluoroelastomers.

Table 32.4.   Comparison of TFE/E/PMVE polymers with other fluoroelastomers

Polymer composition	TFE/E/PMVE	TFE/VF$_2$/HFP	TFE/Propylene
% Hydrogen	1.1	1.1	4
T_g (°C)	−17	−7	0
Compound viscosity MS/1221 °C	26	51	50
Tensile properties			
M$_{100}$ (MPa)	8.5	6.3	9.3
T_b (MPa)	18.3	18.3	20.3
Compression set (%) O-rings			
22 h/200 °C	21	22	31
22 h/0 °C	61	65	84
Fluid resistance % Volume swell			
Skydrol LD, 70 h/121 °C	11	80	33
Toluene, 70 h/25 °C	4	2	51
Relative permeation rate, toluene, 25 °C	5	2	1500

Table 32.5. TFE/E/PMVE fluoroelastomer exposure to severe environments (three days at 150 °C)

Fluid	% Volume swell
30% KOH	12
Sour brine (10% H_2S, 5% amine)	17
Wet sour oil (10% H_2S, 5% amine)	12

12.2 TFE/VF2/HFP/ETHYLENE FLUOROELASTOMERS

Ausimont offers a polymer based on VF_2. HFP, TFE and ethylene [57]. The polymer contains about 5% ethylene which reduces the acidity of the VF_2 hydrogen, making the polymer less susceptible to base attack. The polymer does not respond to bisphenol cures. A bromoperfluoroalkylvinyl ether cure site monomer is used to afford peroxide curability. Exposure of peroxide cured vulcanizates to 1% benzylamine in ASTM oil no. 3 for three days at 160 °C reduced the tensile strength by 11% and the elongation at break by 4%, compared to 67 and 47% respectively for a standard, Ausimont high-fluorine, peroxide curable polymer. Reasonable polymerization rates are obtained by a special microemulsion technique which produces a large number of seed particles [58].

12.3 THERMOPLASTIC FLUOROELASTOMERS

Thermoplastic fluoroelastomers are being marketed by Daikin. They are an extension of their diiodo technology [59,60]. In the first stage iodine-terminated TFE/VF_2/HFP terpolymers are synthesized by emulsion polymerization, using $I(CF_2)_n I$ as the diiodo compound. The polymerization of the hard segment component takes place in the presence of the iodine-terminated terpolymer emulsion. One type contains E/TFE/HFP, the other contains poly-VF_2 as the hard segment. These thermoplastic fluoroelastomers do not need curatives, metal oxides, fillers or process aids, which encourage their use in medical applications. They require radiation curing when compression ratios larger than 10% are used. The polymers can also be chemically cured, alone or in combination with other fluoroelastomers. Typical properties are given in Table 32.6.

12.4 BLENDS

Fluoroelastomers are often blended with other fluoroelastomers for specific purposes. For instance, green strength and tear strength can be improved by blending a high- and a low-viscosity fluoroelastomer. Such blends also show improved mill mixing. Cost considerations were the driving force for early blending studies with other elastomers, research which has found expression in a number of patents [61–63] and publications [64].

Blending with other elastomers is also undertaken to improve a specific property of fluoroelastomers. For instance silicone elastomers have superior

Table 32.6. Thermoplastic fluoroelastomers

Hard segment (Dai-El)	E/TFE/HFP (T-530)	Poly-VF$_2$ (T-630)
Melting point (°C)	220	160
Tensile strength, (MPa)	12	2
Elongation (%)	650	>1000
Compression set		
24 h at 50 °C (%)	11	80
After radiation, 20 Mrad		
Tensile strength (MPa)	11.3	9.5
Elongation (%)	230	240
Compression set		
70 h at 150 °C (%)	23	26
Volume swell, 7 days/40 °C		
Methanol	<5%	<10%

low-temperature properties compared to fluoroelastomers and several attempts are known to improve the low-temperature performance of fluoroelastomers by blending with silicone rubber [65,66]. The two polymers are very incompatible and it has been difficult to develop blends which provide the right balance of low temperature and other desirable properties.

The high-temperature performance of fluoroelastomers is far superior over that of the next best available elastomer, HNBR, a performance not always needed. It would be advantageous for such cases to have an elastomer with intermediate performance at intermediate prices. One recently introduced offering by DuPont is Advanta* specialty elastomer, an alloy based on fluoroelastomers and an acrylic component. It can be used at temperatures up to 175 °C [67,68].

Blends of elastomers with fluoroelastomers, mixed under dynamic curing conditions have also been described. Such blends can be designed to have thermoplastic properties. However, it is difficult to maintain the outstanding properties of fluoroelastomers, using this approach. This is probably the reason why such polymer blends so far have not found widespread applications [69,70].

13 MANUFACTURE

13.1 VF$_2$-BASED FLUOROELASTOMERS

Fluoroelastomers containing VF$_2$ are normally prepared by continuous or semi-batch free-radical emulsion polymerization at pressures up to 7 MPa and temperatures between 50 and 130 °C [71,72]. In a continuous process all ingredients are fed, and the resulting emulsion removed continuously. Such a process is eminently suited for the large scale production of high-quality fluoroelastomers. The emulsion is discharged though a let-down valve into a degasser. Unreacted monomers are recycled; VF$_2$ and TFE are very reactive monomers so that their concentration in the recycle stream is low. The emulsion is coagulated by adding high-valent metal salts of, for instance, calcium or aluminum. The wet crumb thus obtained is dewatered by centrifugation or in a dewatering extruder, then further dried. Many commercial polymers are sold with curatives and/or process aids mixed in. This can be done on a two roll mill, in an internal mixer or mixing extruder.

In a semi-batch process monomers are added continuously to the reactor, optionally also other ingredients. This type of process is more flexible since every charge can be different. The emulsion polymerization is generally not carried out beyond 25% solids in continuous operations; in semi-batch operations higher solids levels can be obtained.

Initiators include ammonium and potassium persulfate. Polymers thus prepared have a high level of carboxylate end groups through the hydrolysis of the sulfate intermediates. They can also be polymerized in combination with redox agents such as sulfites, especially at lower temperatures, where the thermal decomposition of persulfate is too slow. In such cases both carboxylate and sulfonate end groups are produced. Strong acid end groups give high compound viscosities, which harms processing. The polymerization rate can be increased through the use of variable valence metal salts such as ferrous sulfate.

Surfactants, where used, are preferably completely fluorinated in order to avoid chain transfer; most sources quote the use of perfluorooctanoic acid. The pH can be controlled through the addition of base and/or phosphates; the polymerization proceeds well in any event over a wide pH range. Molecular weight can be controlled by adjusting the ratio of persulfate and monomer feeds. Chain transfer agents are used routinely to control molecular weight. They are also beneficial in that they reduce the level of strong acid end groups [19,72–75]. Examples include chlorinated species such as carbon tetrachloride; hydrocarbon-based alcohols such as methanol, ethanol and i-propanol; esters such as diethylmalonate.

Peroxide curable fluoroelastomers require the presence of iodine and/or bromine cure sites. Bromine can be incorporated through the use of small amounts of bromine-containing cure-site monomers such as bromotrifluoroethylene, bromotetrafluorobutene and bromoperfluoroalkylvinyl ether [35,36]. The bromine is randomly distributed along the polymer chain. Bromine and/or iodine are most commonly introduced through the use of bromine- or iodine-containing chain transfer agents. The cure sites are generated in this case only at the end of the polymer chains. Bromine-containing chain transfer agents include KBr. Iodine is normally introduced by polymerizing in the presence of, for example, KI, methylene iodide or diiodoperfluorobutane [38,39,76,77]. Chain transfer agents containing both iodine and bromine have also been described [78].

Free-radical solution polymerization is practiced only on a small scale for the production of low molecular weight polymers, which are primarily used as process aids.

13.2 TFE/PROPYLENE FLUOROELASTOMERS

TFE/P polymers are produced in semi-batch operations by emulsion polymerization in the presence of fluorinated surfactants. Excessive chain transfer takes place at temperatures normally employed for other fluoroelastomers, producing polymers with low molecular weight. Special persulfate redox systems were introduced in 1981 for the production of high molecular weight polymers, allowing the polymerization to proceed at reasonable rates at temperatures between 25 and 35 °C [79].

Microemulsion technology for emulsion polymerization has been described by Ausimont to facilitate the incorporation of ethylene in some of their polymers [57,58].

Theories of emulsion polymerization are not as well developed as for most other emulsion polymerization systems [80]. One reason is that only major producers have the experience of safely handling TFE under pressure.

13.3 QUALITY CONTROL

Quality control includes the measurement of Mooney viscosity, inherent viscosity and polymer composition. Polymer composition can be determined from mass balance, by NMR or by infrared spectroscopy; with infrared methods plant samples are compared to laboratory-produced reference samples.

13.4 TEST METHODS AND SPECIFICATIONS

Fluoroelastomers, their compounds and vulcanizates are tested according to ASTM or DIN procedures, common to the rubber industry. Numerous automotive (ASTM, SAE) and military (AMS) specifications are in circulation.

13.5 SAFETY AND ENVIRONMENTAL CONSIDERATIONS

All manufacturers have brochures available with specific recommendations for the safe handling and disposal of fluoroelastomers and related chemicals [81]. It is highly recommended reading for anyone active in this area. Fluoroelastomers are non-toxic and non-irritating under normal conditions. They can be compounded and cured with standard precautions, such as temperature control, adequate ventilation and so on.

SUMMARY

The fluoroelastomer market has seen a steady growth from the time of its inception. The current market size is estimated to be around 23MM pounds, growing at a rate of approximately 8%. Fluoroelastomers sell for about $20/lb for commodity types, up to $70/lb and sometimes higher for specialty types. The market continues to grow since for many applications only fluoroelastomers will perform satisfactorily. They are solving sealing and other problems in demanding applications in industries such as aircraft and aerospace, automotive, chemical processing and transportation, off-highway and heavy-duty equipment, petroleum refining and transportation. Major uses include bonded seals, radial lip seals, caulks, coatings, vibration dampeners, expansion joints, gaskets, hoses, O-rings, piston seals, custom shapes, stock rod and sheet.

REFERENCES

1. Cooper, J. R., *High Polymers*, vol. XXIII, *Polymer Chemistry of Synthetic Elastomers*, Wiley-Interscience, p. 273 (1968).
2. Arnold, R. G., Barney, A. L., Thompson, D. C., *Rubber Chem. Technol.* **46**, 619 (1973).
3. Vlasak, R. L., Vogel, H. A., Paper no. 23, 'Production of high performance fluoroelastomers', presented at the 109th Meeting of the Rubber Division, American Chemical Society, Minneapolis, Minnesota, April 27–30, 1976.
4. England, D. C., Uschold, R. E., Starkweather, H., Pariser, R., *Proceedings of The Robert A. Welch Conferences on Chemical Research XXVI, Synthetic Polymers*, **1982**, November 15–17, Houston, Texas.
5. Lynn, M. M., Worm, A. T., in *Encyclpedia of Polymer Science and Engineering*, Wiley, New York, vol. 7, p. 257 (1987).
6. Carlson, D. P., Schmiegel, W. W., *Fluoropolymers, Organic*, vol. A11, *Ullmann's Encyclopedia of Industrial Chemistry*, 5th edn, p. 393 (1988).
7. Logothetis, A. L., 'Chemistry of fluorocarbon elastomers', *Prog. Polym. Sci.*, **14**, 251 (1989).
8. Crenshaw, L. E., Tabb, D. L., *The Vanderbilt Rubber Handbook*, 13th edn. p. 211 (1990).
9. Logothetis, A. L., 'Fluoroelastomers', in *Organofluorine Chemistry: Principles and Commercial Applications*, edited by R. E. Banks *et al.*, Plenum Press, New York, p. 373 (1994).
10. Uschold, R. E., *Polymer Journal* (Japan), **17**, 253 (1985).
11. DuPont, *Viton* Bulletin H-22833*, June 1990.
12. Dittman, A. L., Passino, A. J., Wrightson, J. M., US patent 2 689 241 (1954), to M. W. Kellogg.
13. Dixon, S., Rexford, D. R., Rugg, J. S., *Ind. Eng. Chem.*, **49**, 1687 (1957).
14. Rexford, D. R., US patent 3 051 677 (1962), to DuPont.
15. Pailthorp, J. R., Schroeder, H. E., US patent 2 968 649 (1961), to DuPont.
16. DuPont, *Viton* B Bulletin VT-230.B*, March 1988
17. Van Cleeff, A., Paper presented at Scandinavian Rubber Conference, Copenhagen, Denmark, June 10–12 (1985).
18. Van Cleeff, A., Kautsch., *Gummi, Kunstst.*, **39**, 196 (1986).
19. Van Cleeff, A., Kautsch., *Gummi, Kunstst.*, **42 (2)**, 111–114 (1989).
20. Smith, J. F., *J. Polym. Sci.*, Part A-2, **10**, 133 (1972).
21. Schmiegel, W. W., *Die Angewandte Makromolekulare Chemie*, **76/77**, 39 (1979).
22. Patel, K. U., Maier, J. E., US patent 3 655 727 (1972), to 3M.
23. Patel, K. U., Maier, J. E., US patent 3 712 877 (1973), to 3M.
24. De Brunner, M. R., US patent 3 752 787 (1973), to DuPont.
25. Kometani, Y., *et al*, US patent 3 857 807 (1974), to Daikin.
26. Kometani, Y., *et al*, US patent 3 864 298 (1975), to Daikin.
27. Ceccato, G., Geri, S., Colombo, S, L., US patent 3 920 620 (1975), to Montedison.

28. Carlson, D. P., Schmiegel, W. W., US patent 4 957 975 (1990), to DuPont.
29. Grootaert, W. M. A., Kolb, R. E., US patent 4 912 171 (1990), to 3M.
30. Bauerle, J. G., Paper presented at the 1995 SO$_2$ Control Symposium, March 28-31, 1995, Miami, Florida.
31. Baird, R. L., MacLachlan, J. D., French patent 225 9849 (1975), to DuPont.
32. Baird, R. L., MacLachlan, J. D., French 2347 389 (1975), to DuPont.
33. Albano, M., Arcella, V., Brinati, G., Chiodini, G., Minutillo A., EP 525 685 A2 (1993), to Ausimont.
34. Albano, M., Arcella, V., Brinati, G., Chiodini, G., Minutillo A., EP 525 687 A2 (1993), to Ausimont.
35. Apotheker, D., Krusic, P. J., US patent 4 035 565 (1977), to DuPont.
36. Apotheker, D., Krusic, P. J., US patent 4 214 060 (1980), to DuPont.
37. Nudelman, Z. N., Lavrova, L. N., Dontsov, A. A, Kauchuk i Rezina, **4**, 41 (1984).
38. Moore, A. L., US patent 4 973 633 (1990), to DuPont.
39. Moore, A. L., US patent 5 032 6655 (1991), to DuPont.
40. Albano, M., Arcella, V., Brinati, G., Giannetti, E., US patent 5 173 553 (1992), to Ausimont
41. Albano, M., Arcella, V., Brinati, G., Giannetti, E., US patent 5 219 964 (1993), to Ausimont
42. Suzuki, T., Tatemoto, M., Tomoda, M., US patent 4 243 770 (1981), to Daikin.
43. Apotheker, D., Finlay, J. B., Krusic, P. J., Logothetis, A. L., *Rubber Chem. Technol.* **55**, 1004 (1982)
44. Schmiegel, W. W., Logothetis, A. L., ACS Symposium Series No. 260, *Polymers for Fibers and Elastomers*, **10**, 159 (1984)
45. Finlay, J. B., Hallenbeck, A., MacLachlan, J. D., *J. Elastomers Plast.* **10**, 3 (1978).
46. Brasen, W. R., Cleaver, C. S., US patent 3 467 635, to DuPont.
47. Hisasue, M., Kojima, G., Kojima, H., Yamakage, M., Japanese patent JS 52041662 (1977), to Asahi Glass.
48. Hisasue, M., Kojima, G., Kojima, H., Wachi, H., Japanese patent J85019324 (1985), to Asahi Glass.
49. Tabata, Y. *et al. J. Polym. Sci.*, Part **A2**, 2235 (1964).
50. Kojima, G., Tabata, Y., *J. Macromol. Sci., Chem.* **A5**, 1087 (1971).
51. Kojima, G., *et al.*, *Rubber Chem. Technol.*, **50**, 403 (1977).
52. Kojima, G., Wachi, H., *Rubber Chem. Technol.*, **51(5)** 940 1978.
53. Hisasue, M., Kojima, G., Kojima, H., Yamakage, M., Japanese patent J85019325 (1985), to Asahi Glass.
54. Kojima, G., Wachi, H., Int. Rubber Conf., Kyoto, Oct. 15-18, paper no. 16A18 (1985)
55. Moore, A. L., US patent 4 694 045 (1987), to DuPont.
56. Moore, A. L., *Elastomerics*, September 1986
57. Albano, M., Arcella, V., Brinati, G., Chiodini, G., Minutillo, A., US patent 5 264 509 (1993), to Ausimont.
58. Giannetti, E; Visca, M, US patent 4 864 006 (1989), to Ausimont.
59. Nakagawa, T., Tatemoto, M., US patent 4 158 678 (1979), to Daikin.
60. Kawachi, S., GAK 39, 162 (1986).
61. Tomoda, M., Ueta, U., US patent 4 251 399 (1981), to Daikin.
62. Tatemoto, M., Tomoda, M., Ueta, Y., US patent 4 260 698 (1981), to Daikin.
63. Buding, H., Szentivany, Z., Thormer. J., US patent 4 565 614 (1986), to Bayer.
64. Tripathy, A. R., Ghosh, M. K., Das, C. K., *Intern. J. Polymeric Mater.*, **17**, 77 (1992)
65. Yoshida, H., Watanabe, J., Paper no. 122, presented at Rubber Division Meeting, ACS, Detroit, Michigan, October 17-20, 1989.

66. Alberts, H., Huggins, J., Krueger, R., Langstein, G., Moretto, H., EP 582,841 A1 (1994), to Bayer.
67. Tabb, D. L., US patent 5 412 034 (1995), to DuPont.
68. Thomas, E. W., Kotz, D. A., Tabb, D. L., Paper no. 194, presented at Rubber Division. Meeting ACS, Pittsburgh, PA, October 11–14, 1994.
69. Rees, R. W., US patent 5 006 594 (1991), to DuPont.
70. Coran, A. Y., US patent 5 051 480 (1991), to Monsanto.
71. Weaver, S. D., US patent. 3 845 024 (1974), to DuPont.
72. A. L. Moore, US patent 3 839 303 (1974), to DuPont.
73. Gallagher, G. A., US patent. 3 069 401 (1962), to DuPont.
74. Sandberg, C. L., US patent 3 080 347 (1963), to 3M.
75. Gladding, E. K., Nyce, J. L., US patent. 3 707 529 (1972), to DuPont.
76. Arcella, V., Brinati, G., Bonardelli, P., Tommasi, G., US patent US 4 745 165 (1988), to Ausimont.
77. Albano, M., Arcella, V., Brinati, G., Giannetti, E., US patent 5 173 553 (1992), to Ausimont.
78. Abe, M., Naraki, A., Okabe, J., Okamoto, S., US patent 4 943 622 (1990), to Nippon Mektron.
79. Kojima, G., Hisasue, M., *Macromol. Chem.*, **182**, 1429 (1981).
80. 'Handling precautions for Viton* and related chemicals,' *DuPont Bulletin VT.100.1*, April 1987
81. Richards, J. R., Congalidis, J. P., Gilbert, R. G., *J. Appl. Pol. Science*, **37**, 2727 (1989).

Index